# Wheat Stress Management

Ratan Tiwari • Binay Kumar Singh
Ravindra Kumar
Editors

# Wheat Stress Management

## Emerging Solutions and Techniques

Springer

*Editors*
Ratan Tiwari
Indian Institute of Wheat and Barley Research
Karnal, Haryana, India

Binay Kumar Singh
Indian Institute of Wheat and Barley Research
Karnal, Haryana, India

Ravindra Kumar
Indian Institute of Wheat and Barley Research
Karnal, Haryana, India

ISBN 978-981-92-0901-9 ISBN 978-981-92-0902-6 (eBook)
https://doi.org/10.1007/978-981-92-0902-6



This Springer imprint is published by the registered company Springer Nature Singapore Pte Ltd.
The registered company address is: 152 Beach Road, #21-01/04 Gateway East, Singapore 189721, Singapore

# Foreword

The global population is expanding at an unprecedented rate, placing mounting pressure on agricultural systems to deliver sufficient, nutritious, and stable food supplies. Wheat, as a principal staple crop, plays a central role in global food and nutritional security, particularly in countries such as India. However, wheat productivity is increasingly threatened by climate volatility, erratic rainfall, rising temperatures, soil degradation, emerging virulent pathogens, and escalating biotic and abiotic stresses. Wheat crops face a wide spectrum of biotic stresses, including spike diseases, foliar blights, rusts, stem and root pathogens, and emerging complexes of soil-borne organisms. They also confront major abiotic constraints, including drought, terminal heat, salinity, waterlogging, lodging, micronutrient imbalances, metal and metalloid toxicities, and various physiological disorders. Sustainable management of these stresses requires integrated, science-based approaches combining host resistance, improved cultural practices, precision agronomy, early diagnostic tools, and predictive modeling. The volume *Wheat Stress Management: Emerging Solutions and Techniques*, developed by ICAR-Indian Institute of Wheat and Barley Research (IIWBR), addresses these challenges holistically and with a forward-looking approach. The editorial team, Drs. Ratan Tiwari, Binay Kumar Singh, and Ravindra Kumar, has systematically presented the scientific foundations and recent advances in breeding resilient cultivars through both conventional and modern approaches. The chapters critically examine the diversity, underlying mechanisms, diagnosis, and management of major wheat stresses, with strong emphasis on translational research and field-level applicability. This bridges the gap between laboratory discoveries and farm-level implementation. Contributions from leading researchers in wheat genomics, physiology, and pathology provide valuable multidisciplinary perspectives throughout the volume. This book is intended for graduate and postgraduate students of plant sciences, crop physiology, plant pathology, genetics and breeding, biotechnology, and agronomy, as well as for researchers, academicians, policymakers, and professionals involved in agricultural development and agro-industries. As the stress dynamics of wheat cultivation continue to evolve rapidly under the pressures of climate change and global trade, the editors' efforts in presenting this comprehensive and authoritative compendium are highly commendable. It is anticipated that this book will serve as a valuable reference and guiding blueprint for strengthening wheat resilience, stabilizing productivity, and ensuring sustainable food security in the future.

Deputy Director General (Crop Science) D. K. Yadava
Indian Council of Agricultural Research
Dr Rajendra Prasad Road
Krishi Bhavan, New Delhi, India
5 March 2026

# Preface

The global population has surpassed 8 billion and is projected to reach 10–11 billion by the end of this century. This demographic trend poses significant challenges for agricultural systems in providing sufficient, nutritious, and stable food supplies. Wheat, one of the most widely cultivated cereal crops, contributes substantially to daily caloric and protein intake, especially in South Asia and other wheat-dependent regions. In countries like India, wheat plays a vital role in national food security and supports rural livelihoods. However, the productivity and stability of wheat cultivation are under severe threat from climate variability, rising temperatures, frequent heat waves, unpredictable rainfall, soil degradation, the emergence of aggressive pathogen strains, and intensified biotic and abiotic stresses. Given the limited potential to expand cultivable land, increasing yield per unit area through scientifically sound, climate-resilient strategies remains the most practical solution for maintaining self-sufficiency. Consequently, strengthening stress management strategies to protect wheat productivity under changing environmental conditions is essential.

Wheat faces a wide range of biotic stresses, including spike diseases, foliar blights, rusts, stem and root pathogens, and complex soil-borne infections. At the same time, abiotic stresses such as drought, terminal heat, salinity, waterlogging, lodging, micronutrient imbalances, metal and metalloid toxicities, and various physiological disorders further threaten production systems. The combined effects of these stresses, often exacerbated by mono-cropping, intensive input use, and climate change, create complex challenges that require integrated solutions. Stress management does not aim to eliminate stressors entirely, as this is impractical under fluctuating agro-ecological conditions. Instead, it involves regulating these factors to manageable levels through integrated, science-based strategies. Effective management requires a thorough understanding of plant–pathogen and plant–environment interactions, along with the integration of host resistance, molecular breeding, precision agronomy, early diagnostic tools, marker-assisted detection, predictive modeling, nanotechnology interventions, and artificial intelligence-driven decision-support systems.

The book *Wheat Stress Management: Emerging Solutions and Techniques* offers a thorough, future-oriented approach to addressing these complex challenges in a holistic way. Edited by a multidisciplinary team of scientists, including Dr. Ratan Tiwari, Dr. Binay Kumar Singh, and Dr. Ravindra Kumar, the volume systematically covers advances in biotic and abiotic stress biology, the breeding of stress-resistant cultivars using both traditional and genomics-assisted methods, management of spike, foliar, stem, and root diseases, and tackling drought, heat, salinity, waterlogging, lodging, and metal stresses. It also discusses diagnosing and correcting physiological disorders. Dedicated chapters explore emerging areas such as plant biomarkers for early stress detection, nanotechnology applications to enhance tolerance, and the use of artificial intelligence for predictive stress analysis and precision management. The book critically examines the impact of climate change on wheat stress dynamics and suggests adaptive and mitigation strategies tailored to Indian production systems. Contributions from renowned researchers provide multidisciplinary perspectives that connect laboratory research with field practice. Intended for graduate and postgraduate students, researchers, educators, policymakers, and professionals in agricultural development and agro-industries, this volume serves as a reputable and comprehensive resource. By combining core principles with innovative solutions, the book aims to increase wheat resilience, stabilize productivity in stress-affected environments, and promote sustainable food security amid climate change and global challenges.

Karnal, Haryana, India

Ratan Tiwari
Binay Kumar Singh
Ravindra Kumar

# Contents

# Editors and Contributors

## About the Editors

**Ratan Tiwari** is a distinguished agricultural scientist with over 30 years of experience in genomics and molecular breeding for enhancing wheat resistance to biotic and abiotic stresses. He began his career at the ICAR-Directorate of Pulses Research, Kanpur, and now serves as Director at the ICAR-Indian Institute of Wheat and Barley Research (IIWBR), Karnal. Dr. Tiwari holds a PhD and MSc in Genetics and Plant Breeding, and a BSc in Agriculture from Banaras Hindu University (BHU), Varanasi. He has played a key role in developing eight wheat varieties, cultivated on approximately 6.9 million hectares, and eight genetic stocks. He has published over 100 research and review articles in peer-reviewed journals and reviewed more than 50 manuscripts for reputed international publishers. He has supervised five PhD and five MSc students, established academic collaborations and MoUs with leading institutions, and facilitated student training. He has also been invited to serve as an expert at international meetings in Mexico, the UK, France, and Morocco.

**Binay Kumar Singh** is Principal Scientist at ICAR-Indian Institute of Wheat and Barley Research (IIWBR), Karnal, with over 18 years of research experience in genomics and molecular breeding for stress resilience in rice, wheat, and mustard. He holds a PhD and MSc in Molecular Biology and Biotechnology from ICAR-Indian Agricultural Research Institute (IARI), New Delhi, and a BSc in Agriculture from Banaras Hindu University (BHU), Varanasi. He has held key roles, including Head of the Division of Crop Science at the ICAR-Research Complex for North Eastern Hill Region, Meghalaya; Scientist at the Directorate of Rapeseed Mustard Research, Bharatpur; and Senior and Principal Scientist at the ICAR-Indian Institute of Agricultural Biotechnology (IIAB), Ranchi. He has contributed to seven crop varieties and registered three genetic stocks. He has supervised eight MSc students and co-supervised several postgraduate and doctoral scholars. He has established over 20 institutional MoUs and collaborations with start-ups to advance research and innovation. His publications include 80 research papers, 16 popular articles, 4 books, 18 book chapters, and several technical bulletins. He has also edited and reviewed over 30 manuscripts for high-impact journals.

**Ravindra Kumar** is Senior Scientist (Plant Pathology) at ICAR-Indian Institute of Wheat and Barley Research (IIWBR), Karnal, with over 14 years of experience in agricultural research, particularly wheat pathology. He holds a PhD and MSc in Plant Pathology from Banaras Hindu University (BHU), Varanasi, and a BSc (Ag) from Chaudhary Charan Singh University, Meerut. His research focuses on plant disease management, biological control, seed pathology, and plant–microbe interactions. He has contributed to the development of two wheat varieties, DBWH221 and DBW 377. He has published 46 research articles in reputed journals, authored 2 textbooks, edited 3 books, contributed 24 book chapters, and written numerous technical bulletins and popular articles. He is an active member of several professional societies and serves as an Academic and Technical Editor for multiple international journals.

## Contributors

**Saumya Agrawal** Department of Biological Sciences, BITS- Pilani, K.K. Birla Goa Campus, Goa, India

**Rakesh Kumar Bairwa** ICAR-Indian Institute of Wheat and Barley Research, Karnal, Haryana, India

**Poonam Bansal** MMDU University, Ambala, Haryana, India

**Alka Bharati** ICAR Research Complex for NEH Region, Umiam, Meghalaya, India

**Pradeep Kumar Bhati** Borlaug Institute for South Asia (BISA), NASC Complex, New Delhi, India

**Sridhar Bhavani** International Maize and Wheat Improvement Centre (CIMMYT), Global Wheat Program, Nairobi, Kenya

**Tilak Chandra** ICAR-Indian Agricultural Statistics Research Institute, New Delhi, India

**Abhishek Dadsena** ICAR – National Institute for Plant Biotechnology, New Delhi, India

**Meghasree Devaiah** University of Agricultural Sciences, Dharwad, Karnataka, India

**Thokchom Malemnganbi Devi** ICAR-Indian Institute of Wheat and Barley Research, Karnal, Haryana, India

Karnal Hub of IARI Mega University, Karnal, Haryana, India

**Meenu Dua** K.L. Mehta Dayanand College for Women, Faridabad, Haryana, India

**Priti Duhan** K.L. Mehta Dayanand College for Women, Faridabad, Haryana, India

**Vikas Gupta** ICAR-Indian Institute of Wheat and Barley Research, Karnal, Haryana, India

**Gurudatt M. Hegde** University of Agricultural Sciences, Dharwad, Karnataka, India

**M. A. Iquebal** ICAR-Indian Agricultural Statistics Research Institute, New Delhi, India

**Priyanka Jadhav** University of Agricultural Sciences, Dharwad, Karnataka, India

**Sandeep Jaiswal** ICAR Research Complex for NEH Region, Umiam, Meghalaya, India

**Sarika Jaiswal** ICAR-Indian Agricultural Statistics Research Institute, New Delhi, India

**Vikas Joon** ICAR-Central Plantation Crops Research Institute, Kasaragod, Kerala, India

**Arun Kumar Joshi** Borlaug Institute for South Asia (BISA), NASC Complex, New Delhi, India

**Jayanth Kallugudi** ICAR-Indian Institute of Wheat and Barley Research, Shimla, Himachal Pradesh, India

**Gurpreet Kaur** ICAR-Central Soil Salinity Research Institute, Karnal, Haryana, India

**Malemnganbi Keisham** ICAR Research Complex for NEH Region, Umiam, Meghalaya, India

**Kiran** ICAR-Indian Institute of Wheat and Barley Research, Karnal, Haryana, India

**Zennah Kosgey** Kenya Agricultural and Livestock Research Organisation (KALRO), Food Crops Research Centre, Njoro, Kenya

**Neeraj Kulshreshtha** ICAR-Central Soil Salinity Research Institute, Karnal, Haryana, India

**Anita Kumari** Department of Plant and Soil Sciences, Texas Tech University, Lubbock, TX, USA

**Naresh Kumar** Department of Chemistry and Biochemistry, Eternal University, Baru Sahib, HP, India

**Ravindra Kumar** ICAR-Indian Institute of Wheat and Barley Research, Karnal, Haryana, India

**Santosh Kumar** ICAR – Indian Agricultural Research Institute, Gauria Karma, Jharkhand, India

**Sudheer Kumar** ICAR-Indian Institute of Pulses Research, Regional Research Centre, Bikaner, Rajasthan, India

**Vishwakarma Manish Kumar** Borlaug Institute for South Asia (BISA), NASC Complex, New Delhi, India

**Charu Lata** ICAR-Indian Institute of Wheat and Barley Research, Karnal, Haryana, India

**H. M. Mamrutha** ICAR-National Institute of Seed Science and Technology, Regional Station, Bengaluru, Karnataka, India

**Jitendra Kumar Meena** ICAR-Central Research Institute for Jute and Allied Fibres, Barrackpore, West Bengal, India

**Rajesh Mehrotra** Department of Biological Sciences, BITS- Pilani, K.K. Birla Goa Campus, Goa, India

**Sandhya Mehrotra** Department of Biological Sciences, BITS- Pilani, K.K. Birla Goa Campus, Goa, India

**Ian Moot** Department of Plant Science and Landscape Architecture, University of Maryland, College Park, MD, USA

**Nidhi** ICAR-Indian Institute of Wheat and Barley Research, Karnal, Haryana, India

**M. D. Oak** Genetics and Plant Breeding Group, Agharkar Research Institute, Pune, Maharashtra, India

**Vanita Pandey** ICAR-Indian Institute of Wheat and Barley Research, Karnal, Haryana, India

**Nitish Ranjan Prakash** ICAR – Central Soil Salinity Research Institute, Karnal, Haryana, India

**V. Rajeswari** ICAR-Indian Institute of Wheat and Barley Research, Karnal, Haryana, India

**J. M. Rajwade** Nanobioscience Group, Agharkar Research Institute, Pune, Maharashtra, India

**Jagadish Rane** ICAR-Central Institute for Arid Horticulture, Bikaner, Rajasthan, India

**Preety Rani** ICAR-Indian Institute of Wheat and Barley Research, Karnal, Haryana, India

**Rinki** ICAR-Indian Institute of Wheat and Barley Research, Karnal, Haryana, India

**M. S. Saharan** Division of Plant Pathology, ICAR-Indian Agricultural Research Institute, New Delhi, India

**Arpita Sharma** ICAR-Indian Institute of Wheat and Barley Research, Karnal, Haryana, India

Thapar Institute of Engineering and Technology, Patiala, Punjab, India

**Ravindra Singh Shekhawat** ICAR-Indian Institute of Wheat and Barley Research, Karnal, Haryana, India

**Sonia Sheoran** ICAR-Indian Institute of Wheat and Barley Research, Karnal, Haryana, India

**Binay K. Singh** ICAR-Indian Institute of Wheat and Barley Research, Karnal, Haryana, India

**Charan Singh** ICAR-Indian Institute of Wheat and Barley Research, Karnal, Haryana, India

**Gyanendra Singh** ICAR-Indian Institute of Wheat and Barley Research, Karnal, Haryana, India

**Ravi P. Singh** International Maize and Wheat Improvement Centre (CIMMYT), Texcoco, Mexico

**Tanya Singh** ICAR-Indian Institute of Wheat and Barley Research, Karnal, Haryana, India

Karnal Hub of IARI Mega University, Karnal, Haryana, India

**Vikram Singh** ICAR-Indian Institute of Wheat and Barley Research, Karnal, Haryana, India

**Yukti Singh** Department of Biological Sciences, BITS- Pilani, K.K. Birla Goa Campus, Goa, India

**Jyotsana Tilgam** ICAR – National Bureau of Agriculturally Important Microorganisms, Mau, Uttar Pradesh, India

**Ratan Tiwari** ICAR-Indian Institute of Wheat and Barley Research, Karnal, Haryana, India

**Vijay Tiwari** Department of Plant Science and Landscape Architecture, University of Maryland, College Park, MD, USA

**Zeenat Wadhwa** ICAR-Indian Institute of Wheat and Barley Research, Karnal, Haryana, India

# An Overview of Biotic and Abiotic Challenges to Wheat Cultivation in India

1

Binay K. Singh, Ravindra Singh Shekhawat,
Ravindra Kumar, Ratan Tiwari, and Ravi P. Singh

**Abstract**

Wheat (*Triticum aestivum* L.) is the second most important cereal crop in India, providing nearly 20% of its protein and 19% of its calorie intake. Since the Green Revolution of the 1960s, wheat production in India has grown significantly due to the adoption of semidwarf, high-yielding varieties, improved irrigation infrastructure, and advances in agronomic practices. Despite these gains, wheat cultivation still faces various biotic stresses, including rusts, powdery mildew, blast, spot blotch, and insect pests such as aphids and termites. Abiotic stresses such as drought, heat, salinity, metal toxicity, and waterlogging are also widespread and exacerbated by climate change and environmental variability. These challenges lead to considerable reductions in yield and grain quality, posing both economic and nutritional concerns. Over the years, management approaches in wheat have shifted from traditional cultural practices and extensive pesticide application to integrated disease and pest management (IDPM), precision nutrient management, conservation tillage, and breeding initiatives targeting climate resilience through genomics, molecular markers, and gene-editing technologies. Recent innovations, such as artificial intelligence (AI)-enabled phenotyping, remote sensing, and drone-based monitoring, improve early disease detection and stress monitoring. However, the emergence of new threats, including virulent pathogen races such as *Puccinia graminis* f. sp. *tritici* Ug99, pesticide-resistant pest biotypes, and changing climate patterns, necessitates continuous surveillance, adaptive breeding programs, and policies supporting climate-smart agriculture.

B. K. Singh · R. S. Shekhawat · R. Kumar · R. Tiwari
ICAR-Indian Institute of Wheat and Barley Research, Karnal, Haryana, India

R. P. Singh (✉)
International Maize and Wheat Improvement Centre (CIMMYT), Texcoco, Mexico
e-mail: r.singh@cgiar.org

 
R. Tiwari et al. (eds.), *Wheat Stress Management*,
https://doi.org/10.1007/978-981-92-0902-6_1

Increasing genetic diversity through landraces, wild relatives, and participatory breeding, combined with multisectoral collaboration and farmer-centered strategies, is crucial for maintaining adaptive capacity. Policies that promote scientific innovation, environmental protection, extension services, risk mitigation, and equitable access to resources are essential to developing sustainable, climate-resilient wheat production systems.

**Keywords**

Wheat stress management · Biotic and abiotic stress · Climate-resilient wheat breeding · Integrated pest and disease management · Precision nutrient application · Sustainable wheat cultivation

## 1.1 Introduction

### 1.1.1 Significance of Wheat in Indian Agriculture and Food Security

Wheat (*Triticum aestivum* L.) is the backbone of Indian agriculture and food security, ranking second only to rice as a staple crop. It is one of the most affordable food grains and provides 20% of protein and 19% of calories for the Indian population (Ramadas et al. 2019). Wheat is mainly grown during the *rabi* (winter) season, covering nearly 32.76 million hectares and yielding over 117.9 million tonnes (Ministry of Agriculture and Farmers' Welfare, Government of India 2025), making India the second-largest wheat producer in the world after China and contributing about 13–14% to global wheat production (FAO 2024). The crop holds immense economic and agronomic importance, contributing roughly 14% to the gross cropped area, about 4.78% to agricultural GDP (Ministry of Statistics and Program Implementation, Government of India 2025), and supporting rural livelihoods, especially in the Indo-Gangetic Plains. Wheat plays an important role in India's national food security and is a key staple commodity in the Public Distribution System (PDS) implemented under the National Food Security Act (NFSA), 2013. Under the NFSA, approximately 813.5 million beneficiaries—about 67% of the population—receive subsidized foodgrains, with wheat accounting for nearly 45–50% of the total foodgrain allocation in 2023–24 (Annual Report 2023–24 2024; Press Information Bureau (2024). The Green Revolution, launched in the 1960s, caused a paradigm shift in wheat cultivation by introducing high-yielding semidwarf varieties such as "Kalyan Sona" and "Sonalika." These varieties, along with 546 other improved cultivars released under the All India Coordinated Wheat Improvement Project (AICWIP) since 1965, increased productivity from approximately 887 kg $ha^{-1}$ in 1966–67 to over 3587 kg $ha^{-1}$ today (Directorate of Economics and Statistics, Ministry of Agriculture and Farmers Welfare, Government of India n.d.). Recently, significant advances in wheat genomics, including the development of pangenomic resources (IWGSC 2018), marker-assisted selection (MAS), genomic selection (GS), high-throughput phenotyping, speed breeding, and gene editing, have accelerated breeding efforts to develop climate-resilient, high-yielding cultivars with improved nutritional and agronomic traits (Waqas et al. 2025).

### 1.1.2 Overview of Major Biotic and Abiotic Stresses and Physiological Disorders Affecting Wheat Productivity

Wheat productivity in India and worldwide is limited by various biotic and abiotic stresses, as well as physiological disorders, which considerably reduce yield stability and grain quality. Understanding these factors is essential for developing integrated breeding and management strategies within climate-resilient agricultural systems.

#### 1.1.2.1 Biotic Stresses

##### 1.1.2.1.1 Major Diseases Affecting Wheat

The wheat crop is affected by more than 100 diseases caused by various phytopathogens, including fungi, bacteria, viruses, and nematodes. Among these, more than 20 diseases are considered highly damaging to wheat production (Goyal et al. 2021). Some of the most economically important wheat diseases are illustrated in Fig. 1.1 and described in detail in the following sections.

Rusts

Since wheat was domesticated, rust diseases have persisted as a constant threat to global wheat production. These are caused by *Puccinia* species, including stripe rust (*P. striiformis* f. sp. *tritici*), leaf rust (*P. triticina*), and stem rust (*P. graminis* f. sp. *tritici*). Worldwide, wheat rusts are estimated to cause annual losses of US$ 4.3–5.0 billion (Prasad et al. 2020). Stripe rust is highly prevalent in temperate regions with cool, wet weather and appears as bright yellow-orange pustules (sori)

**Fig. 1.1** Some important diseases of wheat: (**a**) stripe rust, (**b**) leaf rust, (**c**) stem rust, (**d**) powdery mildew, (**e**) leaf blight, (**f**) Fusarium head blight (FHB), (**g**) karnal bunt, and (**h**) loose smut

that form stripes on the leaves. These pustules produce yellow-orange urediniospores that rupture the epidermis, mainly on the upper leaf surface. Currently, stripe rust is the most economically important wheat rust disease, causing yield losses up to 100% in susceptible cultivars (Chen 2005; Figueroa et al. 2018). Unlike stripe rust, leaf rust produces scattered pustules instead of striped patterns, mainly on the upper leaf surface. Leaf rust is the most common and widely distributed of the three wheat rust diseases, affecting over 90% of global wheat production (Chai et al. 2020). Under favorable environmental conditions—moderate temperatures of 10–25 °C and moist weather—leaf rust can reduce yields by up to 50% in susceptible cultivars (Huerta-Espino et al. 2011; Chai et al. 2020). Stem rust, another economically important rust disease, primarily develops on stems and leaf sheaths as large, dark-orange to brick-red, elongated pustules. When these pustules rupture, they release a powdery mass of brick-red urediniospores that break through the host epidermis. Under favorable conditions—temperatures of 15–35 °C and high humidity—stem rust can cause yield losses of up to 100% in susceptible cultivars (Leonard and Szabo 2005). A particularly concerning development is the emergence of the highly virulent *P. graminis* race Ug99. It was first detected in Uganda in 1998, and it has since spread across eastern, southern, and northern Africa, as well as into western Asia (Singh et al. 2015).

#### Powdery Mildew

Powdery mildew of wheat is caused by *Blumeria graminis* f. sp. *tritici* (syn. *Erysiphe graminis* f. sp. *tritici*). The disease appears as superficial, greyish-white, cottony patches on leaves, leaf sheaths, stems, and even spikes. These patches may occur on both leaf surfaces, and in severe cases, the white, cottony growth can cover the entire leaf and other aerial parts of the plant. As the disease progresses, the patches turn brownish to dull tan late in the season due to the development of cleistothecia. Under moderate conditions, powdery mildew can cause yield losses of about 10%, whereas under severe infection, losses of up to 62% have been reported (Costamilan 2005).

#### Wheat Blast

Wheat blast disease, caused by *Magnaporthe oryzae* pathotype *Triticum*, has not yet been reported from India. However, this destructive wheat disease was first identified in 2016 in neighboring Bangladesh. It is also common in several South American countries, including Brazil (first reported in 1985), Bolivia, Paraguay, Argentina, and Uruguay, as well as in Zambia. The wheat blast pathogen can infect all aerial parts of the plant, including the spike, leaf, stem, peduncle, glume, awn, and seed. Early symptoms on leaves appear as spindle- or boat-shaped, water-soaked, brown necrotic spots. The most distinctive and severe symptoms, however, occur on the spikes. Infected spikes may be partially or completely bleached, with bleaching usually starting from a blackish-grey infection point at the rachis or the base of the spike. Depending on the site of infection, spikes may partially or fully dry out (Singh et al. 2021). Wheat blast is known for its destructive effects and can cause yield losses of up to 100% (Duveiller et al. 2016; Cruz and Valent 2017).

### Blotches/Blights

Spot blotch of wheat, caused by *Bipolaris sorokiniana* (also known as *Helminthosporium sativum*; teleomorph *Cochliobolus sativus*), is a major disease in warm, humid wheat-growing areas. In India's Eastern Gangetic Plains, Bangladesh, and Nepal, it occurs alongside tan spot, caused by *Pyrenophora tritici-repentis* (Died.) Drechs. (anamorph *Drechslera tritici-repentis* (Died.) Shoemaker), and this disease complex is often referred to as Helminthosporium leaf blight (HLB) (Duveiller et al. 2005). Leaf blight significantly limits wheat production in the Eastern Gangetic Plains, the core wheat-growing region of India, Bangladesh, and Nepal (Duveiller and Sharma 2009). Globally, around 25 million hectares (mha) of wheat are affected, with about 10 mha in the Eastern Gangetic Plains (Roy et al. 2023). Yield losses can reach up to 50% under ideal infection conditions (Sharma and Duveiller 2006). In addition to directly damaging tissues, the pathogen produces host-specific toxins such as helminthosporol, sorokinianin, and bipolaroxin, which increase its virulence (Roy et al. 2023).

Fusarium head blight (FHB), also called wheat head scab or ear blight, is an important disease affecting wheat and other cereals worldwide. Although more than a dozen *Fusarium* species can cause the disease depending on region and climate, *Fusarium graminearum* is recognized as the main pathogen globally (Summerell et al. 2010; Karlsson et al. 2021). The disease is marked by early aging of spikes, starting with the bleaching of infected spikelets while nearby spikelets stay green. As the pathogen advances through the rachis, additional spikelets above and below the original infection site may also bleach, and in severe cases, pinkish-orange spore masses can be seen on infected tissues (Figueroa et al. 2018; Kashyap et al. 2022). *F. graminearum* is a seed-borne pathogen (Kumar et al. 2020) that causes significant yield losses by producing shriveled, lightweight, and poor-quality grains. More importantly, the pathogen contaminates kernels with trichothecene mycotoxins such as deoxynivalenol (DON), which pose serious health risks to humans and animals and render the grain unfit for consumption and trade (McMullen et al. 2012).

### Bunts and Smuts

Bunts and smuts of wheat are caused by basidiomycete fungi. The major diseases include Karnal bunt (caused by *Tilletia indica*, syn. *Neovossia indica* (Mitra) Mundkur), loose smut (caused by *Ustilago segetum* f. sp. *tritici*), flag smut (caused by *Urocystis agropyri*), and hill bunt or stinking/common bunt (caused by *Tilletia caries* and *Tilletia foetida*). Since these diseases rarely occur in epidemic form, their overall economic importance is generally limited, except for Karnal bunt (Iquebal et al. 2021). Field diagnosis of Karnal bunt is difficult, as not all kernels in a spike are infected, and most infected kernels remain asymptomatic until maturity. Infected kernels are typically dark in color and emit a characteristic fishy odor; symptoms usually appear at the germ end, although in some cases the entire kernel may be affected. The darkening results from the fungus converting kernel tissue into a teliospore mass (Kumar et al. 2020). Owing to its quarantine relevance, strict regulations are enforced in many countries to prevent the introduction of the Karnal bunt pathogen. At several instances, the disease has also led to the rejection of Indian wheat consignments in international trade; for instance, in 1996, Turkey, Morocco, and Poland detained shipments, and a cargo worth US$ five million was denied unloading in Poland due to Karnal bunt infection (Kumar et al. 2015).

#### 1.0.2.1.2 Insect Pests

Among insect pests, foliar aphids (*Rhopalosiphum maidis* L., *Rhopalosiphum padi* L., *Sitobion avenae* F.), termites (*Odontotermes obesus* Rambur, *Microtermes obesi* Holmgren), root aphid (*Rhopalosiphum rufiabdominalis* (Sasaki)), pink stem borer (*Sesamia inferens* Walker), and armyworm (*Mythimna separata* Walker) are the main pests affecting the wheat crop.

##### Foliar Aphids

More than 11 aphid species have been reported to attack wheat, including four predominant species—*Sitobion avenae* (Fabricius), *Sitobion miscanthi*, *Rhopalosiphum padi*, and *Rhopalosiphum maidis*. Among these, the corn leaf aphid, *R. maidis*, is the most serious pest in the North-Western Plains of India (Singh and Jasrotia 2020). This small, green aphid infests wheat crops starting at the seedling stage, initially appearing in October–November and reaching peak populations during the ear-ripening stage in February–March. During the vegetative stage, aphids feed on leaves by sucking cell sap, causing chlorosis. Foliar aphids can directly reduce yield by 20–30% through sap extraction (Singh et al. 2020).

##### Termites

About 16 species of termites have been reported to damage wheat crops in India, among which *Odontotermes obesus* (Rambur) and *Microtermes obesi* (Holm) are the dominant species (Chhillar et al. 2006). These small to medium-sized, soft-bodied, slender insects range in color from creamy or dirty white to dark brown and live in underground colonies. Termites are classic pests of wheat under rainfed conditions, and sandy loam soils are particularly conducive to their infestation (Sharma et al. 2004). They attack wheat at all stages of development, from seedlings to maturity. Severely damaged plants appear wilted and dried, can be easily uprooted, and eventually die. Yield losses due to termite damage in wheat have been reported to range from 43% to 80% (Chhillar et al. 2006; Kambrekar et al. 2016).

##### Root Aphid

Root aphids (*Rhopalosiphum rufiabdominalis* (Sasaki)) are stout, pale green insects that attack wheat, barley, oats, and other cereals. They are often not taken seriously because they feed primarily on plants' belowground stems and roots. Root aphid infestations are particularly common in rainfed wheat-growing areas. These aphids feed in the root zone, usually during the seedling stage, especially in light soils. By sucking sap from the roots, they cause the leaves of affected plants to dry up (Kindler et al. 2004). When infested plants are uprooted, aphid colonies can be observed clinging to the root system. Ants are often seen around the collars of plants, feeding on aphid honeydew and carrying aphids to noninfested fields.

Pink Stem Borer

The caterpillar is the damaging stage of the pink stem borer (*Sesamia inferens* Walker). It is small, pink, and smooth, with dark spots on its body and 2- to 2.5-mm-long hairs. During the early stages of crop growth, insect feeding can kill the entire plant, resulting in the formation of "dead hearts." Affected plants turn yellow, and stems can be easily pulled out, often with excreta at the base. Early-stage damage makes larvae easy to observe at the base of affected stems. In later stages, individual tillers die, forming white "dead hearts." Wheat sown in zero-tillage fields within the rice–wheat cropping system is especially vulnerable to pink stem borer infestation. This pest can cause yield losses exceeding 11% in wheat (Baladhiya et al. 2018).

Army Worm

The larval stage of the armyworm (*Mythimna separata* Walker) is the only destructive stage. Newly hatched larvae feed on tender leaves, while older larvae consume developing grains, leaves, flag leaves, awns, and glumes. During the day, the larvae typically hide in soil cracks and feed at night or in the early morning, although in wet, humid conditions, they may also feed during the day. These larvae are voracious feeders, and under favorable conditions, entire fields can be defoliated in a short period. Armyworms persist in rice stubbles before the wheat crop and survive the summer on subsequent crops such as rice (Singh and Sarao 2020).

### 1.0.2.2 Abiotic Stresses

#### 1.0.2.2.1 Drought Stress

Drought stress severely limits wheat cultivation in India, particularly in the semiarid and rainfed regions. It affects wheat at multiple growth stages, including germination, tillering, flowering, and grain filling, resulting in significant reductions in plant growth, biomass accumulation, and grain yield. Physiologically, drought reduces leaf area, chlorophyll content, photosynthetic rate, seedling vigor, and nutrient uptake, while morphologically it decreases root and shoot length and alters root-to-shoot ratios to improve water-uptake efficiency. Additionally, drought stress accelerates leaf senescence and shortens the grain filling period, directly impairing both yield and grain quality. These effects translate into economic losses, with yield reductions in drought-prone states like Rajasthan, Haryana, and Madhya Pradesh ranging from 20% to 60% (Nyaupane et al. 2024; Gaikwad et al. 2025).

#### 1.0.2.2.2 Heat Stress

Heat stress is also a critical abiotic factor that adversely affects wheat (*Triticum aestivum* L.) cultivation in India, especially in semiarid and subtropical regions such as the Indo-Gangetic Plains, Rajasthan, Haryana, and Punjab. Elevated temperatures during key developmental stages, mainly flowering and grain filling, significantly reduce yield by accelerating crop maturation, decreasing grain number and size, and impairing grain quality. Terminal heat stress, occurring when temperatures exceed 31 °C during grain filling, causes premature senescence, disrupts photosynthesis by damaging chloroplasts, and increases oxidative stress through reactive oxygen species (ROS) accumulation. Moreover, high temperatures inhibit enzymatic activities essential for seed germination, stunt root and shoot growth at the seedling stage, and cause pollen sterility, which drastically reduces fertilization success and grain set. Empirical data suggest that a 1 °C increase in temperature can reduce wheat yield by approximately 6%, with late-sown varieties experiencing up to 33% yield losses (Dubey et al. 2020; Abidin et al. 2024). The occurrence of extreme heatwaves, with temperatures exceeding 40 °C during critical periods in March and April, has exacerbated these impacts, advancing wheat maturity by up to two weeks and further reducing productivity.

#### 1.0.2.2.3 Salinity Stress

Salinity stress significantly hampers wheat cultivation in India, particularly in arid and semiarid regions with high soil salinity and sodicity. It reduces growth, development, and yield by causing osmotic stress, ion toxicity, and nutrient imbalances (El Sabagh et al. 2021; Ur Rehman et al. 2025). Physiologically, salinity impairs photosynthesis, nutrient uptake, water relations, seed germination, and root/shoot growth, while accelerating phenological development and reducing fertile tillers, spikelets per spike, kernel weight, and final grain yield, sometimes by up to 45% (Seleiman et al. 2022). Reactive oxygen species (ROS) induced by salt toxicity cause oxidative damage to cellular components (Rani et al. 2018). The impact varies by genotype and growth stage, with greater yield reduction during anthesis and grain filling (Khedr et al. 2022).

#### 1.0.2.2.4 Metal Toxicity

Metal toxicity is an emerging abiotic stress affecting wheat cultivation in India. It is caused primarily due to the accumulation of heavy metals like cadmium (Cd), lead (Pb), chromium (Cr), nickel (Ni), and cobalt (Co) in agricultural soils through industrial pollution, wastewater irrigation, and excessive fertilizer and pesticide use (Rizvi et al. 2020; Muhammad et al. 2025). These heavy metals inhibit wheat growth by inducing root growth inhibition, altering root architecture, and reducing shoot length and biomass. For instance, Cd toxicity leads to stunted growth, disrupted membrane permeability, and excessive ROS production, causing oxidative damage (Maqbool et al. 2025). Chromium causes chlorosis, wilting, and impaired photosynthesis by damaging photosystem II, while Pb reduces photosynthetic pigments and biomass (Singh et al. 2013). Nickel and Co pose dose-dependent toxic and stimulatory effects (Morkunas et al. 2018). Such metal-induced phytotoxicity results in reduced seed germination, poor nutrient uptake, compromised physiological functions, and significant yield losses. Furthermore, the accumulation of heavy metals in wheat grains poses food safety risks. Recent studies show that metal-tolerant plant growth-promoting rhizobacteria (PGPR) can alleviate metal stress by improving nutrient supply and secreting protective biomolecules, offering a sustainable mitigation strategy. Despite growing awareness, the impacts of metal toxicity on wheat and effective remediation approaches have been minimally investigated, jeopardizing yields and food safety (Rizvi et al. 2020; Muhammad et al. 2025).

#### 1.0.2.2.5 Waterlogging

Waterlogging stress significantly limits wheat cultivation in India, especially in sodic and poorly drained soils prevalent in the Indo-Gangetic Plains and canal-irrigated areas. Waterlogging creates hypoxic or anoxic soil conditions that restrict root respiration and nutrient uptake, leading to impaired root growth and diminished shoot development (Marashi and Chinchanikar 2010; Singh 2023). This stress reduces chlorophyll content, photosynthetic efficiency, spike number, kernel number per spike, and 1000-kernel weight, causing substantial yield losses that may exceed 40%, depending on stress duration and growth stage at exposure. Early growth stages such as tillering are particularly sensitive, with yield reductions of up to 18–20% reported for waterlogging lasting 7–14 days (Nóia Júnior et al. 2025). Waterlogged conditions also induce biochemical changes, such as increased proline accumulation and aerenchyma formation in roots, which help plants tolerate oxygen deficiency (Singh 2023). Genotypic variability exists in waterlogging tolerance, with some wheat varieties maintaining biomass and yield under short-term inundation.

### 1.0.2.3 Physiological Disorders

#### 1.0.2.3.1 Lodging

Lodging refers to the permanent displacement of wheat plants from their upright position, commonly caused by weak stem strength and adverse weather conditions such as wind, rain, and hail. It is an important biotic stress in India, particularly in the Indo-Gangetic Plains (Mondal 2020). Lodging primarily occurs near harvest maturity and can reduce wheat grain yield by 12% to 66%, and, in severe cases, up to 80% globally (Khobra et al. 2019). The consequences of lodging include reduced photosynthetic capacity due to altered canopy architecture, increased susceptibility to pests and fungal diseases due to favorable moist microclimates, poor grain filling, and decreased grain quality. Lodging also reduces test weight and Hagberg Falling Number, impairing bread-making quality. Yield losses and quality degradation are more severe if lodging occurs during early grain filling. The advent of semidwarf wheat varieties with dwarfing *Rht* genes significantly improved lodging resistance and overall productivity in India, marking a key milestone in the wheat revolution (Chai et al. 2021). However, as yields and plant stature have changed over time, lodging risk persists, particularly under high-fertility and input conditions. Agronomic practices such as timely sowing, balanced fertilization, and foliar application of growth regulators are employed to manage lodging. Breeding for lodging tolerance, focused on improving stem and root structural integrity, is crucial for sustaining high wheat yields in India (Mondal 2020; Khobra et al. 2019).

#### 1.0.2.3.2 Preharvest Sprouting (PHS)

Preharvest sprouting (PHS) is an important problem affecting wheat cultivation in India, especially in regions prone to late-season rains and high humidity during grain maturity. Preharvest sprouting occurs when physiologically mature wheat grains germinate on the spike before harvest, typically under wet, humid environmental conditions that break seed dormancy (Nakamura 2018). This precocious germination degrades starch and protein in the grain, resulting in poor grain quality, reduced yield, and lower economic returns for farmers. The damage manifests as increased alpha-amylase activity, which correlates with lower Hagberg Falling Number values, a key indicator of grain quality affecting bread-making potential (Ral et al. 2016). Genetic factors controlling seed dormancy and germination sensitivity, such as quantitative trait loci (QTL) and genes like the *Mother of Flowering Time (MFT)*, play crucial roles in PHS resistance (Yang et al. 2025). Breeding efforts focus on identifying and introgressing alleles that enhance dormancy without affecting rapid germination postharvest. Effective management strategies include timely harvesting to avoid exposure to moisture, growing resistant cultivars, and understanding the genetic mechanisms underlying PHS (Singh et al. 2021). Ongoing research is elucidating the molecular and environmental interactions that influence *PHS*, with the aim of developing varieties better suited to India's climate variability and reducing yield and quality losses caused by preharvest sprouting (Kumar et al. 2023a, b).

#### 1.0.2.3.3 Zinc and Iron Deficiency

Iron (Fe) and zinc (Zn) deficiencies are widespread micronutrient constraints that adversely affect wheat cultivation across India, particularly in the Indo-Gangetic Plains and semiarid regions, where soils are often calcareous, alkaline, and nutrient-depleted (Debnath et al. 2021). These deficiencies limit wheat growth by impairing enzymatic functions, chlorophyll synthesis, and photosynthesis, resulting in reduced tillering, poor root development, and decreased grain yield (Ram et al. 2024). Deficiency symptoms include chlorosis, stunted growth, and delayed maturity. Micronutrient-poor soils not only reduce crop productivity but also lower Fe and Zn concentrations in wheat grains, negatively impacting human nutrition and exacerbating micronutrient malnutrition (hidden hunger) prevalent in India (Gupta et al. 2024). Studies reveal that modern high-yielding wheat varieties often have diminished grain Fe and Zn content due to the dilution effect of increased yields (Gul et al. 2023). Foliar and soil application of Fe and Zn fertilizers have shown promising results in improving both wheat yield and grain micronutrient density. Breeding biofortified wheat varieties with enhanced Fe and Zn accumulation is recognized as a sustainable approach to address both agronomic and nutritional challenges. Integrated soil fertility management combining balanced fertilization, organic amendments, and cultivar improvement is essential to mitigate Fe and Zn deficiencies and secure both wheat productivity and nutritional quality in Indian agroecosystems (Morton et al. 2023).

### 1.0.3 Impact of Climate Change and Environmental Variability on Wheat Cultivation

Wheat cultivation is highly sensitive to climatic and environmental changes, and the accelerating pace of climate change significantly threatens its yields, grain quality, and the regions suitable for cultivation. As a temperature-sensitive crop with a narrow optimal temperature range during key growth stages, wheat is especially vulnerable to higher average temperatures, more frequent and intense heatwaves, unpredictable rainfall, and terminal droughts. These environmental stresses, whether alone or combined, disrupt vital physiological processes such as photosynthesis, nutrient distribution, pollen health, and grain development, leading to major yield reductions and poorer end-use quality. Additionally, although higher atmospheric $CO_2$ levels can boost photosynthesis and plant growth, they often cause nutrient dilution, lowering grain protein and micronutrient levels such as Zn and Fe, which affects human nutrition and food security.

#### 1.0.3.1 Rising Temperature and Heat Stress

Rising global temperatures and frequent heatwaves are increasingly threatening wheat yields and quality, especially in heat-prone regions such as Northern and Central India. The IPCC projects that global temperatures will exceed 2 °C above preindustrial levels by 2050 (and nearly 5 °C by 2100) (Field et al. 2014). Wheat is highly vulnerable to heat stress during anthesis and early grain filling, with yield losses reaching up to 44% due to reduced spikelet fertility, impaired pollen function, and shortened grain-filling duration (Djanaguiraman et al. 2020). High daytime temperatures disrupt photosynthesis and pollen viability, while high night temperatures reduce biomass and grain weight through elevated respiration and impaired sink activity (Posch et al. 2022). Grain filling is particularly sensitive to temperatures above 22 °C (Farooq et al. 2011), and every 1 °C rise in global temperature shortens grain-filling by 2.8 days and reduces yield by 6% (Zhao et al. 2017). Consequently, terminal heat stress, especially when coinciding with flowering and grain development, significantly compromises wheat yield and grain quality (Zampieri et al. 2017).

#### 1.0.3.2 Altered Precipitation and Drought

Changing precipitation patterns have caused more frequent and severe droughts in both irrigated and rainfed wheat-growing areas. Although about 80% of the wheat crop in India is grown under irrigated conditions (Reynolds et al. 1999), approximately 66% of the crop receives only partial irrigation (Kang et al. 2017), making the crop vulnerable to moisture stress. In rainfed regions such as Central and Peninsular India, drought stress during anthesis and grain filling can reduce yield by 20–50% (Reynolds et al. 2011). Irregular rainfall also disrupts sowing schedules, delays germination, and causes water stress during early vegetative stages, while sudden rain during maturity may increase lodging, preharvest sprouting (PHS), and disease incidence. Conversely, excessive rainfall or poor drainage can cause waterlogging, especially in lowland areas, which negatively affects root health and oxygen uptake (Setter and Waters 2003).

#### 1.0.3.3 Increased $CO_2$ and Nutrient Imbalance

Since the onset of the Industrial Revolution, atmospheric $CO_2$ concentrations have increased significantly from around 280 ppm to over 400 ppm, mainly due to human activities such as burning fossil fuels and large-scale deforestation (Mishra et al. 2024). Projections suggest that without substantial mitigation efforts, $CO_2$ levels could surpass 550 ppm by 2050, effectively doubling preindustrial levels (IPCC 2023). This rise in $CO_2$ has serious consequences for the global climate, ecosystems, and agriculture. In $C_3$ crops such as wheat, higher atmospheric $CO_2$ has a dual effect: it boosts yields while reducing grain quality. Free-air $CO_2$ enrichment (FACE) experiments have shown yield increases of up to 26% at approximately 602 ppm $CO_2$, mainly due to increased grain number per unit area (Broberg et al. 2017). However, these gains are often balanced by significant drops in grain protein content, baking quality (Panozzo et al. 2014), and vital nutrients, including macro- and microelements and amino acids, raising concerns about the nutritional and end-use quality of wheat, especially in high-protein varieties (Blandino et al. 2020). Other studies indicate that elevated $CO_2$ levels decrease Zn, Fe, and protein concentrations in wheat grains, further threatening nutritional security (Myers et al. 2014). Additionally, $CO_2$-driven stomatal closure may reduce transpirational cooling, raise canopy temperature, and worsen heat stress in field conditions.

#### 1.0.3.4 Elevated Ozone and Air Pollution

Ground-level ozone ($O_3$) pollution represents a serious and escalating threat to wheat production in India, particularly across the Indo-Gangetic Plains. Recent assessments estimate mean wheat yield losses of approximately 14.18% during 2008–2012, corresponding to nearly 15 million tonnes of annual production loss, largely attributable to elevated surface ozone concentrations (Mills et al. 2018). Ozone enters leaves through stomata and induces oxidative stress, leading to damage of chlorophyll, inhibition of key enzymatic activities, and disruption of photosynthetic processes, ultimately reducing grain yield, grain number, and grain size (Feng et al. 2008; Ainsworth et al. 2012). Irrigated wheat is generally more vulnerable than rainfed wheat due to higher stomatal conductance and increased ozone uptake under well-watered conditions (Mills et al. 2018). Expansion of fully irrigated wheat cultivation is therefore expected to exacerbate ozone-induced yield penalties, particularly in arid and semiarid states such as Gujarat, Telangana, and Rajasthan, where additional yield reductions of 9–12% have been projected. Beyond yield losses, ozone exposure adversely affects grain quality by impairing starch biosynthesis, altering protein composition, reducing grain hardness and bulk density, and modifying mineral content and dough rheological properties, thereby compromising nutritional value and end-use quality, despite occasional increases in grain protein concentration associated with stress-induced premature senescence (Feng et al. 2011; Pleijel and Uddling 2012).

#### 1.0.3.5 Climate-Driven Pest and Disease Dynamics

Climate change significantly impacts wheat production by increasing the incidence, severity, and geographic distribution of diseases and insect pests. Rising temperatures, elevated atmospheric $CO_2$ concentrations, altered precipitation patterns, and more frequent extreme weather events create conditions that favor the growth, survival, and virulence of major wheat pathogens, including *Puccinia graminis* f. sp.

*tritici* and *Fusarium* spp. (Chakraborty and Newton 2011; Juroszek and von Tiedemann 2013). Simultaneously, climate-induced abiotic stresses such as heat, drought, and salinity compromise wheat physiological performance and immune responses, which increase host susceptibility to pathogen infection (Atkinson and Urwin 2012). As a result, diseases such as rusts, spot blotch, and Fusarium head blight are occurring more frequently and with greater intensity in climate-vulnerable agroecologies, particularly in the Indo-Gangetic Plains (Sharma et al. 2016; Chakrabarti et al. 2025). Climate change also alters insect pest dynamics by accelerating development rates, increasing the number of generations per season, and extending the seasonal activity of key wheat pests, including aphids (*Schizaphis graminum*), Hessian fly (*Mayetiola destructor*), and wheat stem sawfly (*Cephus cinctus*), while reducing overwintering mortality (Deutsch et al. 2018; Bale et al. 2002). Furthermore, invasive species such as the Russian wheat aphid (*Diuraphis noxia*) are expanding into previously unsuitable wheat-growing regions due to warming climates (Trnka et al. 2007). Elevated $CO_2$ concentrations can decrease nitrogen content in wheat tissues, thereby stimulating compensatory feeding by phloem-feeding insects and exacerbating crop damage (Robinson et al. 2012). Abiotic stress conditions further weaken host defenses, increasing vulnerability to pests such as the wheat curl mite (*Aceria tosichella*), a vector for viral diseases (Murphy and Burrows 2021). Climate variability can also disrupt the synchrony between pests and their natural enemies, reducing the effectiveness of biological control and destabilizing integrated pest management systems (Thomson et al. 2010). Taken together, these interacting factors undermine the durability of conventional disease and pest management strategies, including fungicide applications and host resistance breeding (Bebber et al. 2013).

#### 1.0.3.6 Shifting Agroclimatic Suitability and Crop Calendar

Climate change is significantly shifting agroclimatic suitability and altering the crop calendar for wheat, posing a critical challenge to its sustainable production. Wheat, being a thermosensitive crop, is highly responsive to variations in temperature, rainfall patterns, and seasonal duration (Asseng et al. 2019). Rising minimum and maximum temperatures, especially during the reproductive and grain-filling stages, are shortening the growing season and reducing yield potential in traditional wheat-growing regions such as the Indo-Gangetic Plains (Lobell et al. 2011; Ortiz et al. 2008). Projections suggest that wheat-growing areas in northern and central India may become less suitable due to frequent heatwaves and terminal heat stress, while relatively cooler regions in eastern and northeastern India might become more suitable, requiring a geographical reorientation of wheat cultivation (Jat et al. 2016; Soora et al. 2011). Altered precipitation patterns and increased frequency of extreme weather events, such as early or late rains, droughts, and hailstorms, further disrupt sowing and harvesting windows, heightening vulnerability to crop failure (Aggarwal et al. 2004). The traditional wheat crop calendar—typically sown in November and harvested in March or April—is under pressure, necessitating earlier sowing dates to avoid terminal heat and crop-duration modifications through short-duration or heat-tolerant varieties. Furthermore, climate-induced changes in agroclimatic zones impact soil moisture availability, irrigation scheduling, and pest and disease incidence, thereby affecting overall crop management practices. These shifts require region-specific redefinition of wheat-growing zones, climate-resilient agronomic planning, and adaptive research to help farmers optimize planting schedules and cultivar choices amid changing climatic scenarios (Ortiz et al. 2008; Asseng et al. 2019).

## 1.1 Historical Perspectives on Wheat Cultivation and Stress Management

### 1.1.1 References to Wheat Cultivation in Ancient Indian Texts

Archaeobotanical evidence from excavations at Indus Valley sites such as Harappa (mature period) (Punjab province, Pakistan; ~2600–1900 B.C.), Mohenjo-daro (Sindh province, Pakistan; ~2600–1900 B.C.), Mehrgarh (Baluchistan, Pakistan; 7000–5500 B.C.), and Chanhu-daro (Sindh province, Pakistan; ~4000–1700 B.C.), along with Chalcolithic sites in India including Atranjikhera (Uttar Pradesh, Etah; ~1200–600 B.C.), Chirand (Bihar, Saran; Neolithic ~2500 B.C. to Chalcolithic and Early Iron Age), Kayatha (Ujjain, Madhya Pradesh; 2400–2000 B.C.), Maheshwar (Khargone, Madhya Pradesh; 1600–1300 B.C.), Navdatoli (Maheshwar, Madhya Pradesh; 1600–1300 B.C.), Sonegaon (Maharashtra, Vidarbha; 700–200 B.C.), Ter (Maharashtra, Osmanabad; ~2nd–third century B.C.), and Inamgaon (Pune, Maharashtra; 1600–700 B.C.) provides clear signs of early wheat cultivation and its integration into the diet of the northwestern Indian subcontinent during the later part of the Neolithic era (7000–1000 B.C.). However, wheat is not mentioned in the *Rigveda*, composed around 1500 B.C. (Prakash 1961). This suggests that wheat cultivation and consumption declined with earlier cultures. Wheat adoption appears to have gradually revived after more than 1500 years, possibly through contact between the Aryans, who migrated to the Indian subcontinent between 1800 and 1500 B.C., and non-Aryan groups known to have used wheat, as revealed by excavations at sites as old as 7300 B.C. (Prakash 1961).

By the later Vedic period (1500–800 B.C.), wheat seems to have regained some popularity, as evidenced by references in the *Yajurveda* (Yajur Ved Vājasmeyi Saṃhitā 1959) and the *Śatapatha Brāhmaṇa*, where it was used to prepare groats. However, unlike rice and barley, it was not yet used in offerings, as indicated in the *Sūtras* (800–300 B.C.), implying limited adoption. Early Buddhist and Jain texts, such as the *Vinaya Piṭaka* (1st–4th century A.D.), mention the occasional use of wheat in sugar-coated cakes (Prakash 1961). Wheat gained a more prominent role among cereals during the Maurya and Śuṅga periods (300 B.C.–75 A.D.). The *Jātakas* (100–200 A.D.) mention mathri (Jatakas I–VII 1877), a popular wheat dish.

Medical treatises by Caraka and Suśruta (3rd century A.D.) recorded two cultivars, Madhulika and Nandimukhi, though wheat still ranked below rice and barley in dietary importance (Charak Samhita 1941; Bhishagratna 1963). The renowned physician Dhanvantari described wheat as food for the degraded, reflecting its low status among grains (Prakash 1961); rice and barley remained more prominent. A similar view appeared in the Gupta period (300–750 A.D.) in the Matsya Purana (1972). After the Gupta era (750–1200 A.D.), wheat began to gain popularity, with new recipes (kasar, suhali, phalika, mandaka), and the *Viṣṇudharmottara* (9th–11th century A.D.) (Prakash 1961) listed wheat among five common grains (Randhawa 1980).

Invasions in the 10th century disrupted the lives of the people, but wheat likely became more integrated into the diet, as seen in culinary texts such as *Mānsolāsā* (1126–1135 A.D.) and *Bhojan Kutūhala* (1600 A.D.). From its earliest cultivation in the Neolithic period to the early 1900s, farmers primarily grew diverse landraces and mixed local cultivars ("sorts"). A systematic wheat improvement program started in 1905—and, more notably, the Green Revolution of the late 1960s, driven by the introduction of Mexican semidwarf germplasm with Norin dwarfing genes—shifted breeding focus toward higher yields in native tall landraces and led to significant productivity increases in India, with mean yields rising from 653 kg $ha^{-1}$ in 1951 to about 3587 kg $ha^{-1}$ in 2025; concurrently, national production exceeded 117 million tonnes from approximately 32.76 million hectares.

### 1.1.2 Early Domestication, Adaptation, and Spread of Wheat Across Diverse Agroecological Regions in India

The family Poaceae (grasses) originated approximately 50–70 million years ago. The subfamily Pooideae, containing wheat (*Triticum*), barley (*Hordeum*), and oats (*Avena*), diverged around 20 million years ago (Kellogg 2001). Within *Triticum*, two wild diploid species are recognized: *T. boeoticum* (AbAb) and *T. urartu* (AuAu). The first domesticated diploid wheat, einkorn (*T. monococcum*), arose in the Fertile Crescent from the wild progenitor *T. boeoticum*. In contrast, *T. urartu*, although widespread across the Fertile Crescent (Zohary et al. 2000), was never domesticated but contributed the A genome to polyploid wheats. An ancient allopolyploidization event occurred between *T. urartu* ($2n = 2x = 14$, AuAu) and a B-genome donor from the Sitopsis lineage—most closely related to *Aegilops speltoides* ($2n = 2x = 14$, SS)—approximately 300,000–500,000 years ago, resulting in wild emmer wheat, *T. dicoccoides* ($2n = 4x = 28$, AuAuBB) (Huang et al. 2002). Wild emmer was domesticated more than 10,000 years ago in the northern Fertile Crescent/northern Levant (Turkey), producing cultivated emmer, *T. dicoccum* ($2n = 4x = 28$, AABB). A subsequent hybridization between tetraploid emmer (AABB) and *Aegilops tauschii* ($2n = 2x = 14$, DD) around 9000 years ago generated early hexaploid spelt, *T. spelta* ($2n = 6x = 42$, AABBDD) (Matsuoka and Nasuda 2004). Finally, natural mutations affecting spike architecture and threshability were selected about 8500 years ago to produce the modern free-threshing lineages: durum wheat, *T. durum* ($2n = 4x = 28$, AABB), and bread wheat, *T. aestivum* ($2n = 6x = 42$, AABBDD). Figure 1.2 presents a schematic flow diagram showing the evolution of wheat, from primitive Stone Age grasses to modern durum (macaroni) and bread wheat.

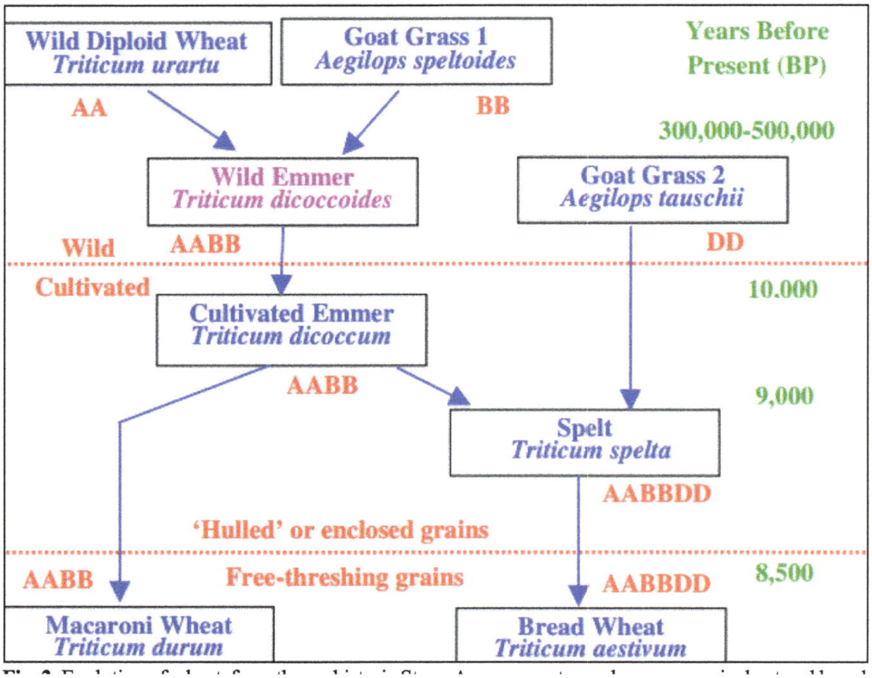


**Fig. 1.2** Evolution of wheat, from the prehistoric Stone Age grasses to modern macaroni wheat and bread wheat. (*Source*: Peng et al. 2011)

Archaeobotanical evidence indicates that the diffusion of wheat from its center of domestication into South Asia occurred through multiple, staggered routes over time. The primary pathway is generally considered to have followed the Iranian plateau and Helmand–Baluchistan corridor, associated with the eastward movement of Southwest Asian agricultural traditions during 7000–5000 B.C., as evidenced by early finds of hulled wheats, particularly emmer, in aceramic and ceramic Neolithic contexts at sites such as Mehrgarh (Kachi Plain, Baluchistan, Pakistan) (7000–5500 B.C.) (Fuller 2006). A secondary dispersal route, potentially linked to later Bronze Age interactions, likely followed a more northerly route via Central Asia and the Indus–Oxus interaction sphere, facilitating the movement of free-threshing wheats into the Indus Valley by 3000–2000 B.C. (Possehl 1997; Spengler 2015).

Within the northwestern Indian subcontinent, the Harappan and post-Harappan phases (2600–1500 B.C.) show increasing archaeobotanical presence of both hulled and free-threshing wheats at sites such as Harappa, Mohenjo-daro, and Rakhigarhi (Weber 2024; Saraswat 2004), suggesting parallel cultivation systems possibly adapted to different irrigation regimes and soil types. Eastward expansion into the Ganga Plain, from the early to mid-2000–1000 B.C., marked a key stage in the ecological diversification of Southwest Asian cereals in India. Archaeobotanical assemblages from Kunal (Fatehabad, Haryana), Atranjikhera (Etah, Uttar Pradesh), Senuwar (Rohtas, Bihar), Chirand (Saran, Bihar), and Taradih (Gaya, Bihar) document the introduction of *T. aestivum* and *T. durum* into mixed farming systems with barley, rice (*Oryza sativa*), and small millets (*Panicum miliaceum*, *Setaria italica*) (Harvey 2006; Fuller 2006; Pokharia et al. 2014). Initial frequencies were low, indicating experimental cultivation and niche adaptation before large-scale adoption.

In contrast to the semiarid Indus–Ghaggar region, the middle and lower Ganga Plains experienced humid subtropical conditions with high monsoonal rainfall, fertile alluvial soils, and short, mild winters, initially limiting the productivity of winter cereals suited to drier regimes (Fuller and Madella 2001; Pokharia 2011). Wheat cultivation persisted mainly on well-drained levees and higher floodplain surfaces, supported by adjusted sowing dates, early-maturing landraces, and integration into multi-cropping systems with pulses (*Lens culinaris*, *Vigna radiata*) to optimize nutrient cycling and manage climatic risks (Harvey and Fuller 2005). Isotopic analyses from Senuwar and Chirand indicate primarily rainfed cultivation using residual soil moisture after monsoon withdrawal (Kingwell-Banham and Fuller 2012). By late 2000–1000 B.C., wheat was a well-established winter crop in the upper and middle Ganga Plains, although it remained less common than rice in eastern zones due to climatic and cultural factors.

Southward dispersal of wheat into peninsular India was later, slower, and patchier, with hulled emmer (locally known as Khapli) dominating over free-threshing wheats. Its adoption, beginning in the mid to late 3000–2000 B.C., is best documented in well-sampled Chalcolithic and later settlements of the Deccan and western Maharashtra, notably Inamgaon, where systematic flotation revealed barley, pulses, and a consistent, though subordinate, presence of wheat (largely emmer) across Chalcolithic phases (Kajale 1988; Dhavalikar et al. 1988). Similar evidence from Daimabad reports wheat alongside millet- and legume-dominated assemblages (Kajale 1977; Vishnu-Mittre et al. 1986). In contrast to the Indus–Ganga *rabi*-cereal core, Deccan agroeconomies remained millet- and pulse-based, with wheat and barley confined to specific niches—raised levees, valley bottoms with residual winter moisture, irrigated plots, and gardens near settlements—where tank irrigation, shallow wells, or flood-recession farming enabled reliable winter harvests (Fuller 2006).

Morphometric and chaff analyses indicate a predominance of hulled types, whose tough glumes provide storability under hot, humid conditions and reduce postharvest losses, while their traits require less intensive threshing—a combination favoring their persistence (Damania 2008; Kajale 1988). Assemblages from other peninsular sites (Hallur, Karnataka; Naikund, Maharashtra; Navdatoli, Madhya Pradesh), from the mid–late 3000–2000 B.C. to the Iron Age, consistently show wheat as a minor crop but one that was integrated across multiple phases, indicating stable incorporation into local cropping systems rather than sporadic introductions (Vishnu-Mittre 1976; Fuller 2006; Korisettar et al. 2001).

### 1.1.3 Indigenous Traditional Knowledge (ITK) Systems for Wheat Stress Management

Indigenous Traditional Knowledge (ITK) systems have historically provided farmers with context-specific, ecologically sound practices for sustaining wheat cultivation amid diverse environmental and biological stresses. Rooted in centuries of observation and experimentation, many of these practices are well documented in ancient Indian texts. The *Rigveda* (8000 B.C.), regarded as the oldest text compiled by humankind, makes references to conservation agriculture in its verses and emphasizes harmony among natural elements—earth, water, air, fire, and akasa (Nene and Sadhale 1997). This ecological worldview laid the foundation for sustainable agricultural practices later elaborated in texts such as *Krishi-Parashara* (400 B.C.), *Kashyapiyakrishisukti* (800 A.D.), *Vrikshayurveda* (1000 A.D.), *Artha-sastra* (321–296 B.C.), and *Krishi Gita* (1500 A.D.).

Since domestication, indigenous farmers have conserved and improved locally adapted wheat landraces that are inherently tolerant of drought, heat, and prevalent diseases (Singh 2020; Manjula and Rengalakshmi 2015). Practices such as seed selection before harvesting, favoring early-maturing earheads from healthy tillers, have long been common, as farmers believed these selections ensured pest-free, vigorous seed stocks. Line sowing of seeds against the wind was employed to reduce lodging and wind damage. Presowing treatments such as soaking wheat seeds in cow dung slurry were widely used, perceived as improving germination, protecting against pests, and providing nutrients, along with insecticidal and fungicidal benefits. Agronomic measures such as mulching, agroforestry-based field buffers, and complex cropping systems mitigated moisture stress, reduced erosion, and stabilized the microclimate (Altieri 1996; Basu 2014).

The *Kashyapiyakrishisukti* vividly emphasized rainwater harvesting and conservation through reservoirs, ponds, wells, and canals, and recommended second cropping supported by fertility enhancement using goat dung, cow dung, and green manure (Ayachit 2002). Similarly, *Krishi-Parashara* stressed the importance of soil and seed health and the uniform use of seed, highlighting efforts to conserve water for crops during Ashwin (October) and Kartika (November) (Sadhale 1999). Community-managed irrigation systems, such as the phad of Maharashtra and the kulhs of Himachal Pradesh, still reflect these principles.

In traditional wheat farming, biotic stress management relied on cultural, physical, and botanical interventions. Deep summer plowing was common to destroy insect and pathogen resting stages, suppress weeds, loosen soil, and improve moisture conservation; its benefits are also emphasized in the *Krishi Gita* (Kumar 2008). *Vrikshayurveda* highlighted seed treatment for vigor, while soil conditioning with ash and salt was used to suppress grubs (Sadhale 1996; FAO 1999). Burning crop residues was a typical practice for rapid field sanitation, pest overwintering control, weed elimination, and nutrient enrichment. Ecological practices such as intercropping and trap cropping minimized pest incidence and maintained ecological balance (Altieri 1996), an approach explicitly mentioned in Kautilya's *Artha-sastra* (Shamasastry 1961).

Manual pest control measures included bonfires, light traps, destruction of egg masses, and handpicking larvae. Farmers also used botanical formulations as biopesticides, employing neem (*Azadirachta indica*), pongamia, garlic, turmeric, cow urine, and fermented plant extracts (Gupta 1999). Ash dusting was particularly effective against a broad range of wheat pests, especially bugs. Other practices included placing red ant nests to foster biological control, encouraging spider webs, spraying garlic and asafoetida solutions, applying lime or oilseed cakes (groundnut, neem) to bunds, mechanically removing larvae, and placing kerosene-soaked gunny bags at water inlets as repellents.

For vertebrate pests, physical deterrents such as scarecrows, white flags, reflective devices, crackers, and drumming were used. Rodent control involved constructing strong, well-plastered narrow bunds to prevent burrowing, blowing smoke into holes to suffocate rodents, and pouring cow dung slurry into burrows. Some farmers prepared bait by mixing crushed leaves and seeds of *Gliricidia sepium* with wheat, which caused natural rodent deaths. Fixing inverted coconut petioles to resemble owls was another innovative method for scaring rats. These measures were especially important during germination and grain-maturation stages, when crops are most vulnerable.

Postharvest protection relied on eco-friendly, low-cost methods: sun drying seeds, mixing stored wheat with turmeric, neem leaves, *Datura* leaves, or garlic, coating seeds with red soil paste, and plastering storage bins with clay and cow dung to regulate moisture and inhibit pests (Meena and Dudi 2014; Sahu and Singh 2016).

## 1.2 Evolution of Scientific Approaches in Wheat Stress Management

### 1.2.1 Influence of Colonial Agricultural Policies on Wheat Farming in India

Between the mid-nineteenth century and 1947, colonial agricultural interventions, such as land revenue and tenancy rules, irrigation schemes, rail-port integration, research and extension institutions, credit reforms, and trade and price controls in the 1930s–40s, significantly restructured wheat production in India. These policies decisively shifted wheat cultivation toward irrigated regions in the northwestern part of the country, especially in Punjab and the canal-served areas of the United Provinces (UP). They also linked *rabi* production to global markets through ports like Karachi and Bombay, while professionalizing research and extension through institutions such as the Imperial Agricultural Research Institute (IARI) and the Imperial Council of Agricultural Research (ICAR). The outcome was a striking expansion in wheat acreage and marketed surplus, but yields remained persistently low by global standards, with increased vulnerability to volatile world prices and entrenched regional inequalities that later influenced postindependence strategies, most notably the Green Revolution (Blyn 2016; Donaldson 2018; Talbot 1982).

Colonial land settlements did not specifically target wheat, but they shaped agrarian incentives in ways that directly influenced the geography of wheat production. In north India's wheat heartlands, Mahalwari-type settlements, where village communities or peasant proprietors were assessed directly, dominated, while in western and southern India, the Ryotwari system, which involved individual assessments of cultivators, was more prevalent. Both systems were characterized by rigid cash-revenue demands that pushed cultivators toward commercial crops and increased market reliance, especially where irrigation or transport infrastructure was available. Empirical research links such tenure regimes to long-term differences in agricultural investment and performance: peasant-proprietor regions often experienced greater provision of public goods and higher investment levels compared to areas affected by zamindari landlordism (Banerjee and Iyer 2005). Punjab's unique legal system, especially the Punjab Tenancy Act (1887) and the Land Alienation Act (1900), aimed to stabilize a base of peasant owners and prevent transfers to moneylenders. Though their effects are debated, these measures complemented the colonial strategy of settling canal colonies with "agricultural" castes and military pensioners. Along with revenue policies, they promoted wheat cultivation under assured irrigation (Whitcombe 1971; Talbot 1982).

The single most important factor driving the regional concentration of wheat was canal irrigation. From the Ganges Canal in UP, opened in the mid-nineteenth century, to the extensive expansion of Punjab's canal colonies from the 1880s onward, covering areas such as the Chenab, Jhelum, Lower Bari Doab, and others, the state transformed semiarid scrub into cultivated land. These colonies were highly planned: land and water were legally allocated, crop rotations were designated, and wheat was prioritized as the leading *rabi* crop. By design, they created wheat-exporting frontiers tightly controlled by state engineers and administrators. The results were striking. Canal-irrigated land in Punjab increased from about 3 million acres in 1885 to over 14 million by 1947, turning the province into India's wheat hub. By the 1920s, Punjab produced nearly one-third of India's wheat, most of which was marketed through Karachi. In UP, canal and well irrigation similarly increased the share of *rabi* crops and stabilized yields across the doabs, although on a smaller scale (Whitcombe 1971; Talbot 1982; Blyn 2016).

Transport integration reinforced these trends. Railways and telegraphs linked interior wheat regions to Karachi, Bombay, and Calcutta, while state-backed carriage contracts and falling transport costs aligned provincial wheat prices with international markets. The railway network boosted trade volumes and more closely connected rural producers to global demand (Donaldson 2018; Hurd 1975). By the late 19th and early 20th centuries, India had become a significant Asian wheat exporter. This commercial integration increased incomes in good years but also introduced risks: during the interwar collapse and the Great Depression, wheat prices plummeted while cultivators' fixed obligations—rents, irrigation fees, and debt payments—remained rigid. The commercialization enabled by railways thus brought both opportunity and risk, exposing cultivators to the fluctuations of global capitalism (Whitcombe 1971).

Alongside these structural changes, colonial agricultural science and credit reforms gained momentum in the early twentieth century. The IARI, established at Pusa in 1905, and ICAR (1929) formalized cereal research under government support. Albert and Gabrielle Howard's *Wheat in India* (1909) remains a key work, cataloging varietal diversity, documenting rust diseases, and outlining improvement principles based on experimental station methods rather than farmer selection. Provincial agricultural colleges, such as Lyallpur (1906), expanded this new agronomic approach, though yield improvements remained modest (Howard et al. 1979; Government of India 1928). To address indebtedness and encourage investment, the government introduced the Agriculturists' Loan Acts in the late nineteenth century and, more notably, the Cooperative Credit Societies Act (1904). Cooperative societies grew rapidly in Punjab and UP, financing wells, seeds, and seasonal inputs, although governance varied and moneylending persisted. Marketing reforms lagged; most wheat still moved through private mandis, and systematic state procurement began only during wartime (Blyn 2016).

The Depression of the 1930s and the challenges of the 1940s marked a turning point. Price collapses devastated farmers, while World War II brought rationing, procurement, and strict interprovincial movement controls. The *Famine Inquiry Commission Report* (1945) criticized the failures and delays in wheat procurement policies, while the House of Lords debated imports and emergency reserves in London. Punjab imposed export restrictions in 1942 to retain surpluses, highlighting intense provincial competition. These wartime controls laid the groundwork for the procurement, rationing, and distribution systems of postcolonial India (Famine Inquiry Commission, 1945). Regionally, Punjab and UP led wheat production, while the Deccan remained focused on millet. Even as land area and irrigation expanded, yields remained low, limited by low availability of fertilizers, rust epidemics, and slow progress in varietal improvement (Blyn 2016). Socially, commercialization exacerbated inequality: canal colonies favored certain groups, while in UP, tenancy conflicts and cycles of debt persisted. Punjab achieved higher marketed surpluses, but risk burdens remained unevenly distributed (Whitcombe 1971; Talbot 1982).

At independence, India inherited a wheat economy focused on irrigated northwestern regions, supported by transport networks linking production to ports and a developing but established research infrastructure. Although these legacies did not overcome yield stagnation during colonial rule, they were crucial in enabling the adoption of high-yielding varieties in the 1960s. Therefore, the Green Revolution directly built on the colonial foundations of irrigation, research institutions, and procurement mechanisms, strengthening the central role of wheat in India's postcolonial agricultural system (Blyn 2016; Government of India 1928; Talbot 1982).

### 1.2.2 Introduction of Improved Wheat Varieties and Agronomic Innovations During the Green Revolution

The introduction of improved wheat varieties and agronomic innovations during the Green Revolution fundamentally transformed India's approach to wheat stress management, boosting national productivity and strengthening resilience to both biotic and abiotic stresses. The breakthrough was driven by the adoption of semidwarf, high-yielding wheat varieties such as Lerma Rojo, Sonora 64, Kalyan Sona, and Sonalika, which were bred with Norin 10 dwarfing genes from Japanese germplasm. This resulted in shorter, sturdier plants that could resist lodging and efficiently use increased fertilizer and irrigation inputs (Smale 1997). These cultivars adapted broadly across the wheat belts of Punjab, Haryana, and western Uttar Pradesh, showing high resistance to rust diseases and greater responsiveness to irrigation, turning wheat from a risky crop into one capable of rapid yield gains even with intensification (Pingali 2012; CIMMYT 2023). Success relied not just on better genetics but also on the spread of agronomic innovations: the Green Revolution saw major investments in canal and tube well irrigation, mechanization, balanced fertilizer use suited to new varieties, and promotion of timely sowing, optimal seed rates, and intensive crop rotations, making double-cropping (rice–wheat or maize–wheat) both feasible and profitable (John and Babu 2021). These advances were supported by strong extension systems, with ICAR, IARI, and collaborations with CIMMYT providing scientific support, participatory breeding, and multilocation trials to ensure variety suitability and stress adaptability across different agroecological zones (CIMMYT 2023; Smale 1997). Backed by government procurement, credit, and price policies, wheat yields more than doubled in a decade, freeing India from chronic import dependence and food insecurity, and laying the foundation for ongoing stress management research using molecular tools, integrated pest and pathogen strategies, and continued adaptation to climate variability (Gulati and Juneja 2018; Pingali 2012; Chand and Singh 2023).

### 1.2.3 Development of Integrated Disease and Pest Management (IDPM) Strategies

The development of integrated disease and pest management (IDPM) strategies in wheat cultivation in India represents an important scientific achievement, combining interdisciplinary approaches to sustainably address the complex biotic stresses that limit wheat productivity. Initially, pest and disease control in Indian wheat farming relied heavily on reactive, often indiscriminate use of chemical pesticides, leading to the buildup of resistance and environmental concerns. Recognizing these challenges, scientists and policymakers gradually moved toward integrated methods that combine cultural, biological, and chemical techniques based on agroecosystem analysis (AESA) and field monitoring to optimize pest control while minimizing ecological damage. The IDPM system emphasizes early pest and disease diagnosis, population monitoring, the establishment of economic thresholds for intervention, and the use of targeted measures only when necessary, thereby reducing reliance on broad-spectrum chemicals (Satyagopal et al. 2014). Central to the IDPM strategy has been the use of disease-resistant and tolerant wheat varieties developed through systematic breeding programs at institutes like IARI and the Directorate of Wheat Research (DWR), which incorporated resistance genes against rusts, loose smut, and other major pathogens, thus providing foundational plant immunity. Biological control agents, such as *Trichoderma viride*, have been used in seed treatments to suppress soil-borne diseases like loose smut, complemented by careful fungicide applications to delay resistance development. For insect pests, the adoption of seed treatments with systemic insecticides such as thiamethoxam and fipronil, soil applications of chlorpyrifos for termite control, and selective foliar sprays timed with pest population surges constitute the chemical component of integrated management, employed alongside measures that conserve natural enemies. Agronomic practices like crop rotation, intercropping with pulses, and residue management further disrupt pest life cycles and enhance biological control, thereby reducing pest pressure sustainably (Kaur et al. 2021). Recent community-level trials have demonstrated yield benefits of 9–15% with AESA-based IPM modules, encouraging widespread adoption in the northwestern wheat belts, where rice–wheat cropping systems are dominant (Kumar et al. 2018). Additionally, extension activities have focused on farmer education in pest scouting, resistance management, and balanced pesticide use to promote IDPM adoption. Ongoing research is advancing precision pest management through spectral monitoring and predictive models to forecast outbreaks.

### 1.2.4 Role of Crop Rotation, Intercropping, and Soil Management Practices in Mitigating Stress Impacts

Crop rotation, intercropping, and soil management practices play an important role in reducing various stresses on wheat cultivation in India. These sustainable, ecosystem-based strategies enhance resilience against abiotic stresses such as drought, heat, and soil degradation, as well as biotic pressures from pests and diseases. Crop rotation, especially the rice–wheat system common in northwestern India, boosts soil fertility and breaks pest and pathogen cycles, thereby reducing stress-related yield losses. Studies show that diversified rotations increase organic matter, soil microbial health, and nutrient availability, leading to better grain quality and stable yields under stress (Adesanya et al. 2024; Dhanda et al. 2022). Intercropping wheat with legumes such as chickpea, mung bean, or faba bean helps reduce disease incidence, including powdery mildew and stripe rust, by breaking monoculture patterns that promote pathogen buildup. It also minimizes pest populations by promoting habitat diversity and the presence of natural enemies, thereby reducing pesticide use (Luo et al. 2021; Lopes et al. 2016; Kaur et al. 2021). Soil management practices support these biological approaches by maintaining soil health and reducing stress. Adding compost, green manures, and crop residues improves soil structure, moisture retention, nutrient cycling, and microbial activity (Kingra et al. 2019). Conservation agriculture methods, such as minimum or zero tillage, protect soil structure and biodiversity, reduce erosion, and improve water-use efficiency, helping mitigate terminal heat and drought stresses common in India's wheat-growing regions (Adesanya et al. 2024; Sharma and Acharya 1987). Furthermore, optimizing sowing dates and managing crop calendars within these rotations enhances climate resilience by avoiding critical stress phases during grain filling (McDonald et al. 2022).

## 1.3 Advances in Wheat Stress Management Strategies

### 1.3.1 Breeding for Disease Resistance: Traditional vs. Biotechnology-Led Approaches

Breeding for disease resistance in wheat has advanced significantly in India, shifting from traditional phenotypic selection to sophisticated biotechnology-based approaches, while both remain essential for managing biotic stresses effectively. Traditional breeding, which has driven wheat improvement for decades, depends on identifying and selecting resistant germplasm through field screening under natural or artificial disease pressure, pedigree and backcross techniques, and conventional hybridization with landraces and wild relatives to introduce resistance genes, especially targeting rusts (stem rust, leaf rust, stripe rust), powdery mildew, Karnal bunt, and spot blotch (Singh et al. 2020; Ren et al. 2024). These methods have produced numerous high-yielding, rust-resistant Indian wheat varieties over the years, mainly by deploying major race-specific genes and slow-rusting adult-plant resistance genes, either singly or in gene pyramids, to enhance resistance durability. However, the rapid emergence of virulent pathogen races has often compromised major gene resistance alone, highlighting the limitations of conventional breeding in providing long-lasting protection. To overcome such limitations, molecular breeding and biotechnological tools have been increasingly adopted in India since the early 2000s, revolutionizing wheat disease resistance strategies. Marker-assisted selection (MAS) uses linked DNA markers to accelerate the introgression of resistance genes into elite lines with greater accuracy, thereby avoiding cumbersome phenotypic screening (Ren et al. 2024; Yadava et al. 2022). Gene editing technologies, such as CRISPR-Cas9, enable precise modifications to susceptibility or resistance genes, creating novel resistant alleles without foreign DNA insertion, promising rapid deployment of broad-spectrum resistance (Kou et al. 2023). Transgenic approaches, although more regulated, have also provided opportunities to incorporate resistance genes from distant species, while chromosome engineering and wide hybridization bring in new diversity from wild relatives, expanding the genetic base beyond domesticated varieties (Savadi et al. 2018). The Indian National Genebank conserves extensive germplasm resources, including landraces and synthetics, which serve as reservoirs of new resistance genes and are increasingly used in genomics-assisted breeding to combine multiple quantitative trait loci for durable resistance (Ren et al. 2024). Despite notable progress, challenges persist in fully integrating multi-omics data with phenotyping for complex traits and ensuring the deployment of durable resistance in farmer-preferred cultivars. Thus, while traditional breeding remains the backbone of wheat improvement due to its proven success and adaptability, biotechnology-driven methods complement and enhance its capabilities by increasing speed, precision, and resistance durability—crucial factors in the context of evolving pathogen populations and climate change impacts on India's wheat production systems (Cowger et al. 2022; Ren et al. 2024). The future of wheat disease resistance breeding in India will depend on the synergistic integration of these approaches to develop sustainable, high-yielding, and resilient wheat cultivars that support food security.

### 1.3.2 Development of Climate-Resilient Wheat Cultivars with Enhanced Drought and Heat Tolerance

The development of climate-resilient wheat cultivars with improved drought and heat tolerance is crucial to reducing the negative effects of abiotic stresses exacerbated by climate change in wheat-growing areas. Wheat yields in India are especially affected by terminal heat stress during grain filling and occasional drought. Traditional breeding initially focused on selecting stable, high-yielding genotypes suited to various environments; however, the rising frequency of extreme temperatures and water shortages has shifted attention towards specific stress-tolerance traits. Recent progress includes extensive germplasm screening, including ancient and native wheat landraces such as *Triticum sphaerococcum* (Indian dwarf wheat), which naturally tolerate heat and drought through physiological stability and genetic factors such as the *BIN2* gene and *TaSG-D1E286K*, known for heat resistance (Gaikwad et al. 2025). Modern breeding combines high-throughput phenotyping of traits such as canopy temperature reduction, membrane stability, osmotic adjustment, and root system traits that enhance drought resilience, with genomic selection using molecular markers linked to QTL for heat and drought tolerance. The use of multi-environment trials to study genotype × environment (G × E) interactions helps precisely identify cultivars like NL 1369 and Sonalika, showing reliable performance under heat and drought stress, evaluated by indices such as the stress susceptibility index (SSI), mean productivity (MP), and geometric mean productivity (GMP) (Bhandari et al. 2024). In addition to conventional methods, marker-assisted breeding, genome editing (CRISPR/Cas9), and transgenic technology support the integration and stacking of multiple tolerance genes to create cultivars with durable, multi-stress resistance (Kou et al. 2023). Moreover, physiological and molecular research identifies key gene networks that regulate heat shock proteins (HSPs), antioxidant enzymes, and hormone signaling pathways involved in stress adaptation and recovery (Singh et al. 2023; Abidin et al. 2024; Sarkar et al. 2021). Efforts are also increasing to expand the genetic base by utilizing wild relatives and landraces, supported by national gene banks and international collaborations such as CIMMYT, to improve resilience under future climate scenarios.

### 1.3.3 Soil Health Management Through Precision Nutrient Application and Conservation Tillage

Soil health management through precision nutrient application and conservation tillage has become a key part of modern wheat stress management strategies in India. It provides sustainable solutions to maintain soil fertility, optimize crop nutrition, and reduce abiotic stresses like drought, nutrient deficiency, and soil degradation. Precision nutrient application involves detailed soil and plant nutrient diagnostics, often supported by technologies such as GPS-guided fertilizer application, leaf color charts, and sensor-based soil testing. These tools help match nutrient supply precisely to crop needs at critical growth stages, improving input-use efficiency. This targeted method reduces environmental pollution from overfertilization, lowers production costs, and enhances nutrient uptake, boosting wheat resilience to nutrient stress and supporting soil microbial activity, crucial for nutrient cycling (Majumdar et al. 2013). Additionally, conservation tillage practices—ranging from zero tillage (ZT) to minimum tillage—minimize soil disturbance, reduce erosion, improve water infiltration, and help retain organic matter. These practices support healthier, more resilient soils, which are especially important for wheat grown in India's variable climate zones (Adesanya et al. 2024). Zero tillage combined with residue retention is especially effective in conserving soil moisture and temperature, reducing heat stress on wheat roots, and promoting beneficial microbial communities that aid nutrient mineralization and disease resistance (Saurabh et al. 2021; Dong et al. 2024). Long-term field trials show that combining precision nutrient management with conservation tillage improves wheat yield stability, boosts soil enzymatic activity, and enhances the plant's ability to withstand drought and heat stress by promoting root growth and overall plant health (Adesanya et al. 2024). This integrated approach aligns with India's goal of sustainable intensification of wheat production by conserving limited soil resources while meeting the nutritional demands of a growing population.

## 1.4 Technological Interventions in the Modern Hi-Tech Era

### 1.4.1 Artificial Intelligence in Wheat Stress Monitoring and Management

Artificial intelligence (AI), especially deep learning-based methods, is becoming increasingly important for monitoring and managing biotic and abiotic stresses in wheat (Barbedo 2025). In disease detection, deep learning models—particularly Convolutional Neural Networks (CNNs)—have shown significant promise in image-based diagnostics (Xu et al. 2023). However, the morphological features of wheat, especially its narrow, vertically oriented leaves, challenge symptom visualization during early disease stages. This has led to a strategic shift toward imaging wheat spikes and kernels, which provide clearer visual signs of infection (Feng et al. 2024). While ground-based imaging remains the primary data-collection method due to its high spatial accuracy, Unmanned Aerial Vehicles (UAVs) are increasingly used for rapid, large-scale, and repeatable assessments under diverse field conditions (Sankaran et al. 2015). The effectiveness of AI models depends largely on extensive, well-annotated datasets. To address this, techniques such as data augmentation, Generative Adversarial Networks (GANs), and Frequency Domain Adaptation (FDA) are utilized (Barbedo 2025). A persistent challenge in AI-based disease classification is the phenotypic similarity among various stress symptoms—both biotic and abiotic—which often leads to diagnostic confusion and high misclassification rates, especially for unseen or mixed stress conditions (Xu et al. 2023). Although classification strategies that use binary schemes with additional "other" classes have been suggested to reduce this problem, their success remains limited by symptom overlap. Few-shot and one-shot learning approaches are promising for identifying rare and emerging diseases, but they are still underused in the wheat disease system (Sun et al. 2023). Moreover, most AI models assume single infections, ignoring the common reality of coinfections in the field. This issue is being addressed through multi-label classification and lesion-level segmentation models (Ahmad et al. 2023). The need for early disease detection is also hindered by the delay between infection and the appearance of visible symptoms in typical RGB imaging. To overcome this, advanced sensing methods—such as hyperspectral, multispectral, thermal, and chlorophyll fluorescence imaging—are used to detect physiological changes that occur before visible symptoms appear (Sankaran et al. 2015). These are often combined with multimodal learning and sensor fusion techniques that merge spectral, thermal, and structural data to improve detection accuracy and environmental adaptability. In weed management, AI models face challenges due to the high morphological and spectral similarity between wheat and various weed species, as well as complications from occlusion and interplant overlap (Li et al. 2024). To improve discrimination, species-specific classification schemes covering multiple weed types are increasingly adopted. However, these models struggle to generalize to untrained weed species and are highly affected by class imbalance, which reduces performance on less common classes (de Camargo et al. 2021). Data augmentation remains the most common strategy for increasing dataset diversity in AI-based weed management, with platforms that include both ground-based and UAV-mounted sensors (Sankaran et al. 2015). Satellite imagery is generally avoided because its resolution is insufficient for intrafield weed discrimination. In pest detection, challenges include the small size, pose variability, and often camouflaged appearance of insect pests on plants (Li et al. 2024). Most existing models focus on single-species detection, with limited work on multispecies classification. Alternative methods, such as near-infrared spectroscopy and electronic nose (E-nose) systems, are increasingly used to detect plant physiological responses—like changes in volatile organic compound (VOC) profiles—as indirect signs of infestation. Nonetheless, these techniques are limited by cultivar-specific VOC signatures and low specificity for pest types, reducing diagnostic accuracy (Fuentes et al. 2021). Research on abiotic stresses like drought and chemical stress is less extensive. Estimating evapotranspiration and monitoring drought are complex because water stress is multifaceted, involving interactions among precipitation, soil moisture, temperature, and crop health. While AI models can integrate multiple data sources—including climate, soil, and vegetation information—for better drought prediction, their success often depends on the availability, resolution, and synchronization of spatial and temporal data. Lodging in wheat is usually assessed using UAV-based RGB or multispectral imaging, often combined with Digital Surface Models (DSMs) to detect canopy deformation (Zhang et al. 2020). Multispectral data has proven more effective in distinguishing lodged from healthy canopies.

### 1.4.2 Genome Editing for Stress Tolerance in Wheat

The CRISPR/Cas-based system is a highly versatile genome editing tool for improving both biotic and abiotic stress resistance in wheat. For biotic stress resistance, a key strategy involves the targeted disruption of susceptibility (S) genes. Unlike resistance (R) gene-mediated defenses, which are often overcome by rapidly evolving pathogen effectors, S gene knockouts can offer durable and broad-spectrum resistance. One of the earliest examples in wheat is the CRISPR/Cas9-mediated deletion of *TaMLO-A1* (Wang et al. 2014) and *TaMLO-B1* (Li et al. 2022), which provided strong resistance to powdery mildew without yield penalties. Subsequently, CRISPR/Cas-mediated knockouts of other S genes, such as *TaEDR1*, *TaMKP1*, *TaWRKY19*, and *TaGW2*, significantly increased resistance to powdery mildew, stripe rust, and leaf rust in wheat, with minimal agronomic trade-offs (Zhang et al. 2017; Liu et al. 2024a; Wang et al. 2022b; Liu et al. 2024b). The CRISPR/Cas-based system has also been used to achieve broad-spectrum resistance to stripe rust by editing *TaCIPK14* (He et al. 2023) and *TaPsIPK1* (Wang et al. 2022a). In the case of Fusarium head blight (FHB), CRISPR/Cas9-mediated deletion of *TaHRC* has been highly effective in conferring *Fhb1*-mediated resistance (Su et al. 2019). Similarly, editing *TaNFXL1*, a gene linked to sensitivity to the Fusarium mycotoxin DON, notably increased resistance (Brauer et al. 2020). For necrotrophic pathogens, knockouts of *TaTsn1* and *TaSnn5* reduced sensitivity to host-selective toxins ToxA and Tox5, thus alleviating diseases such as Septoria nodorum blotch, tan spot, and spot blotch (Karmacharya et al. 2023; Poddar et al. 2023). Genome editing has also enabled the development of virus-resistant wheat by disrupting S genes, such as *TaeIF4E*, thereby improving resistance to wheat yellow mosaic virus (WYMV) without reducing yield (Kan et al. 2023). Additionally, CRISPR/Cas9 editing of wheat orthologues of *HvPDIL5–1*, *HvIMP-α*, and *HvSDN1* has provided resistance to both WYMV and barley yellow dwarf virus (BYDV) (Jin et al. 2022). Beyond S gene knockouts, innovative genome-editing approaches such as Cas-Exo, TATSI, and PrimeRoot show great potential for site-specific deployment of R genes. Base and prime editing can effectively introduce functional SNPs into resistance genes like *Lr67* and *Lr34*, where one or two nucleotide polymorphisms differentiate susceptible and resistant alleles (Milne et al. 2019). Moreover, engineering nucleotide-binding leucine-rich repeat (NLR) receptors through domain swapping, effector reprogramming, and reactivation of pseudogenized genes such as *lr21Ψ* is now feasible in wheat. Genome editing of integrated domains in NLRs, along with modifications to effector-targeted decoy or guardee proteins monitored by NLRs, holds great promise for enhancing effector recognition and broadening disease resistance in wheat (Waites et al. 2025).

In addition to biotic stresses, genome editing has effectively been used to improve wheat tolerance to abiotic stresses such as drought, salinity, cold, and heat. CRISPR/Cas9-based overexpression or knockout of important genes like *TaCER1–6A*, *TaHAG1*, *TaPGK*, *TaHSFA6e*, and *TaPHT1;9* has successfully boosted cuticular wax production, antioxidant activity, energy metabolism, and Arsenic (As) resistance in wheat (Haber et al. 2024). Regulatory genes such as *TaSnRK2.9* and various transcription factors—DREB, bZIP, MYB, and NAC—have also been targeted to influence broad-spectrum stress response pathways (Ahmed et al. 2020). Functional studies involving negative regulators like DIW1 (*TaPP2C158*) show that gene knockouts can increase drought tolerance (Wang et al. 2023), while editing *TaRPK1* has improved root structure and water absorption under stress, with a mutation efficiency of 41.93% (Abdul Rahim et al. 2024). Additionally, advanced genome-editing methods—including promoter editing, base editing, and EvolvR-mediated directed evolution—expand the options for accelerating genetic improvements and boosting climate resilience in wheat.

### 1.4.3 Automated Phenotyping and Aerial Platforms for Wheat Stress Monitoring

The integration of automated high-throughput phenotyping (HTP) platforms and unmanned aerial vehicles (UAVs or drones) has revolutionized the assessment of abiotic and biotic stresses in wheat by enabling precise, rapid, and nondestructive monitoring of plant responses in both field and controlled environments. HTP platforms with multispectral, hyperspectral, thermal, fluorescence, and RGB imaging sensors enable real-time observation of stress-related physiological changes, such as canopy temperature depression, chlorophyll fluorescence, water-use efficiency, and stay-green traits, providing a better understanding of genotype-by-environment interactions and stress-adaptation mechanisms (Yang et al. 2020). Ground-based systems like the LemnaTec Scanalyzer and PhenoTron facilitate accurate temporal monitoring of traits including wilting, biomass accumulation, and senescence, while aerial platforms increase scalability by capturing spatial variability across large field trials, especially under drought, heat, and salinity stress conditions (Araus and Cairns 2014; Sankaran et al. 2015). UAV-based thermal imaging has proven particularly effective in detecting early heat and water stress by measuring canopy temperature as an indicator of stomatal conductance, while multispectral indices such as NDVI, PRI, and CWSI are strongly correlated with photosynthetic performance and yield stability under stress (Deery et al. 2014; Crain et al. 2016). Additionally, 3D imaging and LiDAR technologies have enhanced the measurement of plant architectural traits—such as leaf angle, height, and tiller number—which are essential for stress resilience and yield improvement. These phenomics tools not only speed up trait analysis and QTL mapping for stress tolerance but also support genomic selection by providing high-dimensional, temporal phenotypic data for predictive modeling (Shakoor et al. 2017). As climate variability increases, integrating drone-assisted phenotyping with genomic and environmental data will be key to precision breeding strategies focused on developing climate-resilient wheat cultivars.

## 1.5 Statistical Insights on Wheat Cultivation and Stress Impact

### 1.5.1 Trends in Wheat Production and Yield Fluctuations Over the Decades

The journey of wheat production in India from 1965 to 2025, detailed in Table 1.1, reflects a remarkable transformation in the country's agricultural sector. The Green Revolution of the late 1960s marked the first major leap, driven by the introduction of high-yielding varieties (HYVs), expansion of irrigation networks, and increased use of fertilizers, leading to rapid production gains. In 1965–66, India produced about 12.3 million tonnes of wheat from 13.4 million hectares, with yields averaging less than 1 t $ha^{-1}$. During the 1990s and early 2000s, improved HYVs, expanded irrigation, and greater mechanization propelled production to over 76 million tonnes by 1999–2000, from an area of around 27 million hectares, while yields rose to about 2.8 t $ha^{-1}$. Growth during this period was largely driven by productivity gains alongside the expansion of cultivated area. Since 2000, wheat output has risen steadily, supported by advances in crop breeding, integrated nutrient and water management, and policy support, although occasional setbacks have occurred due to extreme weather events, pest outbreaks, and market fluctuations. By 2010–11, production had crossed 85 million tonnes, with yields exceeding 3 t $ha^{-1}$. In recent years, the adoption of climate-resilient and biofortified varieties, precision farming practices, and improved postharvest handling has further strengthened resilience.

**Table 1.1** Journey of wheat post Green Revolution (1965–2025)

| Period/year | Production growth | Trend |
|---|---|---|
| 1965–1994 | Production gain due to yield improvement (varietal improvement and agronomic practices) | From ~12.3 to ~62 million tonnes |
| 1995–2014 | Production gain due to acreage and yield increment | From ~65 million tonnes to ~95 million tonnes |
| 2015–2024 | Yield improvement | Reached ~113 million tonnes; yield ~3.5 t/ha |
| 2025 forecast | Yield improvement over time | Projected 117.51 million tonnes (record high) |
| 1951–2025 | Production, yield, and acreage growth | From 653 kg/ha to ~3587 kg/ha |

DA&FW, Govt. of India

By 2024–25, wheat production surpassed 117 million tonnes from about 32.76 million hectares, with yields around 3.6 t $ha^{-1}$. The projected 2025 production of approximately 117.51 million tonnes emphasizes the sector's adaptability, driven by technological advancements, farmer innovation, and sustained government support.

### 1.5.2 Data on Yield Losses Caused by Major Biotic and Abiotic Stress Factors

Wheat production is limited by a wide range of biotic and abiotic stresses that act individually or together to reduce yield and grain quality across various agroecological zones. Among biotic factors, fungal diseases are the most economically important, with rusts—leaf (brown), stem (black), and stripe (yellow)—present in nearly all wheat-growing regions and causing yield losses of 10%–70% during epidemics (Singh et al. 2016); leaf rust typically causes 5–20% loss, with severe outbreaks reaching about 50% (Kolmer 2005). Stripe rust can lead to 50–60% losses and even total crop failure in susceptible cultivars (Chen 2005), while stem rust may result in 20–40% losses, with localized epidemics historically destroying entire harvests (Singh et al. 2016). Other significant foliar and spike diseases include powdery mildew, Karnal bunt, spot blotch, Septoria leaf blotch, and Fusarium head blight, the latter of which can contaminate grain with harmful mycotoxins that are harmful to humans and livestock. Insect pests such as sap-feeding aphids, which can transmit viral diseases, the Hessian fly, armyworms, and termites cause notable damage, especially in rainfed systems. Weeds like *Phalaris minor*, *Avena ludoviciana*, *Chenopodium album*, and *Convolvulus arvensis* reduce yields through resource competition, with herbicide-resistant *P. minor* posing additional management challenges. Abiotic stresses are equally damaging, with drought—both terminal and intermittent—reducing biomass and grain weight, resulting in global yield losses of about 27.5%, rising to approximately 32% under severe conditions (Daryanto et al. 2016). Terminal heat stress during anthesis and grain filling accelerates aging, shortens grain filling, and decreases grain size, resulting in yield reductions of 51–81% and up to 89% in vulnerable cultivars. Cold and frost damage can cause seedling mortality or floral sterility, especially in high-altitude or early-sown crops, with severe events causing near-total yield loss (Araus et al. 2008). Soil salinity and sodicity impair germination, tillering, and nutrient absorption, accounting for 40% losses in the Indo-Gangetic Plain and between 50% and 88% in severely affected areas (Munns and Tester 2008; Sharma et al. 2023). Waterlogging in heavy, poorly drained soils induces root hypoxia and can reduce yields by 20–50% (Setter and Waters 2003). Nutrient deficiencies, especially nitrogen, zinc, and sulfur, restrict canopy growth, cause chlorosis, and lower grain protein content. Physical damage from hailstorms, strong winds, and lodging—often worsened by excessive nitrogen use—also contributes to harvest losses.

### 1.5.3 Economic and Social Consequences of Stress-Related Yield Declines

The economic and social consequences of stress-related declines in wheat yields are significant. Abiotic stresses such as drought, heat, frost, salinity, and waterlogging can reduce yields by 10–50%, while biotic stresses, including rusts, smuts, aphids, and termites, cause additional losses of 5–30%, depending on severity and management gaps (Shiferaw et al. 2013; Oerke 2006; Lesk et al. 2016). Such reductions not only lower farm incomes but also decrease national grain supplies, leading to price volatility and food insecurity. In India, a 10% loss in wheat yield due to heat stress alone could result in annual economic losses exceeding USD 1.5 billion, with cascading effects on rural livelihoods, wage levels, and nutritional outcomes. Socially, repeated crop failures weaken farmer resilience, increase debt, and worsen rural distress, sometimes resulting in distress migration from affected areas (Morton 2007). These stresses also impact gender equity, as women often shoulder increased labor and household food provisioning during production shortfalls. Over time, unaddressed stress impacts threaten to undermine rural economies, national self-sufficiency goals, and climate adaptation efforts. Yield losses reduce farm incomes, raise per-unit production costs, and increase vulnerability to food price swings, especially in smallholder systems. At the national level, stress-related yield reductions in wheat lead to lower buffer stocks, greater dependence on imports, and potential inflation (FAO 2021). Chronic crop failures worsen rural poverty, drive distress migration, and pose nutritional risks by decreasing staple grain availability (Lobell et al. 2011). The combined effect of abiotic and biotic stresses diminishes resilience in wheat farming systems, making targeted adaptation strategies—such as breeding stress-tolerant varieties, improving resource management, and implementing integrated pest and disease control—crucial for maintaining productivity and livelihoods. Yield instability increases production risks, discourages investment in better seeds and inputs, and leads smallholders to adopt low-input, low-risk cropping systems that further limit productivity. Reduced wheat supply pushes up market prices, disproportionately affecting low-income consumers who rely on wheat-based diets. In regions where wheat provides a large part of household calories—such as North India, Pakistan, and parts of North Africa—these stresses threaten both farm incomes and nutritional security, potentially increasing malnutrition and social vulnerability. Persistent heat-related production shortfalls disrupt buffer stocks and lead to export restrictions, tightening domestic availability. These shocks push up food prices, weaken government efforts to ensure food security, and increase reliance on imports. Significant income losses reduce farmers' ability to invest in resilient cultivars and necessary inputs, while ongoing yield volatility heightens debt, migration, and nutritional insecurity, ultimately challenging the government's capacity to maintain buffer stocks, control prices, and ensure food self-sufficiency.

### 1.5.4 Government Policies and Interventions for Wheat Production Enhancement

1. *Minimum Support Price (MSP) and Procurement Measures*: The government guarantees a Minimum Support Price (MSP) for wheat, protecting farmers from price drops and encouraging investment in quality inputs and practices. The MSP is recommended annually by the Commission for Agricultural Costs and Prices (CACP) based on production costs and demand–supply dynamics. The Food Corporation of India (FCI) and state agencies procure wheat at the MSP under the Central Pool for the Public Distribution System (PDS) and for maintaining buffer stocks. During the Rabi Marketing Season (RMS) 2025–26, 300.35 lakh metric tonnes of wheat were procured for the central pool, benefiting more than 25 lakh farmers.
2. *The All India Coordinated Research Project on Wheat & Barley (AICRPW&B)*: Operational since 1965 and led by the Indian Institute of Wheat and Barley Research (IIWBR) in Karnal, focuses on breeding high-yielding, disease-resistant, and heat-tolerant varieties. It also aims to develop region-specific cultivars tailored for irrigated, rainfed, and late-sown conditions.
3. *National Food Security Mission—Wheat (NFSM-Wheat)*: Launched in 2007 and funded by the central government, NFSM aims to increase the production of rice, wheat, and pulses through sustainable methods. Its wheat-specific initiatives focus on boosting productivity, restoring soil fertility, and promoting resource-efficient practices.
4. *Crop Insurance and Farmer Income Protection*: The Pradhan Mantri Fasal Bima Yojana (PMFBY), launched in 2016, offers comprehensive crop insurance coverage at all stages of crop failure, including damages from weather events and pest attacks, thus supporting farmer income stability.
5. *Export Management and Buffer Stocking*: To ensure domestic food security, the government enforces export restrictions during periods of low supply and adjusts import tariffs as needed. For example, in 2022, after heatwave-induced yield losses, wheat exports were restricted, procurement targets were revised, and the use of early-maturing, heat-tolerant varieties like HD 2967, HD 3086, and DBW 187 was encouraged.

Additionally, the government has implemented targeted interventions during crisis years, such as special measures to contain the rust epidemics in Punjab during 2009–10.

## 1.6 Future Challenges and Prospects for Sustainable Wheat Cultivation

### 1.6.1 Climate Change and Its Projected Impact on Wheat Production Systems in India

Climate change poses a complex challenge to sustainable wheat cultivation in India, threatening to disrupt established production systems through rising temperatures, changing precipitation patterns, more frequent extreme weather events, and higher atmospheric $CO_2$ levels. Wheat is very sensitive to terminal heat stress during grain filling and to drought during both vegetative and reproductive stages, which are likely to become more severe under climate change scenarios. Studies utilizing climate models, crop simulations, and vulnerability assessments predict yield declines of between 10% and over 30% by mid-century if current farming practices remain unchanged, with impacts varying across regions and depending on climate model assumptions (Kumar et al. 2014; Kaur et al. 2023; Sonkar et al. 2019). Rising temperatures shorten the grain-filling period and reduce biomass, resulting in lighter, lower-quality grains. Additionally, heat and drought stress increase the prevalence of pests and diseases such as rusts and aphids, further reducing yields. Variability in monsoon patterns and groundwater depletion also threaten irrigation resources crucial for maintaining wheat production in frost-free, semiarid regions (Jain et al. 2021; Bowden et al. 2023). At the same time, elevated atmospheric $CO_2$ may partially counteract negative effects by boosting photosynthesis and water-use efficiency, though this benefit is often offset by nutrient deficiencies and heat stress (Chavan et al. 2019). Adaptive measures are crucial, including developing and adopting climate-resilient wheat varieties with improved heat and drought tolerance, adjusting planting schedules to avoid heat periods, optimizing water and nutrient management, and implementing integrated pest management designed for emerging threats (Kumar et al. 2022; Gaikwad et al. 2025). The integration of advanced molecular breeding, precision agriculture, and real-time climate forecasting can strengthen system resilience. Policy measures supporting farmer education, infrastructure development, and climate-smart agriculture are critical to buffering yield shocks and ensuring food security. The intricate impacts of climate change on India's wheat systems highlight the need for interdisciplinary research, continuous monitoring, and proactive adaptation strategies to address future challenges sustainably.

### 1.6.2 Emerging Threats from New Pathogens, Insect Pests, and Environmental Stressors

Emerging threats from new pathogens, insect pests, and environmental stressors increasingly threaten the sustainability of wheat cultivation in India, amplifying traditional challenges and highlighting the need for proactive research and adaptive management strategies. Rapidly evolving fungal pathogens, such as new virulent races of stripe rust (*Puccinia striiformis* f. sp. *tritici*) and stem rust (*Puccinia graminis* f. sp. *tritici*), including the notorious Ug99 lineage and variants adapted to warmer climates, are expanding their geographic range, facilitated by rising temperatures and altered microclimates, endangering the genetic resistance in existing wheat varieties (Prasad et al. 2023; Pequeno et al. 2024). Similarly, new biotypes of insect pests, such as wheat aphids (*Schizaphis graminum*) with increased pesticide resistance and broader host ranges, along with invasive species such as the wheat bulb fly, increase pest management challenges (Rathee and Dalal 2018). Environmental stressors, primarily driven by climate change, create complex pressures including more frequent droughts, irregular monsoon patterns, higher heat stress, and soil salinization, all of which heighten plant vulnerability to pests and diseases and directly reduce yield potential (Sharma et al. 2023; Adesanya et al. 2024). Additionally, intensification of cropping and monoculture systems disrupts ecological balances, weakening natural pest enemies and accelerating resistance development in pests and pathogens (Kelly and Guo 2007). The proliferation of mycotoxin-producing fungi due to climate variability also poses food safety risks. Addressing these emerging threats requires integrated surveillance using remote sensing, genomic diagnostics for early identification of new pathogen strains, and landscape-level pest management approaches that improve agroecological resilience (Mohan et al. 2025). Advances in molecular breeding focused on durable resistance gene pyramiding, combined with biocontrol agents and precision pesticide applications based on agroecosystem monitoring, form the technological frontline (Li et al. 2019). Strengthening national policies to promote coordinated research, farmer education, and resilient seed systems is essential to preemptively address these evolving threats and ensure sustainable wheat production amid increasing biotic and abiotic challenges.

### 1.6.3 Strategies for Enhancing Genetic Diversity and Adaptive Capacity in Wheat Breeding

Enhancing genetic diversity and adaptive capacity in wheat breeding is a key strategy to tackle current and future challenges in sustainable wheat cultivation in India, especially as biotic and abiotic stresses are changing with climate change. Genetic diversity offers the essential material for breeders to create cultivars with broad and durable resistance to pathogens, pests, and environmental extremes, thereby supporting yield stability and food security. Historically, Indian wheat breeding mainly relied on a narrow genetic base, dominated by a few elite varieties, which heightened vulnerability to new stresses due to genetic uniformity (Kumar et al. 2022). To address this, modern breeding programs increasingly utilize the diverse germplasm stored in India's National Genebank and international repositories such as CIMMYT, focusing on varied landraces, synthetic wheats from crosses between *Triticum durum* and wild relatives, and wild *Triticeae* species in the primary gene pool that contain novel resistance genes and stress tolerance traits (Yadava et al. 2022; Kou et al. 2023). Marker-assisted selection and genomic tools accelerate the identification and integration of beneficial alleles into elite backgrounds through prebreeding efforts (Ren et al. 2024). Pyramiding multiple resistance and tolerance genes through molecular breeding enhances cultivars' ability to withstand multiple stresses simultaneously, reducing the risk of resistance failure (Cowger et al. 2022). Advanced genomic selection models take into account polygenic traits related to drought, heat, and disease tolerance with higher accuracy, enabling faster genetic improvements (Xiong et al. 2024). Participatory breeding involving farmers also helps capture locally adapted genotypes with traits suited to specific micro-environments. Maintaining and expanding genetic diversity combines traditional germplasm use with modern biotechnological approaches, allowing the wheat breeding programs to develop resilient wheat cultivars with broad adaptive capacity. Collaboration among national and international research networks, along with policies supporting germplasm conservation, equitable access, and capacity building, will be crucial to unlocking the full potential of genetic diversity and securing India's wheat future.

### 1.6.4 Policy Recommendations for Sustainable and Climate-Smart Agriculture

Policy recommendations for sustainable and climate-smart agriculture are vital to securing the future of wheat cultivation in India, especially given the growing challenges posed by climate change, resource constraints, and evolving biotic stresses. A coherent policy framework should focus on promoting climate-resilient practices that combine advanced scientific innovations with traditional knowledge systems, thereby boosting productivity, environmental sustainability, and farmer livelihoods. Key recommendations include prioritizing investments in research and development to speed up the breeding and dissemination of climate-resilient wheat varieties with improved heat, drought, and disease tolerance through genomics-assisted approaches and precision agronomy (Kumar et al. 2022; Gaikwad et al. 2025). Policies should support the adoption of conservation agriculture, including zero tillage and residue retention, along with precision nutrient management, to sustain soil health, improve water efficiency, and reduce greenhouse gas emissions, and be reinforced by capacity-building initiatives tailored for smallholders (Jat et al. 2025; Adesanya et al. 2024). Strengthening extension services and digital platforms is crucial for delivering real-time climate advisories, pest and disease forecasts, and best management practices to farmers across various agroecological zones (Kumar et al. 2023a, b). An emphasis on integrated pest management (IPM) and biocontrol methods will reduce reliance on pesticides, promote ecological balance, and lower input costs. Additionally, establishing climate risk insurance schemes, promoting water-saving technologies, and encouraging diversified cropping systems will enhance farmers' resilience and income stability (Bindu 2024; Singh et al. 2023). Overcoming infrastructure barriers, ensuring fair access to quality seeds and credit, and fostering stakeholder collaboration across the public and private sectors are crucial to scaling up climate-smart agriculture (CSA) technologies. Ultimately, policy frameworks that combine scientific progress with socioeconomic and environmental considerations are essential to transform Indian wheat production systems toward sustainability and resilience amid climate change.

## 1.7 Conclusion and Recommendations

The comprehensive analysis of wheat stress management in India, covering historical development, biotic and abiotic stresses, and modern technological and policy responses, highlights the critical importance of integrated, multidimensional strategies to ensure sustainable wheat production amid increasing pressures. India's wheat sector, crucial to national food security and rural livelihoods, has undergone significant transformations since the Green Revolution, driven by high-yielding varieties, expanded irrigation, and improved agronomic practices. However, ongoing and intensifying stresses, such as fungal diseases like rusts and spot blotch, insect pests, and abiotic factors like drought, heat, salinity, as well as emerging metal toxicity and waterlogging, expose the complex vulnerabilities in wheat production systems. Advances in breeding, from traditional methods to biotechnological approaches such as marker-assisted selection and genome editing, have enhanced disease resistance and abiotic stress tolerance. Meanwhile, integrated disease and pest management, precision nutrient application, conservation tillage, and crop diversification practices bolster resilience at the agroecosystem level. New threats from emerging pathogens, evolving pest biotypes, and climate-driven environmental variability require continuous monitoring, swift adaptation, and innovative management solutions. Future prospects depend on harnessing genomic diversity, deploying climate-smart cultivars, improving automated phenotyping and AI-based monitoring, and establishing strong policy frameworks that combine scientific innovation, extension services, social equity, and environmental sustainability. These cohesive efforts are vital to mitigating the impacts of climate change, protecting wheat productivity, ensuring food and nutritional security, and supporting rural livelihoods in India's dynamic agricultural sector, reaffirming wheat's essential role as a staple crop amid twenty-first-century challenges.

## References


Abdul Rahim A, Uzair M, Rehman N et al (2024) CRISPR/Cas9 mediated *TaRPK1* root architecture gene mutagenesis confers enhanced wheat yield. J King Saud Univ Sci 36:103063. https://doi.org/10.1016/j.jksus.2023.103063

Abidin ZU, Mahmood A, Najeeb Alawadi HF et al (2024) Heat stress responses and mitigation strategies in wheat: an updated and comprehensive review. Not Bot Horti Agrobo 52(3):13636. https://doi.org/10.15835/nbha52313636

Adesanya T, Francis Z, Myriam F et al (2024) Crop rotation diversity and tillage effects on soil and wheat grain nutrient concentration in an organically-managed system. J Agric Food Res 18:101411. https://doi.org/10.1016/j.jafr.2024.101411

Aggarwal PK, Joshi PK, Ingram JSI et al (2004) Adapting food systems of the Indo-Gangetic plains to global environmental change: key information needs to improve policy formulation. Environ Sci Pol 7(6):487–498. https://doi.org/10.1016/j.envsci.2004.07.006

Ahmad A, Saraswat D, El Gamal A (2023) A survey on using deep learning techniques for plant disease diagnosis and recommendations for development of appropriate tools. Smart Agric Technol 3:100083. https://doi.org/10.1016/j.atech.2022.100083

Ahmed RF, Irfan M, Shakir HA et al (2020) Engineering drought tolerance in plants by modification of transcription and signalling factors. Biotechnol Biotechnol Equip 34:781–789. https://doi.org/10.1080/13102818.2020.1805359

Ainsworth EA, Yendrek CR, Sitch S et al (2012) The effects of tropospheric ozone on net primary productivity and implications for climate change. Annu Rev Plant Biol 63:637–661. https://doi.org/10.1146/annurev-arplant-042110-103829

Altieri MA (1996) Agroecology: the science of sustainable agriculture, 2nd edn. CRC Press. https://doi.org/10.1201/9780429495465

Annual report 2023–24 (2024) Department of Food and Public Distribution, Ministry of Consumer Affairs, Food and Public Distribution, Government of India. https://dfpd.gov.in

Araus JL, Cairns JE (2014) Field high-throughput phenotyping: the new crop breeding frontier. Trends Plant Sci 19(1):52–61. https://doi.org/10.1016/j.tplants.2013.09.008

Araus JL, Slafer GA, Royo C et al (2008) Breeding for yield potential and stress adaptation in cereals. Crit Rev Plant Sci 27(6):377–412. https://doi.org/10.1080/07352680802467736

Asseng S, Martre P, Maiorano A et al (2019) Climate change impact and adaptation for wheat protein. Glob Chang Biol 25(1):155–173. https://doi.org/10.1111/gcb.14481

Atkinson NJ, Urwin PE (2012) The interaction of plant biotic and abiotic stresses: from genes to the field. J Exp Bot 63:3523–3543. https://doi.org/10.1093/jxb/ers100

Ayachit SM (Tr.) (2002) Kashyapiyakrishisukti (a treatise on agriculture by Kashyapa). Agri history bulletin no. 4. Asian Agri-History Foundation, Secunderabad, p 158

Baladhiya HC, Sisodiya DB, Pathan NP (2018) A review on pink stem borer, *Sesamia inferens* Walker: a threat to cereals. J Entomol Zool Stud 6:1235–1239

Bale JS, Masters GJ, Hodkinson ID et al (2002) Herbivory in global climate change research: direct effects of rising temperature on insect herbivores. Glob Chang Biol 8:1–16. https://doi.org/10.1046/j.1365-2486.2002.00451.x

Banerjee A, Iyer L (2005) History, institutions and economic performance: the legacy of colonial land tenure systems in India. Am Econ Rev 95(4):1190–1213

Barbedo JGA (2025) A review of artificial intelligence techniques for wheat crop monitoring and management. Agronomy 15:1157. https://doi.org/10.3390/agronomy15051157

Basu JP (2014) Agroforestry, climate change mitigation and livelihood security in India. N Z J For Sci 44(Suppl 1):S11. https://doi.org/10.1186/1179-5395-44-S1-S11

Bebber DP, Ramotowski MAT, Gurr SJ (2013) Crop pests and pathogens move polewards in a warming world. Nat Clim Chang 3:985–988. https://doi.org/10.1038/nclimate1990

Bhandari R, Paudel H, Nyaupane S et al (2024) Climate resilient breeding for high yields and stable wheat (*Triticum aestivum* L.) lines under irrigated and abiotic stress environments. Plant. Stress 11:100352. https://doi.org/10.1016/j.stress.2024.100352

Bhishagratna KL (Trans. & Ed.) (1963) The Sushruta Samhita (Vols. I–III; Sutra Sthana and other sections). Chaukhamba Sanskrit Series Office, Varanasi

Bindu MS (2024) Climate change and crop insurance: a risk mitigation tool for climate-resilient agriculture—significance, initiatives and prospects in India. Online article. https://www.linkedin.com/pulse/climate-change-crop-insurance-risk-mitigation-tool-pradhan-mantri-ihyrc

Blandino M, Badeck FW, Giordano D et al (2020) Elevated $CO_2$ impact on common wheat (*Triticum aestivum* L.) yield, wholemeal quality, and sanitary risk. J Agric Food Chem 68:10574–10585. https://doi.org/10.1021/acs.jafc.0c02975

Blyn G (2016) Agricultural trends in India, 1891–1947: output, availability, and productivity. University of Pennsylvania Press, Philadelphia

Bowden C, Foster T, Parkes B (2023) Identifying links between monsoon variability and rice production in India through machine learning. Sci Rep 13:2446. https://doi.org/10.1038/s41598-023-27752-8

Brauer EK, Balcerzak M, Rocheleau H et al (2020) Genome editing of a deoxynivalenol-induced transcription factor confers resistance to Fusarium graminearum in wheat. Mol Plant-Microbe Interact 33:553–560. https://doi.org/10.1094/MPMI-11-19-0332-R

Broberg MC, Högy P, Pleijel H (2017) $CO_2$-induced changes in wheat grain composition: meta-analysis and response functions. Agronomy 7:32. https://doi.org/10.3390/agronomy7020032

Chai Y, Pardey PG, Hurley TM et al (2020) A probabilistic bio-economic assessment of the global consequences of wheat leaf rust. Phytopathology 110:1886–1896. https://doi.org/10.1094/PHYTO-02-20-0032-R

Chai S, Yao Q, Zhang X et al (2021) The semi-dwarfing gene *Rht-dp* from dwarf Polish wheat (*Triticum polonicum* L.) is the "green revolution" gene *Rht-B1b*. BMC Genomics 22:63. https://doi.org/10.1186/s12864-021-07367-x

Chakrabarti B, Bhatia A, Deo A et al (2025) Climatic stresses and adaptation options for south Asian wheat: a systematic literature review. Front Agron 7:1670235. https://doi.org/10.3389/fagro.2025.1670235

Chakraborty S, Newton AC (2011) Climate change, plant diseases and food security: an overview. Plant Pathol 60:2–14. https://doi.org/10.1111/j.1365-3059.2010.02411.x

Chand R, Singh J (2023) From Green Revolution to Amrit Kaal: lessons and way forward for Indian agriculture. https://doi.org/10.13140/RG.2.2.24291.12320

Charak Samhita (1941) Caraka. Caraka Saṃhitā. Translated and with commentary by Sharma RK, Dash B (1976–2002). 7 vols. Chowkhamba Sanskrit Series Office, Varanasi

Chavan SG, Duursma RA, Tausz M et al (2019) Elevated $CO_2$ alleviates the negative impact of heat stress on wheat physiology but not on grain yield. J Exp Bot 70:6447–6459. https://doi.org/10.1093/jxb/erz386

Chen X (2005) Epidemiology and control of stripe rust [*Puccinia striiformis* f. sp. *tritici*] on wheat. Can J Plant Pathol 27:314–337. https://doi.org/10.1080/07060660509507230

Chhillar BS, Saini RK, Roshanlal K (2006) Emerging trends in economic entomology. CCSHAU Press, Hisar

CIMMYT (2023) India transforms wheat for the world. https://www.cimmyt.org/news/india-transforms-wheat-for-the-world/

Costamilan L (2005) Variability of the wheat powdery mildew pathogen *Blumeria graminis* f. sp. *tritici* in the 2003 crop season. Fitopatol Bras 30. https://doi.org/10.1590/S0100-41582005000400015

Cowger C, Bulli P, Chen F et al (2022) Advances in breeding for wheat disease resistance. Front Plant Sci 13:1034506. https://doi.org/10.3389/fpls.2022.1034506

Crain J, Wei Y, Barker J et al (2016) Development and deployment of a portable field phenotyping platform. Crop Sci 56:1–12. https://doi.org/10.2135/cropsci2015.05.0290

Cruz CD, Valent B (2017) Wheat blast disease: danger on the move. Trop Plant Pathol 42:210–222. https://doi.org/10.1007/s40858-017-0159-z

Damania AB (2008) History, achievements, and current status of genetic resources conservation. Agron J 100:27–39. https://doi.org/10.2134/agronj2005.0239cs

Daryanto S, Wang L, Jacinthe PA (2016) Global synthesis of drought effects on maize and wheat production. PLoS One 11:e0156362. https://doi.org/10.1371/journal.pone.0156362

de Camargo T, Schirrmann M, Landwehr N et al (2021) Optimized deep learning model as a basis for fast UAV mapping of weed species in winter wheat crops. Remote Sens 13:1704. https://doi.org/10.3390/rs13091704

Debnath S, Mandal B, Saha S et al (2021) Are the modern-bred rice and wheat cultivars in India inefficient in zinc and iron sequestration? Environ Exp Bot 189:104535. https://doi.org/10.1016/j.envexpbot.2021.104535

Deery D, Jimenez-Berni J, Jones H et al (2014) Proximal remote sensing buggies and potential applications for field-based phenotyping. Agronomy 4:349–379. https://doi.org/10.3390/agronomy4030349

Deutsch CA, Tewksbury JJ, Tigchelaar M et al (2018) Increase in crop losses to insect pests in a warming climate. Science 361:916–919. https://doi.org/10.1126/science.aat3466
Dhanda S, Yadav A, Yadav DB et al (2022) Emerging issues and potential opportunities in the rice–wheat cropping system of North-Western India. Front Plant Sci 13:832683. https://doi.org/10.3389/fpls.2022.832683
Dhavalikar MK, Sankalia HD, Ansari ZD (1988) Excavations at Inamgaon, vol 1. Deccan College Postgraduate and Research Institute, Pune, p 1
Directorate of Economics and Statistics, Ministry of Agriculture and Farmers Welfare, Government of India (n.d.) Agricultural Statistics of various reports. https://desagri.gov.in/
Djanaguiraman M, Narayanan S, Erdayani E et al (2020) Effects of high temperature stress during anthesis and grain filling on photosynthesis, lipids and grain yield in wheat. BMC Plant Biol 20:268. https://doi.org/10.1186/s12870-020-02479-0
Donaldson D (2018) Railroads of the raj: estimating the impact of transportation infrastructure. Am Econ Rev 108(4–5):899–934. https://doi.org/10.1257/aer.20101199
Dong F, Wang L, Xu T et al (2024) Multi-omics analysis of soil microbiota and metabolites in dry-land wheat fields under different tillage methods. Sci Rep 14:24066. https://doi.org/10.1038/s41598-024-74620-0
Dubey R, Pathak H, Chakrabarti B et al (2020) Impact of terminal heat stress on wheat yield in India and options for adaptation. Agric Syst 181:102826. https://doi.org/10.1016/j.agsy.2020.102826
Duveiller E, Sharma RC (2009) Genetic improvement and crop management strategies to minimize yield losses in warm non-traditional wheat growing areas due to spot blotch pathogen *Cochliobolus sativus*. J Phytopathol 157:521–534. https://doi.org/10.1111/j.1439-0434.2008.01534.x
Duveiller E, Kandel YR, Sharma RC et al (2005) Epidemiology of foliar blights (spot blotch and tan spot) of wheat in the plains bordering the Himalayas. Phytopathology 95:248–256. https://doi.org/10.1094/PHYTO-95-0248
Duveiller E, He X, Singh PK (2016) Wheat blast: an emerging disease in South America potentially threatening wheat production. In: Bonjean A, van Ginkel M (eds) World wheat book: a history of wheat, vol 3. Lavoisier, Paris
El Sabagh A, Islam MS, Skalicky M et al (2021) Salinity stress in wheat (*Triticum aestivum* L.) in the changing climate: adaptation and management strategies. Front Agron 3:661932. https://doi.org/10.3389/fagro.2021.661932
FAO (1999) Farmers' practices in managing pests and diseases. In: FAO, Chapter on plant protection practices. FAO, Rome. https://www.fao.org/4/X5672E/x5672e07.htm
FAO (2021) The impact of disasters and crises on agriculture and food security 2021. Food and Agriculture Organization of the United Nations, Rome. https://doi.org/10.4060/cb3673en
Farooq M, Bramley H, Palta JA et al (2011) Heat stress in wheat during reproductive and grain-filling phases. Crit Rev Plant Sci 30:491–507. https://doi.org/10.1080/07352689.2011.615687
Feng Z, Kobayashi K, Ainsworth EA (2008) Impact of elevated ozone concentration on growth, physiology, and yield of wheat (*Triticum aestivum* L.): a meta-analysis. Glob Chang Biol 14:2696–2708. https://doi.org/10.1111/j.1365-2486.2008.01673.x
Feng Z, Pang J, Kobayashi K et al (2011) Differential responses in two varieties of winter wheat to elevated ozone concentration under fully open-air field conditions. Glob Chang Biol 17:580–591. https://doi.org/10.1111/j.1365-2486.2010.02184.x
Feng G, Gu Y, Wang C et al (2024) Wheat Fusarium head blight automatic non-destructive detection based on multi-scale imaging: a technical perspective. Plants 13:1722. https://doi.org/10.3390/plants13131722
Field CB, Barros VR, Mach KJ et al (2014) Technical summary. In: Field CB, Barros VR, Dokke DJ et al (eds) Climate change 2014: impacts, adaptation, and vulnerability. Part A: global and sectoral aspects. Contribution of Working Group II to the Fifth Assessment Report of the Intergovernmental Panel on Climate Change. Cambridge University Press, Cambridge and New York, pp 35–94
Figueroa M, Hammond-Kosack KE, Solomon PS (2018) A review of wheat diseases—a field perspective. Mol Plant Pathol 19:1523–1536. https://doi.org/10.1111/mpp.12618

Food and Agriculture Organization (FAO) (2024) Agricultural production statistics. FAOSTAT analytical brief 96

Fuentes S, Tongson E, Unnithan RR et al (2021) Early detection of aphid infestation and insect–plant interaction assessment in wheat using a low-cost electronic nose (E-nose), near-infrared spectroscopy and machine learning modeling. Sensors 21:5948. https://doi.org/10.3390/s21175948

Fuller DQ (2006) Agricultural origins and frontiers in South Asia: a working synthesis. J World Prehist 20:1–86. https://doi.org/10.1007/s10963-006-9006-8

Fuller DQ, Madella M (2001) Issues in Harappan archaeobotany: retrospect and prospect. In: Settar S, Korisettar R (eds) Indian archaeology in retrospect. Protohistory, vol II. Manohar Publishers, New Delhi, p 317

Gaikwad KB, Mazumder AK, Kumar M et al (2025) Evaluating heat and drought resilience in ancient Indian Dwarf wheat *Triticum sphaerococcum* Percival using stress tolerance indices. Sci Rep 15:18970. https://doi.org/10.1038/s41598-025-02502-0

Government of India (1928) Royal Commission on agriculture in India: report. Government of India Press, Bombay

Goyal L, Sharma CM, Singh A et al (2021) Leaf and spike wheat disease detection and classification using an improved deep convolutional architecture. Inform Med Unlocked 25:100642. https://doi.org/10.1016/j.imu.2021.100642

Gul R, Gupta M, Kour S et al (2023) Influence of zinc and iron nutrition on growth and quality of wheat (*Triticum aestivum* L.) in Shivalik foothills. Pharma Innov J 12(3):4325–4328

Gulati A, Juneja R (2018) Innovations and revolutions in Indian agriculture: a review. J Agric Sci Technol B 8:1–12. https://doi.org/10.17265/2161-6264/2018.08.001

Gupta A (1999) Science, sustainability and social purpose: barriers to effective articulation, dialogue and utilization of formal and informal science in public policy. Indian Institute of Management Ahmedabad, Research and Publication Department, IIMA Working Papers

Gupta OP, Singh A, Pandey V et al (2024) Critical assessment of wheat biofortification for iron and zinc: a comprehensive review of conceptualization, trends, approaches, bioavailability, health impact, and policy framework. Front Nutr 10:1310020. https://doi.org/10.3389/fnut.2023.1310020

Haber Z, Sharma D, Selvaraj KSV et al (2024) Is CRISPR/Cas9-based multi-trait enhancement of wheat forthcoming? Plant Sci 341:112021. https://doi.org/10.1016/j.plantsci.2024.112021

Harvey EL (2006) Early agricultural communities in northern and eastern India: an archaeobotanical investigation. PhD thesis. UCL, London

Harvey E, Fuller DQ (2005) Investigating crop processing using phytolith analysis: the example of rice and millets. J Archaeol Sci 32:739–752. https://doi.org/10.1016/j.jas.2004.12.010

He F, Wang C, Sun H et al (2023) Simultaneous editing of three homoeologues of *TaCIPK14* confers broad-spectrum resistance to stripe rust in wheat. Plant Biotechnol J 21:354–368. https://doi.org/10.1111/pbi.13956

Howard A, Howard GLCM, Howard M (1979) Wheat in India: its production, varieties, and improvement. Periodical Expert Book Agency, New Delhi

Huang S, Sirikhachornkit A, Su X et al (2002) Genes encoding plastid acetyl-CoA carboxylase and 3-phosphoglycerate kinase of the Triticum/Aegilops complex and the evolutionary history of polyploid wheat. Proc Natl Acad Sci USA 99:8133–8138. https://doi.org/10.1073/pnas.072223799

Huerta-Espino J, Singh RP, German S et al (2011) Global status of wheat leaf rust caused by *Puccinia triticina*. Euphytica 179:143–160. https://doi.org/10.1007/s10681-011-0361-x

Hurd J (1975) Railways and the expansion of markets in India, 1861–1921. Explor Econ Hist 12:263–288

Intergovernmental Panel on Climate Change (IPCC) (2023) Climate change 2021: the physical science basis. Working group I contribution to the sixth assessment report of the IPCC. Cambridge Univ Press, Cambridge

Iquebal MA, Mishra P, Maurya R et al (2021) Centenary of soil- and airborne wheat Karnal bunt disease research: a review. Biology (Basel) 10:1152. https://doi.org/10.3390/biology10111152

Jain M, Fishman R, Mondal P et al (2021) Groundwater depletion will reduce cropping intensity in India. Sci Adv 7:eabd2849. https://doi.org/10.1126/sciadv.abd2849
Jat ML, Dagar JC, Sapkota TB et al (2016) Climate change and agriculture: adaptation strategies and mitigation opportunities for food security in South Asia and Latin America. Adv Agron 137:127–235. https://doi.org/10.1016/bs.agron.2015.12.005
Jat RK, Meena VS, Pazhanisamy S et al (2025) Enhancing wheat resilience in subtropical agroecosystems through climate-resilient agriculture strategies. Front Agron 7:1535701. https://doi.org/10.3389/fagro.2025.1535701
Jatakas I–VII (1877) Trubner & Co., London. Reprinted 1968, Luzac & Co., London
Jin H, Han X, Wang Z et al (2022) Barley *GRIK1 SnRK1* kinases subvert a viral virulence protein to upregulate antiviral RNAi and inhibit infection. EMBO J 41:e110521. https://doi.org/10.15252/embj.2021110521
John DA, Babu GR (2021) Lessons from the aftermaths of Green Revolution on food system and health. Front Sustain Food Syst 5:644559. https://doi.org/10.3389/fsufs.2021.644559
Juroszek P, von Tiedemann A (2013) Climate change and potential future risks through wheat diseases: a review. Eur J Plant Pathol 136:21–33. https://doi.org/10.1007/s10658-012-0144-9
Kajale MD (1977) On the botanical findings from excavations at Daimabad, a Chalcolithic site in western Maharashtra, India. Curr Sci 46:818–819
Kajale MD (1988) Ancient plant economy at Chalcolithic Tuljapur Garhi, district Amraoti, Maharashtra. Curr Sci 57:377–379
Kambrekar DN, Halagalimath SP, Somanagouda G (2016) Management of termites in wheat with fipronil 3% G—new insecticide molecule. J Exp Zool India 19:185–189
Kan J, Cai Y, Cheng C et al (2023) CRISPR/Cas9 guided knockout of *eIF4E* improves wheat yellow mosaic virus resistance without yield penalty. Plant Biotechnol J 21:893–895. https://doi.org/10.1111/pbi.14002
Kang S, Hao X, Du T et al (2017) Improving agricultural water productivity to ensure food security in China under changing environment: from research to practice. Agric Water Manag 179:5–17. https://doi.org/10.1016/j.agwat.2016.05.007
Karlsson I, Persson P, Friberg H (2021) Fusarium head blight from a microbiome perspective. Front Microbiol 12:628373. https://doi.org/10.3389/fmicb.2021.628373
Karmacharya A, Li D, Leng Y et al (2023) Targeting disease susceptibility genes in wheat through wide hybridization with maize expressing Cas9 and guide RNA. Mol Plant-Microbe Interact 36:554–557. https://doi.org/10.1094/MPMI-01-23-0004-SC
Kashyap PL, Kumar S, Shekhawat PS et al (2022) Field disease scoring guide for wheat and barley research workers. ICAR-Indian Institute of Wheat and Barley Research, Karnal, Technical Bulletin No 32, p 101
Kaur G, Gupta G, Hooda K (2021) Intercropping systems in wheat (*Triticum sativum* L.) for insect pests and disease management—a review. J Pharm Res Int 33:122–129. https://doi.org/10.9734/jpri/2021/v33i53B33688
Kaur P, Sandhu S, Kothiyal S et al (2023) An assessment of adaptation measures to enhance wheat productivity under climate change during early, mid and end of the 21st century in Indian Punjab. J Agric Sci 161:1–37. https://doi.org/10.1017/S0021859623000436
Kellogg EA (2001) Evolutionary history of the grasses. Plant Physiol 125:1198–1205. https://doi.org/10.1104/pp.125.3.1198
Kelly M, Guo Q (2007) Integrated agricultural pest management through remote sensing and spatial analyses. In: Ciancio A, Mukerji KG (eds) General concepts in integrated pest and disease management. Springer, Dordrecht, pp 161–181. https://doi.org/10.1007/978-1-4020-6061-8_7
Khedr RA, Sorour SGR, Aboukhadrah SH et al (2022) Alleviation of salinity stress effects on agro-physiological traits of wheat by auxin, glycine betaine, and soil additives. Saudi J Biol Sci 29:534–540. https://doi.org/10.1016/j.sjbs.2021.09.027
Khobra R, Sareen S, Meena BK et al (2019) Exploring the traits for lodging tolerance in wheat genotypes: a review. Physiol Mol Biol Plants 25:589–600. https://doi.org/10.1007/s12298-018-0629-x

Kindler D, Hesler L, Elliott N et al (2004) Cereal and grass hosts of the rice root aphid, *Rhopalosiphum rufiabdominalis* (Sasaki), and a description of an efficient greenhouse rearing technique. J Agric Urban Entomol 21:51–59

Kingra PK, Kaur J, Majumder R (2019) Management strategies for sustainable wheat (*Triticum aestivum* L.) production under climate change in South Asia—a review. J Agric Phys 19:21–34

Kingwell-Banham E, Fuller DQ (2012) Shifting cultivators in South Asia: expansion, marginalisation and specialisation over the long term. Quat Int 249:84–95. https://doi.org/10.1016/j.quaint.2011.05.025

Kolmer JA (2005) Tracking wheat rust on a continental scale. Curr Opin Plant Biol 8:441–449. https://doi.org/10.1016/j.pbi.2005.05.001

Korisettar R, Venkatasubbaiah PC, Fuller DQ (2001) Brahmagiri and beyond: the archaeology of the Southern Neolithic. In: Settar S, Korisettar R (eds) Indian archaeology in retrospect, vol 1: prehistory, archaeology of South Asia. Indian Council of Historical Research, New Delhi, p 151

Kou H, Zhang Z, Yang Y et al (2023) Advances in the mining of disease resistance genes from *Aegilops tauschii* and their utilization in wheat. Plants 12:880. https://doi.org/10.3390/plants12040880

Kumar BM (2008) Krishi Gita (agricultural verses). Agri-history bulletin no. 7. Asian Agri-History Foundation, Secunderabad, p 112

Kumar N, Soora NK, Aggarwal PK et al (2014) Vulnerability of wheat production to climate change in India. Clim Res 59:173–187. https://doi.org/10.3354/cr01212

Kumar A, Gupta A, Atwal SS et al (2015) Post-harvest management of Karnal bunt, a quarantine disease in wheat. Plant Pathol J 14:23–30. https://doi.org/10.3923/ppj.2015.23.30

Kumar V, Jat HS, Sharma PC et al (2018) Can productivity and profitability be enhanced in intensively managed cereal systems while reducing the environmental footprint of production? Assessing sustainable intensification options in the breadbasket of India. Agric Ecosyst Environ 252:132–147. https://doi.org/10.1016/j.agee.2017.10.006

Kumar R, Gupta A, Srivastava S et al (2020) Diagnosis and detection of seed-borne fungal phytopathogens. In: Kumar R, Gupta A (eds) Seed-borne diseases of agricultural crops: detection, diagnosis and management. Springer, Singapore. https://doi.org/10.1007/978-981-32-9046-4_5

Kumar S, Jacob SR, Mir RR et al (2022) Indian wheat genomics initiative for harnessing the potential of wheat germplasm resources for breeding disease-resistant, nutrient-dense, and climate-resilient cultivars. Front Genet 13:834366. https://doi.org/10.3389/fgene.2022.834366

Kumar A, Sharma R, Singh B et al (2023a) Predicting the incidence and severity of wheat aphids and development of a web-enabled forewarning system in India. J Environ Biol 44:83–90. https://doi.org/10.22438/jeb/44/1/MRN-5058

Kumar M, Kumar S, Sandhu KS et al (2023b) GWAS and genomic prediction for pre-harvest sprouting tolerance involving sprouting score and two other related traits in spring wheat. Mol Breed 43:14. https://doi.org/10.1007/s11032-023-01357-5

Leonard KJ, Szabo LJ (2005) Stem rust of small grains and grasses caused by *Puccinia graminis*. Mol Plant Pathol 6:99–111. https://doi.org/10.1111/j.1364-3703.2005.00273.x

Lesk C, Rowhani P, Ramankutty N (2016) Influence of extreme weather disasters on global crop production. Nature 529:84–87. https://doi.org/10.1038/nature16467

Li F, Upadhyaya NM, Sperschneider J et al (2019) Emergence of the Ug99 lineage of the wheat stem rust pathogen through somatic hybridisation. Nat Commun 10:5068. https://doi.org/10.1038/s41467-019-12927-7

Li S, Lin D, Zhang Y et al (2022) Genome-edited powdery mildew resistance in wheat without growth penalties. Nature 602:455–460. https://doi.org/10.1038/s41586-022-04395-9

Li Z, Wang D, Yan Q et al (2024) Winter wheat weed detection based on deep learning models. Comput Electron Agric 227:109448. https://doi.org/10.1016/j.compag.2024.109448

Liu S, Liu H, Guo M et al (2024a) Knockout of *GRAIN WIDTH2* has a dual effect on enhancing leaf rust resistance and increasing grain weight in wheat. Plant Biotechnol J 22:2007–2009. https://doi.org/10.1111/pbi.14320

Liu S, Zhang F, Su J et al (2024b) CRISPR-targeted mutagenesis of mitogen-activated protein kinase phosphatase 1 improves both immunity and yield in wheat. Plant Biotechnol J 22:1–3. https://doi.org/10.1111/pbi.14312

Lobell DB, Schlenker W, Costa-Roberts J (2011) Climate trends and global crop production since 1980. Science 333:616–620. https://doi.org/10.1126/science.1204531

Lopes T, Hatt S, Xu Q et al (2016) Wheat (*Triticum aestivum* L.) based intercropping systems for biological pest control. Pest Manag Sci 72:2193–2202. https://doi.org/10.1002/ps.4332

Luo C, Ma L, Zhu J et al (2021) Effects of nitrogen and intercropping on the occurrence of wheat powdery mildew and stripe rust and the relationship with crop yield. Front Plant Sci 12:637393. https://doi.org/10.3389/fpls.2021.637393

Majumdar K, Jat MI, Pampolino M et al (2013) Nutrient management in wheat: current scenario, improved strategies and future research needs in India. J Wheat Res 4:1–10

Manjula M, Rengalakshmi R (2015) Seasonal climate information for ensuring agricultural sustainability and food security of smallholder rainfed farmers: experience from India. In: Proceedings of 16th annual global development network conference, Casablanca, Morocco, June 11–13, 2015

Maqbool S, Ramzan T, Haider A et al (2025) Impact of thiourea on wheat's morpho-physiological and ionic attributes (*Triticum aestivum* L.) under lead stress: reducing the translocation of lead from soil to roots, shoots, and grains. ACS Omega 10(3):3054–3066. https://doi.org/10.1021/acsomega.4c09941

Marashi SK, Chinchanikar GS (2010) Effect of waterlogging on yield and yield components of wheat (*Triticum aestivum* L.). Int J Agric Eng Res 5:561–567

Matsuoka Y, Nasuda S (2004) Durum wheat as a candidate for the unknown female progenitor of bread wheat: an empirical study with a highly fertile $F_1$ hybrid with *Aegilops tauschii* Coss. Theor Appl Genet 109:1710–1717. https://doi.org/10.1007/s00122-004-1791-x

Matsya Purana (1972) (Tr) Vasu SC; (Ed) Akhtar JD. Oriental Publ, Delhi

McDonald AJ, Balwinder-Singh, Keil A et al (2022) Time management governs climate resilience and productivity in the coupled rice–wheat cropping systems of eastern India. Nat Food 3:542–551. https://doi.org/10.1038/s43016-022-00549-0

McMullen M, Bergstrom G, de Wolf E et al (2012) Unified effort to fight an enemy of wheat and barley: Fusarium head blight. Plant Dis 96:1712–1728. https://doi.org/10.1094/PDIS-03-12-0291-FE

Meena ML, Dudi A (2014) Indigenous technological practices followed by wheat growers in Rajasthan. Indian J Extn Educ R D 22:30–35

Mills G, Sharps K, Simpson D et al (2018) Ozone pollution will compromise efforts to increase global wheat production. Glob Chang Biol 24:3560–3574. https://doi.org/10.1111/gcb.14157

Milne RJ, Dibley KE, Schnippenkoetter W et al (2019) The wheat *Lr67* gene from the sugar transport protein 13 family confers multipathogen resistance in barley. Plant Physiol 179:1285–1297. https://doi.org/10.1104/pp.18.00945

Ministry of Agriculture and Farmers' Welfare, Government of India (2025) Final estimates of production of foodgrains, oilseeds and other commercial crops for 2024–25. https://agriwelfare.gov.in

Ministry of Statistics and Programme Implementation, Government of India (2025) Agricultural statistics at a glance 2024–25. https://www.mospi.gov.in

Mishra AK, Gupta GS, Agrawal SB et al (2024) Understanding the impact of elevated $CO_2$ and $O_3$ on growth and yield in Indian wheat cultivars: implications for food security in a changing climate. Environ Pollut 362:124990. https://doi.org/10.1016/j.envpol.2024.124990

Mohan R, Singh VK, Chetan KK et al (2025) Multiple patho-phenotyping and molecular analysis to characterize wide-spectrum durable leaf rust resistance in wheat collections from India. Front Microbiol 16:1596282. https://doi.org/10.3389/fmicb.2025.1596282

Mondal T (2020) Lodging in wheat: its causes, ill effects, and management for higher productivity and profitability. Agri 2:AL202100

Morkunas I, Woźniak A, Mai VC et al (2018) The role of heavy metals in plant response to biotic stress. Molecules 23(9):2320. https://doi.org/10.3390/molecules23092320

Morton JF (2007) The impact of climate change on smallholder and subsistence agriculture. Proc Natl Acad Sci USA 104:19680–19685. https://doi.org/10.1073/pnas.0701855104

Morton CM, Pullabhotla H, Bevis L et al (2023) Soil micronutrients linked to human health in India. Sci Rep 13:13591. https://doi.org/10.1038/s41598-023-39084-8

Muhammad A, Waseem K, Sh F et al (2025) Mitigating heavy metal pollution in agriculture: a multi-omics and nanotechnology approach to safeguard global wheat production. J Hazard Mater Adv 100584. https://doi.org/10.1016/j.hazadv.2024.100584

Munns R, Tester M (2008) Mechanisms of salinity tolerance. Annu Rev Plant Biol 59:651–681. https://doi.org/10.1146/annurev.arplant.59.032607.092911

Murphy CY, Burrows ME (2021) Management of the wheat curl mite and wheat streak mosaic virus with insecticides on spring and winter wheat. Front Plant Sci 12:682631. https://doi.org/10.3389/fpls.2021.682631

Myers SS, Zanobetti A, Kloog I et al (2014) Increasing $CO_2$ threatens human nutrition. Nature 510:139–142. https://doi.org/10.1038/nature13179

Nakamura S (2018) Grain dormancy genes responsible for preventing pre-harvest sprouting in barley and wheat. Breed Sci 68:295–304. https://doi.org/10.1270/jsbbs.17138

Nene YL, Sadhale N (1997) Agriculture and biology in Rigveda. Asian Agri-Hist 1:177–190

Nóia Júnior R, Stocca V, Martre P et al (2025) Enabling modeling of waterlogging impact on wheat. Field Crop Res 333:110090. https://doi.org/10.1016/j.fcr.2025.110090

Nyaupane S, Poudel M, Panthi B et al (2024) Drought stress effect, tolerance, and management in wheat—a review. Cogent Food Agric 10:2296094. https://doi.org/10.1080/23311932.2023.2296094

Oerke EC (2006) Crop losses to pests. J Agric Sci 144:31–43. https://doi.org/10.1017/S0021859605005708

Ortiz R, Sayre KD, Govaerts B et al (2008) Climate change: can wheat beat the heat? Agric Ecosyst Environ 126:46–58. https://doi.org/10.1016/j.agee.2008.01.019

Panozzo JF, Walker CK, Partington DL et al (2014) Elevated carbon dioxide changes grain protein concentration and composition and compromises baking quality: a FACE study. J Cereal Sci 60:461–470. https://doi.org/10.1016/j.jcs.2014.08.011

Peng J, Sun D, Nevo E (2011) Wild emmer wheat, *Triticum dicoccoides*, occupies a pivotal position in wheat domestication process. Aust J Crop Sci 5:1127–1143

Pequeno DNL, Ferreira TB, Fernandes JMC et al (2024) Production vulnerability to wheat blast disease under climate change. Nat Clim Chang 14:178–183. https://doi.org/10.1038/s41558-023-01902-2

Pingali PL (2012) Green revolution: impacts, limits, and the path ahead. Proc Natl Acad Sci USA 109:12302–12308. https://doi.org/10.1073/pnas.0912953109

Pleijel H, Uddling J (2012) Yield vs. quality trade-offs for wheat in response to carbon dioxide and ozone. Glob Chang Biol 18:596–605. https://doi.org/10.1111/j.1365-2486.2011.02489.x

Poddar S, Tanaka J, Running KLD et al (2023) Optimization of highly efficient exogenous-DNA-free Cas9 ribonucleoprotein mediated gene editing in disease susceptibility loci in wheat (Triticum aestivum L.). Front Plant Sci 13:1084700. https://doi.org/10.3389/fpls.2022.1084700

Pokharia AK (2011) Palaeoethnobotany at Lahuradewa: a contribution to the 2nd millennium BC agriculture of the Ganga Plain, India. Curr Sci 101:1569–1578

Pokharia AK, Kharakwal JS, Srivastava A (2014) Archaeobotanical evidence of millets in the Indian subcontinent with some observations on their role in the Indus civilization. J Archaeol Sci 42:442–455. https://doi.org/10.1016/j.jas.2013.11.029

Posch BC, Zhai D, Coast O et al (2022) Wheat respiratory $O_2$ consumption falls with night warming alongside greater respiratory $CO_2$ loss and reduced biomass. J Exp Bot 73(3):915–926. https://doi.org/10.1093/jxb/erab454

Possehl GL (1997) The transformation of the Indus civilization. J World Prehist 11:425–472. https://doi.org/10.1007/BF02220556

Prakash O (1961) Foods and drinks in ancient India. Munshi Lai Manohar Lai, Delhi

Prasad P, Bhardwaj SC, Gangwar OP et al (2020) Wheat rust research: impact, thrusts, and road-map to sustained wheat production. In: Sareen S, Sharma P, Singh C et al (eds) Improving cereal productivity through climate smart practices. Woodhead Publishing, pp 177–203

Prasad P, Thakur R, Bhardwaj SC et al (2023) Virulence and genetic analysis of *Puccinia graminis* f. sp. *tritici* in the Indian sub-continent from 2016 to 2022 and evaluation of wheat varieties for stem rust resistance. Front Plant Sci 14:1196808. https://doi.org/10.3389/fpls.2023.1196808

Press Information Bureau (2024) Ministry of Consumer Affairs, Food & Public Distribution, Government of India. Free foodgrains to 81.35 crore beneficiaries under National Food Security Act: Cabinet Decision. https://www.pib.gov.in/PressReleasePage.aspx?PRID=1886215

Ral JP, Whan A, Larroque O et al (2016) Engineering high α-amylase levels in wheat grain lowers falling number but improves baking properties. Plant Biotechnol J 14(1):364–376. https://doi.org/10.1111/pbi.12390

Ram S, Malik VK, Gupta V et al (2024) Impact of foliar application of iron and zinc fertilizers on grain iron, zinc, and protein contents in bread wheat (*Triticum aestivum* L.). Front Nutr 11:1378937. https://doi.org/10.3389/fnut.2024.1378937

Ramadas S, Kiran TM, Singh G (2019) Wheat production in India: trends and prospects. IntechOpen. https://doi.org/10.5772/intechopen.86341

Randhawa MS (1980) A history of agriculture in India. ICAR, New Delhi

Rani R, Prasad S, Jaiswal B et al (2018) Screening of wheat varieties (*Triticum aestivum* L.) under natural salt stress condition for yield and yield related traits. Int J Curr Microbiol Appl Sci Special Issue-7:2746–2751

Rathee M, Dalal PK (2018) Emerging insect pests in Indian agriculture. Indian J Entomol 80:267. https://doi.org/10.5958/0974-8172.2018.00043.3

Ren R, Zhou X, Feng J (2024) Editorial: wheat disease resistance: diagnosis, germplasm mining, and molecular breeding. Front Plant Sci 15:1500414. https://doi.org/10.3389/fpls.2024.1500414

Reynolds MP, Rajaram S, Sayre KD (1999) Physiological and genetic changes of irrigated wheat in the post–green revolution period and approaches for meeting projected global demand. Crop Sci 39:1611–1621. https://doi.org/10.2135/cropsci1999.3961611x

Reynolds M, Bonnett D, Chapman SC et al (2011) Raising yield potential of wheat. I. Overview of a consortium approach and breeding strategies. J Exp Bot 62:439–452. https://doi.org/10.1093/jxb/erq311

Rizvi A, Zaidi A, Ameen F et al (2020) Heavy metal induced stress on wheat: phytotoxicity and microbiological management. RSC Adv 10:38379–38303. https://doi.org/10.1039/d0ra05610c

Robinson EA, Ryan GD, Newman JA (2012) A meta-analytical review of the effects of elevated $CO_2$ on plant–arthropod interactions highlights the importance of interacting environmental and biological variables. New Phytol 194:321–336. https://doi.org/10.1111/j.1469-8137.2012.04074.x

Roy C, He X, Gahtyari NC et al (2023) Managing spot blotch disease in wheat: conventional to molecular aspects. Front Plant Sci 14:1098648. https://doi.org/10.3389/fpls.2023.1098648

Sadhale N (Tr.) (1996) Surapala's Vrikshayurveda (the science of plant life by Surapala). Agri-history bulletin no. 1. Asian Agri-History Foundation, Secunderabad, p 94

Sadhale N (Tr.) (1999) Krishi-Parashara (agriculture by Parashara). Agri-history bulletin no. 2. Asian Agri-History Foundation, Secunderabad, p 94

Sahu O, Singh A (2016) Indigenous practices for eco-friendly storage of food grains and seeds. Adv Plants Agric Res 3:00101

Sankaran S, Khot LR, Zúñiga Espinoza C et al (2015) Low-altitude, high-resolution aerial imaging systems for row and field crop phenotyping: a review. Eur J Agron 70:112–123. https://doi.org/10.1016/j.eja.2015.07.004

Saraswat KS (2004) Plant economy of early farming communities. In: Singh BP (ed) Early farming communities of the Kaimur (excavations at Senuwar). Publication Scheme, Jaipur, p 416

Sarkar S, Islam AKMA, Barma NCD et al (2021) Tolerance mechanisms for breeding wheat against heat stress: a review. S Afr J Bot 138:262–277. https://doi.org/10.1016/j.sajb.2021.01.003

Satyagopal K, Sushil SN, Jeyakumar P et al (2014) AESA-based IPM package for wheat. Extension Publication, p 58

Saurabh K, Rao KK, Mishra JS et al (2021) Influence of tillage-based crop establishment and residue management practices on soil quality indices and yield sustainability in rice–wheat cropping system of eastern Indo-Gangetic Plains. Soil Tillage Res 206:104841. https://doi.org/10.1016/j.still.2020.104841

Savadi S, Prasad P, Kashyap PL et al (2018) Molecular breeding technologies and strategies for rust resistance in wheat (*Triticum aestivum*) for sustained food security. Plant Pathol 67:771–791. https://doi.org/10.1111/ppa.12802

Seleiman MF, Aslam MT, Alhammad BA et al (2022) Salinity stress in wheat: effects, mechanisms and management strategies. Phyton-Int J Exp Bot 91(4):667–694. https://doi.org/10.32604/phyton.2022.017365

Setter T, Waters I (2003) Review of prospects for germplasm improvement for waterlogging tolerance in wheat, barley and oats. Plant Soil 253:1–34. https://doi.org/10.1023/A:1024573305997

Shakoor N, Lee S, Mockler T (2017) High throughput phenotyping to accelerate crop breeding and monitoring of diseases in the field. Curr Opin Plant Biol 38:184–192. https://doi.org/10.1016/j.pbi.2017.05.006

Shamasastry R (1961) Kautilya's Arthasastra, 7th edn. Mysore Printing and Publishing House, Mysore, p 488

Sharma PD, Acharya CL (1987) Effect of soil management on rain-fed wheat in Northern India. II. Nutrient uptake, plant growth and yield. Soil Tillage Res 9:79–89. https://doi.org/10.1016/0167-1987(87)90052-3

Sharma RC, Duveiller E (2006) Spot blotch continues to cause substantial grain yield reductions under resource-limited farming conditions. J Phytopathol 154:482–488. https://doi.org/10.1111/j.1439-0434.2006.01134.x

Sharma AK, Babu KS, Nagarajan S et al (2004) Distribution and status of termite damage to wheat crop in India. Indian J Entomol 66:235–237

Sharma RC, Duveiller E, Ortiz-Ferrara G (2016) Progress and challenge towards reducing wheat spot blotch threat in the Eastern Gangetic Plains of South Asia: is climate change already taking its toll? Field Crop Res 103:109–118. https://doi.org/10.1016/j.fcr.2007.05.004

Sharma K, Dhanda S, Leharwan M et al (2023) Challenges of herbicide-resistant weeds in wheat in India and its management: a review. Agric Rev 44:536–541. https://doi.org/10.18805/ag.R-2233

Shiferaw B, Smale M, Braun HJ et al (2013) Crops that feed the world 10. Past successes and future challenges to the role played by wheat in global food security. Food Sec 5:291–317. https://doi.org/10.1007/s12571-013-0263-y

Singh S (2020) Farmers' perception of climate change and adaptation decisions: a micro-level evidence from Bundelkhand region. India Ecol Indic 116:106475. https://doi.org/10.1016/j.ecolind.2020.106475

Singh A (2023) Genetic variation in waterlogging tolerance of wheat in sodic soil. Plant Arch 23:248–252

Singh B, Jasrotia P (2020) Impact of integrated pest management (IPM) module on major insect pests of wheat and their natural enemies in North-Western plains of India. J Cereal Res 12:114–119

Singh B, Sarao P (2020) Efficacy of biopesticides and insecticides against armyworm *Mythimna separata* (Walker). Indian J Entomol 82:369. https://doi.org/10.5958/0974-8172.2020.00082.6

Singh HP, Mahajan P, Kaur S et al (2013) Chromium toxicity and tolerance in plants. Environ Chem Lett 11:10. https://doi.org/10.1007/s10311-013-0407-5

Singh RP, Hodson DP, Jin Y et al (2015) Emergence and spread of new races of wheat stem rust fungus: continued threat to food security and prospects of genetic control. Phytopathology 105:872–884. https://doi.org/10.1094/PHYTO-01-15-0030-FI

Singh RP, Singh PK, Rutkoski J et al (2016) Disease impact on wheat yield potential and prospects of genetic control. Annu Rev Phytopathol 54:303–322. https://doi.org/10.1146/annurev-phyto-080615-095835

Singh B, Simon A, Halsey K et al (2020) Characterisation of bird cherry-oat aphid (*Rhopalosiphum padi* L.) behaviour and aphid host preference in relation to partially resistant and susceptible wheat landraces. Ann Appl Biol 177:184–194
Singh C, Kamble U, Gupta V et al (2021) Pre-harvest sprouting in wheat: current status and future prospects. J Cereal Res 13:114484. https://doi.org/10.25174/2582-2675/2021/114484
Singh K, Jhorar RK, Sidhpuria MS et al (2023) Energy conservation prospects in water intensive paddy–wheat cropping system for groundwater pumping in the semi-arid region of Haryana. PeerJ 11:e14815. https://doi.org/10.7717/peerj.14815
Smale M (1997) The Green Revolution and wheat genetic diversity: some unfounded assumptions. World Dev 25:1257–1269. https://doi.org/10.1016/S0305-750X(97)00038-7
Sonkar G, Mall R, Banerjee T et al (2019) Vulnerability of Indian wheat against rising temperature and aerosols. Environ Pollut 255:113114. https://doi.org/10.1016/j.envpol.2019.07.114
Soora NK, Aggarwal PK, Swaroopa R et al (2011) Impact of climate change on crop productivity in Western Ghats, coastal and northeastern regions of India. Curr Sci 101:332–341
Spengler RN (2015) Agriculture in the Central Asian Bronze Age. J World Prehist 28:215–253. https://doi.org/10.1007/s10963-015-9087-3
Su Z, Bernardo A, Tian B et al (2019) A deletion mutation in *TaHRC* confers *Fhb1* resistance to Fusarium head blight in wheat. Nat Genet 51:1099–1105. https://doi.org/10.1038/s41588-019-0425-8
Summerell BA, Laurence MH, Liew ECY et al (2010) Biogeography and phylogeography of Fusarium: a review. Fungal Divers 44:3–13. https://doi.org/10.1007/s13225-010-0060-2
Sun J, Cao W, Fu X et al (2023) Few-shot learning for plant disease recognition: a review. Agron J 116:1–14. https://doi.org/10.1002/agj2.21285
Talbot I (1982) Punjab and the Raj, 1849–1947
The International Wheat Genome Sequencing Consortium (IWGSC), Appels R, Eversole K, Stein N et al (2018) Shifting the limits in wheat research and breeding using a fully annotated reference genome. Science 361:eaar7191. https://doi.org/10.1126/science.aar7191
Thomson LJ, Macfadyen S, Hoffmann AA (2010) Predicting the effects of climate change on natural enemies of agricultural pests. Biol Control 52:296–306. https://doi.org/10.1016/j.biocontrol.2009.01.022
Trnka M, Muška F, Semerádová D et al (2007) European corn borer life stage model: regional estimates of pest development and spatial distribution under present and future climate. Ecol Model 207:61–84. https://doi.org/10.1016/j.ecolmodel.2007.04.014
Ur Rehman S, Yang J, Zhang J et al (2025) Salt stress in wheat: a physiological and genetic perspective. Plant Stress 16:100832. https://doi.org/10.1016/j.stress.2025.100832
Vishnu-Mittre S (1976) Setaria in ancient plant economy of India. J Palaeosci 25:559–564. https://doi.org/10.54991/jop.1976.1042
Vishnu-Mittre S, Sharma A, Chanchala (1986) Ancient plant economy at Daimabad (appendix II). In: Dali SA (ed) Daimabad 1976–79. Memoirs of the Archaeological Survey of India, Government of India Central Publication Branch, Calcutta, p 588
Waites J, Achary VMM, Syombua ED et al (2025) CRISPR-mediated genome editing of wheat for enhancing disease resistance. Front Genome Ed 7:1542487. https://doi.org/10.3389/fgeed.2025.1542487
Wang Y, Cheng X, Shan Q et al (2014) Simultaneous editing of three homoeoalleles in hexaploid bread wheat confers heritable resistance to powdery mildew. Nat Biotechnol 32:947–951. https://doi.org/10.1038/nbt.2969
Wang N, Tang C, Fan X et al (2022a) Inactivation of a wheat protein kinase gene confers broad-spectrum resistance to rust fungi. Cell 185:2961–2974.e19. https://doi.org/10.1016/j.cell.2022.06.027
Wang N, Fan X, He M et al (2022b) Transcriptional repression of *TaNOX10* by *TaWRKY19* compromises ROS generation and enhances wheat susceptibility to stripe rust. Plant Cell 34:1784–1803. https://doi.org/10.1093/plcell/koac001

Wang J, Li C, Li L et al (2023) *DIW1* encoding a clade IPP2C phosphatase negatively regulates drought tolerance by de-phosphorylating *TaSnRK1.1* in wheat. J Integr Plant Biol 65:1918–1936. https://doi.org/10.1111/jipb.13504

Waqas M, Rasheed A, Peng J (2025) Wheat genomics frontiers for gene discovery and breeding applications. WheatOmics 1:3. https://doi.org/10.1007/s44412-025-00002-6

Weber SA (2024) Plants and Harappan subsistence, an example of stability and change from Rojdi, 1st edn. Routledge, London

Whitcombe E (1971) Agrarian conditions in Northern India, vol 1. Thomson Press, New Delhi

Xiong W, Reynolds MP, Montes C et al (2024) New wheat breeding paradigms for a warming climate. Nat Clim Chang 14:869–875. https://doi.org/10.1038/s41558-024-02069-0

Xu L, Cao B, Zhao F et al (2023) Wheat leaf disease identification based on deep learning algorithms. Physiol Mol Plant Pathol 123:101940. https://doi.org/10.1016/j.pmpp.2022.101940

Yadava DK, Hossain F, Choudhury PR et al (2022) Crop cultivars developed through molecular breeding, 2nd edn. Indian Council of Agricultural Research, New Delhi

Yajurv Ved Vajasmeyi Samhita (1959) (Tr) Devi Chand. All India Dayanand Salvation Mission, Hoshiyarpur

Yang W, Feng H, Zhang X et al (2020) Crop phenomics and high-throughput phenotyping: past decades, current challenges, and future perspectives. Mol Plant 13:187–214. https://doi.org/10.1016/j.molp.2020.01.008

Yang R, Harrison M, Yang Y et al (2025) Pre-harvest sprouting in cereals: global incidence, impacts and mitigation strategies. Field Crop Res 333:110111. https://doi.org/10.1016/j.fcr.2025.110111

Zampieri M, Ceglar A, Dentener F et al (2017) Wheat yield loss attributable to heat waves, drought and water excess at the global, national and subnational scales. Environ Res Lett 12:064008. https://doi.org/10.1088/1748-9326/aa723b

Zhang Y, Bai Y, Wu G et al (2017) Simultaneous modification of three homoeologs of *TaEDR1* by genome editing enhances powdery mildew resistance in wheat. Plant J 91:714–724. https://doi.org/10.1111/tpj.13599

Zhang Z, Flores P, Igathinathane C et al (2020) Wheat lodging detection from UAS imagery using machine learning algorithms. Remote Sens 12:1838. https://doi.org/10.3390/rs12111838

Zhao C, Liu B, Piao S et al (2017) Temperature increase reduces global yields of major crops in four independent estimates. Proc Natl Acad Sci USA 114:9326–9331. https://doi.org/10.1073/pnas.1701762114

Zohary D, Hopf M, Weiss E (2000) Domestication of plants in the Old World, 4th edn. Oxford University Press, Oxford

# Advances and Challenges in Developing Stress-Resilient Wheat Cultivars in India

2

Vishwakarma Manish Kumar, Pradeep Kumar Bhati, and Arun Kumar Joshi

**Abstract**

Wheat production in India is under mounting pressure from escalating abiotic stresses, most notably heat, drought, salinity, and waterlogging, as well as from fast-evolving biotic challenges, including rusts, spot blotch, and newly emerging pathogens. Developing stress-resilient cultivars has therefore become crucial to maintaining yield stability under climate variability. This chapter synthesizes progress in India's wheat improvement pipeline, tracing advances from early selection and Green Revolution dwarfing genes to contemporary genomics-assisted and systems-level breeding approaches. We highlight key genetic, physiological, and biochemical mechanisms underlying abiotic and biotic stress tolerance, including major resistance genes, quantitative trait loci, transcriptional regulators, and adaptive traits such as efficient root systems, canopy temperature depression, and stay-green behaviour. Recent innovations, including genomic selection, speed breeding, high-throughput phenotyping, CRISPR–Cas genome editing, and microbiome-mediated resilience, are accelerating genetic gain and enabling more precise trait introgression. Case studies of recent multi-stress-tolerant cultivars demonstrate the successful deployment of these technologies in farmer-preferred backgrounds. Remaining challenges include breeding for complex polygenic stress combinations, strengthening seed systems, countering new pathogen races, and integrating climate-informed predictive breeding. Overall, the convergence of modern genomics, phenomics, and digital agriculture provides a strong platform for delivering resilient wheat varieties suited to future agro-climatic scenarios.

**Keywords**

Wheat breeding · Stress resilience · Abiotic and biotic stresses · Genomics-assisted selection · High-throughput phenotyping · Climate-smart cultivars

V. M. Kumar (✉) · P. K. Bhati · A. K. Joshi
Borlaug Institute for South Asia (BISA), NASC Complex, New Delhi, Delhi, India
e-mail: m.vishwakarma@cgiar.org



R. Tiwari et al. (eds.), *Wheat Stress Management*,
https://doi.org/10.1007/978-981-92-0902-6_2

## 2.1 Introduction

Bread wheat (*Triticum aestivum*) is the primary staple food crop in India, accounting for over 35% of total food grain production and providing approximately 20–25% of the total calorie intake (ICAR-IIWBR 2023). With production exceeding 112 million tons in the 2023–2024 crop year, India is the world's second-largest producer, contributing substantially to national food security. However, wheat production in India is continually challenged by multiple factors, including climate change and various biotic and abiotic stresses.

Abiotic stresses such as terminal heat and drought pose significant threats to wheat productivity and have become increasingly severe over the past two decades due to climate change. Projections indicate that temperature increases of 1.5–4.0 °C by 2100 could result in yield declines of 15–20% in North-West India, the country's major wheat-producing region and a key part of the Indo-Gangetic Plains (Aggarwal 2008). Simultaneously, declining groundwater resources and unpredictable monsoon patterns are intensifying water stress, a critical constraint for successful wheat cultivation. Soil salinity and waterlogging further complicate production in certain agro-ecologies. Biotic stresses constitute another major threat. Among these, the rapidly evolving wheat rusts (stem, leaf, and stripe rust), spot blotch, and newly emerging diseases such as wheat blast (reported in neighbouring Bangladesh) are of particular concern. In the absence of appropriately resistant varieties, these diseases can cause yield losses ranging from 70% to complete crop failure (Singh et al. 2016).

These challenges clearly illustrate that developing stress-resilient wheat cultivars is an urgent national priority. Such efforts are essential for safeguarding food security and ensuring a stable supply of wheat for a growing population. Stress-resilient cultivars are equally important for stabilizing farmer incomes and protecting the livelihoods of more than 100 million wheat-growing farmers who face substantial production risks. Developing varieties that require fewer inputs (water, fertilizers, and pesticides) and can withstand existing and emerging pathogen races will be the key to achieving these goals. Successful wheat production will enable India to meet domestic demand while maintaining its potential as a global exporter.

### 2.1.1 Traditional Resistance Breeding Approaches and Emerging Advances

The food shortages of the 1950s compelled Indian scientists to prioritize grain yield as the primary breeding objective under well-managed production conditions. The introduction of semi-dwarf wheat varieties from CIMMYT, Mexico, triggered a dramatic surge in wheat production and laid the foundation of the Green Revolution. Within just 4 years, India's wheat harvest doubled to 20 million tons, enabling the country to achieve self-sufficiency in wheat (Joshi and Kropff 2018). Once self-sufficiency was attained, breeding strategies shifted toward enhancing yield stability, resilience to biotic and abiotic stresses, and superior grain quality across diverse management conditions. These advances were made possible through the intelligent integration of traditional breeding knowledge with emerging scientific innovations, supported by strong national and international collaborations.

Conventional breeding relied either on selection within exotic germplasm introductions or on segregating generations derived from hybridization among available genetic materials (Fig. 2.1). Mutagen-treated populations were occasionally exploited by some institutions to identify superior variants. The most successful approach, however, involved controlled hybridization among parents with complementary traits (e.g., high yield × rust resistance), followed by rigorous multigenerational selection within target environments such as drought-prone regions and disease nurseries (Joshi et al. 2007). For several decades, grain yield and rust resistance remained the core breeding priorities in India. To improve tolerance to abiotic stresses, breeders began incorporating physiological traits conferring resilience, such as canopy temperature depression (an indicator of transpiration cooling and water-use efficiency), stay-green behaviour (delayed senescence under heat), and

**Fig. 2.1** Evolution of breeding techniques

favourable root architecture (for enhanced water and nutrient extraction from deeper soil layers) (Reynolds et al. 2021).

Over the past two decades, molecular breeding tools, particularly marker-assisted selection (MAS) and gene pyramiding, have become central to breeding programs. These approaches facilitated precise and rapid introgression of key genes (e.g., *Yr36* for heat tolerance, the pleiotropic slow-rusting locus *Lr34/Sr57/Yr18/Pm38*, and *Sr26* for stem rust resistance) into elite cultivars, thereby reducing reliance on labour-intensive phenotypic screening in early generations (Vishwakarma et al. 2022). More recently, genomic selection (GS) has been deployed at selected centres through international partnerships. GS enables prediction of genomic estimated breeding values (GEBVs) for complex, polygenic traits, including multiple biotic and abiotic stresses, using genome-wide marker data, substantially shortening breeding cycles by allowing selection based solely on genetic merit (Crossa et al. 2017).

Emerging breeding advances such as speed breeding have further accelerated genetic improvement by enabling up to six wheat generations per year under controlled conditions with extended photoperiods and optimized temperatures (Watson et al. 2018; Vishwakarma et al. 2024). High-throughput phenotyping (HTP) platforms employing drones, satellites, and ground-based sensors equipped with RGB, multispectral, hyperspectral, and thermal imaging capabilities have revolutionized phenotyping by providing non-destructive, large-scale measurements of plant health, water status, and biomass. These datasets can be integrated with genomic information to enhance the accuracy of selection (Araus et al. 2018). Today, the convergence of classical breeding, molecular tools, genomics, speed breeding, and advanced phenotyping is yielding a new generation of climate-resilient wheat cultivars, equipped to meet the challenges of twenty-first-century agriculture.

## 2.2 Historical Perspectives on Stress-Resilient Wheat Breeding in India

India's progress in developing stress-resilient wheat reflects a continuous process from traditional landrace selection to systematic, research-based breeding. The accumulated diversity, breeder expertise, and coordinated programs conducted over the decades enabled breakthroughs such as semi-dwarf cultivars, molecular breeding, and genomics-assisted selection, which continue to shape current and future breeding priorities.

### 2.2.1 Evolution of Wheat Breeding Strategies for Stress Resistance

Wheat is a crop of ancient origin, cultivated since around 3000 BC (Tomar et al. 2004). Generations of Indian farmers practised selection long before formal breeding programs were established. Their early methods were the first forms of selective breeding, primarily targeting two traits: tolerance to stresses and grain quality, both external and related to taste or cooking properties. These practices evolved through necessity, observation, and adaptation to local environments. Owing to the broad agro-ecological variability across India, the history of wheat improvement for stress resistance reflects a continuous journey of scientific innovation and adaptive breeding. Early wheat improvement in India can broadly be grouped into three categories: (i) selection or hybridization within locally collected or available materials, (ii) exotic germplasm introduction, and (iii) hybridization with exotic-adapted germplasm to achieve sustainable yield and enhanced resistance.

**Wheat Cultivation Up to the Early 1900s**

Before modern breeding methods were introduced, Indian farmers cultivated native, heterogeneous landraces and mixtures known as 'sorts', which were classified based on grain characteristics such as colour and texture rather than botanical traits (Pal 1966). The wheat sorts cultivated in India carried generic local names, such as Sharbati, Dara, Pissi, Safed Pissi, Karachi Choice, Lal Pissi, and Lal Kanak, representing mixtures of landraces rather than genetically uniform varieties. Despite their low yield potential, these landraces were well adapted to local disease pressures and variable abiotic stresses, including drought, heat, and challenging soil conditions.

**Early Modern Breeding (1905–1962)**

Systematic wheat improvement efforts in India began in 1905 at Lyallpur (now Faisalabad), Punjab, and at the Imperial Agricultural Research Institute (IARI) in Pusa, Bihar. This period can be divided into three phases: (i) pure-line selection from local mixtures, (ii) hybridization to combine desirable characters, and (iii) targeted development of disease-resistant cultivars.

One of the earliest systematic studies was conducted by Albert and Gabrielle Howard at the Imperial Agricultural Research Institute's Phipps Laboratory (Naulakha) in Pusa, Bihar, established in 1904. Their mission was to elevate wheat to the same level of importance as rice by improving and popularizing it, marking the beginning of modern wheat breeding in India. Their work involved surveying, collecting, purifying, characterizing, and classifying landraces, common varieties, and wheat types (B 1925). According to their estimates, 7.6 million tons of wheat were harvested from 10.2 million hectares in 1909 in undivided India. Remarkably, wheat yields stagnated at around 750 kg/ha for nearly 80 years, from 1885 to 1965 (Smale et al. 2008).

With the assistance of Mia Abdul Rehman, Howards carried out extensive selection efforts across India and developed varieties such as Pusa 4, Pusa 6, and Pusa 12. Concurrently, D. Milne selected Type 11, which was recommended for northern India by the Government Agricultural College, Lyallpur. Later, in 1919, Punjab 8A and Punjab 9D replaced Type 11, being recommended for timely sown irrigated and rainfed conditions, respectively. Varieties developed at IARI, Pusa, were designated with the prefix 'P' (e.g., P4, P12), while those developed at the new IARI campus in New Delhi used the prefix 'NP' (New Pusa), such as NP 770. Varieties developed at Lyallpur were designated PbC (Punjab Cross) and later shortened to 'C' (e.g., C

306). Additional series included AO (Powarkheda, Madhya Pradesh), Hy (Hyderabad), and NI (Niphad). These series dominated Indian wheat breeding until the introduction of semi-dwarf Mexican genotypes in the 1960s.

Among early Indian wheat varieties, NP 4 was the first to receive a distinct name. Subsequent improved types, such as NP 6 and NP 12, resulted from systematic selection and purification. By 1955, the cultivar C 591 had spread to nearly 80% of the Punjab region (Pal 1966). Among India's early wheat breeders, Chaudhary Ramdhan Singh was one of the most prominent. After working briefly with the Howards at IARI (1914–1919), he joined the Government College of Agriculture, Lyallpur, in 1926. In 1933–34, he developed the important varieties '591 Gandam' (C 591 or Pb 591) and Pb 518. He pioneered cross-breeding to create variability and later contributed varieties such as C 217, C 228, C 280, C 253, C 273, C 281, and C 285. His influence continued even after retirement; the widely popular variety C 306, released in 1965, originated from his breeding material. Similarly, WH 147, released in 1975, was derived from his contributions.

Parallel efforts at other centres yielded important varieties through pure-line selection: AO 13, AO 85, AO 88, and AO 90 at Powarkheda (Madhya Pradesh); Bansi 202, Motia, and Gulab at Niphad; and K 13, K 46, K 49, K 53, and K 54 at Kanpur (1928–1958). Prof. K. C. Mehta and Dr. B. P. Pal also made major contributions by developing cultivars resistant to rust and smut diseases, thereby addressing yield instability. Notable among these were NP 783 and NP 784 (brown rust resistance), NP 785 and NP 786 (yellow rust resistance), and NP 789 and NP 790 (stem rust resistance), all released in 1935. NP 809 was later developed as a cultivar resistant to loose smut and all three rusts (Krishnappa et al. 2022; Gupta et al. 2023).

During the pre–Green Revolution period, wheat improvement efforts primarily targeted disease resistance and yield enhancement in tall, traditional genotypes. However, as India's population grew, reports of food shortages increased. The yield potential of tall varieties had plateaued, and they were prone to lodging under high irrigation and fertilizer inputs, making further production gains impossible without fundamentally new plant types.

### 2.2.2 Role of Traditional Selection Methods and Early Breeding Programs

The development of stress-resilient wheat in India has its origin in the foundational work of early-twentieth-century scientists and the traditional knowledge of farmers who often managed their plant materials with the skill of trained breeders. This early work laid the groundwork for establishing the institutional framework and genetic resources that later enabled major advancements. Pioneering institutions such as the Indian Agricultural Research Institute (IARI), Pusa, New Delhi, under the leadership of scientists like Dr. B. P. Pal, systematically collected, evaluated, and selected from local landraces, followed by pure-line selection and progeny testing. A significant milestone was the release of NP 809 in 1954, the first wheat variety resistant to all three rusts (Smale et al. 2008). Although these early breeding programs were limited in scope, they established a crucial foundation of genetic diversity and a scientific network. They demonstrated that traits for local adaptation could be identified and fixed through meticulous selection. The first scientific efforts using pure-line selection from landraces led to varieties such as 'Pusa 4' and 'NP 4' (Singh et al. 2022). Overall, the early breeding initiatives and conventional selection methods did not produce the high-yielding, stress-resistant cultivars seen today, but they provided valuable genetic material, scientific expertise, and institutional support that enabled the Green Revolution and subsequent advances.

### 2.2.3 Impact of the Green Revolution in Enhancing Wheat Productivity and Stress Resilience

The mid-1960s marked the beginning of the Green Revolution, made possible through well-planned collaboration between India and CIMMYT, Mexico. This transformative period resulted from coordinated efforts between ICAR and CIMMYT, led by Dr. Norman E. Borlaug (CIMMYT) and Dr. M. S. Swaminathan in India. Crucially, the political will of the Government of India, particularly the decisive actions of the then Agriculture Minister, Mr. C. Subramaniam—facilitated the large-scale introduction of dwarf wheat lines from Mexico. Several Indian scientists played a major role in evaluating these materials and releasing them as varieties for farmers. The major constraint of lodging, a significant stress under high-input conditions, was effectively addressed through the incorporation of dwarfing genes (*Rht1* and *Rht2*) derived from 'Norin 10' and introduced via CIMMYT lines (Singh et al. 2022). Early introductions such as 'Kalyan Sona' and 'Sonalika' represented the first generation of high-yielding, stress-resilient cultivars; they were rust-resistant and well suited to the agro-climatic needs of Indian farmers across the Gangetic plains. 'Sonalika', being early maturing, was able to escape terminal heat stress and remained dominant in farmers' fields until the 1990s, especially in the eastern Gangetic plains (Smale et al. 2008).

Once the semi-dwarf, high-yielding varieties from CIMMYT were established, Indian breeders launched extensive hybridization and backcrossing programs that produced a series of outstanding cultivars. Subsequently, many bread wheat varieties grown in India and globally have descended from the Veery and Attila crosses (Smale et al. 2008), and significant synthetic wheat germplasm has contributed to the improvement of modern breeding pools. The Green Revolution thus transformed India from a 'ship-to-mouth' nation to a food self-sufficient country by linking high yield with resistance to major pests and diseases (Swaminathan 2013). The mid-twentieth century marked the introduction of alien germplasm and the beginning of systematic hybridization efforts, but it was the influx of semi-dwarf germplasm from CIMMYT in the 1960s that catalyzed the true revolution. This period marked a shift from stress avoidance strategies (e.g., earliness) to combining high-yield potential with intrinsic stress tolerance (e.g., lodging resistance conferred by dwarfing genes). Targeted breeding following the Green Revolution led to cultivars adapted to specific stress environments, such as 'HI 1500' for drought and 'KRL 1–4' for salinity (Paroda et al. 2013).

With the advent of molecular breeding in the twenty-first century, precision improvement using genomics-assisted selection and gene editing has become possible. This transition, from exploiting natural variation to creating intentional variation, illustrates how wheat breeding in India has continually evolved its methodologies to meet the dynamic demands of a challenging agricultural system.

### 2.2.4 Contributions of ICAR, CIMMYT, and Other Research Organizations

The tripartite collaboration among ICAR, CIMMYT, and the State Agricultural Universities (SAUs) represents one of the most successful models of international agricultural partnership globally. Its impact is clearly reflected in the tremendous progress in India's wheat production, which increased from 5.6 million tons in 1947–48 to 113.3 million tons in 2023–24.

**ICAR (Indian Council of Agricultural Research)** The All India Coordinated Wheat and Barley Improvement Project (AICW&BIP), coordinated by the Indian Institute of Wheat and Barley Research (IIWBR), Karnal, forms the backbone of India's wheat improvement efforts. It enables multi-location testing across diverse agro-climatic zones, ensuring broad adaptation and stress resilience. Major institutions actively engaged in wheat breeding include IARI, New Delhi; PAU, Ludhiana; HAU, Hisar; GBPUAT, Pantnagar; BHU, Varanasi; ARI, Pune; UAS, Dharwad; and JNKVV, Jabalpur, while centres such as CSSRI focus on salinity research.

**CIMMYT (The International Maize and Wheat Improvement Center)** It has been an indispensable partner, serving as a global source of novel germplasm, a provider of advanced training for Indian scientists, and a collaborator in strategic research initiatives, including efforts to combat Ug99 stem rust and the emerging threat of wheat blast. The CIMMYT–India office ensures a continuous flow of elite breeding lines into the national program (Joshi et al. 2007). To accelerate genetic gains and develop advanced breeding materials, CIMMYT adopted modern strategies such as selected bulk, rapid-cycle genomic selection (RCGS), and rapid-cycle recurrent selection (RCRS), combined with nursery selection incorporating genomic selection followed by final-stage phenotypic evaluation (Kumar et al. 2025).

**BISA (The Borlaug Institute for South Asia)** It was established in 2011 as a joint ICAR–CIMMYT initiative. It strengthens efforts toward environmental rehabilitation and food, nutrition, and livelihood security in South Asia. Through the ICAR–CIMMYT–BISA collaboration, 52 wheat varieties have been released between 2014 and 2024 (Fig. 2.2).

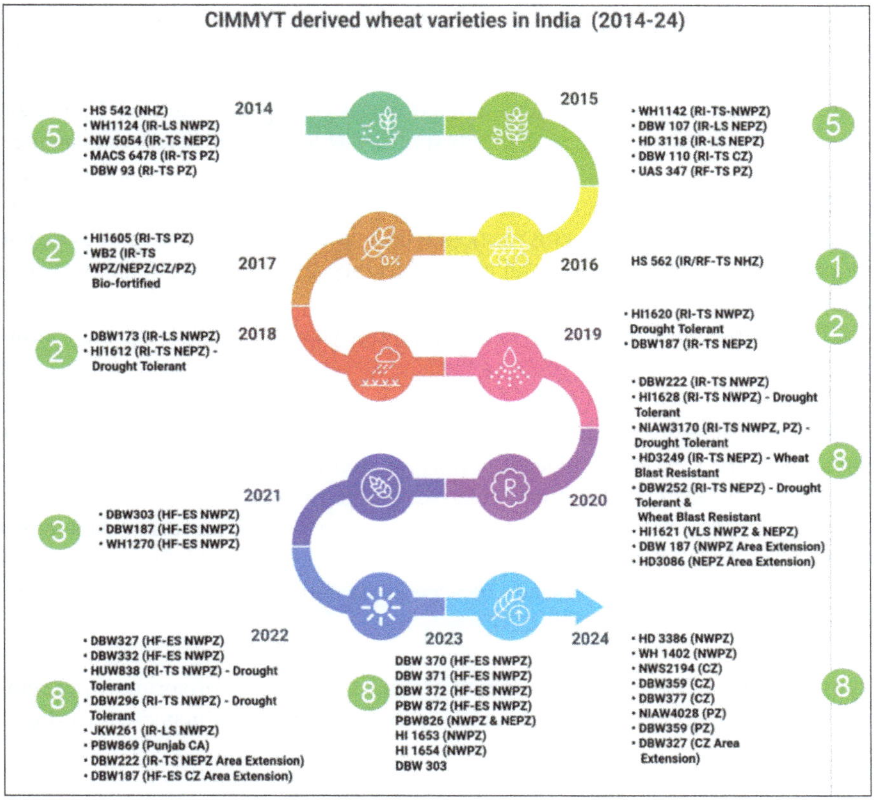


**Fig. 2.2** ICAR–CIMMYT–BISA collaboration from 2014 to 2024

## 2.3 Foundational Principles and Advances in Stress-Resilient Wheat Breeding

As breeding objectives have expanded to address increasingly complex stress environments, the methods used to develop resilient wheat varieties have also evolved. To appreciate the role of modern genomic and physiological tools, it is first necessary to revisit the classical breeding techniques that established the core principles of recombination, selection, and trait introgression.

### 2.3.1 Classical Breeding and Hybridization Techniques for Stress Tolerance

Traditional approaches remain the mainstay of breeding for stress resistance. Under controlled-stress conditions (such as drought-stressed fields or rust nurseries), these approaches involve generating genetic diversity through crossing, followed by pedigree selection, the modified bulk technique, or single-seed descent (SSD). A particular stress-resistance gene is often transferred from a donor parent to a superior but susceptible recurrent parent through backcross breeding. To explore novel resilience genes, absent from the cultivated wheat gene pool, extensive hybridization with wild relatives, such as *Aegilops* spp. and *Thinopyrum elongatum,* is undertaken (Ahtisham et al. 2025). While modern biotechnology offers sophisticated tools, traditional breeding and hybridization remain the most widely used and foundational methods for developing resilient wheat cultivars. These techniques enable the combination of high yield and stress tolerance by harnessing natural genetic diversity through recombination and selection.

Classical breeding follows a systematic, multi-year cycle involving the following steps:

**Germplasm Screening** Identifying sources of tolerance from genetically diverse pools is the first step in any breeding effort for stress tolerance. Germplasm collections, including landraces and wild relatives with strong resistance to diseases and drought, such as *Aegilops tauschii* and *Triticum dicoccoides,* serve as valuable reservoirs of useful alleles. International germplasm banks at CIMMYT and ICARDA are key centres housing thousands of accessions and readily share material with wheat breeders worldwide (Fig. 2.3).

**Hybridization (Crossing)** To combine desirable traits in the offspring ($F_1$ generation), a high-yielding, adaptable parent is deliberately cross-pollinated with a stress-

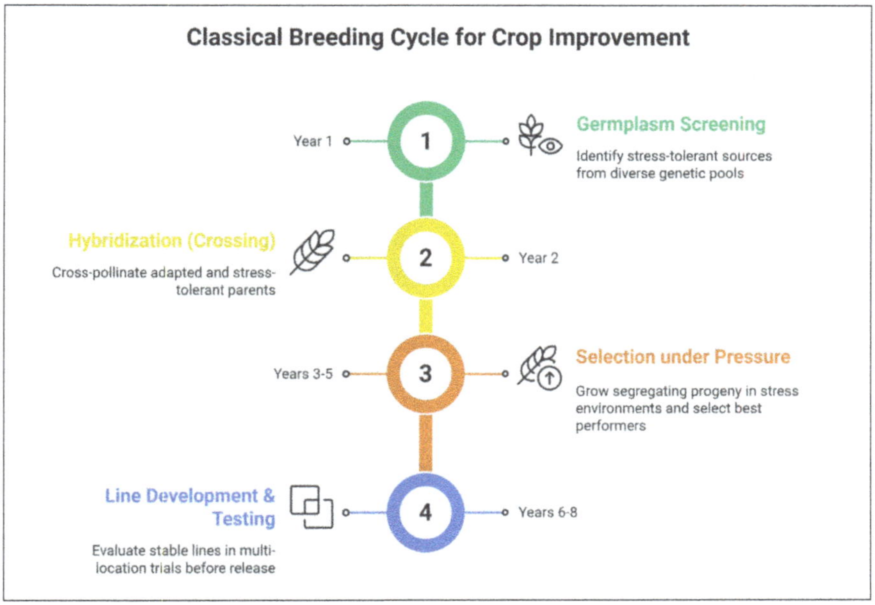


**Fig. 2.3** Approach followed for wheat improvement for stress tolerance

tolerant donor parent. The progenies are advanced according to the breeding objectives, and promising lines are selected for subsequent testing and potential release.

***Backcross Breeding*** This method is ideal for introducing a major gene into an elite variety, such as rust resistance. Through repeated backcrossing to the elite recurrent parent, most of the elite genome is recovered while incorporating the target trait from the donor. Numerous race-specific rust resistance genes have been deployed using this approach. Limited backcrossing, often involving a single backcross, has frequently been used to develop high-yielding genotypes with new traits. A recent notable example is the transfer of the biological nitrification inhibition (BNI) trait in wheat (Vishwakarma et al. 2025).

***Wide Hybridization*** To broaden the genetic base and introduce new genes, wheat is crossed with distant wild relatives. Following the Green Revolution, CIMMYT conducted major efforts to explore crosses between spring and winter wheat (Rajaram et al. 2002). Hybridizations in the late 1970s between semidwarf spring and winter wheat lines were notably successful, producing numerous promising advanced lines. The 1BL/1RS rye translocation, providing rust resistance, yield stability, and abiotic stress tolerance, is a classic example of successful wide hybridization.

**Selection Under an Appropriate Environment** Once a hybrid ($F_1$) is selfed in wheat, genetic segregation occurs in the $F_2$ and subsequent generations, and transgressive segregants may appear due to novel gene combinations. Progenies are evaluated under targeted stress conditions (such as drought-prone fields or disease nurseries), and only the best-performing plants are selected each generation. This ensures that the environment effectively identifies the hardiest genotypes.

***Recurrent Selection*** For complex, polygenic traits such as heat tolerance, recurrent selection is employed. A base population is created, superior plants are selected under stress, and these are intercrossed to gradually accumulate favourable alleles. The process is repeated across cycles to produce progressively improved populations.

**Line Development and Testing** Stable and uniform lines are evaluated in multilocation trials to assess their performance and adaptability before release as new varieties. Through established breeding methods such as pedigree selection or the selected bulk approach, segregating generations ($F_2$–$F_6$) are handled to isolate superior, stress-tolerant lines. Multiple sister lines can be derived from a single cross, each potentially suited to different regions or management conditions. The Veery lines are a well-known example, where around 10 advanced lines from one breeding effort were released in multiple countries due to their broad adaptability and superior performance.

### 2.3.2 Marker-Assisted Selection (MAS) and Its Role in Targeted Trait Improvement

Over the last few decades, MAS has improved the accuracy and efficiency of breeding for simply inherited traits. By using DNA-based markers, breeders can save time and resources by selecting for important genes at the seedling stage, such as *Lr34* for durable rust resistance and *Yr36* for heat tolerance. Multiple resistance genes can be pyramided into a single variety to provide more durable and broad-spectrum resistance. This approach also helps reduce linkage drag while facilitating the effective introgression of genes from wild relatives. The use of markers for genes such as *Lr24/Sr24*, *Lr19/Sr25*, *Yr36*, and the pleiotropic slow-rusting gene *Lr34/Yr18/Sr57/Pm38* represents a notable achievement in India (Vishwakarma et al. 2024).

### 2.3.3 Understanding the Genetic Basis of Stress Resilience: Key Genes and Pathways

The complexity of the wheat genome poses a significant challenge in elucidating the genetic basis of stress resilience. The three subgenomes (A, B, and D) of bread wheat (*Triticum aestivum*) originate from three distinct ancestral species, resulting in a hexaploid species. Consequently, the genome is enormous (~17 Gb), highly repetitive, and contains three homologous copies of many genes. Although this complexity provides genetic buffering, it also complicates efforts to determine the precise functions of individual genes. Nonetheless, the completion of the high-quality wheat reference genome (IWGSC 2018) has significantly facilitated the accurate discovery of genes.

#### 2.3.3.1 Key Genes and Pathways for Abiotic Stress Resilience

Abiotic stress resilience in wheat is not governed by a single 'magic bullet' gene but is a complex, quantitative trait shaped by interconnected regulatory networks. Drought tolerance involves activation of multiple physiological and molecular pathways, with key regulators such as the DREB/CBF transcription factors binding to DRE/CRT elements to induce a suite of stress-responsive genes (Agarwal et al. 2017). Overexpression of *TaDREB1*, for example, enhances tolerance to drought, heat, and salinity through broad transcriptional reprogramming. Other major contributors include NAC transcription factors such as *TaNAC69*, which modulate genes associated with water retention and root architecture (Guérin et al. 2022), and Late Embryogenesis Abundant (LEA) proteins like *TaLEA3* that safeguard cellular structures during dehydration (Liu et al. 2019). Osmoprotectant biosynthesis genes (*P5CS* for proline and *BADH* for glycine betaine) help maintain turgor, while aquaporins, particularly PIP family proteins, regulate water flux under stress (Trono and Pecchioni 2022). Heat resilience similarly relies on coordinated genetic responses, with heat shock proteins (HSP70, HSP90) functioning as molecular chaperones activated by heat shock transcription factors such as *TaHsfA6f* (Vishwakarma et al. 2018; Trono and Pecchioni 2022). Heat also induces antioxidant genes encoding APX, CAT, and SOD, which detoxify reactive oxygen species to prevent oxidative injury, while heat-stable variants of *Rubisco activase* help maintain photosynthetic efficiency. Salinity tolerance is mediated by genes that restrict sodium uptake or sequester it into vacuoles, including high-affinity $K^+$ transporters (*TmHKT1;4-A2*, *TaHKT1;5-D*) that remove $Na^+$ from xylem sap (James et al. 2011), vacuolar $Na^+/H^+$ exchangers such as *TaNHX2*, and the Salt Overly Sensitive (SOS) pathway comprising SOS1, SOS2, and SOS3, which maintains ionic homeostasis (Ding et al. 2019). These multilayered genetic mechanisms indicate the inherently polygenic nature of stress resilience in wheat and highlight the need for integrated breeding strategies to enhance tolerance to multiple abiotic challenges.

#### 2.3.3.2 Key Genes and Pathways for Abiotic Stress Resilience

Gene-for-gene interactions, in which a plant's resistance (R) gene product recognizes a specific pathogen avirulence (Avr) effector and triggers a defence response, are fundamental to biotic stress resistance (Flor 1956). Disease-resistance genes in wheat predominantly encode nucleotide-binding site leucine-rich repeat (NBS–LRR) proteins, and more than 2000 such genes have been identified in the wheat genome, the highest number reported in any plant species to date. Bread wheat is affected by numerous diseases, including powdery mildew, stripe rust, stem rust, and epidemic leaf rust, many of which are caused by rapidly evolving fungal pathogens that frequently give rise to new virulent isolates. Several R genes conferring resistance to diverse diseases have been identified and cloned from common wheat and its wild relatives. Notable examples include the leaf rust resistance genes *Lr1* (CC–NBS–LRR) (Cloutier et al. 2007), *Lr10* (CC–NBS–LRR) (Feuillet et al. 2003), *Lr21* (CC–NBS–LRR) (Huang et al. 2009), and *Lr34* (ABC transporter) (Krattinger et al. 2009); the stripe rust resistance gene *Yr10* (LZ–START) (Temel et al. 2008); and the stem rust resistance gene *Sr33* (CC–NBS–LRR) (Periyannan et al. 2013; Saintenac et al. 2013). Despite the identification of numerous R genes, no single gene is sufficient to provide comprehensive protection against the wide range of wheat pathogens. Important rust resistance genes include *Sr22*, *Sr35*, *Sr45*, and *Sr46* (for stem rust); *Lr34*, *Lr67*, and *Lr68* (for leaf rust); and *Yr5*, *Yr15*, and *Yr36* (for stripe rust). Among these, *Lr34* and *Lr67* are notable as 'slow-rusting' adult plant resistance (APR) genes that confer durable, broad-spectrum resistance.

Fusarium head blight (FHB) resistance is an especially complex, quantitative trait. Nearly 500 QTL have been reported since the first FHB-resistance locus was described by Waldron et al. (1999). Among these, *Fhb1* on chromosome 3BS is the most important and consistent QTL, imparting resistance through mechanisms such as enhanced cell wall thickening and mycotoxin detoxification. R-gene-mediated responses typically lead to a hypersensitive reaction and systemic acquired resistance (SAR), accompanied by the induction of pathogenesis-related (PR) proteins, including chitinases and glucanases, which directly inhibit pathogen growth.

### 2.3.4 Physiological and Biochemical Adaptations Associated with Abiotic and Biotic Stress Tolerance

Under stressful conditions, a plant's ability to survive and thrive depends on a complex repertoire of physiological and biochemical adaptations. These adaptations are interconnected responses designed to maintain homeostasis, conserve resources, and directly counteract stress rather than functioning as isolated events. Some of the important strategies employed by wheat to withstand these challenges are as follows:

#### 2.3.4.1 Abiotic Stress Tolerance

Abiotic stressors frequently cause cellular damage, disrupt water balance, and hinder essential functions such as photosynthesis.

**Drought Stress** Under drought conditions, wheat employs a combination of physiological and biochemical strategies to maintain water status and cellular integrity. Plants develop deeper and more robust root systems, with key traits such as xylem vessel diameter, root length density, and root angle, allowing access to moisture in deeper soil layers. Stomatal closure, regulated primarily by abscisic acid (ABA), minimizes transpirational water loss but also restricts $CO_2$ uptake required for photosynthesis. Additional protective features include leaf rolling and a thick epicuticular wax layer that reduces evaporative losses. At the cellular level, osmotic adjustment through the accumulation of compatible solutes (proline, glycine betaine, and sugars such as sucrose and trehalose) helps maintain turgor pressure and metabolic activities. Because drought induces oxidative stress, wheat activates an extensive antioxidant defence system comprising enzymatic (SOD, CAT, APX, GR) and non-enzymatic (ascorbate, glutathione, carotenoids) components to detoxify reactive oxygen species (ROS).

**Heat Stress** Heat stress triggers physiological adjustments that enable wheat plants to prevent overheating and maintain cellular stability. Increased transpiration enhances cooling, while heat-tolerant genotypes maintain membrane integrity by increasing the proportion of saturated fatty acids, which reduces membrane fluidity and leakage. Plants may also use paraheliotropism, modifying leaf orientation to reduce light absorption during peak temperatures and protect Photosystem II, the most heat-sensitive component of the photosynthetic machinery. Biochemically, heat tolerance depends largely on heat shock proteins (HSP70, HSP90, and small HSPs), which act as molecular chaperones to refold denatured proteins and prevent aggregation. Similar to drought, heat stress generates ROS, prompting the upregulation of the antioxidant system to mitigate oxidative damage.

**Salinity Stress** Salinity imposes both ionic and osmotic challenges, prompting wheat plants to adopt mechanisms that limit ion toxicity and maintain water balance. Tolerant cultivars restrict the uptake of sodium ($Na^+$) and chloride ($Cl^-$) at the root interface and rely on specialized transporters to sequester $Na^+$ into vacuoles, preventing harmful cytosolic accumulation. Some wild wheat relatives possess salt-secreting glands, although this trait is rare in modern cultivars. Biochemically, salinity induces osmotic stress, which is counteracted by the synthesis of compatible solutes such as proline and glycine betaine. ABA accumulation under saline conditions promotes stomatal closure and activates downstream stress-responsive genes, enabling plants to conserve water and adjust metabolism for sustained growth under high-salt environments.

#### 2.3.4.2 Biotic Stress Tolerance: Defence and Deterrence

Biotic stress responses are generally divided into two major categories: pre-formed and induced defences. Pre-formed (constitutive) defences include physical barriers such as a thickened cell wall reinforced with hydroxyproline-rich glycoproteins, lignin, and callose, which restrict pathogen penetration and insect feeding. The cuticle and wax layer also inhibit fungal spore adhesion and germination. In addition, phytoanticipins, pre-existing biochemical inhibitors such as phenolics and saponins, suppress or deter pests and pathogens.

Induced defences are activated when wheat detects a pathogen or insect. Structural and physiological responses often begin with the hypersensitive response (HR), a rapid, localized form of programmed cell death at the infection site that restricts pathogen spread by depriving it of living tissue. Cell wall reinforcement follows, involving the deposition of additional callose, lignin, and phenolic compounds at the site of attack. Biochemically, an oxidative burst, the rapid production of reactive oxygen species (ROS), such as $H_2O_2$, is one of the earliest responses, providing both direct antimicrobial activity and serving as a signal to activate downstream defences. Pathogen recognition also triggers the de novo synthesis of phytoalexins, including terpenoids and flavonoids, which enhance resistance.

A crucial component of induced resistance is the accumulation of pathogenesis-related (PR) proteins, such as β-1,3-glucanases and chitinases, which degrade fungal cell wall components. Proteinase inhibitors further contribute to defence by impairing digestive enzymes in insect herbivores, thereby reducing feeding and growth. Hormonal signalling networks play a central regulatory role: the salicylic acid (SA) pathway mediates defence against biotrophic pathogens (e.g., rusts) and activates systemic acquired resistance (SAR), a long-lasting, whole-plant immune state. In contrast, the jasmonic acid (JA) and ethylene (ET) pathways are critical for resistance against insect herbivores and necrotrophic pathogens that kill host tissue for nutrition.

### 2.3.5 The Interplay and Trade-Offs

The relationship between biotic and abiotic stress responses, often mediated by shared signalling molecules, is a key concept in plant stress biology. Crosstalk between pathways can produce either synergistic or antagonistic effects. For example, in drought-stressed wheat, abscisic acid (ABA), which plays a central role in abiotic stress responses, can antagonize the salicylic acid (SA) pathway, potentially increasing susceptibility to certain biotrophic pathogens. Conversely, reactive oxygen species (ROSs), generated under both abiotic and biotic stresses, can function as a common signalling hub that coordinates multiple defence pathways. Fitness costs also play an important role. Many defence-related modifications are energetically expensive: the synthesis of osmoprotectants, antioxidants, and pathogenesis-related (PR) proteins diverts energy and carbon that would otherwise support growth and yield, contributing to a 'yield penalty'. The objective of modern breeding is to optimize this balance by fine-tuning stress responses so that plants maintain resilience without compromising productivity under stress.

### 2.3.6 Integration of Soil Microbiome and Plant–Microbe Interactions in Stress Management

The rhizosphere microbiome functions as a 'second genome' for plants, as highlighted by recent research. Plant growth-promoting rhizobacteria (PGPR) and arbuscular mycorrhizal fungi (AMF) are two key groups of beneficial microbes that enhance stress tolerance by improving nutrient and water uptake. They also help modulate plant hormone levels, including the production of growth regulators such as IAA and enzymes like ACC deaminase that reduce stress-induced ethylene. When plants interact with certain beneficial microbes, particularly PGPR, these microorganisms can activate a plant-wide defence mechanism known as induced systemic resistance (ISR). ISR strengthens the plant's general defence capacity, making it more resistant to subsequent pathogen and pest attacks. An emerging strategy for enhancing crop resilience involves breeding wheat cultivars with an improved ability to attract and sustain these beneficial microbial communities (Kaushal and Wani 2016).

## 2.4 Recent Innovations in Developing Stress-Resilient Wheat Cultivars

The rapid evolution of breeding techniques has opened new avenues for developing wheat cultivars capable of withstanding multiple and increasingly severe stresses. Some of the recent innovations that complement traditional breeding and significantly accelerate the development of climate-resilient wheat are as follows:

### 2.4.1 CRISPR–Cas Genome Editing

CRISPR–Cas9 provides an unprecedented ability to precisely and specifically modify the wheat genome without introducing foreign DNA. Its applications include the elimination of susceptibility (S) genes, such as editing *TaMLO* to confer resistance to powdery mildew, and the enhancement of stress tolerance through targeted modification of genes associated with heat response, nutritional quality, and grain size, including *TaGW2* and *TaPDS*. Another emerging strategy is the engineering of immunity by rewriting promoter regions of resistance genes to achieve more robust and durable defence responses. Although CRISPR technology is still in its early stages in India, it holds tremendous promise for rapidly generating novel genetic diversity for improving stress resilience in wheat (Elsharawy and Refat 2023).

### 2.4.2 High-Throughput Phenotyping and AI-Driven Models for Stress-Tolerant Variety Selection

The phenotyping bottleneck in wheat breeding is being progressively addressed through the integration of high-throughput phenotyping (HTP) systems with machine learning (ML) and artificial intelligence (AI) tools (Table 2.1). Advanced sensor-equipped drones can scan hundreds of field plots in a single day, capturing detailed information on plant height, biomass, canopy temperature, and chlorophyll content (NDVI). AI and ML models analyse these large datasets to forecast yield and stress-tolerance performance, identify subtle phenotypic patterns that are not easily detectable by the human eye, and correlate spectral signatures with underlying genetic markers. Together, these technologies significantly enhance the efficiency and accuracy of selecting stress-tolerant wheat varieties.

**Table 2.1** High-throughput phenotyping platforms, sensors, and key stress-related traits measured in wheat

| Phenotyping platform | Sensors used | Key measurable traits for stress tolerance in wheat |
|---|---|---|
| Ground-based rovers | RGB, hyperspectral, thermal, LiDAR | Canopy cover, biomass, plant height (3D structure), and early seedling vigour |
| Unmanned aerial vehicles (UAVs/drones) | RGB, multispectral, hyperspectral, thermal | *Drought/heat:* Canopy temperature (CT, as a proxy for stomatal conductance), normalized difference vegetation index (NDVI), water band index (WBI)<br>*Nutrient stress:* Chlorophyll-related indices<br>*Disease:* Detection of disease patches via spectral signatures |
| Controlled environment (greenhouses) | Automated imaging systems (RGB, fluorescence) | Root architecture (with specialized systems), early stress response, and germination rates |

### 2.4.3 Advances in Gene Stacking and Multi-trait Engineering for Resistance to Multiple Stresses

Creating cultivars capable of withstanding multiple stresses is essential for future wheat improvement. This includes pyramiding genes for various abiotic stresses, such as heat and drought, as well as for diseases such as rusts and spot blotch. Priority should also be given to pleiotropic genes that confer partial resistance to multiple rusts and even some abiotic stressors, such as *Lr34*. Transcriptional regulators, including DREB genes, offer broad-spectrum tolerance by activating suites of stress-responsive pathways.

### 2.4.4 Use of Omics Technologies in Wheat Improvement

**Genomics** Next-generation sequencing (NGS) has enabled the identification of thousands of markers and candidate genes and has facilitated the development of high-quality wheat reference genomes.

**Transcriptomics** RNA–Seq studies reveal key biological pathways by identifying all genes that are upregulated or downregulated in response to specific stress stimuli.

**Proteomics and Metabolomics** These approaches identify proteins and metabolites, such as carbohydrates, amino acids, and antioxidants, that accumulate under stress. Such information provides valuable selection indicators and deepens our understanding of the biochemical processes underlying stress tolerance.

### 2.4.5 Case Studies of Recently Released Stress-Resilient Wheat Varieties in India

A detailed variety-wise description is beyond the scope of this chapter. Readers seeking comprehensive information on wheat varieties released over the past two decades may refer to Gupta et al. (2018, 2023). A concise summary of major varietal developments from 1960 to 2020 is presented in Table 2.2. Some of the latest varieties bred for targeted environments and widely adopted across different zones are described in Table 2.3.

**Table 2.2** Era-wise leading wheat varieties and their maximum cultivation extent in India

| Era | Leading variety | Developer | Peak coverage (Mha) |
|---|---|---|---|
| 1970s–1980s | Sonalika, Kalyan Sona | IARI | ~5–6 (each, at their peak) |
| 1990s–2000s | PBW 343 | PAU, Ludhiana | ~7–8 |
| 2010s | HD 2967 | IARI, New Delhi | ~8–9 (highest ever) |
| 2020s (current) | HD 3086, DBW 187 | IARI & IIWBR | ~4–5 (each, currently) |

**Table 2.3** List of some recent varieties bred for targeted environments and widely adopted across different zones

| | |
|---|---|
| HD 3226 | High-yielding cultivar with strong resistance to all three rusts and tolerance to terminal heat stress |
| DBW 187 | Early-maturing cultivar resistant to foliar blights and stripe rust, suitable for late planting to escape terminal heat |
| H 1105 | Performs well under limited irrigation and is resistant to brown and yellow rust |
| KRL 283 | Specifically adapted to saline–alkaline soils |
| HD 3410 | Landmark heat- and drought-tolerant variety for late-sown, irrigated NWPZ conditions. It escapes terminal heat by completing key growth stages early and ensuring rapid grain filling. It maintains strong photosynthetic efficiency and canopy vigour under stress, yielding up to 7.8 t/ha even with end-December sowing. The variety shows robust resistance to all three rusts and strong seedling resistance to Karnal bunt. Its adaptability is especially valuable in rice–wheat systems where delayed rice harvests force late sowing, helping farmers avoid yield losses from rising late-season heatwaves |
| DBW 332 | Notified in 2022 for CZ and NEPZ. It is a drought- and heat-tolerant variety suited to limited-irrigation conditions. It delivers high yields of up to 6.2 t/ha with only 2–3 irrigations, outperforming older cultivars under water-limited conditions. It is resistant to brown and stripe rust |
| WB 2 | Notified in 2021, it is a salt-tolerant wheat variety developed specifically for saline–alkaline soils. It is a derivative of KRL 1–4 and achieves 4.0–5.0 t/ha under salt-affected conditions. It is moderately early in maturity and resistant to stripe and leaf rusts |
| HI 8805 (Pusa Wheat 8805) - | Notified in 2022, it is a disease-resistant variety bred specifically to counter the highly virulent stem rust race Ug99 and its variants. It combines key race-specific *Sr* genes with adult plant resistance (APR) genes, providing durable, broad-spectrum protection. The variety is also resistant to stripe rust and leaf rust. Suited to irrigated, timely-sown conditions of the NWPZ, HI 8805 delivers high yields of up to 7.5 t/ha and possesses good milling quality with bold grains |
| HD 3385 | Notified in 2022, it is a multi-stress-tolerant variety combining terminal heat tolerance with resistance to all three rusts. It maintains green leaf area during grain filling under heat (stay-green trait) and incorporates durable rust-resistance genes from elite backgrounds. With yields up to 7.6 t/ha, it performs reliably across sowing dates and disease pressures, making it highly suitable for timely to late-sown conditions in the NWPZ |

## 2.5 Technological Interventions for Enhanced Stress Resilience

Technological innovations are increasingly complementing genetic and physiological approaches to strengthen stress resilience in wheat. The scope and contributions of some of the latest technological advances, including digital tools, AI-driven analytics, and precision farming techniques in modern breeding programs, are as follows.

### 2.5.1 Role of Artificial Intelligence (AI) and Machine Learning (ML) in Accelerating Wheat Breeding Programs

AI and ML models are increasingly employed in genomic prediction, using DNA sequences from new lines to accurately forecast their performance. They are also applied in image analysis, where drone-captured images automatically categorize disease symptoms and quantify stress damage. Additionally, AI/ML helps optimize breeding schemes by simulating multiple approaches to identify the most effective strategies (Sandhu et al. 2021).

### 2.5.2 Deployment of Digital Agriculture Tools for Stress Prediction and Mitigation

When integrated with real-time meteorological data, crop simulation models such as the Decision Support System for Agrotechnology Transfer (DSSAT) and the Agricultural Production Systems Simulator (APSIM) can predict yield losses due to impending stressors, such as heatwaves, enabling proactive irrigation scheduling and early warnings. Mobile applications, such as Maharashtra's Crop Pest Surveillance and Advisory Project (CROPSAP), facilitate real-time pest and disease monitoring and provide timely alerts to farmers.

### 2.5.3 Precision Agriculture Techniques for Resource-Efficient Wheat Farming

Technologies such as GPS-guided tractors, drip irrigation, and variable rate technology (VRT) for water and fertilizer application ensure precise input use, minimizing waste and environmental impact while optimizing the genetic potential of stress-resilient varieties.

### 2.5.4 Utilization of Remote Sensing, GIS, and Climate Modelling for Breeding and Stress Monitoring

Remote sensing from satellites allows large-scale monitoring of crop health and identification of stress hotspots. Geographic information systems (GIS) help determine breeding objectives for different mega-environments by mapping soil types, water availability, and historical disease outbreaks. Climate models enable breeders to future-proof their programs by predicting regions likely to experience increased heat or drought stress.

### 2.5.5 Automated Phenotyping Platforms and Drone-Based Field Evaluations

Plant development and stress responses can be continuously monitored using fully automated phenotyping platforms in both greenhouses and fields, equipped with conveyor systems and sensor arrays. Drones (unmanned aerial vehicles, UAVs) provide high-resolution, multispectral data on crop status, making them invaluable tools for high-throughput phenotyping in field conditions.

## 2.6 Statistical Insights on Stress-Resilient Wheat Breeding in India

India's progress in wheat improvement is reflected not only in the release of new cultivars but also in their field performance, adoption patterns, and socioeconomic impact. Evaluating historical trends against current outcomes provides valuable insight into the effectiveness of breeding programs and the challenges that remain. The key indicators of national progress in developing and deploying stress-resilient wheat cultivars are as follows.

### 2.6.1 Historical vs. Current Trends

Wheat production in India has risen sharply from about ten million tonnes in the 1960s to over 117 million tonnes today, shifting the country from food scarcity to self-sufficiency through the combined impact of improved cultivars, enhanced agronomy, expanded irrigation, and coordinated national efforts. However, similar to global trends, the growth rate of wheat productivity has slowed in the past two decades due to more frequent heatwaves, droughts, erratic rainfall, the emergence of new rust races and foliar diseases, declining soil fertility, resource degradation, and delays in sowing, particularly in the rice–wheat system. Periodic extreme climatic events, including severe terminal heat and unseasonal rainfall during grain filling, have caused substantial yield losses across major wheat-growing regions. These patterns highlight the urgent need to accelerate the development and adoption of stress-tolerant, climate-resilient wheat varieties and to strengthen on-farm management practices to safeguard production under increasing climatic uncertainty.

**Success Rate of Breeding for Stress Tolerance** According to IIWBR (2023), more than half of the cultivars released in the past decade exhibit clear tolerance to one or more major stresses, demonstrating the effectiveness of current breeding approaches in the country. This represents a significant advancement compared with earlier periods, when varietal releases were largely driven by yield and disease resistance alone. Modern breeding pipelines, supported by genomics, MAS, HTP phenotyping, and international collaborations, have strengthened the precision and speed of selecting stress-resilient genotypes. The growing number of multi-stress-tolerant cultivars validates the effectiveness of current breeding strategies.

**Increasing Yield** Stress-resilient cultivars consistently outperform older, susceptible varieties by 15–25% under challenging conditions owing to their improved canopy temperature regulation, deeper and more efficient root systems, stable grain filling under terminal heat, multi-gene resistance to rusts and foliar diseases, and enhanced water and nutrient use efficiency. Under optimal management, recently released varieties such as HD 2967, HD 3086, HD 3410, and DBW 187 yield comparable to or even exceed those of traditional high-yielding cultivars, including in late-sown or water-limited environments. These performance gains clearly demonstrate that modern genetic improvements can buffer the negative impacts of climate variability and help sustain production stability.

**Adoption Rate** Adoption of recently developed resilient varieties is slower in Eastern India due to socioeconomic constraints and smaller landholdings, compared with regions having strong extension systems, such as Punjab and Haryana. Nevertheless, the overall wheat seed replacement rate has doubled over the last decade, accelerating the uptake of new varieties compared with the post-Green Revolution period. One factor contributing to this faster replacement is the use of pre-release seed multiplication, initially implemented to fast-track Ug99-resistant wheat varieties (Joshi et al. 2011).

**Socioeconomic Impact** The cultivation of stress-resilient wheat cultivars has delivered substantial economic and livelihood benefits, with rust-resistant varieties saving farmers billions of rupees by preventing major crop losses and reducing reliance on fungicides, while drought-tolerant cultivars have improved production stability in rainfed regions, lowering risks for resource-poor farmers. The adoption of early sown, heat-tolerant varieties has also allowed farmers to shift sowing to October, resulting in yield gains of nearly 1 t/ha compared with traditional November planting, a major advantage in rice, wheat systems where delayed rice harvests often constrain sowing time. Collectively, these advances strengthen national food security and enhance climate resilience and economic stability for millions of wheat-growing households.

## 2.7 Future Prospects and Global Challenges

Ensuring the sustainability of wheat production in the coming decades will depend on how effectively breeding programs respond to rapidly changing environments, evolving pathogens, and shifting policy landscapes.

### 2.7.1 Challenges Caused by Climate Change and the Need for Adaptive Breeding Techniques

As climate unpredictability increases, reactive breeding must give way to proactive approaches. This involves designing ‘future climates’ using climate models and screening germplasm under controlled conditions, such as polyhouses, to select lines suited for anticipated stress scenarios.

### 2.7.2 New Pest Infestations and Changing Pathogen Threats

Ongoing evolution of pathogens, such as the Ug99 race group of stem rust and new virulent pathotypes of yellow rust, necessitates continuous surveillance, rapid gene discovery, and swift deployment of resistant cultivars through marker-assisted selection (MAS) and gene-editing techniques. Emerging threats, such as wheat blast, also demand increased research investment.

### 2.7.3 Funding and Policy Issues in Stress-Resilient Wheat Research

Long-term breeding programs require sustained financial investment. Policies should promote funding for advanced genotyping and phenotyping infrastructure and provide a simplified legal system for the assessment and distribution of genetically modified crops. Such measures would incentivize public–private partnerships and the integration of complementary expertise.

### 2.7.4 The Need for International and Multidisciplinary Collaborations

The complexity of stresses in wheat requires breeders to collaborate with pathologists, physiologists, soil scientists, climate scientists, data analysts, and social scientists. International initiatives such as the Borlaug Global Rust Initiative (BGRI) and the International Wheat-Genome Sequencing Consortium (IWGSC) are more critical than ever for sharing knowledge and resources.

### 2.7.5 Systems Biology, Epigenetics, and Advanced Gene-Editing Approaches

Gene-editing technologies are evolving from simple gene deletions to precise nucleotide modifications and targeted gene insertions. Understanding stress-induced epigenetic changes, such as DNA methylation and histone modifications, can reveal persistent, heritable 'stress memories' that prime plants for future stress, providing new avenues for breeding resilient wheat varieties.

## 2.8 Conclusion and Way Forward

The development of stress-resilient wheat in India has progressed from simple selection to a sophisticated science integrating genetics, phenomics, and data analytics, with collaboration among ICAR, CIMMYT, and SAUs playing a critical role in producing high-yielding, heat-, drought-, and rust-tolerant cultivars that enhance food security. Nevertheless, significant challenges remain, including breeding for complex polygenic stresses and their interactions, harnessing the root microbiome, slow adoption of modern technologies in some regions, and the need for faster varietal replacement to keep pace with evolving pathogens. Addressing these gaps requires increased investment in both basic and applied stress-resilience research, modernization of infrastructure through a nationwide network of genotyping and high-throughput phenotyping platforms, and the establishment of a predictable, science-based regulatory system for gene-edited products. Strengthening seed systems to ensure efficient production and delivery, promoting conservation agriculture practices that enhance water use efficiency and soil health, and adopting precision farming technologies are equally critical. By integrating these approaches, India can build a resilient, sustainable, and productive wheat farming system, secure a steady supply of genetically diverse, multi-stress-tolerant cultivars, and continue to play a major role in global food security.

## References

Agarwal PK, Gupta K, Lopato S et al (2017) Dehydration-responsive element binding transcription factors and their applications for the engineering of stress tolerance. J Exp Bot 68(9):2135–2148. https://doi.org/10.1093/jxb/erx118

Aggarwal PK (2008) Global climate change and Indian agriculture: impacts, adaptation and mitigation. Indian J Agric Sci 78(11):911–919

Ahtisham M, Obaid Z, Kisana MTT (2025) Harnessing wild relatives for crop improvement: genetic resources, breeding strategies, and applications in enhancing yield, quality, and resilience. Premier J Environ Sci 4:100019. https://doi.org/10.70389/PJES.100019

Araus JL, Kefauver SC, Zaman-Allah M et al (2018) Translating high-throughput phenotyping into genetic gain. Trends Plant Sci 23(5):451–466. https://doi.org/10.1016/j.tplants.2018.02.001

B A (1925) Crop production in India: a critical survey of its problems. Nature 116:4–5. https://doi.org/10.1038/116004a0

Cloutier S, McCallum BD, Loutre C et al (2007) Leaf rust resistance gene *Lr1*, isolated from bread wheat (*Triticum aestivum* L.), is a member of the large psr567 gene family. Plant Mol Biol 65(1–2):93–106. https://doi.org/10.1007/s11103-007-9201-8

Crossa J, Pérez-Rodríguez P, Cuevas J et al (2017) Genomic selection in plant breeding: methods, models, and perspectives. Trends Plant Sci 22(11):961–975. https://doi.org/10.1016/j.tplants.2017.08.011

Ding H, Ma D, Huang X et al (2019) Exogenous hydrogen sulphide alleviates salt stress by improving antioxidant defenses and the salt overly sensitive pathway in wheat seedlings. Acta Physiol Plant 41:123. https://doi.org/10.1007/s11738-019-2918-6

Elsharawy H, Refat M (2023) CRISPR/Cas9 genome editing in wheat: enhancing quality and productivity for global food security—a review. Funct Integr Genomics 23(3):265. https://doi.org/10.1007/s10142-023-01190-1

Feuillet C, Travella S, Stein N et al (2003) Map-based isolation of the leaf rust disease resistance gene *Lr10* from the hexaploid wheat (*Triticum aestivum* L.) genome. Proc Natl Acad Sci USA 100(25):15253–15258. https://doi.org/10.1073/pnas.2435133100

Flor HH (1956) The complementary genetics systems in flax and flax rust. Adv Genet 8:29–54. https://doi.org/10.1016/S0065-2660(08)60498-8

Guérin C, Dupuits C, Mouzeyar S et al (2022) Insights into four NAC transcription factors involved in grain development and in response to moderate heat in the Triticeae tribe. Int J Mol Sci 23(19):11672. https://doi.org/10.3390/ijms231911672

Gupta A, Singh C, Kumar V et al (2018) Wheat varieties notified in India since 1965. ICAR-Indian Institute of Wheat & Barley Research, Karnal, p 101

Gupta A, Kumar V, Kumar P et al (2023) Compendium of wheat varieties notified in India during 2018–2023. Research Bull No. 51. ICAR-Indian Institute of Wheat & Barley Research, Karnal, p 32

Huang L, Brooks S, Li W et al (2009) Evolution of new disease specificity at a simple resistance locus in a crop-weed complex: reconstitution of the *Lr21* gene in wheat. Genetics 182(2):595–602. https://doi.org/10.1534/genetics.108.099614

ICAR-IIWBR (2023) Project coordinator's report: all India coordinated research project on wheat and barley. Indian Institute of Wheat and Barley Research, Karnal

International Wheat Genome Sequencing Consortium (IWGSC) (2018) Shifting the limits in wheat research and breeding using a fully annotated reference genome. Science 361(6403):eaar7191. https://doi.org/10.1126/science.aar7191

James RA, Byrt CS, Blake C et al (2011) Major genes for $Na^+$ exclusion, *Nax1* and *Nax2* (wheat *HKT1;4* and *HKT1;5*), decrease $Na^+$ accumulation in bread wheat leaves under saline and waterlogged conditions. J Exp Bot 62(8):2939–2947. https://doi.org/10.1093/jxb/err003

Joshi AK, Kropff M (2018) CIMMYT continues to sustainably address global hunger. Indian Farming 68(10):13–17. https://epubs.icar.org.in/index.php/IndFarm/article/view/84460

Joshi AK, Mishra B, Chatrath R et al (2007) Wheat improvement in India: present status, emerging challenges and future prospects. Euphytica 157:431–446. https://doi.org/10.1007/s10681-007-9385-7

Joshi AK, Azab M, Mosaad M et al (2011) Delivering rust resistant wheat to farmers: a step towards increased food security. Euphytica 179:187–196. https://doi.org/10.1007/s10681-010-0314-9

Kaushal M, Wani SP (2016) Plant-growth-promoting rhizobacteria: drought stress alleviators to ameliorate crop production in drylands. Ann Microbiol 66:35–42. https://doi.org/10.1007/s13213-015-1112-3

Krattinger SG, Lagudah ES, Spielmeyer W et al (2009) A putative ABC transporter confers durable resistance to multiple fungal pathogens in wheat. Science 323(5919):1360–1363. https://doi.org/10.1126/science.1166453

Krishnappa G, Tyagi B, Gupta V et al (2022) Wheat breeding. In: Yadava DK, Dikshit HK, Mishra GP, Tripathi S (eds) Fundamentals of field crop breeding. Springer, Singapore. https://doi.org/10.1007/978-981-16-9257-4_2

Kumar A, Kumar A, Pandey D et al (2025) Breeding and selection strategies to accelerate genetic gains in wheat for South Asia [version 1]. VeriXiv 2:110. https://doi.org/10.12688/verixiv.1139.1

Liu H, Xing M, Yang W et al (2019) Genome-wide identification of and functional insights into the late embryogenesis abundant (LEA) gene family in bread wheat (*Triticum aestivum*). Sci Rep 9:13375. https://doi.org/10.1038/s41598-019-49759-w

Pal BP (1966) Wheat (Monograph). Indian Council of Agricultural Research, ICAR, New Delhi

Paroda RS, Dasgupta B, Singh SS et al (2013) Proceedings of the regional consultation on improving wheat productivity in Asia, Bangkok, 26–27 April 2012, p 224

Periyannan S, Moore J, Ayliffe M et al (2013) The gene *Sr33*, an ortholog of barley *Mla* genes, encodes resistance to wheat stem rust race Ug99. Science 341(6147):786–788. https://doi.org/10.1126/science.1239028

Rajaram S, Borlaug NE, van Ginkel M (2002) CIMMYT international wheat breeding. In: Curtis BC, Rajaram S, Gomez Macpherson H (eds) Bread wheat improvement and production, Plant production and protection series No. 30, pp 103–117

Reynolds MP, Lewis JM, Ammar K et al (2021) Harnessing translational research in wheat for climate resilience. J Exp Bot 72(14):5134–5157. https://doi.org/10.1093/jxb/erab256

Saintenac C, Zhang W, Salcedo A et al (2013) Identification of wheat gene *Sr35* that confers resistance to Ug99 stem rust race group. Science 341(6147):783–786. https://doi.org/10.1126/science.1239022

Sandhu K, Patil SS, Pumphrey M et al (2021) Multitrait machine- and deep-learning models for genomic selection using spectral information in a wheat breeding program. Plant Genome 14(3):e20119. https://doi.org/10.1002/tpg2.20119

Singh RP, Singh PK, Rutkoski J et al (2016) Disease impact on wheat yield potential and prospects of genetic control. Annu Rev Phytopathol 54:303–322. https://doi.org/10.1146/annurev-phyto-080615-095835

Singh GP, Kumar PR, Khan H et al (2022) The wheat revolution: historical milestones in wheat breeding in India. Breeding Front Wheat 1:1–18

Smale M, Singh J, Di Falco S et al (2008) Wheat breeding, productivity and slow variety change: evidence from the Punjab of India after the green revolution. Aust J Agric Resour Econ 52(4):419–432. https://doi.org/10.1111/j.1467-8489.2008.00435.x

Swaminathan MS (2013) Genesis and growth of the yield revolution in wheat in India: lessons for shaping our agricultural destiny. Agric Res 2(3):183–188

Temel A, Senturk-Akfirat F, Ertugrul F et al (2008) *Yr10* gene polymorphism in bread wheat varieties. Afr J Biotechnol 7:2328–2332

Tomar SMS, Vinod, Singh B (2004) Distant hybridization in wheat. ICAR, New Delhi

Trono D, Pecchioni N (2022) Candidate genes associated with abiotic stress response in plants as tools to engineer tolerance to drought, salinity and extreme temperatures in wheat: an overview. Plants 11(23):3358. https://doi.org/10.3390/plants11233358

Vishwakarma H, Junaid A, Manjhi J et al (2018) Heat stress transcripts, differential expression, and profiling of heat stress tolerant gene *TaHsp90* in Indian wheat (*Triticum aestivum* L.) cv C306. PLoS One 13(6):e0198293. https://doi.org/10.1371/journal.pone.0198293

Vishwakarma MK, Yadav PS, Rai VP et al (2022) Molecular markers and genomics-assisted breeding for improving crop plants. In: Samuel J, Kumar A, Singh J (eds) Relationship between microbes and the environment for sustainable ecosystem services, vol 1. Elsevier, pp 303–334. https://doi.org/10.1016/B978-0-323-89938-3.00014-1

Vishwakarma MK, Kumar U, Bhati PK et al (2024) Harnessing cooler nights in the central zone of India to achieve speed breeding under field conditions. Rom Agric Res 41:261–270. https://www.incda-fundulea.ro/rar/nr41fol/rar41.24.pdf

Vishwakarma MK, Kumar U, Bhati PK et al (2025) Molecular markers for the identification of BNI capacity in chromosome addition lines and their use in marker-assisted breeding. Euphytica 221:188. https://doi.org/10.1007/s10681-025-03624-2

Waldron BL, Moreno-Sevilla B, Anderson JA et al (1999) RFLP mapping of QTL for Fusarium head blight resistance in wheat. Crop Sci 39(3):805–811. https://doi.org/10.2135/cropsci1999.0011183X003900030032x

Watson A, Ghosh S, Williams MJ et al (2018) Speed breeding is a powerful tool to accelerate crop research and breeding. Nat Plants 4:23–29. https://doi.org/10.1038/s41477-017-0083-8

# Wheat Spike Diseases Under Changing Climatic Scenario with Special Emphasis on Fusarium Head Blight

3

Ravindra Kumar, Sudheer Kumar, M. S. Saharan, and Ratan Tiwari

**Abstract**

Fusarium head blight (FHB) poses a significant threat to global food security by reducing yields, diminishing grain quality, and causing mycotoxin contamination. This chapter synthesizes current knowledge on the aetiology, epidemiology, and management of FHB, with emphasis on climate-driven disease dynamics. FHB is caused by a complex of *Fusarium* species, among which *Fusarium graminearum* is the most aggressive and widely distributed, exhibiting adaptability across diverse agroecological zones. Environmental factors such as temperature, humidity, and rainfall during anthesis are critical for host infection, pathogen dispersal, and survival, thereby increasing epidemic risk under climate variability and conservation agriculture. Mycotoxin production, including deoxynivalenol (DON), nivalenol (NIV), and zearalenone (ZEN), exacerbates the socio-economic impact by compromising food and feed safety. In India, FHB is transitioning from a minor to a potentially significant disease due to shifts in climate, agronomic practices, and pathogen aggressiveness, highlighting the need for enhanced surveillance and targeted interventions. Considerable variability in pathogen populations and aggressiveness, coupled with limited genetic resistance in current germplasm, presents persistent challenges. Integrated disease management strategies, including cultural practices, deployment of resistant cultivars, fungicide application, biological control, and forecasting systems, are

R. Kumar · R. Tiwari
ICAR-Indian Institute of Wheat and Barley Research, Karnal, Haryana, India

S. Kumar
ICAR-Indian Institute of Pulses Research, Regional Research Centre, Bikaner, Rajasthan, India

M. S. Saharan (✉)
Division of Plant Pathology, ICAR-Indian Agricultural Research Institute, New Delhi, India
e-mail: mssaharan@iari.res.in



R. Tiwari et al. (eds.), *Wheat Stress Management*,
https://doi.org/10.1007/978-981-92-0902-6_3

assessed, with attention to their complementary roles and inherent limitations. Advances in molecular breeding, quantitative trait locus (QTL) mapping, genomic selection, and gene editing offer promising avenues for developing durable resistance. However, challenges remain due to the polygenic nature of resistance and genotype-by-environment interactions. A climate-resilient, multidisciplinary approach integrating epidemiology, genomics, and sustainable management practices is essential to address the escalating threat of FHB in wheat production systems.

**Keywords**

Fusarium head blight (FHB) · *Fusarium graminearum* · Climate change · Mycotoxins · Wheat disease epidemiology · Integrated disease management

## 3.1 Introduction

Wheat (*Triticum* spp.) is susceptible to a range of biotic stresses caused by fungi, bacteria, nematodes, and viruses. Many of these diseases can affect the most economically significant part of the crop, the spikes, in addition to producing primary symptoms on other parts of the wheat plant. Spot blotch, caused by *Bipolaris sorokiniana*, can significantly impact wheat spikes, resulting in reduced size, low weight, shrivelled, and discoloured kernels. Some wheat diseases specifically target the spikes or primarily produce symptoms on the spikes. In addition to spot blotch, wheat crop is affected by Karnal bunt, loose smut, tan spot, foot rot, hill bunt, and Fusarium head blight or head scab. In this chapter, our emphasis is on Fusarium head blight (FHB), which is becoming an important disease affecting wheat spikes in many countries under a changing climate.

## 3.2 Fusarium Head Blight (FHB)

FHB, also known as wheat head scab or ear blight, is a spike disease of significant concern due to its adaptability to the climatic conditions prevalent across major wheat-producing regions worldwide. FHB substantially reduces grain yield and quality through spike infection and the accumulation of toxic secondary metabolites. The disease affects key cultivated wheat species, including *T. aestivum* L. and *T. durum* Desf., and is recognized as one of the most complex and economically damaging diseases of small grain cereals. In India, FHB was first reported in Arunachal Pradesh in 1973 (Roy 1974) and was historically considered a minor disease compared to rusts and spot blotch. However, recent evidence indicates an increasing epidemiological significance, driven by changing climatic and agronomic conditions. Systematic surveys across diverse agro-climatic zones have confirmed the presence and dynamic distribution of multiple *Fusarium* species associated with FHB (Hegde et al. 2023). Environmental factors such as high humidity, frequent rainfall, and moderate temperatures during anthesis strongly influence disease development and favour infection and epidemic outbreaks. Under favourable conditions, yield losses may exceed 50–80% (Kaul et al. 2022). A defining feature of FHB is the production of mycotoxins, including deoxynivalenol (DON), nivalenol (NIV), and zearalenone (ZEN), which pose significant threats to food safety, animal health, and international trade. The resurgence of FHB epidemics in major wheat-growing regions, including North America, Europe, China, and South America, has been attributed to climate variability, increased maize–wheat rotations, and conservation agriculture practices that enhance inoculum carryover on crop residues (Jung et al. 2022). Climate change is projected to further intensify FHB risk by altering pathogen life cycles, expanding geographic distribution, and increasing the frequency of weather conditions conducive to infection (Jung et al. 2022). Among FHB-infected samples, *F. graminearum* is consistently identified as the dominant pathogen and demonstrates high adaptability to variable environments (Kaul et al. 2022). New pathogen reports, such as the detection of *F. poae* causing FHB in Karnataka during the 2021–2022 cropping season, further illustrate expanding pathogen diversity and the potential emergence of new threats in India (Devaiah et al. 2023). The increasing adoption of conservation agriculture practices in the Indo-Gangetic Plains, combined with changing weather patterns characterized by unseasonal rainfall and elevated humidity during wheat flowering, is expected to enhance inoculum persistence and disease incidence. A comprehensive understanding of FHB dynamics under changing climatic scenarios is therefore essential to strengthen resistance breeding, surveillance, and integrated disease management strategies for sustainable wheat production. During epidemics, FHB can reduce yields by 10–70%, depending on cultivar susceptibility and environmental conditions (Wegulo et al. 2013). Severe outbreaks in the United States during the 1990s resulted in cumulative economic losses estimated at billions of US dollars, highlighting the considerable threat this disease poses to global wheat production (Nganje et al. 2004). Ongoing climate change, characterized by rising temperatures, altered precipitation patterns, and more frequent extreme weather events, is projected to intensify FHB risk by promoting pathogen survival, dispersal, and infection efficiency (Jung et al. 2022).

### 3.2.1 Symptomatology

FHB typically presents as premature senescence of wheat spikes and is recognized as one of the most significant cereal diseases globally (Figueroa et al. 2018). Initial symptoms include bleaching of infected spikelets, while adjacent healthy spikelets remain green. As the fungal phytopathogen progresses along the rachis, spikelets above and below the primary infection site may also bleach. Pinkish orange masses of fungal spores are often visible on infected spikelets upon close examination (Kashyap et al. 2022). FHB causes substantial reductions in wheat yield, as affected grains are small, shrivelled, possess inferior bulk and quality (tombstone appearance), and are contaminated with mycotoxins (McMullen et al. 2012). Infected

**Fig. 3.1** (**a**) Symptoms of FHB in wheat; (**b**) FHB-infected grain exhibiting a characteristic 'tombstone' appearance; (**c**) pure culture of *F. graminearum* grown on potato dextrose agar (PDA); and (**d**) macroconidia of *F. graminearum* observed under a compound microscope. (Photograph courtesy of Ravindra Kumar)

grains are frequently contaminated with mycotoxins such as DON, which pose serious health risks to humans and livestock and render the grain unsuitable for consumption and trade. Figure 3.1 presents the symptoms, pathogen morphology, and cultural characteristics of *Fusarium graminearum* associated with FHB in wheat.

### 3.2.2 Disease Aetiology

FHB is a complex disease involving a diverse array of pathogenic species, with 15–17 fungal species associated across various agroecological regions. Among these, *F. graminearum* Schwabe (teleomorph: *Gibberella zeae*), an ascomycetous fungus, is widely regarded as the most aggressive and globally dominant causal agent of head blight in wheat and other small-grain cereals (Gilchrist and Dubin 2000; Saharan and Sharma 2009; Summerell et al. 2010; Karlsson et al. 2021). Additional contributors to the FHB complex include *F. culmorum*, *F. avenaceum*, *F. poae*, *F. verticillioides* (syn. *F. moniliforme*), *F. oxysporum*, and the snow mould pathogen *Microdochium nivale* (Molnár et al. 2024). The prevalence of these species is influenced by climatic conditions and cropping systems (Xu et al. 2021).

In India, initial investigations identified *F. avenaceum* and *F. graminearum* as causal agents of ear blight and scab in widely cultivated wheat varieties, including Sonalika, Kalyan Sona, Safed Lerma, and Lerma Rojo, based on samples from the Siang district of Arunachal Pradesh. Subsequent studies confirmed the presence of *F. graminearum* in various wheat genotypes, such as Sonalika, C 306, and several other cultivars and breeding lines at the IARI Regional Station in Wellington, demonstrating its adaptability to diverse agro-climatic conditions. Broader surveys in northwestern India documented a wider range of associated species, including

*M. nivale*, *F. compactum*, *F. subglutinans*, *F. pallidoroseum* (syn. *F. semitectum* complex), and *F. oxysporum*, particularly in Punjab and adjacent regions (Singh and Aujla 1994; Kumar et al. 2021). Systematic isolations further identified multiple *Fusarium* species, including *F. graminearum*, *F. verticillioides*, *F. oxysporum*, *F. equiseti*, *F. solani*, and *F. semitectum*, from infected wheat spikes. *F. graminearum* was consistently predominant in locations such as Lahaul Valley, Punjab, and Wellington (Saharan et al. 2003; Kumar et al. 2021). This persistent predominance aligns with global trends, where *F. graminearum* is recognized as the principal FHB pathogen in major wheat-producing areas (Alisaac and Mahlein 2023).

### 3.2.3 Disease Cycle

Although FHB is generally classified as a monocyclic disease, its epidemiology is governed by a dynamic and environmentally responsive infection cycle (Alisaac et al. 2020). The pathogen persists between cropping seasons as ascospores within perithecia or as asexual spores (macroconidia or microconidia) on infected crop residues, such as wheat and maize stubble, where it survives saprophytically. These propagules constitute the primary inoculum, and a wide range of gramineous and non-gramineous weeds further contribute to inoculum carryover and pathogen persistence (Manstretta and Rossi 2015).

Favourable environmental conditions, particularly prolonged moisture and moderate temperatures during the wheat anthesis stage, promote the dispersal of ascospores and conidia by wind or rain splash onto exposed floral tissues, which are highly susceptible to infection (Karlsson et al. 2021). Following deposition, spores germinate on the spikelet surface and form germ tubes that facilitate adhesion and colonization. Initial fungal growth occurs epiphytically on floral structures, including the lemma, palea, and ovary. At this stage, early induction of trichothecene mycotoxin biosynthesis has been documented prior to host penetration, suggesting a role in undermining host defences.

The pathogen invades host tissues through natural openings or by direct mechanical and enzymatic action, initiating a biotrophic phase characterized by predominantly intercellular hyphal growth. This phase rapidly transitions to a necrotrophic stage, which involves extensive inter- and intracellular colonization of spike tissues and leads to spikelet bleaching, kernel shrivelling, and significant yield and quality losses (Alisaac and Mahlein 2023). During the necrotrophic phase, high concentrations of mycotoxins such as deoxynivalenol (DON) accumulate in infected tissues and developing grains (Stein et al. 2009). These compounds contaminate the harvest and function as virulence factors by inhibiting host protein synthesis, suppressing defence responses, and conferring a competitive advantage to the pathogen over other microorganisms (Hao et al. 2023). At the cellular level, *F. graminearum* forms specialized infection structures, including lobate appressoria and multicellular infection cushions, which facilitate host penetration. These structures are closely associated with localized trichothecene biosynthesis, highlighting the coordination between morphogenesis and pathogenicity (Boenisch and Schäfer 2011).

### 3.2.4 Mycotoxins and Socio-Economic Implications of FHB

*Fusarium* species responsible for FHB produce a diverse array of mycotoxins that pose significant risks to food safety and grain quality. Among these, type B trichothecenes, particularly DON and NIV, are the most prevalent and economically important. Other toxins, such as ZEN, fumonisins, fusaric acid, and a growing number of emerging mycotoxins, including enniatins, beauvericin, moniliformin, and fusaproliferin, are also frequently detected in contaminated grain. DON, commonly known as 'vomitoxin', serves as a major virulence factor for *F. graminearum* by inhibiting eukaryotic protein synthesis and suppressing host defence mechanisms. Consumption of DON-contaminated food or feed can result in immunotoxic, neurotoxic, and gastrointestinal disorders in both humans and livestock. NIV, which is structurally similar to DON, is generally considered more toxic because of its pronounced immunosuppressive and cytotoxic properties (Drakulic et al. 2016). Exposure to these mycotoxins has been linked to various pathological conditions, including vomiting, intestinal inflammation, dermal irritation, hemorrhagic lesions, and immune dysfunction (Olariu et al. 2025). FHB and the associated mycotoxins, particularly DON or 'vomitoxin', represent significant social, economic, and health challenges globally. Besides the severe yield losses in small-grain cereals, viz., wheat, barley, oats, the contaminating FHB mycotoxins pose direct risks to human and animal health, leading to strict regulatory, financial, and food security implications. In highly ideal conditions, it may be expected that the FHB-contaminated wheat grains should not be allowed to enter the food or feed chain to avoid extremely harmful effects on human and cattle's health due to this disease. Therefore, a segregation mechanism is essential to divert such grain from edible use to non-edible industrial purposes, as social responsibility is linked to FHB.

### 3.2.5 Variation in *Fusarium* spp./Isolates Causing Head Scab of Wheat in India

Systematic surveys conducted in ecologically diverse wheat-growing regions of India, such as Dalang Maidan in the Lahaul Valley (Himachal Pradesh) and Wellington in the Nilgiri Hills (Tamil Nadu), have consistently demonstrated the predominance of *F. graminearum* in head scab–infected wheat spikes (Kumar et al. 2021). This finding highlights the notable ecological plasticity of the pathogen across both temperate high-altitude and humid subtropical environments. Twenty-nine isolates obtained from infected ear heads were confirmed as *F. graminearum* using an integrated approach of morphological, cultural, and molecular methods. Pathogenicity assays on a diverse panel of wheat genotypes (UP 2338, PBW 343, Sonalika, HD 2967, HD 3086, HD 29, MACS 5049, HS 645, and VL 1013) revealed substantial variability in aggressiveness. Isolates such as Fg-W10 and Fg-W24 (Wellington) exhibited high virulence at 7- and 14-day post-inoculation, while Fg-W7 and Fg-W26 showed comparatively low pathogenic potential (Kaul et al. 2022). This variability likely reflects differential adaptation and selection pressures imposed by local environmental factors, including temperature, humidity, and host diversity.

Molecular characterization using 23 simple sequence repeat (SSR) markers from the *F. graminearum* PH-1 reference genome yielded successful amplification for 21 markers, demonstrating their reliability for population-level analysis. Cluster analysis of SSR polymorphism identified two main genetic groups: Group A, comprising two distinct isolates (Fg-W27 from Wellington and Fg-L2 from Lahaul-Spiti), and Group B, comprising the remaining 27 isolates (Kaul et al. 2022). This distribution suggests a largely homogeneous population with isolated pockets of divergence. Such a pattern may reflect ongoing gene flow and localized adaptation, potentially influenced by climatic heterogeneity and cropping practices (Wang et al. 2011). The presence of high pathogenic variability alongside limited genetic divergence suggests that minor genetic changes or regulatory differences may significantly impact virulence traits (Kaul et al. 2022). In the context of climate change, these adaptable pathogen populations are likely to respond rapidly to changing environmental conditions, increasing the risk of more severe epidemics (Kaul et al. 2022). These results highlight the importance of continuous surveillance, the incorporation of diverse sources of resistance, and the implementation of region-specific breeding strategies to effectively manage FHB in India.

### 3.2.6 Epidemiology

FHB exhibits marked sensitivity to environmental conditions, especially during anthesis and early grain development, when both host susceptibility and pathogen aggressiveness reach their peak. The primary causal agent, *F. graminearum*, generates sexual spores (ascospores) within perithecia and asexual macroconidia in sporodochia, both of which contribute to epidemic development under favourable conditions (Alisaac and Mahlein 2023). Wind-dispersed ascospores are considered the main primary inoculum, facilitating long-distance spread and initiating infections on exposed floral tissues (Markell and Francl 2003). The pathogen persists between cropping seasons primarily on infected crop residues, where perithecia and sporodochia form and persist, creating a strong epidemiological link between residue retention and subsequent disease outbreaks. Environmental factors play a critical role in regulating these processes. Perithecia can form at soil temperatures as low as 6–10 °C, with optimal development occurring between 15 and 20 °C (Gilchrist and Dubin 2000). High soil moisture and relative humidity significantly enhance fruiting body formation and maturation (Andersen et al. 2015). As a result, humid conditions in late summer or early autumn promote inoculum accumulation, increasing the risk of FHB epidemics in the following season (Al-Hashimi et al. 2025). In spring, rising temperatures and intermittent rainfall events synchronize ascospore release, with optimal production for *F. graminearum* reported at approximately 29 °C (Manstretta and Rossi 2016). Successful infection of wheat spikes requires extended surface wetness, and at least 48 hours of high humidity immediately after inoculation is essential for disease establishment. Under these favourable conditions, characterized by warm temperatures and sustained moisture, visible blight symptoms can appear within 2–4 days post-infection (Wegulo et al. 2013). The close interplay among temperature, moisture, and crop phenology highlights the strong dependence of FHB epidemics on climate and helps explain the increasing disease incidence observed under climate variability and intensified conservation agriculture (Jung et al. 2022).

### 3.2.7 Management

*F. graminearum*, the primary causal agent of FHB, is highly adapted to temperate and warm-humid environments. It is considered one of the most significant plant-pathogenic fungi due to its global distribution, substantial economic impact, and well-characterized pathogenicity mechanisms. Historically, FHB epidemics have caused extensive damage in major wheat-producing regions. The frequency and severity of these outbreaks have increased significantly in recent decades across North America, Europe, Asia, and South America, primarily due to changes in climate, cropping systems, and pathogen population dynamics (Karlsson et al. 2021). This trend highlights the need for integrated and climate-resilient management strategies.

The management of FHB presents significant challenges due to complex interactions among pathogen biology, environmental factors, and agricultural practices. The increased use of maize–wheat rotations and the adoption of conservation agriculture, especially reduced- or zero-tillage with residue retention, have led to greater persistence of infected crop debris (Landschoot et al. 2013). This persistence enhances inoculum buildup, contributing to more frequent FHB epidemics. A primary obstacle in FHB control is the latent infection process, as visible symptoms on wheat spikes emerge only at advanced stages, by which time mycotoxin contamination has typically occurred (Beccari et al. 2019). This timing reduces the efficacy of curative interventions. Consequently, effective management necessitates a preventive, integrated approach centred on three objectives: reducing inoculum sources, limiting inoculum dispersal, and protecting susceptible tissues during anthesis (Ashiq et al. 2022). These goals can be addressed through a combination of cultural practices, deployment of resistant cultivars, chemical control, and biological methods. Nevertheless, overreliance on fungicides is limited by economic constraints, environmental risks, the potential for resistance development, and restricted access for smallholder farmers, particularly in developing regions (Powell and Vujanovic 2021). Table 3.1 provides a comparative overview of various FHB management strategies.

#### 3.2.7.1 Cultural Management

Cultural practices form the basis of FHB management by reducing inoculum and modifying the agroecosystem to limit conditions favourable to disease development. Tillage methods that incorporate or bury crop residues accelerate decomposition and decrease the survival of *Fusarium* species (Dill-Macky and Jones 2000).

**Table 3.1** Comparative overview of management strategies for FHB

| Management strategy | Key components | Effectiveness | Limitations | Best use scenario |
|---|---|---|---|---|
| Cultural practices | Crop rotation (non-maize), residue management, tillage, optimized sowing time, balanced nutrition, and irrigation management | Moderate (20–50% disease reduction) | Labour-intensive; residue burning has environmental concerns; limited effectiveness under high inoculum pressure | Baseline strategy in all production systems |
| Host resistance | Use of resistant/ moderately resistant cultivars (e.g., Sumai 3-derived lines) | Moderate to high (30–70% reduction) | No complete resistance; polygenic inheritance; genotype × environment interaction | Core strategy for sustainable management |
| Fungicide application | Triazoles (tebuconazole, propiconazole, metconazole) application at anthesis | Moderate to high (40–80% reduction when timed correctly) | Timing critical; incomplete control; cost; resistance development; environmental impact | High-risk years and regions with forecasted outbreaks |
| Biological control (General) | Microbial BCAs (*Bacillus*, *Pseudomonas*, *Trichoderma*, etc.) | Variable (20–60% reduction) | Inconsistent field performance; sensitivity to environmental conditions; limited commercial products | Integrated systems and organic farming |
| Fungal antagonists | *Trichoderma* spp., *Clonostachys rosea*, *Gliocladium* spp. | Moderate (30–60% reduction) | Variable efficacy across environments; formulation and shelf-life issues | Supplement to cultural and chemical control |
| Bacterial antagonists | *Bacillus*, *Pseudomonas*, *Streptomyces* | Moderate to high (30–70% reduction; strong DON suppression in some cases) | Field inconsistency; strain-specific effects; regulatory challenges | Integrated disease management and microbiome-based approaches |
| Plant extracts/ natural compounds | Flavonoid-rich extracts (e.g., *Zanthoxylum bungeanum*) | Low to moderate (20–50% reduction) | Limited field validation; scalability issues; variable efficacy | Eco-friendly niche applications |

(continued)

**Table 3.1** (continued)

| Management strategy | Key components | Effectiveness | Limitations | Best use scenario |
|---|---|---|---|---|
| Forecasting models | Weather-based risk prediction, decision support systems | Indirect but high impact (improves timing and efficiency) | Requires accurate weather data and infrastructure; regional calibration needed | Precision agriculture and advisory services |
| Integrated disease management (IDM) | Combination of resistance + cultural + fungicide + forecasting | High (up to 70–90% combined reduction) | Requires coordination, knowledge, and resources | Most effective and recommended approach |

The application of fertilizers or green manures increases microbial activity, which promotes competitive suppression of the pathogen (Todorović et al. 2024). Additional approaches, such as residue removal by baling or destruction by controlled burning, can further reduce inoculum loads, though these methods require careful consideration of environmental trade-offs (Dill-Macky and Jones 2000). Crop rotation, especially the exclusion of maize as a preceding crop, is particularly effective in reducing inoculum carryover (Dill-Macky and Jones 2000). Other agronomic strategies, which provide moderate but cumulative benefits, include early sowing, selection of early-maturing cultivars to avoid peak infection periods, optimized plant density, balanced nutrient management, and irrigation practices that prevent excessive canopy humidity during flowering (Mueller et al. 2024). Effective weed management also reduces alternative hosts (Suproniene et al. 2019). The integration of these cultural measures with disease forecasting systems further enhances their effectiveness by supporting informed decision-making based on predicted risk levels for subsequent crops (McMullen et al. 2008).

#### 3.2.7.2 Chemical Control

Fungicides are a critical component of integrated FHB management, particularly when applied preventively at the optimal growth stage. Demethylation inhibitor (DMI) fungicides, especially triazoles such as tebuconazole, metconazole, and propiconazole, have demonstrated efficacy in reducing disease severity and mycotoxin accumulation when applied at early anthesis (Edwards 2022). Field studies consistently report significant reductions in FHB severity and deoxynivalenol levels, as well as yield improvements under favourable conditions (Shittu et al. 2025). Combinations such as propiconazole plus tebuconazole generally provide superior control compared to single active ingredients (Paul et al. 2019). However, fungicide efficacy depends on precise application timing, which is often complicated by asynchronous flowering among tillers and unpredictable weather conditions. Certain fungicide classes, particularly quinone outside inhibitors (QoIs), have been associated with increased accumulation of DON, limiting their suitability for FHB management (Paul et al. 2018). Although fungicides offer substantial benefits, they rarely provide complete control and must be integrated with additional management practices. The sporadic occurrence of FHB epidemics further reduces the economic reliability of fungicide applications, especially in regions with variable disease pressure (Salgado et al. 2014).

#### 3.2.7.3 Biological Control and Microbiome-Based Approaches

Biological control offers an environmentally sustainable alternative to chemical interventions and is receiving increasing attention in FHB management. Recent advances in microbiome research highlight the essential role of plant-associated microbial communities in suppressing pathogens and enhancing host resistance (Herrera-Vásquez et al. 2025). Numerous bacterial and fungal biocontrol agents have been studied, including *Bacillus*, *Pseudomonas*, and *Streptomyces*, as well as fungal antagonists such as *Trichoderma*, *Clonostachys rosea*, and *Sphaerodes mycoparasitica* (Vujanovic et al. 2017). These organisms mitigate disease severity through mechanisms such as competition for nutrients and space, production of antimicrobial compounds, mycoparasitism, and induction of systemic resistance in the host (Petrucci et al. 2025). Although experimental results are promising, commercial adoption of biocontrol agents remains constrained by inconsistent field performance and environmental variability (Nguyen et al. 2025). Integration of microbiome-based strategies with breeding programs and agronomic practices represents a promising approach to sustainable, climate-resilient FHB management.

##### 3.2.7.3.1 Fungal Antagonists

Fungal biocontrol agents demonstrate considerable potential for managing FHB due to their rapid growth, adaptability, and persistence under diverse environmental conditions. Their target-specific properties further enhance their effectiveness. Moreover, fungi can persist in the environment without a host by shifting from parasitism to saprotrophy, thereby supporting long-term sustainability. Species including *Trichoderma harzianum*, *T. viride*, and *Clonostachys rosea* have been extensively studied for antagonistic activity against *Fusarium* spp. These fungi employ mechanisms such as mycoparasitism, secretion of cell wall–degrading enzymes, and production of antifungal metabolites. Experimental studies report significant inhibition of *F. graminearum* growth and reduced disease severity in both in vitro and greenhouse settings (Saharan et al. 2007; Zanon et al. 2024). For example, *Clonostachys rosea* suppresses sporulation of *F. graminearum* on wheat residues, reducing inoculum potential (Kulik et al. 2024). Foliar application of *Trichoderma* spp., alone or in combination, results in notable reductions in FHB severity (Panwar et al. 2014). Recent studies have also identified novel fungal antagonists, including *Sarocladium strictum*, *Anthracocystis floculosa*, and *Penicillium olsonii*, with promising biocontrol potential (Rojas et al. 2020).

##### 3.2.7.3.2 Bacterial Antagonists and Plant-Derived Compounds

Bacterial biocontrol agents represent a critical component of integrated FHB management. Genera such as *Bacillus*, *Pseudomonas*, *Streptomyces*, and *Paenibacillus* exhibit strong antagonistic activity against *Fusarium* spp. through mechanisms including competitive exclusion, antibiotic production, secretion of lytic enzymes, and induction of host defence responses (Shi et al. 2014). Multiple bacterial strains significantly reduce FHB severity and mycotoxin accumulation, with notable decreases in DON levels under control conditions (Zhao et al. 2014). *Bacillus subtilis* produces antifungal lipopeptides such as fengycins, while *Pseudomonas fluorescens* strains are effective in reducing disease severity and improving grain quality. Actinobacteria, including *Streptomyces rimosus* and *Rhodococcus* spp., also show potential in suppressing FHB and associated mycotoxins (Kovaleski et al. 2025), although efficacy may be affected by interactions with co-occurring pathogens. In addition to microbial agents, plant-derived compounds are emerging as promising eco-friendly alternatives. Extracts from medicinal plants such as *Zanthoxylum bungeanum*, which are rich in flavonoids including quercetin and kaempferol derivatives, demonstrate antifungal activity by reducing pathogen biomass and mycotoxin accumulation in wheat spikes (Abbas and Yli-Mattila 2022). These strategies present new opportunities for sustainable, residue-free disease management.

#### 3.2.7.4 FHB Resistance with Emphasis on Molecular Breeding and Genomic Approaches

According to Schroeder and Christensen (1963), the wheat plant's growth stage influences FHB infection. The flowering period affects susceptibility to *Fusarium* infection. Passive or active are the two types of resistance against head blight or head scab. Active resistance is conferred by physiological traits. Whereas morphological and phenological characteristics, including plant height, the presence or absence of awns, floral opening size and duration, and anther retention, are key drivers of passive resistance. To elaborate, increased plant height helps keep wheat spikes out of raindrops that may carry pathogens from topsoil and crop debris. Awns in wheat spikes can trap disease-causing agents, increasing the probability of natural infection, while awnless spikes receive reduced infection. A short, narrow floral opening increases resistance by limiting floret exposure to pathogens. Anthers and pollens with a long retention period can trap pathogens and promote host-penetration by facilitating germination of micro- or macroconidia. Mechanisms of FHB resistance have been classified into five types: Type I (resistance to initial infection), Type II (resistance to pathogen proliferation within the plant), Type III (resistance to kernel infection), Type IV (yield tolerance), and Type V (resistance to mycotoxin production). Later, two more types have also been proposed: Type VI (resistance to late blighting) and Type VII (resistance to blighting above the inoculation point). These mechanisms collectively contribute to FHB resistance across diverse wheat germplasm. Sirangelo (2024) claimed that out of the total released wheat varieties, not a single cultivar has been proven immune or absolutely resistant to FHB. The polygenic nature of the disease and the influence of environmental factors make this task highly complex. Quantifying FHB resistance is resource-intensive and time-consuming, and outcomes are often affected by weather variability, necessitating repeated evaluations under diverse conditions.

Molecular markers facilitate the identification of genes and quantitative trait loci linked with FHB resistance in breeding populations. Among the several QTLs identified, a major QTL located on chromosome 3BS has demonstrated a sizable contribution to resistance. Singh (1995) identified additive interactions among at least three minor genes in Frontana wheat that contribute to FHB resistance. The authors recommended using the cotton web technique as a reliable inoculation method for evaluating FHB resistance in wheat lines, a recommendation later confirmed by Saharan et al. (2004).

Advancements in molecular genetics, genomics, and high-throughput phenotyping have significantly accelerated the genetic improvement of FHB resistance. FHB resistance is quantitatively inherited and controlled by numerous quantitative trait loci (QTLs) distributed throughout the wheat genome, with their effects influenced by genetic background and environmental conditions. The most extensively characterized locus, *Fhb1* on chromosome 3BS, originates from the Chinese cultivar Sumai 3 and confers strong Type II resistance by restricting pathogen spread within the spike (Neupane et al. 2025). Other important QTLs, including *Fhb2*, *Fhb4*, *Fhb5*, and *Fhb7*, have been identified from diverse germplasm sources such as Asian cultivars and wild relatives, each contributing to distinct resistance components. Marker-assisted selection has been widely used to introgress major QTLs, such as *Fhb1*, into elite wheat backgrounds. However, the complex, polygenic nature of FHB resistance limits the effectiveness of single-gene introgression strategies. As a result, pyramiding multiple QTLs is now favoured to enhance the durability and stability of resistance across environments. The integration of genomic selection has further improved breeding efficiency by enabling the prediction of breeding values from genome-wide marker data, thereby accelerating the development of resistant cultivars without the need for extensive phenotyping in every generation.

Recent progress in next-generation sequencing and genome-wide association studies has enabled the identification of novel loci and candidate genes associated with FHB resistance, including those involved in cell wall reinforcement, detoxification pathways, and defence signalling. The discovery of *Fhb7*, which encodes a glutathione S-transferase that detoxifies trichothecenes, marks a significant advancement in understanding mechanisms of toxin resistance (Hou et al. 2024). Emerging biotechnological approaches, such as gene editing with CRISPR/Cas systems, provide opportunities for precise modification of resistance genes and susceptibility factors. These include editing host genes related to toxin sensitivity or enhancing defence pathways to achieve durable resistance. Transcriptomic and metabolomic analyses are increasingly used to elucidate host–pathogen interactions and identify biomarkers associated with resistance. In India, the integration of molecular breeding tools with conventional selection is still developing. However, the incorporation of validated QTLs such as *Fhb1* into locally adapted germplasm, combined with genomic selection and multi-environment phenotyping, offers considerable potential for developing climate-resilient, FHB-resistant wheat varieties. Future strategies should prioritize the use of diverse germplasm, including landraces and wild relatives, to broaden the genetic base of resistance and ensure long-term sustainability.

## 3.3 Conclusion

FHB represents a significant and emerging threat to wheat production, particularly under changing climatic conditions characterized by altered temperature, humidity, and cropping practices that enhance pathogen survival and infection. Primarily caused by *F. graminearum*, this disease results in substantial yield losses and mycotoxin contamination, thereby threatening food safety and international trade. In India, the significance of FHB is increasing due to rising pathogen diversity and favourable agro-climatic conditions. Effective management necessitates an integrated approach that includes cultural practices, deployment of resistant cultivars, application of fungicides, and utilization of biological control. Furthermore, recent advances in molecular breeding and genomics provide promising avenues for developing durable resistance in wheat.

## References

Abbas A, Yli-Mattila T (2022) Biocontrol of *Fusarium graminearum*, a causal agent of Fusarium head blight of wheat, and deoxynivalenol accumulation: from in vitro to in planta. Toxins 14(5):299. https://doi.org/10.3390/toxins14050299

Al-Hashimi A, Aina O, Daniel AI et al (2025) Critical review on characterization, management, and challenges of Fusarium head blight disease in wheat. Physiol Mol Plant Pathol 136:102557. https://doi.org/10.1016/j.pmpp.2024.102557

Alisaac E, Mahlein AK (2023) Fusarium head blight on wheat: biology, modern detection and diagnosis and integrated disease management. Toxins 15(3):192. https://doi.org/10.3390/toxins15030192

Alisaac E, Rathgeb A, Karlovsky P et al (2020) Fusarium head blight: effect of infection timing on spread of *Fusarium graminearum* and spatial distribution of deoxynivalenol within wheat spikes. Microorganisms 9(1):79. https://doi.org/10.3390/microorganisms9010079

Andersen KF, Madden LV, Paul PA (2015) Fusarium head blight development and deoxynivalenol accumulation in wheat as influenced by post-anthesis moisture patterns. Phytopathology 105(2):210–219. https://doi.org/10.1094/PHYTO-04-14-0104-R

Ashiq S, Edwards S, Watson A et al (2022) Biofumigation for the management of *Fusarium graminearum* in a wheat–maize rotation. Pathogens 11(12):1427. https://doi.org/10.3390/pathogens11121427

Beccari G, Arellano C, Covarelli L et al (2019) Effect of wheat infection timing on Fusarium head blight causal agents and secondary metabolites in grain. Int J Food Microbiol 290:214–225. https://doi.org/10.1016/j.ijfoodmicro.2018.10.014

Boenisch MJ, Schäfer W (2011) *Fusarium graminearum* forms mycotoxin-producing infection structures on wheat. BMC Plant Biol 11:110. https://doi.org/10.1186/1471-2229-11-110

Devaiah M, Hegde GM, Patil PV et al (2023) Predominance and prevalence of Fusarium head blight of wheat in northern parts of Karnataka. J Farm Sci 36(1):62–66

Dill-Macky R, Jones RK (2000) The effect of previous crop residues and tillage on Fusarium head blight of wheat. Plant Dis 84(1):71–76. https://doi.org/10.1094/PDIS.2000.84.1.71

Drakulic J, Kahar MH, Ajigboye O et al (2016) Contrasting roles of deoxynivalenol and nivalenol in host-mediated interactions between *Fusarium graminearum* and *Sitobion avenae*. Toxins 8(12):353. https://doi.org/10.3390/toxins8120353

Edwards SG (2022) Pydiflumetofen co-formulated with prothioconazole: a novel fungicide for Fusarium head blight and deoxynivalenol control. Toxins 14(1):34. https://doi.org/10.3390/toxins14010034

Figueroa M, Hammond-Kosack KE, Solomon PS (2018) A review of wheat diseases—a field perspective. Mol Plant Pathol 19(6):1523–1536. https://doi.org/10.1111/mpp.12618

Gilchrist L, Dubin H (2000) Fusarium head blight. Food and Agriculture Organization. https://www.fao.org/4/y4011e/y4011e0j.htm

Hao G, McCormick S, Tiley H et al (2023) NX trichothecenes are required for *Fusarium graminearum* infection of wheat. Mol Plant-Microbe Interact 36(5):294–304. https://doi.org/10.1094/MPMI-08-22-0164-R

Hegde GM, Devaiah M, Uday G et al (2023) First report of *Fusarium poae* causing Fusarium head blight of wheat in India. Plant Dis 107(10). https://doi.org/10.1094/PDIS-05-23-0879-PDN

Herrera-Vásquez A, Schlechter R, Armijo-Godoy G et al (2025) Editorial: pathogen suppression by plant-associated microbiota. Front Plant Sci 16:1604449. https://doi.org/10.3389/fpls.2025.1604449

Hou B, Wang D, Yan F et al (2024) Fhb7-GST catalyzed glutathionylation effectively detoxifies the trichothecene family. Food Chem 439:138057. https://doi.org/10.1016/j.foodchem.2023.138057

Jung JY, Kim JH, Baek M et al (2022) Adapting to the projected epidemics of Fusarium head blight of wheat in Korea under climate change scenarios. Front Plant Sci 13:1040752. https://doi.org/10.3389/fpls.2022.1040752

Karlsson I, Persson P, Friberg H (2021) Fusarium head blight from a microbiome perspective. Front Microbiol 12:628373. https://doi.org/10.3389/fmicb.2021.628373

Kashyap PL, Kuar S, Shekhawat PS et al (2022) Field disease scoring guide for wheat and barley research workers. ICAR-Indian Institute of Wheat and Barley Research, Karnal, Technical Bulletin No. 32, p 101

Kaul N, Kashyap PL, Kumar S et al (2022) Genetic diversity and population structure of head blight disease causing fungus *Fusarium graminearum* in northern wheat belt of India. J Fungi 8(8):820. https://doi.org/10.3390/jof8080820

Kovaleski M, Maurer TR, Banfi M et al (2025) Biocontrol of plant pathogens by actinomycetes: a bibliometric analysis and systematic review. World J Microbiol Biotechnol 41:243. https://doi.org/10.1007/s11274-025-04422-7

Kulik T, Treder K, Rochoń M et al (2024) Measurement of the effectiveness of *Clonostachys rosea* in reducing Fusarium biomass on wheat straw. J Appl Genet 65:937–947. https://doi.org/10.1007/s13353-024-00906-8

Kumar HMA, Saharan MS, Aggarwal R et al (2021) In vitro mycelium growth variation among *Fusarium graminearum* isolates causing head blight of wheat in India. J Pharmacogn Phytochem 10(1):1417–1419

Landschoot S, Audenaert K, Waegeman W et al (2013) Influence of maize–wheat rotation systems on Fusarium head blight infection and deoxynivalenol content in wheat under low versus high disease pressure. Crop Prot 52:14–21. https://doi.org/10.1016/j.cropro.2013.04.013

Manstretta V, Rossi V (2015) Effects of weather variables on ascospore discharge from *Fusarium graminearum* perithecia. PLoS One 10(9):e0138860. https://doi.org/10.1371/journal.pone.0138860

Manstretta V, Rossi V (2016) Effects of temperature and moisture on development of *Fusarium graminearum* perithecia in maize stalk residues. Appl Environ Microbiol 82(1):184–191. https://doi.org/10.1128/AEM.02436-15

Markell SG, Francl LJ (2003) Fusarium head blight inoculum: species prevalence and *Gibberella zeae* spore type. Plant Dis 87(7):814–820. https://doi.org/10.1094/PDIS.2003.87.7.814

McMullen M, Halley S, Schatz B et al (2008) Integrated strategies for Fusarium head blight management in the United States. Cereal Res Commun 36(Suppl 6):563–568. https://doi.org/10.1556/CRC.36.2008.Suppl.B.45

McMullen M, Bergstrom G, De Wolf E et al (2012) Unified effort to fight an enemy of wheat and barley: Fusarium head blight. Plant Dis 96:1712–1728. https://doi.org/10.1094/PDIS-03-12-0291-FE

Molnár O, Vida G, Puskás K (2024) Fusarium species associated with Fusarium head blight in Hungarian wheat fields. Plant Dis 108(3):558–562. https://doi.org/10.1094/PDIS-06-23-1167-SR

Mueller N, Wegulo S, Brhel J (2024) Fusarium head blight management. CropWatch, University of Nebraska–Lincoln. https://cropwatch.unl.edu/2024/fusarium-head-blight-management/

Neupane A, Tamburic-Llincic L, Brûlé-Babel A et al (2025) Genetic improvement of FHB and DON resistance by combining the *Fhb1* gene with additional resistance QTL in winter wheat population. Plant Genome 18(3):e70084. https://doi.org/10.1002/tpg2.70084

Nganje WE, Kaitibie S, Wilson WW et al (2004) Economic impacts of Fusarium head blight in wheat and barley: 1993–2001. Agribus Appl Econ Rep 538:10.22004/ag.econ.23627

Nguyen HT, Pham TT, Nguyen PT et al (2025) Microbial biocontrol in agriculture: from mechanistic understanding to field application. Discov Plants 2:334. https://doi.org/10.1007/s44372-025-00421-y

Olariu RM, Fiţ NI, Bouari CM et al (2025) Mycotoxins in broiler production: impacts on growth, immunity, vaccine efficacy, and food safety. Toxins 17(6):261. https://doi.org/10.3390/toxins17060261

Panwar V, Aggarwal A, Singh G et al (2014) Efficacy of foliar spray of *Trichoderma* isolates against *Fusarium graminearum* causing head blight of wheat. J Cereal Res 6(1):59–63

Paul PA, Bradley CA, Madden LV et al (2018) Meta-analysis of the effects of QoI and DMI fungicide combinations on Fusarium head blight and deoxynivalenol in wheat. Plant Dis 102(12):2602–2615. https://doi.org/10.1094/PDIS-02-18-0211-RE

Paul PA, Salgado JD, Bergstrom G et al (2019) Integrated effects of genetic resistance and prothioconazole + tebuconazole application timing on Fusarium head blight in wheat. Plant Dis 103(2):223–237. https://doi.org/10.1094/PDIS-04-18-0565-RE

Petrucci A, Cesarini M, Vicente I et al (2025) Exploring the synergistic potential of *Trichoderma gamsii* T6085 and *Clonostachys rosea* IK726 for biological control of Fusarium head blight in wheat. Microbiol Res 296:128152. https://doi.org/10.1016/j.micres.2025.128152

Powell AJ, Vujanovic V (2021) Evolution of Fusarium head blight management in wheat: scientific perspectives on biological control agents and crop genotypes protocooperation. Appl Sci 11(19):8960. https://doi.org/10.3390/app11198960

Rojas EC, Jensen B, Jørgensen HJ et al (2020) Selection of fungal endophytes with biocontrol potential against Fusarium head blight in wheat. Biol Control 144:104222. https://doi.org/10.1016/j.biocontrol.2020.104222

Roy AK (1974) Ear blight and scab of wheat in Arunachal Pradesh (*Fusarium avenaceum*). Curr Sci 43:162

Saharan MS, Sharma AK (2009) Studies on pathogenic variation among *Fusarium* spp./isolates causing head blight/head scab of wheat in India. Indian Phytopathol 62:402

Saharan MS, Kumar J, Tiwari R (2003) Evaluation of spring and winter wheat lines against *Fusarium graminearum* causing head scab. In: Proceedings of the Indian Society of Mycology and Plant Pathology, Jaipur, India

Saharan MS, Naef A, Kumar J et al (2004) Variation among isolates of *Fusarium graminearum* associated with head scab in India. In: Proceedings of the international symposium on Fusarium head blight, Orlando, USA

Saharan MS, Naef A, Kumar J et al (2007) Characterization of variability among isolates of *Fusarium graminearum* associated with head scab of wheat using DNA markers. Curr Sci 92(2):230–235

Salgado JD, Madden LV, Paul PA (2014) Efficacy and economics of integrating in-field and harvesting strategies to manage Fusarium head blight of wheat. Plant Dis 98(10):1407–1421. https://doi.org/10.1094/PDIS-01-14-0093-RE

Schroeder HW, Christensen JJ (1963) Factors affecting resistance of wheat to scab caused by *Gibberella zeae*. Phytopathology 53:831–838

Shi C, Yan P, Li J et al (2014) Biocontrol of *Fusarium graminearum* growth and deoxynivalenol production in wheat kernels with bacterial antagonists. Int J Environ Res Public Health 11(1):1094–1105. https://doi.org/10.3390/ijerph110101094

Shittu O, McFeaters T, Esker P (2025) Fungicide efficacy for Fusarium head blight and foliar disease. Penn State Extension. https://extension.psu.edu/fungicide-efficacy-for-fusarium-head-blight-and-foliar-disease/

Singh R (1995) Genetic analysis of resistance to scab in spring wheat cultivar Frontana. Plant Dis 79:238. https://doi.org/10.1094/PD-79-0238

Singh PJ, Aujla SS (1994) Effect of lodging on development of head scab of wheat. Indian Phytopathol 47:256–257

Sirangelo TM (2024) Molecular investigations to improve Fusarium head blight resistance in wheat: an update focusing on multi-omics approaches. Plants (Basel) 13(16):2179. https://doi.org/10.3390/plants13162179

Stein JM, Osborne LE, Bondalapati KD et al (2009) Fusarium head blight severity and deoxynivalenol concentration in wheat in response to *Gibberella zeae* inoculum concentration. Phytopathology 99(6):759–764. https://doi.org/10.1094/PHYTO-99-6-0759

Summerell BA, Laurence MH, Liew ECY et al (2010) Biogeography and phylogeography of *Fusarium*: a review. Fungal Divers 44:3–13. https://doi.org/10.1007/s13225-010-0060-2

Suproniene S, Kadziene G, Irzykowski W et al (2019) Weed species within cereal crop rotations can serve as alternative hosts for *Fusarium graminearum* causing Fusarium head blight of wheat. Fungal Ecol 37:30–37. https://doi.org/10.1016/j.funeco.2018.10.002

Todorović I, Abrouk D, Fierling N et al (2024) Manure amendments and fungistasis, and relation with protection of wheat from *Fusarium graminearum*. Appl Soil Ecol 201:105506. https://doi.org/10.1016/j.apsoil.2024.105506

Vujanovic V, Daida MA, Daida P (2017) qPCR assessment of aurofusarin gene expression in mycotoxigenic *Fusarium* species challenged with mycoparasitic and chemical control agents. Biol Control 109:51–57. https://doi.org/10.1016/j.biocontrol.2017.03.010

Wang JH, Ndoye M, Zhang JB et al (2011) Population structure and genetic diversity of the *Fusarium graminearum* species complex. Toxins 3(8):1020–1037. https://doi.org/10.3390/toxins3081020

Wegulo SN, Bockus WW, Nopsa JF et al (2013) Integration of fungicide application and cultivar resistance to manage Fusarium head blight in wheat. In: Fungicides – showcases of integrated plant disease management from around the world. InTech. https://doi.org/10.5772/53096

Xu F, Liu W, Song Y et al (2021) The distribution of *Fusarium graminearum* and *Fusarium asiaticum* causing Fusarium head blight of wheat in relation to climate and cropping system. Plant Dis 105(10):2830–2835. https://doi.org/10.1094/PDIS-01-21-0013-RE

Zanon MSA, Cavaglieri LR, Palazzini JM et al (2024) *Bacillus velezensis* RC218 and emerging biocontrol agents against *Fusarium graminearum* and *Fusarium poae* in barley: in vitro, greenhouse and field conditions. Int J Food Microbiol 413:110580. https://doi.org/10.1016/j.ijfoodmicro.2024.110580

Zhao Y, Selvaraj JN, Xing F et al (2014) Antagonistic action of *Bacillus subtilis* strain SG6 on *Fusarium graminearum*. PLoS One 9(3):e92486. https://doi.org/10.1371/journal.pone.0092486

# 4 Advances in Foliar and Stem Disease Management in Wheat: Pathogen Biology, Resistance Breeding, and Integrated Strategies for Durable Control

Sridhar Bhavani and Zennah Kosgey

**Abstract**

Foliar and stem diseases are among the most persistent threats to global wheat production, causing significant yield reductions, instability, and higher production costs across various agro-ecological zones. Rust diseases, such as stem rust, stripe rust, and leaf rust, pose the greatest transboundary risk because of their epidemic potential, rapid evolutionary changes, and ability to disperse over long distances. Meanwhile, non-rust foliar diseases, including Septoria tritici blotch, spot blotch, and tan spot, exert chronic pressure on wheat systems, especially under intensified production and changing climatic conditions. Additionally, emerging and trade-sensitive diseases, such as wheat blast and bunt diseases, complicate management and deployment strategies. This chapter summarizes recent advances in pathogen biology, epidemiology, and evolution, focusing on how these processes impact resistance durability and disease risk at regional and global levels. The discussion emphasizes host resistance as the foundation for sustainable disease management, highlighting progress in adult plant resistance, resistance gene pyramiding, and the use of genomic tools to accelerate breeding for quantitative resistance. The chapter also reviews the roles of phenotyping platforms, surveillance networks, and integrated management approaches in increasing resilience amid climate variability and evolving pathogen populations.

S. Bhavani (✉)
International Maize and Wheat Improvement Centre (CIMMYT), Global Wheat Program, Nairobi, Kenya
e-mail: s.bhavani@cgiar.org

Z. Kosgey
Kenya Agricultural and Livestock Research Organisation (KALRO), Food Crops Research Centre, Njoro, Kenya



R. Tiwari et al. (eds.), *Wheat Stress Management*,
https://doi.org/10.1007/978-981-92-0902-6_4

**Keywords**

Wheat diseases · Foliar and stem diseases · Durable disease resistance · Pathogen population dynamics · Climate change · Resistance breeding · Integrated disease management

## 4.1 Introduction

Wheat (*Triticum aestivum* L.) is the most widely cultivated cereal crop globally and is essential for food and nutritional security. Maintaining wheat production and productivity is becoming increasingly difficult due to climatic variability, intensified agricultural practices, and expansion into marginal environments. Biotic stresses, especially foliar and stem diseases, continue to cause significant yield losses and production instability across wheat-growing regions, despite advances in breeding and crop protection. Global assessments estimate that plant diseases account for 17–21% of yield losses in major food crops, with wheat among the most affected cereals (Savary et al. 2019). These losses result from both episodic epidemics and ongoing disease pressure, which reduce photosynthetic capacity, radiation-use efficiency, and grain filling. The extent of these impacts varies by region, depending on climate, farming practices, cultivar choice, and access to disease management technologies.

Rust diseases are among the most destructive and persistent threats to global wheat production. Stem rust (*Puccinia graminis* f. sp. *tritici*), stripe rust (*P. striiformis* f. sp. *tritici*), and leaf rust (*P. triticina*) exhibit rapid pathogen evolution, long-distance dispersal, and frequent breakdown of host resistance (Kolmer 2005; Singh et al. 2015). Since the early 2000s, the emergence of highly virulent populations, such as the Ug99 stem rust lineage and aggressive stripe rust races adapted to warmer climates, has significantly altered global risk profiles and revealed the vulnerability of genetically uniform cultivars. Beyond rusts, non-rust foliar and stem diseases, including Septoria tritici blotch, spot blotch, and tan spot, impose substantial, yet often underestimated, yield constraints. Additionally, emerging and re-emerging threats such as wheat blast and bunt diseases increasingly shape breeding priorities, deployment strategies, and phytosanitary policies due to their yield impact, transboundary spread, and trade implications (Ceresini et al. 2018; Bishnoi et al. 2020). These disease complexes highlight the limitations of single-gene resistance and emphasize the necessity for integrated, durable disease management strategies.

## 4.2 Global Importance of Foliar and Stem Diseases in Wheat

Foliar and stem diseases are major constraints on global wheat production, negatively affecting yield, grain quality, and costs, and significantly contributing to yield gaps worldwide (Savary et al. 2019). Among these, rust diseases pose the greatest global threat because of their epidemic potential and ability to cross borders. Stripe rust is the most prevalent and damaging, especially as it spreads into warmer, previously unaffected regions (Ali et al. 2017). Although stem rust occurs less frequently, it can cause devastating losses in susceptible cultivars (Singh et al. 2015; Lewis et al. 2024). Leaf rust continues to cause yield reductions in temperate wheat-growing areas. Additionally, non-rust foliar diseases pose important, often overlooked challenges to wheat productivity. Septoria tritici blotch is particularly damaging in temperate regions, especially in Europe, where yield losses can exceed 50% under favorable conditions (Fones and Gurr 2015; Ghaffary et al. 2018). Spot blotch and tan spot are becoming more significant in warmer environments, especially under heat stress, residue retention, and reduced tillage systems that promote pathogen survival and infection (Lamari and Bernier 1991; Gupta et al. 2018).

Besides direct yield losses, several wheat diseases significantly impact grain quality and trade restrictions. Wheat blast and Karnal bunt illustrate how economic damages and regulatory limits increase the overall burden of wheat diseases, highlighting the importance of integrated strategies that include host resistance breeding, disease surveillance, and crop management (Beddow et al. 2015; Ceresini et al. 2018; Garima et al. 2020). These issues are intensified by climate change, which affects pathogen survival, dispersal, and aggressiveness, thereby raising disease pressure and posing new threats. As a result, adaptive and proactive disease management strategies are crucial to ensure the long-term stability of wheat production systems under future climate conditions (Bebber et al. 2016; Chaloner et al. 2021).

## 4.3 Major Foliar and Stem Diseases of Wheat

### 4.3.1 Rust Diseases of Wheat

Rust diseases are the most economically important foliar and stem diseases of wheat worldwide. These diseases, caused by obligate biotrophic fungi of the genus Puccinia, including stem rust, stripe rust, and leaf rust, affect nearly all wheat-growing regions and continue to shape global breeding and monitoring efforts.

*Stem rust*, caused by *Puccinia graminis* f. sp. *tritici*, has historically been the most destructive disease affecting wheat. Severe epidemics lead to extensive lodging, shriveled grain, and catastrophic yield losses when susceptible cultivars are grown in favorable conditions. While stem rust was largely controlled during the latter half of the twentieth century through the deployment of resistance genes and the eradication of alternate hosts in certain areas, it re-emerged as a global concern following the detection of the Ug99 race group in East Africa in the late 1990s (Singh et al. 2015, 2021). Subsequent worldwide surveillance has revealed a more complex epidemiological landscape than initially expected. Multiple variants within the Ug99 lineage have gained additional virulence, and genetically distinct, highly virulent races have appeared independently in Europe and South America (Lewis et al. 2024). Genomic analyses show that mutation, recombination, and migration collectively drive virulence diversification, suggesting that the threat posed by stem rust is not limited to a single lineage or geographic area.

*Stripe rust*, caused by *Puccinia striiformis* f. sp. *tritici*, is the most widespread and damaging rust disease affecting wheat globally. Although it has traditionally been associated with cool, humid environments, the disease has expanded into warmer regions over the past 20 years. This spread is due to the rise of aggressive, high-temperature-adapted pathogen populations (Milus et al. 2009; Ali et al. 2017). The occurrence of sexual recombination and long-distance dispersal has increased genetic diversity and sped up pathogen adaptation, often leading to the failure of race-specific resistance genes. Yield losses from stripe rust usually range between 10% and 40% with moderate to severe infections and can be even higher if epidemics start early and align with vulnerable growth stages (Chen 2020; Singh et al. 2021). Its sensitivity to environmental conditions and rapid adaptation to favorable weather create significant management challenges, especially in the context of increasingly variable climates.

*Leaf rust*, caused by *Puccinia triticina*, is the most consistently prevalent rust disease affecting wheat worldwide. Although it is generally less damaging than stem or stripe rust during individual outbreaks, its widespread presence and persistent occurrence lead to significant cumulative yield losses, particularly in temperate wheat-growing areas. Leaf rust populations show notable virulence diversity, and the frequent failure of resistance due to the widespread use of single major resistance genes highlights the pathogen's rapid evolution and adaptability (Zhang et al. 2019).

### 4.3.2 Non-rust Foliar and Stem Diseases

Besides rusts, various non-rust foliar and stem diseases also significantly affect wheat production, mainly due to ongoing disease development and interactions with abiotic stresses.

*Septoria tritici blotch (STB)*, caused by *Zymoseptoria tritici*, is the most important foliar disease of wheat in temperate regions, especially in Europe. STB reduces green leaf area during grain filling and can lead to yield losses of 10–30%, with losses over 50% under favorable conditions in susceptible cultivars (Fones and Gurr 2015; Ghaffary et al. 2018). The pathogen's long latent period, high genetic diversity, and quick development of fungicide resistance complicate management and increase reliance on host resistance.

*Spot blotch*, caused by *Bipolaris sorokiniana*, poses a major challenge in warmer, moisture-stressed environments, especially in South Asia, East Africa, and parts of Latin America. The disease often interacts with heat stress and terminal drought, resulting in significant yield losses in susceptible cultivars (Gupta et al. 2018; Singh et al. 2021). Resistance to spot blotch is mainly quantitative and is often mixed with plant maturity and adaptation traits, making breeding more difficult.

*Tan spot*, caused by *Pyrenophora tritici-repentis*, has become increasingly significant in residue-retaining and reduced-tillage systems. The disease is characterized by necrotrophic interactions, with host-selective toxins such as ToxA playing a key role in symptom development and host specificity (Lamari and Bernier 1991; Friesen et al. 2006). Although yield losses are usually moderate, tan spot can significantly raise overall disease pressure when it coincides with rusts and STB.

### 4.3.3 Other Globally Important Wheat Diseases Influencing Management and Deployment

*Wheat blast*, caused by *Magnaporthe oryzae* pathotype *triticum*, is one of the most significant emerging threats to global wheat production. After its initial appearance in South America, the disease spread to South Asia and, more recently, to Africa, mainly through the movement of infected seed (Ceresini et al. 2018; Tembo et al. 2020). In warm, humid environments, especially during heading and grain filling, wheat blast can lead to yield losses ranging from moderate reductions to nearly complete crop failure (Ceresini et al. 2018; Singh et al. 2021).

Current genetic resistance to wheat blast mainly relies on the 2NS/2AS translocation from *Aegilops ventricosa*, which offers partial resistance to blast, as well as to several rusts and nematodes (He et al. 2020; Shahinnia et al. 2022). Nonetheless, the rise of virulent pathogen populations that can overcome 2NS-linked resistance highlights the dangers of genetic uniformity and emphasizes the need for diverse resistance sources.

*Bunt diseases*, especially Karnal bunt (*Tilletia indica*), form another group of globally important wheat diseases. Although yield losses are generally modest, Karnal bunt has significant economic impacts due to strict quarantine rules and its negative effects on grain quality (Bishnoi et al. 2020; Garima et al. 2020). Management mainly emphasizes exclusion, surveillance, and accurate diagnostics rather than direct control of diseases in the field. Crown and root diseases, along with plant-parasitic nematodes, cause yield variability, especially in warm, water-scarce areas. These stresses often interact with foliar diseases by speeding up senescence and reducing plant tolerance to canopy infections (Savary et al. 2017; Yavuzaslanoglu et al. 2024).

## 4.4 Pathogen Biology, Epidemiology, and Transboundary Risk

A thorough understanding of pathogen biology and epidemiology is crucial for effectively managing foliar and stem diseases in wheat. The course of disease epidemics depends on pathogen survival between seasons, inoculum production and dispersal, infection development within the crop canopy, and the rate at which pathogen populations adapt to host resistance and control measures. Although wheat pathogens vary in lifestyle, from obligate biotrophs to necrotrophs, many management challenges arise from common epidemiological processes that occur at the field and at the regional scales.

### 4.4.1 Disease Cycles and Inoculum Sources

Rust fungi (*Puccinia* spp.) are obligate biotrophs that require living host tissue and produce large amounts of wind-dispersed urediniospores. These spores can travel hundreds to thousands of kilometers, explaining the transboundary nature of rust outbreaks and the potential for regional epidemics to become international threats (Kolmer 2005). Local persistence is supported by volunteer wheat and green-bridge hosts, which enable early-season establishment under favorable conditions.

In contrast, *Zymoseptoria tritici*, the causal agent of Septoria tritici blotch, persists on infected crop residues. Epidemics are initiated by both splash-dispersed asexual spores and airborne sexual ascospores, facilitating local intensification and regional connectivity (Hailemariam et al. 2020). The extended latent period typical of STB allows for significant pathogen buildup before symptoms appear, making disease monitoring and control timing more difficult.

Wheat blast has a unique epidemiological profile. Besides local spread via airborne conidia, infected seed serves as a key vector for long-distance transmission, highlighting the importance of seed health testing, quarantine, and phytosanitary measures in areas at risk of blast (Cruz and Valent 2017; Tembo et al. 2020). Likewise, bunt diseases mainly spread through contaminated seed and soilborne teliospores, emphasizing the need for exclusion and surveillance-based management strategies (Garima et al. 2020).

### 4.4.2 Canopy Development and Epidemic Progression

Canopy architecture and crop developmental stage significantly influence the progression of foliar disease epidemics. Dense canopies create humid microclimates that encourage infection and prolong leaf wetness, thereby aiding disease development for both rusts and blotch diseases. Rust diseases are polycyclic, undergoing multiple infection cycles per season, allowing rapid exponential increases in disease severity after an epidemic begins.

Stripe rust is especially sensitive to temperature and moisture, but modern pathogen populations include strains with wider temperature tolerance, raising the risk in warmer environments and at lower latitudes (Milus et al. 2009). In contrast, STB progresses more slowly, with staggered lesion development on leaves and a longer epidemic duration caused by successive infection events (Hailemariam et al. 2020).

Necrotrophic pathogens, including *Pyrenophora tritici-repentis* and *Bipolaris sorokiniana*, are significantly influenced by residue management practices and interactions with environmental stressors. Reduced tillage and residue retention help maintain inoculum for tan spot, while heat and moisture stress exacerbate spot blotch severity, especially in South Asia and sub-Saharan Africa (Lamari and Bernier 1991; Gupta et al. 2018).

### 4.4.3 Pathogen Evolution and Implications for Durability

A key feature of wheat foliar and stem diseases is the rapid evolutionary adaptation of pathogen populations in response to selection pressures. In rust pathogens, virulence evolution results from mutation, recombination, and migration, alongside strong selection from widespread deployment of resistance genes. Evidence of sexual recombination in stripe rust populations has clarified the mechanisms underlying the rapid emergence of new virulence combinations and frequent resistance failures.

In the case of STB, large effective population sizes, frequent recombination, and strong fungicide selection pressures have caused rapid changes in sensitivity profiles and pathogen genotypes. The resistance development in *Zymoseptoria tritici* shows how management practices can quickly impact pathogen populations, emphasizing the need to combine fungicide applications with host resistance strategies (Hawkins et al. 2014; Lucas et al. 2015).

Wheat blast and tan spot highlight evolutionary risks. The emergence of blast isolates that are virulent to the widely used 2NS/2AS translocation emphasizes the dangers of genetic uniformity, while the horizontal transfer of the *ToxA* gene in *Pyrenophora tritici-repentis* shows how specific evolutionary events can quickly alter host–pathogen interactions (Friesen et al. 2006; Ceresini et al. 2018).

### 4.4.4 Scale of Risk and the Need for Coordinated Surveillance

The epidemiology of wheat diseases operates across different spatial scales. Rusts are prime examples of pathogens whose long-distance dispersal poses a global risk, requiring coordinated international surveillance and early warning systems (Kolmer 2005). Similarly, wheat blast demands both regional and international preparedness due to its seedborne transmission and potential for rapid establishment in new areas (Cruz and Valent 2017). These challenges have resulted in the development of integrated disease surveillance platforms that combine field observations, pathogen diagnostics, and climate-based risk modeling to support timely management and breeding decisions (Singh et al. 2021; Chen et al. 2025).

## 4.5 Host Resistance as the Foundation of Disease Management

Host resistance is one of the most effective and sustainable methods for controlling foliar and stem diseases in wheat. Unlike chemical or agronomic practices, resistance built into cultivars works continuously during the growing season, functions independently of farmer decisions after deployment, and does not involve ongoing financial or environmental costs.

### 4.5.1 Types of Resistance and Epidemiological Consequences

Resistance to wheat diseases is typically classified as either race-specific (qualitative) or non-race-specific (quantitative). Race-specific resistance, usually conferred by single major-effect genes, can provide near-complete protection against pathogen races lacking the corresponding virulence. Although this form of resistance has been widely utilized against rusts and powdery mildew, it has often become ineffective after extensive deployment (Ellis et al. 2014). Quantitative resistance, often expressed as adult plant resistance (APR), is controlled by multiple loci and helps slow down epidemic development rather than completely preventing infection. APR usually exerts weaker directional selection on pathogen populations and is more durable across different environments and years (Krattinger et al. 2009; Escamilla et al. 2025).

### 4.5.2 Resistance to Rust Diseases

In the rust pathosystem, extensive evidence from historical deployment and recent epidemics shows that single major resistance genes rarely remain effective when widely deployed. While race-specific genes can initially offer nearly complete protection, widespread use exerts strong selective pressure on pathogen populations, accelerating the evolution and spread of virulence. The repeated failure of common rust resistance genes across continents confirms that durable rust control cannot rely on single-gene solutions, regardless of their initial success (Ellis et al. 2014; Singh et al. 2021).

Durable resistance to rust diseases is achieved through integrated resistance architectures, with adult plant resistance (APR) serving as the main component. APR is typically polygenic, expressed at various growth stages, and works by reducing infection efficiency, lesion growth, sporulation, and epidemic development rather than preventing infection entirely. Well-characterized APR loci, such as *Sr2*, *Yr18/Lr34*, *Yr29/Lr46*, *Lr67*, and *Lr68*, provide partial yet consistent protection across diverse environments and pathogen populations. When these loci are combined with additional minor-effect genes, the resistance phenotypes become harder for pathogen populations to overcome, as selection pressure is spread across multiple mechanisms instead of targeting a single one (Krattinger et al. 2009; Ellis et al. 2014; Singh et al. 2021).

Literature also shows that APR-based resistance remains effective even as rust populations evolve through mutation, recombination, and migration. Unlike race-specific genes, APR loci generally maintain activity across multiple pathogen lineages and environmental conditions, making them especially valuable as climate variability increases and disease ranges expand. Importantly, APR does not work alone; its durability is improved when combined with strategically deployed major genes that provide early-season protection, while APR helps limit late-season epidemic development and reduces the risk of catastrophic resistance failure.

CIMMYT-led breeding programs have operationalized these principles by applying sustained selection under intense and genetically diverse rust pressure in global hotspot environments. By consistently exposing breeding material to evolving pathogen populations and combining complementary APR loci within broadly adapted genetic backgrounds, these programs have developed resistance profiles that perform reliably across multiple regions and years. The long-term field performance of CIMMYT-derived wheat varieties in Africa, South Asia, and Latin America demonstrates that APR stacks can maintain effective rust resistance at scale, even under high disease pressure and rapidly evolving pathogen populations (Kolmer 2005; Singh et al. 2015).

These collective experiences highlight a key principle in rust resistance breeding: durability comes not from single genes but from the strategic combination, selection, and deployment of resistance components over time. Breeding strategies that focus on resistance diversity, continuous field validation, and informed deployment are crucial for maintaining effective rust control in global wheat production systems.

### 4.5.3 Resistance to Non-rust and Emerging Diseases

Resistance to STB, spot blotch, tan spot, and bunt diseases is mainly quantitative and affected by environmental factors. While this complexity makes phenotyping and selection more difficult, quantitative resistance tends to be more durable than resistance from a single gene. In the case of wheat blast, dependence on the 2NS/2AS translocation shows both the advantages and disadvantages of major resistance sources, emphasizing the importance of diversifying resistance strategies and avoiding widespread genetic uniformity (He et al. 2020; Shahinnia et al. 2022).

## 4.6 Advances in Breeding, Phenotyping Platforms, and Genomic Tools

Advancements in managing foliar and stem diseases in wheat mainly originate from improved breeding methods, better phenotyping facilities, and the use of genomic tools. Over the past 20 years, disease resistance breeding has shifted from focusing on individual genes to creating robust resistance profiles that remain effective across diverse environments and pathogen populations.

### 4.6.1 Field Phenotyping Platforms and Hotspot Environments

Accurate phenotyping is the most crucial element of disease resistance breeding. For rust diseases, establishing and maintaining high-pressure hotspot environments has been essential to global breeding progress. Regions such as East Africa (notably Kenya and Ethiopia), South Asia, and Mexico naturally experience high and diverse rust pressure, which has been crucial for identifying durable resistance and screening international germplasm (Kolmer 2005; Singh et al. 2015).

These platforms support continuous selection against evolving pathogen populations and enable early detection of new races that could threaten resistance. The success of this approach has led to the development of similar phenotyping networks for other diseases. For wheat blast, specialized precision phenotyping platforms in South Asia and Latin America now provide controlled environments for screening germplasm under consistent, severe disease pressure, aiding resistance discovery and validation (Singh et al. 2021; Shahinnia et al. 2022).

For non-rust diseases such as Septoria tritici blotch (STB), spot blotch, and tan spot, performing multi-location phenotyping across different environments is essential to capture genotype-by-environment interactions and distinguish true resistance from disease escape. Combining managed stress environments with natural field conditions has improved the accuracy of selecting for quantitative resistance (Ghaffary et al. 2018; Gupta et al. 2018).

### 4.6.2 Genetic Mapping and Resistance Gene Discovery

Recent advances in molecular genetics have accelerated the identification of resistance loci for foliar and stem diseases. Techniques such as linkage mapping and genome-wide association studies have been widely used to analyze resistance to rusts, STB, spot blotch, and blast, revealing a complex genetic architecture characterized by many loci with small to moderate effects (Shahinnia et al. 2022; Escamilla et al. 2025). For rust diseases, cloning several resistance genes has provided a better understanding of the underlying mechanisms, including transporter-mediated resistance and pattern recognition pathways (Krattinger et al. 2009). Although gene cloning has improved understanding of resistance biology, its practical breeding impact has been maximized when translated into diagnostic markers that enable efficient pyramiding of multiple loci.

In wheat blast, genetic analyses have confirmed the crucial role of the 2NS/2AS translocation and have identified additional resistance loci that offer opportunities for diversification (He et al. 2020; Shahinnia et al. 2022). In necrotrophic diseases, mapping studies have emphasized the importance of host-sensitivity loci, such as *Tsn1*, highlighting the need to select against susceptibility alleles in resistance breeding (Lamari and Bernier 1991; Friesen et al. 2006).

### 4.6.3 Marker-Assisted Selection and Genomic Selection

Marker-assisted selection is now a common practice in resistance breeding for foliar and stem diseases, especially rusts, where diagnostic markers for major genes and key adult-plant resistance (APR) loci are available. Marker-assisted pyramiding has improved the efficiency and reliability of combining resistance loci, thereby decreasing reliance on phenotypic selection alone (Ellis et al. 2014; Singh et al. 2021).

Genomic selection has transformed breeding for quantitative disease resistance by enabling selection based on genome-wide marker profiles rather than individual loci. This approach is particularly useful for diseases such as STB, spot blotch, and blast, where resistance is polygenic and heavily affected by environmental factors (Voss-Fels et al. 2019). Combining genomic selection with high-quality phenotypic data from hotspot environments has accelerated genetic gains in disease resistance while maintaining yield and adaptation. It has also proven especially effective for rust diseases, where resistance is polygenic and influenced by the environment (Poland et al. 2012; Rutkoski et al. 2014; Shahinnia et al. 2022).

### 4.6.4 Integration with Agronomic and Chemical Management

While host resistance is essential for disease management, its effectiveness improves when combined with other agronomic and chemical strategies. Fungicides remain important in controlling foliar diseases, especially in systems where resistance is partial or epidemics spread quickly. However, the increase in fungicide resistance in pathogens such as *Zymoseptoria tritici* highlights the limits of chemical control and the need to focus on genetic resistance (Hawkins et al. 2014; Lucas et al. 2015).

Agronomic practices such as crop rotation, residue management, and adjustment of planting dates influence disease pressure and interact with genetic resistance. Although these practices alone rarely provide complete disease control, combining them with resistant cultivars results in more consistent disease suppression and decreases selection pressure on both resistance genes and fungicides.

## 4.7 Integrated Approaches to Disease Management and Future Perspectives

Recent progress in managing foliar and stem diseases in wheat shows that durable control cannot be achieved through isolated actions. Sustainable disease management requires combining resistant cultivars, deploying them strategically over time and space, conducting thorough disease surveillance, and using effective agronomic practices. Evidence from rust disease management indicates that the durability of resistance depends on both the traits of resistance genes and their deployment across diverse production areas. Similar vulnerabilities are seen in wheat blast, bunt diseases, and necrotrophic pathogens, where genetic uniformity and inadequate surveillance can quickly undermine control strategies (Ceresini et al. 2018; Bishnoi et al. 2020).

Enhancing global disease preparedness requires sustained investment in advanced phenotyping platforms, coordinated pathogen surveillance, and integration of regional data. Developments in genomics, remote sensing, and predictive modeling offer valuable tools for forecasting disease risks and informing breeding and management strategies. However, the effectiveness of these tools depends on strong collaboration among breeders, pathologists, and national programs. Therefore, progress in managing foliar and stem diseases should be measured by the long-term stability and resilience of wheat production systems under changing pathogen pressures and climate variability, rather than by the complete eradication of disease. Achieving this goal depends on integrating diverse resistance sources, strategically deploying them, and coordinating international efforts.

## 4.8 Role of Fungicides and Integrated Crop Management

Fungicides have long been essential in controlling foliar and stem diseases of wheat, especially in high-input production systems. Applying fungicides at the right time can reduce rust outbreaks and safeguard both yield and grain quality during times of high disease pressure. Triazoles, strobilurins, and, more recently, succinate dehydrogenase inhibitors are frequently used against rusts and foliar blotch diseases, often in combination to improve effectiveness and delay resistance (Hawkins et al. 2014; Oliver 2014). However, reliance on fungicides as the primary method of disease control is increasingly constrained by economic, environmental, and biological challenges. The rapid development and widespread occurrence of fungicide resistance in *Zymoseptoria tritici* populations, especially in Europe, highlight the vulnerability of reliance on chemical management. Documented decreases in sensitivity across multiple modes of action have restricted effective options and increased production costs (Lucas et al. 2015).

Although fungicide resistance has developed more gradually in rust pathogens, chemical control remains limited by logistical challenges, especially in smallholder systems, where high costs, limited access, and suboptimal application timing reduce its effectiveness. Moreover, fungicides are ineffective against seed-borne and quarantine-sensitive diseases such as wheat blast and Karnal bunt. Managing these diseases primarily depends on resistant cultivars, clean seed, and regulatory measures rather than field-level chemical treatments (Cruz and Valent 2017; Garima et al. 2020). Therefore, integrated crop management strategies that combine genetic resistance with complementary agronomic practices are essential for sustainable disease control. Practices like crop rotation, residue management, adjusting planting dates, and optimizing canopy architecture can significantly affect disease development, especially for necrotrophic and residue-borne pathogens, including *Zymoseptoria tritici* and *Pyrenophora tritici-repentis* (Lamari and Bernier 1991; Ghaffary et al. 2018). Although these practices alone rarely provide complete disease control, their combination with resistant cultivars improves system resilience and lessens dependence on chemical inputs.

## 4.9 Impact of Climate Change on Foliar and Stem Diseases of Wheat

Climate change is fundamentally altering the epidemiology and impact of foliar and stem diseases in wheat by changing temperature patterns, rainfall, and the frequency of extreme weather events. These shifts influence pathogen survival, infection efficiency, host susceptibility, and the geographic range of both crops and pathogens. Rising temperatures have allowed the spread of stripe rust into warmer regions and higher altitudes, driven by the emergence of thermotolerant pathogen populations (Milus et al. 2009; Ali et al. 2017). Additionally, climate-driven changes in wind patterns and increased weather variability are expected to modify long-distance dispersal routes and raise epidemic unpredictability (Bebber et al. 2014). Non-rust foliar diseases are also affected: increased rainfall intensity and humidity promote Septoria tritici blotch development, while combined heat and moisture stress worsen spot blotch severity in tropical and subtropical regions (Gupta et al. 2018; Hailemariam et al. 2020). Furthermore, wheat blast risk is closely linked to warm, humid conditions during heading and grain filling, raising concerns about potential disease spread under future climate scenarios (Ceresini et al. 2018).

Climate change also affects host resistance mechanisms. Elevated temperatures can change how resistance is expressed, reduce the effectiveness of specific resistance genes, or alter host–pathogen interactions. These temperature-dependent dynamics highlight the importance of testing resistance in representative target environments and breeding for stability across diverse and changing thermal conditions.

## 4.10 Emerging Tools and Technologies for Disease Management

Recent advances in genomics, phenomics, and data science are transforming strategies for managing foliar and stem diseases in wheat. High-throughput genotyping platforms now enable genome-wide dissection of disease resistance and accelerate marker-assisted and genomic selection in breeding programs (Voss-Fels et al. 2019; Escamilla et al. 2025). Meanwhile, remote sensing and digital phenotyping technologies, including multispectral and hyperspectral imaging combined with machine learning, offer powerful tools for early disease detection, severity assessment, and field-scale evaluation of resistance responses. These methods complement traditional phenotyping and enhance selection accuracy (Mahlein 2016). Furthermore, pathogen genomics and molecular diagnostics improve disease surveillance and early warning systems through rapid race identification, population genomics, and real-time monitoring of virulence evolution, especially for rust pathogens and increasingly for wheat blast (Kolmer 2005; Singh et al. 2021). When integrated with climate-driven risk modeling, these advances enable a shift from reactive disease control to proactive, predictive, and precision-based disease management.

## 4.11 Pathogen Diagnostics and Molecular Surveillance

Accurate and timely pathogen diagnostics are crucial for managing foliar and stem diseases in wheat, especially given the transboundary movement of pathogens and their rapid evolution. While traditional phenotypic race analysis based on differential host responses remains fundamental for rust diseases, it is increasingly complemented by molecular and genomic diagnostic tools that enhance resolution, speed, and consistency across regions. Recent advances in pathogen genomics have enabled diagnostics that can distinguish major lineages, track virulence evolution, and reconstruct migration pathways at continental and global scales. Integrating genome-wide data with coordinated field surveillance has strengthened early warning systems and improved detection of emerging threats before they lead to widespread epidemics (Kolmer 2005). High-resolution population genomic analyses of stem rust show that virulence diversification results from multiple evolutionary processes, including mutation, recombination, and migration. This emphasizes the need for ongoing genomic surveillance alongside phenotypic monitoring, particularly in epidemiological hubs for rust evolution and spread.

Recent innovations have expanded diagnostic capacity in resource-limited settings. Portable, field-deployable sequencing platforms like MARPLE (Mobile and Real-time PLant disease diagnostics) combine targeted sequencing with rapid analysis to enable near real-time identification of wheat rust pathogens directly from field samples. These platforms have been successfully used in Africa and Asia to strengthen local and regional surveillance capacity (Hubbard et al. 2015). For wheat blast and bunt diseases, diagnostics are especially important due to seedborne transmission and quarantine implications. PCR-based and genome-informed assays are increasingly employed to confirm incursions, differentiate pathogenic lineages, and support phytosanitary decisions (Cruz and Valent 2017; Tembo et al. 2020). Despite significant progress, diagnostic capacity remains uneven across regions, particularly in low- and middle-income countries. This disparity highlights the need to improve infrastructure, unify protocols, and integrate diagnostics with phenotyping and surveillance systems to boost global preparedness and ensure that disease management decisions are based on timely, reliable pathogen data.

## 4.12 Early Warning, Forecasting, and Advisory Systems

Early warning, forecasting, and advisory systems are crucial for turning disease surveillance and pathogen diagnostics into actionable insights for breeders, policymakers, and farmers. For wheat foliar and stem diseases, especially rusts, these systems are essential because epidemics can develop and spread rapidly across regions before field-level measures can be implemented. Early warning platforms integrate data from field surveillance, pathogen diagnostics, and weather conditions to identify increased disease risk before major outbreaks happen. Regional rust monitoring networks, supported by real-time reporting, have facilitated early detection of new virulence patterns and unusual disease activity. This has enabled proactive responses such as targeted fungicide applications, varietal advisories, and informed breeding decisions (Kolmer 2005).

Forecasting models improve early warning systems by linking disease development to environmental factors such as temperature, rainfall, and humidity. Epidemiological and climate-based models have been successfully used to predict rust infection risk and epidemic progression across various spatial scales, especially when forecasts are combined with surveillance data rather than used alone (Chen et al. 2025; Bebber et al. 2016). Advisory systems are the final and most important step, turning risk assessments into practical advice for varietal deployment, fungicide timing, and seed movement—particularly in regions prone to transboundary spread. Despite ongoing challenges such as gaps in surveillance, uneven diagnostic capacity, and climate-related uncertainties, strengthening these systems through consistent investment in field networks, data integration, and communication channels, along with closer collaboration among breeders, pathologists, meteorologists, and extension services, can significantly reduce disease risks and improve the durability of resistance deployment in wheat production systems.

## 4.13 Challenges, Knowledge Gaps, and Future Research Priorities

Although significant progress has been achieved, several challenges still limit the effective management of foliar and stem diseases in wheat. A major knowledge gap remains in understanding quantitative resistance mechanisms, especially for non-rust diseases and emerging threats like wheat blast. Limitations in phenotyping accuracy and incomplete analysis of genotype-by-environment interactions further hinder progress. Genetic uniformity continues to be a major vulnerability, as the widespread reliance on a small set of resistance sources, exemplified by the extensive use of the 2NS/2AS translocation for wheat blast resistance, increases the risk of resistance failure and emphasizes the need for diversification at both genetic and deployment levels (Ceresini et al. 2018; He et al. 2020).

Capacity limitations in disease surveillance, diagnostics, and phenotyping infrastructure, especially in low- and middle-income countries, continue to limit global preparedness for transboundary disease threats. Addressing these shortcomings requires strengthening regional and international networks, enhancing data sharing, and increasing investment in training and infrastructure. Future research should focus on interdisciplinary integration, linking breeding, pathology, agronomy, and climate science to develop resistance strategies effective under future climate conditions while maintaining yield and grain quality. Achieving these goals will require ongoing investment and coordinated global cooperation.

## 4.14 Conclusion and Future Outlook

Over the past two decades, advances in foliar and stem disease management in wheat have significantly enhanced understanding of pathogen biology, resistance mechanisms, and breeding strategies. Despite these advances, rust diseases remain the primary global threat. Furthermore, increasing pressure from non-rust foliar diseases, wheat blast, and trade-sensitive pathogens highlights the need for comprehensive, system-level disease management approaches. Achieving durable control in wheat requires combining diverse resistance sources, deploying them strategically, establishing robust surveillance systems, and employing complementary agronomic practices instead of relying solely on single genes or isolated measures. Evidence from rusts, wheat blast, bunt diseases, and necrotrophic pathogens highlights the risks of genetic uniformity and the benefits of resistance diversity. Future success depends on strengthening global phenotyping and surveillance platforms, integrating innovative genomic, phenomic, and data-driven tools into breeding programs, and fostering collaboration across regions and disciplines. Given climate uncertainty and evolving pathogen threats, a continuous commitment to integrated, durable disease management strategies is crucial to sustain wheat production and ensure global food security.

## References


Ali S, Rodriguez-Algaba J, Thach T et al (2017) Yellow rust epidemics worldwide were caused by pathogen races from divergent genetic lineages. Front Plant Sci 8:1057. https://doi.org/10.3389/fpls.2017.01057

Bebber DP, Holmes T, Smith D et al (2014) Economic and physical determinants of the global distributions of crop pests and pathogens. New Phytol 202(3):901–910. https://doi.org/10.1111/nph.12722

Bebber DP, Castillo ÁD, Gurr SJ (2016) Modelling coffee leaf rust risk in Colombia with climate reanalysis data. Philos Trans R Soc Lond Ser B Biol Sci 371(1709):20150458. https://doi.org/10.1098/rstb.2015.0458

Beddow JM, Pardey PG, Chai Y et al (2015) Research investment implications of shifts in the global geography of wheat stripe rust. Nat Plants 1:15132. https://doi.org/10.1038/nplants.2015.132

Bishnoi SK, He X, Phuke RM et al (2020) Karnal bunt: a re-emerging old foe of wheat. Front Plant Sci 11:569057. https://doi.org/10.3389/fpls.2020.569057

Ceresini PC, Castroagudín VL, Rodrigues FÁ et al (2018) Wheat blast: past, present, and future. Annu Rev Phytopathol 56:427–456. https://doi.org/10.1146/annurev-phyto-080417-050036

Chaloner TM, Gurr SJ, Bebber DP (2021) Plant pathogen infection risk tracks global crop yields under climate change. Nat Clim Chang 11:710–715. https://doi.org/10.1038/s41558-021-01104-8

Chen X (2020) Pathogens which threaten food security: *Puccinia striiformis*, the wheat stripe rust pathogen. Food Sec 12:239–251. https://doi.org/10.1007/s12571-020-01016-z

Chen W, Yao L, Zeng Q et al (2025) Spatiotemporal relationship of wheat stripe rust occurrence revealing the key inoculum origin area and migration route of *Puccinia striiformis* f. sp. *tritici* in the Yunnan–Guizhou epidemiological region. Phytopathol Res 7:73. https://doi.org/10.1186/s42483-025-00347-z

Cruz CD, Valent B (2017) Wheat blast disease: danger on the move. Trop Plant Pathol 42:210–222. https://doi.org/10.1007/s40858-017-0159-z

Ellis JG, Lagudah ES, Spielmeyer W et al (2014) The past, present and future of breeding rust resistant wheat. Front Plant Sci 5:641. https://doi.org/10.3389/fpls.2014.00641
Escamilla DM, Li D, Negus KL et al (2025) Genomic selection: essence, applications, and prospects. Plant Genome 18(2):e70053. https://doi.org/10.1002/tpg2.70053
Fones H, Gurr S (2015) The impact of *Septoria tritici* blotch disease on wheat: an EU perspective. Fungal Genet Biol 79:3–7. https://doi.org/10.1016/j.fgb.2015.04.004
Friesen TL, Stukenbrock EH, Liu Z et al (2006) Emergence of a new disease as a result of interspecific virulence gene transfer. Nat Genet 38(8):953–956. https://doi.org/10.1038/ng1839
Garima S, Satish K, Singh G et al (2020) Karnal bunt of wheat: etiology, breeding and integrated management. Crop Prot 139:105376. https://doi.org/10.1016/j.cropro.2020.105376
Ghaffary SMT, Chawade A, Singh PK (2018) Practical breeding strategies to improve resistance to Septoria tritici blotch of wheat. Euphytica 214:122. https://doi.org/10.1007/s10681-018-2205-4
Gupta PK, Chand R, Vasistha NK et al (2018) Spot blotch disease of wheat: the current status of research on genetics and breeding. Plant Pathol 67(3):508–531. https://doi.org/10.1111/ppa.12781
Hailemariam BN, Kidane Y, Ayalew A (2020) Epidemiological factors of Septoria tritici blotch (*Zymoseptoria tritici*) in durum wheat (*Triticum turgidum*) in the highlands of Wollo, Ethiopia. Ecol Process 9:61. https://doi.org/10.1186/s13717-020-00258-1
Hawkins NJ, Cools HJ, Sierotzki H et al (2014) Paralog re-emergence: a novel, historically contingent mechanism in the evolution of antimicrobial resistance. Mol Biol Evol 31(7):1793–1802. https://doi.org/10.1093/molbev/msu134
He X, Kabir MR, Roy KK et al (2020) QTL mapping for field resistance to wheat blast in the Caninde#1/Alondra population. Theor Appl Genet 133(9):2673–2683. https://doi.org/10.1007/s00122-020-03624-x
Hubbard A, Lewis CM, Yoshida K et al (2015) Field pathogenomics reveals the emergence of a diverse wheat yellow rust population. Genome Biol 16:23. https://doi.org/10.1186/s13059-015-0590-8
Kolmer JA (2005) Tracking wheat rust on a continental scale. Curr Opin Plant Biol 8(4):441–449. https://doi.org/10.1016/j.pbi.2005.05.001
Krattinger SG, Lagudah ES, Spielmeyer W et al (2009) A putative ABC transporter confers durable resistance to multiple fungal pathogens in wheat. Science 323(5919):1360–1363. https://doi.org/10.1126/science.1166453
Lamari L, Bernier CC (1991) Genetics of tan necrosis and extensive chlorosis in tan spot of wheat caused by *Pyrenophora tritici-repentis*. Phytopathology 81:1092–1095. https://doi.org/10.1094/Phyto-81-1092
Lewis CM, Morier-Gxoyiya C, Hubbard A et al (2024) Resurgence of wheat stem rust infections in western Europe: causes and how to curtail them. New Phytol 243(2):537–542. https://doi.org/10.1111/nph.19864
Lucas JA, Hawkins NJ, Fraaije BA (2015) The evolution of fungicide resistance. Adv Appl Microbiol 90:29–92. https://doi.org/10.1016/bs.aambs.2014.09.001
Mahlein AK (2016) Plant disease detection by imaging sensors – parallels and specific demands for precision agriculture and plant phenotyping. Plant Dis 100(2):241–251. https://doi.org/10.1094/PDIS-03-15-0340-FE
Milus EA, Kristensen K, Hovmøller MS (2009) Evidence for increased aggressiveness in a recent widespread strain of *Puccinia striiformis* f. sp. *tritici* causing stripe rust of wheat. Phytopathology 99(1):89–94. https://doi.org/10.1094/PHYTO-99-1-0089
Oliver RP (2014) A reassessment of the risk of rust fungi developing resistance to fungicides. Pest Manag Sci 70(11):1641–1645. https://doi.org/10.1002/ps.3767
Poland J, Endelman J, Dawson J et al (2012) Genomic selection in wheat breeding using genotyping-by-sequencing. Plant Genome 5:103–113. https://doi.org/10.3835/plantgenome2012.06.0006
Rutkoski J, Poland J, Singh RP et al (2014) Genomic selection for quantitative adult plant resistance to stem rust in wheat. Plant Genome 7:1–10. https://doi.org/10.3835/plantgenome2014.02.0006
Savary S, Bregaglio S, Willocquet L et al (2017) Crop health and its global impacts on the components of food security. Food Secur 9:10. https://doi.org/10.1007/s12571-017-0659-1

Savary S, Willocquet L, Pethybridge SJ et al (2019) The global burden of pathogens and pests on major food crops. Nat Ecol Evol 3(3):430–439. https://doi.org/10.1038/s41559-018-0793-y

Shahinnia F, Geyer M, Schürmann F et al (2022) Genome-wide association study and genomic prediction of resistance to stripe rust in current Central and Northern European winter wheat germplasm. Theor Appl Genet 135(10):3583–3595. https://doi.org/10.1007/s00122-022-04202-z

Singh RP, Hodson DP, Jin Y et al (2015) Emergence and spread of new races of wheat stem rust fungus: continued threat to food security and prospects of genetic control. Phytopathology 105(7):872–884. https://doi.org/10.1094/PHYTO-01-15-0030-FI

Singh PK, Gahtyari NC, Roy C et al (2021) Wheat blast: a disease spreading by intercontinental jumps and its management strategies. Front Plant Sci 12:710707. https://doi.org/10.3389/fpls.2021.710707

Tembo B, Mulenga RM, Sichilima S et al (2020) Detection and characterization of fungus (*Magnaporthe oryzae* pathotype *Triticum*) causing wheat blast disease on rain-fed grown wheat (*Triticum aestivum* L.) in Zambia. PLoS One 15(9):e0238724. https://doi.org/10.1371/journal.pone.0238724

Voss-Fels KP, Cooper M, Hayes BJ (2019) Accelerating crop genetic gains with genomic selection. Theor Appl Genet 132:669–686. https://doi.org/10.1007/s00122-018-3270-8

Yavuzaslanoglu E, Hekimhan H, Nicol J et al (2024) Plant parasitic nematodes of wheat. In: Zencirci N, Altay F, Baloch FS et al (eds) Advances in wheat breeding. Springer, Singapore. https://doi.org/10.1007/978-981-99-9478-6_7

Zhang LY, Meng QF, Yan HF et al (2019) Virulence and molecular genetic diversity of the *Puccinia triticina* population in Hebei Province of China in 2008 and 2010. Eur J Plant Pathol 153:153–166. https://doi.org/10.1007/s10658-018-1549-x

# Strategies for Effective Management of Root Diseases in Wheat

# 5

Gurudatt M. Hegde, Meghasree Devaiah, and Priyanka Jadhav

**Abstract**

Wheat root diseases remain a persistent challenge to global wheat productivity, threatening both yield stability and food security. Among the most damaging pathogens are *Rhizoctonia solani* and *Bipolaris sorokiniana*, which cause spot blotch and common root rot; *Gaeumannomyces graminis* var. *tritici*, the agent of take-all; *Fusarium pseudograminearum* and *F. culmorum*, responsible for Fusarium crown rot; and several *Pythium* species. Wheat foot rot, caused by *Sclerotium rolfsii*, further diminishes productivity by causing seedling mortality before and after emergence and can lead to 25–50% yield loss through severe seedling infection. These soilborne pathogens attack root and crown tissues, resulting in necrosis, impaired nutrient uptake, and overall plant decline. The development of resistant cultivars has been accelerated by advances in CRISPR-based genome editing, quantitative trait mapping, and marker-assisted selection, while host-induced gene silencing (HIGS), RNA interference (RNAi), and microbial consortia have strengthened integrated disease management. Emerging technologies such as remote sensing, high-throughput phenotyping, and AI-driven advisory platforms enable early detection, precise management, and farmer-centered decision support. Integrating genomics, soil health enhancement, and digital monitoring within an ecosystem-based approach will be crucial for sustained progress, and enhanced collaboration among researchers, policymakers, and farmers is essential to achieving resilient, sustainable wheat production worldwide.

**Keywords**

Wheat root diseases · Foot rot · Take all · Fusarium crown rot · *Rhizoctonia solani* · Molecular diagnostics · Integrated disease management

G. M. Hegde (✉) · M. Devaiah · P. Jadhav
University of Agricultural Sciences, Dharwad, Karnataka, India
e-mail: hegdegm@uasd.in



R. Tiwari et al. (eds.), *Wheat Stress Management*,
https://doi.org/10.1007/978-981-92-0902-6_5

## 5.1 Introduction

Wheat root diseases pose a major challenge to global wheat production, significantly reducing yields and threatening food security. The most damaging root diseases include common root rot caused by *Bipolaris sorokiniana*; Fusarium crown rot mainly caused by *Fusarium pseudograminearum* and *F. culmorum*; Pythium root rot caused by various *Pythium* species; rhizoctonia root rot resulting from *Rhizoctonia solani*; and take-all disease caused by *Gaeumannomyces graminis* (Zhou et al. 2025; Pu et al. 2025). These soilborne pathogens damage crown and root tissues, cause necrosis, disrupt nutrient transport, and weaken plants under environmental stress. Foot rot of wheat caused by *Sclerotium rolfsii* Sacc. has gained increasing attention due to pathogen variability across regions, especially in rainfed areas, posing a serious problem. *Sclerotium rolfsii*—a widespread, highly destructive soilborne fungus—was first described by Rolfs (1892) on tomato in Florida and later named by Saccardo (1911), with its presence in India reported by Shaw (2018). Economically important in tropical and subtropical regions, it infects over 500 plant species and significantly limits wheat production in rainfed areas of Madhya Pradesh, Karnataka, and Gujarat. In major wheat-growing regions, root diseases cause substantial yield losses, with Fusarium crown rot alone reducing annual yields by 6–10% in Australia (over AUD $400 million), and localized losses reaching 35% during severe epidemics (Murray and Brennan 2010; Pu et al. 2025). Similar impacts are observed in the Pacific Northwest of the United States, where *Fusarium* infections cause whiteheads and grain shriveling (Smiley et al. 2005). Worldwide, wheat diseases reduce yields by approximately 7.7%, resulting in multi-billion-dollar economic losses (Friskop et al. 2025).

Effective management of wheat root diseases is crucial not only for maintaining yields but also for improving soil health and ensuring production stability. Integrated disease management—combining resistant cultivars, crop rotation, biological control agents, targeted fungicide applications, and enhanced cultural practices—has proven effective in reducing pathogen inoculum and disease pressure (Zhou et al. 2025; Chaudhary and Pujari 2021). Advances in molecular breeding, including marker-assisted selection and genomic tools for developing resistant varieties, along with remote-sensing-based early detection and precision agriculture, further strengthen modern management strategies (Pu et al. 2025; Murray and Brennan 2010). Gaining a deeper understanding of the interactions among wheat, root pathogens, and soil microbiome dynamics remains essential for designing durable and economically viable disease-control options. This chapter reviews recent progress in managing wheat root diseases, emphasizing integrated management approaches, molecular breeding for resistance, and technological innovations, while highlighting advances within India in the context of global challenges such as climate change and emerging pathogen dynamics. The major wheat root diseases, associated pathogens, yield impact, affected regions, and management focus are indicated in Table 5.1.

**Table 5.1** Major wheat root diseases, associated pathogens, yield impact, affected regions, and management focus

| Disease | Primary pathogen(s) | Yield loss (per cent) | Major affected regions | Primary management focus | References |
|---|---|---|---|---|---|
| Fusarium crown rot (FCR) | *Fusarium pseudograminearum*, *F. culmorum* | 6–10 (up to 35 during epidemics) | Australia, USA, India, China | Use of resistant cultivars, crop rotation, and residue management | Murray and Brennan (2010) |
| Common root rot (CRR) | *Bipolaris sorokiniana* | 5–15 | South Asia, North America, Africa | Seed treatment, resistant cultivars, and biological control | |
| Take-all | *Gaeumannomyces graminis* var. *tritici* | 5–30 | Europe, Australia, New Zealand | Crop rotation, soil health management, and biological control (*Pseudomonas fluorescens*) | Thomashow and Weller (1990) and Cook (2000) |
| Rhizoctonia root rot | *Rhizoctonia solani* (AG-8, AG-2) | 5–12 | USA (Pacific Northwest), India, Australia | Soil health improvement, reduced tillage, biocontrol with *Trichoderma* spp. | Pu et al. (2025) |
| Foot rot | *Sclerotium rolfsii* | 10–30 | India | Soil and seed treatment with biocontrol agents and microbial consortia | Priyanka et al. (2022) |
| Pythium root rot | *Pythium* spp. complex | 3–8 | Temperate wheat-growing regions, irrigated areas | Proper drainage, fungicidal seed treatment, balanced fertilization | Zhou et al. (2025) |
| Spot blotch (root–foliar complex) | *Bipolaris sorokiniana* (dual infection: roots + foliage) | 15–25 (up to 40 in humid regions) | South Asia, East Africa, and parts of Latin America | Resistant genotypes, integrated nutrient and residue management | Bartoš et al. (2018) and Buster et al. (2023) |

## 5.2 Historical Perspectives and Contemporary Understanding of Wheat Root and Foot Rot Diseases

The recognition and management of wheat root diseases have a long history intertwined with the development of agricultural science. Documentation from the late nineteenth century indicates that root rot symptoms in wheat were initially attributed to environmental stress, before soilborne pathogens were identified as the cause. Indigenous agricultural communities recognized the effects of root diseases through empirical observation and developed practical management strategies, such as crop rotation and natural seed treatments, to minimize disease buildup (Mathad et al. 2013; Kumar et al. 2009). In the late 1800s and early 1900s, pioneering plant pathologists began systematic investigations to understand disease etiology. Henry L. Bolley's work in North Dakota laid the foundation by isolating fungal pathogens responsible for root rots and establishing their primary role in yield decline (Figueroa et al. 2018). Around the same time, Lawrence R. Waldron started breeding programs to develop disease-resistant wheat varieties, marking an early shift toward genetic solutions (Waldron and Clark 1936). Take-all disease, first observed in Australia in the 1850s, became one of the earliest root diseases to attract extensive scientific attention. The identification of *Gaeumannomyces graminis* as its causal agent and the discovery of soil suppressiveness, mainly driven by antibiotic-producing *Pseudomonas fluorescens*, provided key insights that shaped biological control strategies (Thomashow and Weller 1990; Cook 2000). During the Green Revolution, the widespread adoption of high-yielding cultivars and intensive inputs increased yields but also revealed vulnerabilities in root system health, as many cultivars lacked strong resistance to root diseases (Bartoš et al. 2018). Extensive monoculture and fertilizer use occasionally worsened disease problems, prompting renewed focus on breeding for root disease resistance and integrating cultural, chemical, and biological management approaches. Advances in soil health management, including maintaining organic matter, balanced fertilization, and appropriate tillage, have since become critical in suppressing soilborne pathogens (Janvier et al. 2007). Modern surveillance systems, which incorporate remote sensing, climate-based forecasting, and pathogen monitoring, now enable timely interventions and mark the transition from traditional management to multifaceted, science-driven strategies (Pu et al. 2025).

Foot rot caused by *Sclerotium rolfsii* is a destructive soilborne disease that primarily attacks the stem near the soil line, although other plant parts may be affected under favorable conditions. The pathogen produces a rapidly spreading white, fluffy mycelial mat and characteristic round, tan- to brown-colored sclerotia. Infected plants show stem girdling, leaf chlorosis, wilting, loss of vigor, and premature death and can often be easily pulled from the soil, revealing white mycelial growth around the crown and roots. The causal fungus, *Athelia rolfsii* (anamorph: *Sclerotium rolfsii*), is a basidiomycete that mainly exists as mycelium and sclerotia. It attacks crown and stem tissues by producing oxalic acid and a range of pectinolytic and cellulolytic enzymes that kill host cells and facilitate colonization. Once established, the pathogen rapidly spreads through plant tissues, forming abundant mycelium and sclerotia, particularly under warm, humid conditions. Susceptible plants develop basal stem lesions, leading to wilting and eventual death.

## 5.3 Scientific Advancements in Wheat Root Disease Management Over the Past Centuries

The management of wheat root diseases has advanced significantly through various scientific developments, beginning with classical breeding techniques aimed at increasing resistance to soilborne pathogens. Early efforts relied on phenotypic selection within diverse wheat germplasm, with landraces and wild relatives such as *Thinopyrum ponticum* and *Secale cereale* serving as important sources of resistance genes (Su et al. 2021; FAO 2020). Wide hybridization and backcrossing enabled the transfer of valuable loci into cultivated wheat, while conventional breeding programs incorporated major genes such as *Sb1* and *Lr34*, leading to improved field performance under high disease pressure (Krattinger et al. 2009; Zhang et al. 2024). The identification of QTLs offering partial resistance through quantitative genetics and pedigree mapping in the late twentieth and early twenty-first centuries greatly contributed to broad-spectrum, durable resistance (Su et al. 2021; Palma-Guerrero et al. 2021). Progress in understanding pathogen biology, plant immunity, rhizosphere ecology, and epidemiology clarified infection mechanisms and informed intervention timing (Gayan and Timothy 2021). Cultural practices such as crop rotation with nonhosts, organic amendments, and conservation agriculture are vital for disrupting pathogen life cycles, increasing microbial diversity, and reducing disease (Jayaraman et al. 2021; Janvier et al. 2007). Meanwhile, advances in seed treatment formulations and fungicide chemistry have enabled control of soilborne pathogens such as sharp eyespot and Fusarium crown rot, with systemic fungicides and combination treatments providing early-season protection and lowering inoculum carryover (Yang et al. 2022). The use of disease forecasting models based on weather data, crop phenology, and GIS has further improved risk prediction and decision-making, promoting timely interventions and sustainable management (Audsley et al. 2005; Savary et al. 2015; Pu et al. 2025). Overall, these developments have created an integrated disease management approach that combines ecological principles, host resistance, chemical and biological controls, and predictive modeling to protect wheat productivity and ensure agricultural resilience.

Extensive research on wheat foot rot caused by *Sclerotium rolfsii* has refined chemical, biological, and integrated management strategies. Under unsterilized soil conditions, fungicides such as Tritisan, Brassicol, and Thiram improved seedling vigor and reduced pre- and post-emergence rot, while Captan and Ceresan wet-minimized seed and seedling mortality. Benodanil drenching and seed treatment with Campogran M were highly effective under controlled conditions, although calixin caused phytotoxicity when used as a seed dressing. Field trials identified Panoram, Brassicol (quintozene), Panoctine-35 (guazatine), Vitavax (carboxin), and Plantavax (oxycarboxin) as effective options, with seed treatments providing up to 35 days of protection even in heavily infested soils (Kalappanavar 2018). Tribunil completely inhibited *Sclerotium rolfsii* across all tested concentrations, whereas 2,4-D showed variable inhibition (Nargund et al. 1983). Biological control with antagonists such as *Trichoderma harzianum*, *Bacillus subtilis*, and *Pseudomonas fluorescens* reduced disease incidence; *Trichoderma harzianum* and *Bacillus subtilis* conferred high seedling survival, whereas *Pseudomonas fluorescens* was less effective. Soil application of Tricho-compost further improved seedling survival and plant growth. Integrated field management demonstrated that combining effective fungicides (carboxin 37.5% + thiram 37.5%), *Trichoderma harzianum*, crude pongamia oil, and indigenous knowledge, such as cow urine, drastically reduced both pre- and post-emergence incidence. Recent studies highlight the superiority of microbial consortia over single inoculants, with combinations involving *Trichoderma harzianum*, *Pseudomonas fluorescens*, *Bacillus subtilis*, and *Neofusicoccum parvum* showing up to 80% inhibition of *Sclerotium rolfsii*. However, the standard chemical check (carboxin 37.5% + thiram 37.5%) still achieved complete inhibition.

## 5.4 Resistant Wheat Sources for *Sclerotium rolfsii*

Nargund et al. (1983) screened 25 wheat genotypes, including varieties and multilines, and identified Lerma Roja as the most resistant and Agra local as highly susceptible to foot rot. Subsequent evaluations revealed that only a few varieties—WL 711, HI 977, K 8705, HP 1660, HUW 342, HD 2189, HS 240, NP 200, UP 2565, HS 420, PBW 343, RAJ 3765, and PBW 373—exhibited marked tolerance to the disease. Similarly, Mishra et al. (1995), in a pot experiment conducted at Sagar, India (1988–1991), found that AKW-619, HD 2270, K 8027, UP 262, and HI 1077 showed no disease symptoms, whereas lines such as PBN-51, PBW 175, K 8006, K 8853, Raj 3232; DWR 163, and HD 2402 were categorized as resistant (0–10% disease incidence), with the remaining genotypes classified as moderately resistant to highly susceptible. Further screening by Harlapur et al. (1993) of forty-six cultivars in artificially infested soil at New Delhi reported no immune lines. However, CPAN 2099 exhibited the highest resistance, with 73% seedling survival, and several cultivars—including DWR 158, HD 2189, DWR 137, DWR 152, WH 416, and Sonalika—recorded survival rates exceeding 60%. Complementing these phenotypic evaluations, Uday et al. (2021) conducted a molecular diversity analysis of *Sclerotium rolfsii* isolates from Karnataka using the ITS5/ITS4 primer pair to amplify the ITS region, which showed high homology with *Athelia rolfsii* sequences reported from other crops; notably, no ITS sequences had previously been reported from wheat. Phylogenetic analysis of these isolates revealed significant variability, grouping them into two distinct clades.

## 5.5 Advances and Technological Innovations in Modern Wheat Root Disease Management

Substantial progress in wheat root disease management has been achieved in recent years, driven by breakthroughs in molecular breeding, biotechnology, and digital agriculture. Limitations of traditional phenotypic selection—particularly its environmental dependence and slow pace—have been mitigated through marker-assisted selection (MAS) and genomics-enabled breeding, which allow efficient identification and introgression of resistance loci into elite cultivars. Molecular markers such as SNPs, SSRs, and DArTs have facilitated the mapping and pyramiding of quantitative trait loci (QTLs) that confer partial but durable resistance to complex soilborne pathogens, including Fusarium crown rot, take-all, and spot blotch (Gupta and Mir 2010; Miedaner and Korzun 2012; Collard and Mackill 2008; Cowger et al. 2022; Palma-Guerrero et al. 2021). Genome-wide association studies (GWASs) and fine-mapping have accelerated the discovery of novel sources of resistance. At the same time, genome editing technologies—particularly CRISPR/Cas9—have enabled functional validation of resistance (R) and susceptibility (S) genes and expanded the potential for engineering broad-spectrum, durable resistance (Lyzenga and Kagale 2021; Wang et al. 2025). Complementing genetic innovations, microbial biocontrol agents such as *Trichoderma*, *Pseudomonas*, and *Bacillus* species have emerged as vital tools for sustainable disease suppression through mycoparasitism, competitive exclusion, antifungal metabolite production, and induced systemic resistance (Meena et al. 2017; Jayaraman et al. 2021; Krishnappa et al. 2023). Microbial consortia, often combined with organic amendments or micronutrients, have demonstrated considerable success in managing *Fusarium*, *Rhizoctonia*, and take-all across field conditions. Eco-friendly approaches such as RNAi and HIGS further expand the toolkit for disrupting essential pathogen genes (Wang et al. 2020; Krishnappa et al. 2023). Together, these innovations reinforce integrated disease management (IDM) strategies that synergize genetic resistance, crop rotation, precision chemical application, cultural practices, biocontrol, and data-driven diagnostics. However, climate change introduces new complexities: rising temperatures, altered precipitation patterns, and shifts in soil microbiomes are increasing the aggressiveness and distribution of major root pathogens, underscoring the need for more resilient and adaptable management approaches (Singh et al. 2023).

Parallel advances in digital and high-tech agriculture have transformed the detection, monitoring, and management of wheat root diseases. Remote sensing technologies—including drone-based hyperspectral imaging, UAV-mounted multispectral sensors, and satellite platforms—now enable large-scale surveillance and early detection, with hyperspectral cameras capable of accurately identifying infections before visible symptoms emerge (Guo et al. 2021; Nigam et al. 2019). Soil health sensors integrated into digital platforms provide real-time data on pathogen load, moisture levels, and microbial activity, enabling rapid diagnosis and intervention. Gene editing, especially CRISPR–Cas systems, has accelerated targeted disruption of susceptibility genes, insertion of broad-spectrum resistance genes, and modification of regulatory pathways for more durable defenses, thereby overcoming the rapid evolution of pathogen populations and the breakdown of single-gene resistance (Greenwood et al. 2023; Waites et al. 2025). High-throughput phenotyping (HTP) using robotics, multispectral imaging, and advanced analytics enables precise evaluation of disease resistance and stress responses across thousands of genotypes, providing robust datasets for dissecting quantitative resistance traits (Bakhshi et al. 2025). Genomic selection and predictive breeding models further improve breeding efficiency by estimating genomic breeding values early in the selection cycle, accelerating the development of varieties resistant to *Fusarium*, *Rhizoctonia*, and take-all (Robertsen et al. 2019; Daetwyler et al. 2014). Finally, digital extension services and mobile-based advisory systems—such as DEWAS—integrate weather data, disease models, and field observations to deliver precise, location-specific recommendations, thereby democratizing access to advanced diagnostics and promoting timely interventions, especially for smallholder farmers (Islam et al. 2023; Islam et al. 2025). Collectively, AI-driven diagnostics, remote sensing, genomic innovations, CRISPR-based editing, HTP, and digital advisory ecosystems are fundamentally reshaping wheat root disease management, thereby boosting precision, sustainability, and resilience amid changing climatic pressures and rapidly evolving pathogen populations.

## 5.6 Statistical Insights on Wheat Root Disease Incidence and Management in India

Wheat remains a critical staple for India, contributing significantly to national food security and rural livelihoods (Ministry of Agriculture and Farmers Welfare; GoI 2025). With a projected production of nearly 114 million metric tons, wheat was cultivated on approximately 31 million hectares in 2024–2025; however, root and crown diseases have increasingly challenged production across several key agroclimatic zones (Sharma et al. 2025). Recent surveillance across India's major wheat-growing regions—including the Indo-Gangetic Plains, central, western, and peninsular zones—reveals considerable regional variation in disease incidence and severity (Sharma et al. 2025). In Punjab and Haryana, historically vulnerable to rust outbreaks, effective management in 2024–2025 limited stripe and yellow rust to localized areas through the widespread adoption of resistant cultivars (PBW 343, DBW 303) and timely fungicide applications. Minor outbreaks of brown rust and leaf blight occurred in Uttar Pradesh and Bihar but remained restricted due to strict extension advisories and the use of resistant cultivars. In central and western India, including Madhya Pradesh and Maharashtra, sporadic cases of spot blotch, Fusarium head blight, and leaf rust were reported—often associated with specific susceptible varieties or inadequate crop rotation (Sharma et al. 2025). While stem rust and take-all remained rare nationwide, Rhizoctonia and Fusarium crown rot emerged more frequently in late-sown, irrigated wheat or fields following rice cultivation.

Across agroclimatic zones, yield losses from major root diseases such as spot blotch, Fusarium crown rot, and common root rot typically range from 5% to 15%. However, they can escalate to 25–35% in epidemic-prone districts with susceptible cultivars and conducive weather (Buster et al. 2023). Economic assessments estimate that annual losses from wheat root and crown diseases across the country exceed ₹3500 crore (~US$425 million), averaged over regions and years. Government interventions—through subsidies, policy support, and targeted extension—have significantly influenced disease outcomes. State governments promote the adoption of resistant cultivars and integrated disease management in hotspot districts, while the National Food Security Mission (NFSM) provides seed subsidies and support for agrochemicals. Impact evaluations show that these programs considerably reduce disease spread and yield losses (Ministry of Agriculture and Farmers Welfare; GoI 2025). Enhanced surveillance and increased farmer awareness, facilitated by digital platforms and mobile advisories, have improved early detection and on-farm responses (Sharma et al. 2025). Despite emerging pathogen strains and challenges from climate variability, India has maintained sustained success in managing wheat root diseases, as evidenced by consistent production gains and a decline in major outbreaks. Continued progress will depend on farmer participation, strategic rotation of resistant cultivars, and strengthened surveillance systems.

## 5.7 Future Projections and Emerging Global Challenges in Wheat Root Disease Management

Climate change presents significant risks to wheat root disease management worldwide, as rising temperatures, more variable precipitation, and changing soil microbial communities increase the frequency and severity of major root pathogens (Prasad et al. 2021). Predictive models forecast that by mid-century, disease caused by the rapidly evolving *Magnaporthe oryzae-Triticum* strain could spread from 6.4 to 13.5 million hectares globally, and under extreme emission scenarios, up to 75% of wheat-growing regions in South America and Africa may become highly susceptible (Pequeno et al. 2024; Prasad et al. 2021). Similar patterns are anticipated for other root pathogens—including *Fusarium* spp., *Rhizoctonia solani*, and take-all—as warming and shifting rainfall patterns boost pathogen virulence, prolong infection periods, and disrupt beneficial soil microbiomes that naturally inhibit disease. The emergence of increasingly virulent pathogen strains, such as wheat blast (*Zymoseptoria tritici*) and various *Fusarium* species, highlights the growing threat. Their genetic flexibility accelerates adaptation to fungicides and resistance genes, leading to rapid failure of single-gene resistance and necessitating continual updates of resistance loci in breeding programs.

Addressing these escalating challenges requires strong global collaboration, rapid-response surveillance systems, and resilient agricultural policies. Coordinated networks for pathogen monitoring, data sharing, and rapid germplasm mobilization—supported by initiatives such as the wheat initiative and the Borlaug Global Rust Initiative—are essential for early detection and mitigation of emerging root disease threats. Policy frameworks increasingly emphasize ecosystem-based and sustainable approaches, including improved crop rotation, enhanced soil health, and the deployment of multi-disease-resistant cultivars. Subsidies should incentivize integrated disease management, climate-resilient practices, and farmer-accessible forecasting platforms, while strengthened surveillance and quarantine systems are needed to curb the spread of transboundary diseases. Precision agriculture and biotechnology provide powerful tools for future management: AI-assisted sensors, satellite mapping, and dynamic advisory systems enable near-real-time outbreak detection, while

CRISPR-based gene stacking accelerates the development of durable resistance (Waites et al. 2025). Together with farmer training, equitable access to technologies, and supportive trade and policy environments, these innovations will help build a more resilient global wheat production system in the face of climate change.

## 5.8 Conclusion and Way Forward

Recent decades have seen significant progress in managing wheat root diseases, driven by advances in genomic selection, biocontrol, molecular breeding, and digital technologies for forecasting and diagnostics. Wheat varieties with strong resistance to major root pathogens such as *Fusarium*, *Rhizoctonia*, and take-all have been rapidly developed through marker-assisted selection, CRISPR gene editing, and high-throughput phenotyping. AI-driven diagnostic platforms and precision agriculture tools now support large-scale surveillance and timely interventions, reducing disease outbreaks and stabilizing yields across various agroclimatic regions. Combining traditional practices—such as crop rotation, residue management, and locally adapted cultivars—with advanced molecular and digital approaches is increasingly vital for sustainable disease control. This integration is strengthened by microbial biocontrol agents, precision agrochemical application, and resistant germplasm that perform well in field conditions. Soil health monitoring and farmer-focused mobile advisories further enable data-driven, locally relevant decision-making. However, substantial research and policy gaps remain, requiring ongoing investment in genetic resource conservation, multi-location trials, and coordinated surveillance networks, especially given complex inheritance patterns, rapid pathogen evolution, and resistance breakdowns. Strong policy support is essential for promoting integrated disease management, encouraging environmentally friendly innovations, and ensuring equitable access to improved seeds, technologies, and extension services. Successful application of scientific advances depends on multidisciplinary collaboration among breeders, pathologists, agronomists, and policymakers. Moving forward, global and national priorities should emphasize cooperation, resilience, and adaptability, with breeders stacking multiple resistance genes and root traits; pathologists and data scientists enhancing disease surveillance through machine learning and real-time data; and policymakers endorsing integrated approaches and international cooperation for rapid responses to emerging threats and fair access to benefits. Through proactive research, institutional support, and ongoing stakeholder engagement, wheat productivity and food security can be protected against current and future root disease challenges.

## References

Audsley E, Milne A, Paveley N (2005) A foliar disease model for use in wheat disease management decision support systems. Ann Appl Biol 147:161–172. https://doi.org/10.1111/j.1744-7348.2005.00023.x

Bakhshi T, Mehrabi R, Sarbarzeh MA et al (2025) Monitoring wheat leaf rust severity using machine learning techniques. Sci Rep 15:42951. https://doi.org/10.1038/s41598-025-24407-8

Bartoš P, Šíp V, Chrpová J et al (2018) Achievements and prospects of wheat breeding for disease resistance. Czech J Genet Plant Breed 38:16–28. https://doi.org/10.17221/6107-CJGPB

Buster M, Simpfendorfer S, Guppy C et al (2023) Impact of *Fusarium* crown rot on root system area and links to genetic variation within commercial wheat varieties. Agronomy 13(12):2955. https://doi.org/10.3390/agronomy13122955

Chaudhary R, Pujari M (2021) Major diseases of wheat and their management: a review. Plant Arch 21. https://doi.org/10.51470/PLANTARCHIVES.2021.v21.no2.037

Collard BC, Mackill DJ (2008) Marker-assisted selection: an approach for precision plant breeding in the twenty-first century. Philos Trans R Soc Lond Ser B Biol Sci 363(1491):557–572. https://doi.org/10.1098/rstb.2007.2170

Cook RJ (2000) Advances in plant health management in the twentieth century. Annu Rev Phytopathol 38:95–116. https://doi.org/10.1146/annurev.phyto.38.1.95

Cowger C, Bulli P, Chen F et al (2022) Editorial: advances in breeding for wheat disease resistance. Front Plant Sci 13:1034506. https://doi.org/10.3389/fpls.2022.1034506

Daetwyler HD, Bansal UK, Bariana HS et al (2014) Genomic prediction for rust resistance in diverse wheat landraces. Theor Appl Genet 127(8):1795–1803. https://doi.org/10.1007/s00122-014-2341-8

FAO (2020) Wheat disease management: a global perspective. Food and Agriculture Organization of the United Nations, Rome

Figueroa M, Hammond-Kosack KE, Solomon PS (2018) A review of wheat diseases—A field perspective. Mol Plant Pathol 19(6):1523–1536. https://doi.org/10.1111/mpp.12618

Friskop AJ, Sisson AJ, Allen TW et al (2025) Estimated yield reductions and economic losses on wheat caused by disease from 2018 through 2021. Plant Health Prog 26(3):325–343. https://doi.org/10.1094/PHP-09-24-0087-RS

Gayan K, Timothy F (2021) Understanding the plant–pathogen interaction associated with *Septoria nodorum* blotch of wheat. In: Achieving durable disease resistance in cereals. Burleigh Dodds Science Publishing. https://doi.org/10.19103/AS.2021.0092.14

Greenwood JR, Zhang X, Rathjen JP (2023) Precision genome editing of crops for improved disease resistance. Curr Biol 33(11):R650–R657. https://doi.org/10.1016/j.cub.2023.05.003

Guo A, Huang W, Dong Y et al (2021) Wheat yellow rust detection using UAV-based hyperspectral technology. Remote Sens 13:123. https://doi.org/10.3390/rs13010123

Gupta P, Mir RR (2010) Marker-assisted wheat breeding: present status and future possibilities. Mol Breed 26:145–161. https://doi.org/10.1007/s11032-009-9359-7

Harlapur SI, Srikant K, Hegde RK et al (1993) Wheat genotypes against foot rot. Curr Res Univ Agric Sci Bangalore 22(2):33–34

Islam MM, Adil MAA, Talukder MA et al (2023) DeepCrop: deep learning-based crop disease prediction with web application. J Agric Food Res 14:100764. https://doi.org/10.1016/j.jafr.2023.100764

Islam M, Azad A, Arman SE et al (2025) PlantCareNet: an advanced system to recognize plant diseases with dual-mode recommendations for prevention. Plant Methods 21(1):52. https://doi.org/10.1186/s13007-025-01366-9

Janvier C, Villeneuve F, Alabouvette C et al (2007) Soil health through soil disease suppression: which strategy from descriptors to indicators? Soil Biol Biochem 39:1–23

Jayaraman S, Naorem A, Lal R et al (2021) Disease-suppressive soils—beyond food production: a critical review. J Soil Sci Plant Nutr 21:1437–1465. https://doi.org/10.1007/s42729-021-00451-x

Kalappanavar IK (2018) Wheat diseases and their management in Karnataka: an overview. J Wheat Res 4(1):11–19

Krattinger SG, Lagudah ES, Spielmeyer W et al (2009) A putative ABC transporter confers durable resistance to multiple fungal pathogens in wheat. Science 323(5919):1360–1363. https://doi.org/10.1126/science.1166453

Krishnappa C, Kavya BS, Kumar AH et al (2023) Endophytic microbes and their role in plant health. In: Mathur P, Kapoor R, Roy S (eds) Microbial symbionts and plant health: trends

and applications for changing climate, rhizosphere biology. Springer, Singapore. https://doi.org/10.1007/978-981-99-0030-5_12
Kumar G, Chhetry N, Belbahri L (2009) Indigenous pest and disease management practices in traditional farming systems in North East India: a review. J Plant Breed Crop Sci 1:28–38
Lyzenga WJ, Kagale S (2021) Wheat improvement using genome editing technology. BioTechniques 71(6):577–579. https://doi.org/10.2144/btn-2021-0067
Mathad RC, Vasudevan SN, Mathad NC et al (2013) Traditional seed treatment and storage methods of Northeastern Region of Karnataka. Asian Agri-History 17(3):233–239
Meena M, Swapnil P, Zehra A et al (2017) Beneficial microbes for disease suppression and plant growth promotion. In: Microbial interventions in agriculture and environment. Springer. https://doi.org/10.1007/978-981-10-6593-4_16
Miedaner T, Korzun V (2012) Marker-assisted selection for disease resistance in wheat and barley breeding. Phytopathology 102(6):560–566. https://doi.org/10.1094/PHYTO-05-11-0157
Ministry of Agriculture and Farmers Welfare, Government of India (2025) Agricultural statistics at a glance 2025. Government of India, New Delhi
Mishra PC, Pathak RK, Jain SK (1995) Incidence of foot-rot disease in wheat in Sagar region. JNKVV Res J 26(2):80
Murray G, Brennan J (2010) Estimating disease losses to the Australian wheat industry. Australas Plant Pathol 39:85–96. https://doi.org/10.1071/AP09064
Nargund VB, Kulkarni S, Hegde RK (1983) Studies on foot rot of wheat caused by *Sclerotium rolfsii* Sacc. in Karnataka. II. Effect of number of irrigations on incidence of foot rot. Curr Res Univ Agric Sci Bangalore 12(7/9):76–77
Nigam R, Bhattacharya BK, Kot R et al (2019) Wheat blast detection and assessment combining ground-based hyperspectral and satellite-based multispectral data. Int Arch Photogramm Remote Sens Spatial Inf Sci XLII-3/W6:473–475
Palma-Guerrero J, Chancellor T, Canning G et al (2021) Take-all disease: new insights into an important wheat root pathogen. Trends Plant Sci 26(8):1360–1385. https://doi.org/10.1016/j.tplants.2021.02.009
Pequeno DNL, Ferreira TB, Fernandes JMC et al (2024) Production vulnerability to wheat blast disease under climate change. Nat Clim Chang 14:178–183. https://doi.org/10.1038/s41558-023-01902-2
Prasad P, Bhardwaj SC, Thakur RK et al (2021) Prospects of climate change effects on crop diseases with particular reference to wheat. J Cereal Res 13(2):118–135. https://doi.org/10.25174/2582-2675/2021/112817
Priyanka J, Hegde GM, Jahagirdar S et al (2022) Evaluation of microbial consortia against *Sclerotium rolfsii Sacc.* causing foot rot of wheat. J Farm Sci 35(02):219–222. https://doi.org/10.61475/jfm.v35i02.77
Pu L, Jin Q, Cai X et al (2025) Crown rot in wheat: pathogen biology, host responses, and management strategies. Stress Biol 5(1):52. https://doi.org/10.1007/s44154-025-00247-4
Robertsen CD, Hjortshøj RL, Janss LL (2019) Genomic selection in cereal breeding. Agronomy 9(2):95. https://doi.org/10.3390/agronomy9020095
Rolfs PH (1892) Tomato blight—some hints. Bull Fla Agric Exp Stn 18
Saccardo PA (1911) Notae mycologicae Series XIII. *Ann Mycol* 9(3):249–257
Savary S, Stetkiewicz S, Brun F et al (2015) Modelling and mapping potential epidemics of wheat diseases—examples on leaf rust and *Septoria tritici* blotch using EPIWHEAT. Eur J Plant Pathol 142:771–790. https://doi.org/10.1007/s10658-015-0650-7
Sharma P, Kashyap P, Kumar R et al (2025) Wheat crop health newsletter (February 2025). Wheat Crop Health Newsl 30(4):1–15
Shaw FJF (2018) The genus rhizoctonia in India. Forgotten Books
Singh BK, Delgado-Baquerizo M, Egidi E et al (2023) Climate change impacts on plant pathogens, food security and paths forward. Nat Rev Microbiol 21:640–656. https://doi.org/10.1038/s41579-023-00900-7

Smiley RW, Gourlie JA, Easley SA et al (2005) Pathogenicity of fungi associated with the wheat crown rot complex in Oregon and Washington. Plant Dis 89(9):949–957. https://doi.org/10.1094/PD-89-0949

Su J, Zhao J, Zhao S et al (2021) Genetics of resistance to common root rot (spot blotch), *Fusarium* crown rot, and sharp eyespot in wheat. Front Genet 12:699342. https://doi.org/10.3389/fgene.2021.699342

Thomashow LS, Weller DM (1990) Role of antibiotics and siderophores in biocontrol of take-all disease of wheat. Plant Soil 129:93–99. https://doi.org/10.1007/BF00011695

Uday G, Priyanka K, Hegde GM et al (2021) Molecular detection and characterization of *Sclerotium rolfsii* Sacc. causing wheat foot rot isolated from northern Karnataka, India. J Plant Pathol 103:991–995. https://doi.org/10.1007/s42161-021-00879-1

Waites J, Achary VMM, Syombua ED et al (2025) CRISPR-mediated genome editing of wheat for enhancing disease resistance. Front Genome Ed 7:1542487. https://doi.org/10.3389/fgeed.2025.1542487

Waldron LR, Clark JA (1936) Breeding rust-resistant spring wheats. Science 83(2144):106–108

Wang M, Wu L, Mei Y et al (2020) Host-induced gene silencing of multiple genes of *Fusarium graminearum* enhances resistance to Fusarium head blight in wheat. Plant Biotechnol J 18(12):2373–2375. https://doi.org/10.1111/pbi.13401

Wang DL, Yang PP, Wang SJ (2025) Stacking of multiple resistance genes in wheat via transgenic approaches. Triticeae Genom Genet 16(3):120–129. https://doi.org/10.5376/tgg.2025.16.0013

Yang X, Zhang Z, Yuan Y et al (2022) Control efficiency of hexaconazole–lentinan against wheat sharp eyespot and wheat crown rot and associated effects on rhizosphere soil fungal community. Front Microbiol 13:1014969. https://doi.org/10.3389/fmicb.2022.1014969

Zhang Y, Chen G, Zang Y et al (2024) *Lr34/Yr18/Sr57/Pm38* confers broad-spectrum resistance to fungal diseases via sinapyl alcohol transport for cell wall lignification in wheat. Plant Commun 5(12):101077. https://doi.org/10.1016/j.xplc.2024.101077

Zhou H, Cao W, Li Y et al (2025) Research progress on the analysis of resistance genes and mechanisms of wheat *Fusarium* crown rot. Front Plant Sci 16:1603842. https://doi.org/10.3389/fpls.2025.1603842

# Diagnosis and Mitigation of Wheat Physiological Disorders

6

Charu Lata, Naresh Kumar, Jayanth Kallugudi, and Gurpreet Kaur

**Abstract**

Wheat (*Triticum aestivum* L.) is the most widely cultivated cereal crop worldwide and plays a vital role in ensuring global food security. However, its productivity is increasingly threatened by physiological disorders caused by non-pathogenic factors such as salinity, heat stress, drought, frost, waterlogging, and nutrient imbalances. Unlike biotic diseases, these disorders arise from environmental extremes and metabolic imbalances, often resulting in persistent damage that is difficult to manage. Such stresses frequently impair photosynthesis, grain filling, spike development, and nutrient translocation, thereby reducing both yield and quality. Precise and early diagnosis is therefore essential to distinguish physiological disorders from biotic diseases and to implement appropriate remedial measures. In the present scenario, diagnostic approaches have progressed from traditional field observations to advanced molecular, imaging, and sensor-based techniques. Mitigation strategies now focus on integrated approaches, including precision nutrient management, development of climate-resilient varieties, application of stress-protective chemicals, improved water management, foliar nutrient supplementation, and the use of precision farming tools such as drones, remote sensing, and IoT-based devices. This chapter pro-

C. Lata (✉)
ICAR-Indian Institute of Wheat and Barley Research, Karnal, Haryana, India
e-mail: charu.sharma@icar.org.in

N. Kumar
Department of Chemistry and Biochemistry, Eternal University, Baru Sahib, HP, India

J. Kallugudi
ICAR-Indian Institute of Wheat and Barley Research, Shimla, HP, India

G. Kaur
ICAR-Central Soil Salinity Research Institute, Karnal, Haryana, India



R. Tiwari et al. (eds.), *Wheat Stress Management*,
https://doi.org/10.1007/978-981-92-0902-6_6

vides a comprehensive understanding of wheat physiological disorders, traces the transition from traditional practices to modern technological innovations, and emphasizes the importance of integrated strategies to safeguard wheat production under changing climatic conditions.

**Keywords**

Wheat · Physiological disorders · Rapid diagnosis · Mitigation · Modern strategies

## 6.1 Introduction

Wheat (*Triticum aestivum* L.) is one of the oldest and most important cereal crops worldwide, cultivated on over 220 million hectares with an annual production exceeding 799 million tons (FAOSTAT 2024). It constitutes a major source of carbohydrates, proteins, minerals, and vitamins for a large segment of the global population. In India, wheat occupied about 31 million hectares during 2023–24 and contributed more than 113 million tons to the national food basket, with the Indo-Gangetic Plains serving as the principal production zone. Wheat thus supports food security, rural livelihoods, and the agro-economy, particularly in temperate and subtropical regions (ICAR-IIWBR 2025). Despite this significance, productivity is constrained by a wide range of biotic and physiological disorders. Biotic stresses are caused by living agents such as fungal pathogens, insect pests, and nematodes, whereas physiological disorders result from non-living factors, including temperature extremes, drought, nutrient imbalances, and soil salinity. A subset of these abiotic stresses manifests as physiological disorders, which, though non-pathogenic, can severely affect plant growth, development, grain quality, and yield potential (Lata et al. 2017). The incidence and severity of such disorders are influenced by adverse environmental conditions, genetic predisposition, and intricate interactions between plant physiology and external stressors.

### 6.1.1 Definition and Overview of Physiological Disorders in Wheat

Physiological disorders are defined as non-pathogenic disruptions in normal plant development and metabolic functions that arise from unfavorable environmental conditions or genetic abnormalities rather than infections by living organisms. These disorders are noninfectious in nature and result from disturbances in the plant's normal physiological and metabolic processes. They may be induced by abiotic factors (Fig. 6.1) such as heat, drought, salinity, nutrient imbalances, moisture fluctuations, or soil toxicities (Levitt 1980).

Unlike diseases caused by pathogens, physiological disorders do not spread from plant to plant and usually affect entire fields homogeneously if the causative abiotic factor is widespread. Common examples in wheat include spikelet sterility, heat

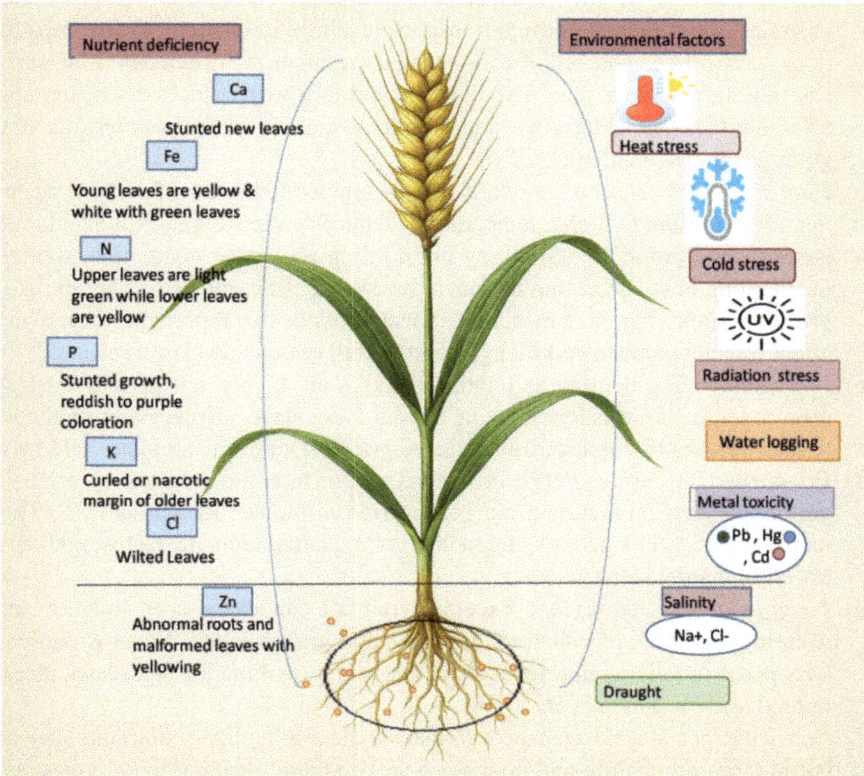


**Fig. 6.1** Schematic representation of physiological disorders in wheat and their impacts on plant growth and metabolism

stress–induced grain shriveling, pre-harvest sprouting, frost injury to developing spikes, chlorosis from micronutrient deficiencies, terminal heat stress, lodging due to excessive nitrogen application, sterility caused by temperature fluctuations during anthesis, poor grain set from micronutrient deficiencies such as zinc or copper, and premature senescence. These disorders impair fundamental physiological processes such as photosynthesis, grain development, nutrient translocation, and respiration. Because they are directly influenced by environmental conditions, their frequency and severity are expected to increase if the global surface temperature rises to approximately 1.5 °C by 2052, as projected by the IPCC (2012). In recent years, with increasing climate variability, physiological disorders have gained significant attention due to their impact on wheat productivity and food security (Atta et al. 2023). Each disorder is strongly influenced by environmental conditions, cultivar tolerance, and crop management practices, making integrated monitoring and mitigation strategies essential for stable wheat productivity. Some key examples are outlined below:

(a) *Drought-induced sterility*: Inadequate soil moisture during tillering, flowering, or grain filling impairs nutrient transport, resulting in incomplete fertilization, poor seed filling, and reduced biomass accumulation. Water deficit at early stages lowers survival rates and tillering, at mid-growth reduces dry matter and tiller numbers, and at terminal stages limits assimilate production, fertility, and grain weight (Le Gall et al. 2015).
(b) *Heat stress*: Wheat requires relatively low temperatures (14–15 °C) for ripening, and exposure to higher temperatures reduces grain weight. Prolonged high temperatures (up to 35 °C) during grain filling accelerate maturation, shorten the duration of starch accumulation, reduce kernel size, and cause shriveling of grains by impairing enzyme activity. Heat stress is also reported to negatively affect spikelet number, spike length, and overall grain yield (Tiwari et al. 2017).
(c) *Frost injury*: The booting and anthesis stages are highly sensitive to frost. A drop in temperature below freezing during these stages damages reproductive tissues, leading to floret sterility, reduced grain set, or even complete yield loss.
(d) *Pre-harvest sprouting (PHS)*: Prolonged wet or humid conditions before harvest can trigger premature grain germination in the ear, known as PHS. This increases α-amylase activity, degrading grain quality, reducing test weight, and lowering market value.
(e) *Lodging and waterlogging*: Excessive rainfall can cause crop lodging and waterlogging, both of which reduce yield and grain quality. Wheat is particularly prone to lodging during reproductive stages, making it a significant threat to productivity (Singh et al. 2017).
(f) *Nutrient deficiencies*: Imbalances or inadequate availability of nutrients such as boron (causing sterility and poor grain set) and zinc (leading to chlorosis and stunted growth) adversely affect photosynthesis, enzyme function, and reproductive development.

### 6.1.2 Distinction Between Physiological Disorders and Biotic Diseases

A critical component of effective crop management is the ability to distinguish between physiological disorders and biotic diseases. Biotic diseases, caused by living organisms such as fungi, bacteria, viruses, and nematodes, can often be managed through resistant varieties, fungicide applications, or biological control (Lata et al. 2022). In contrast, physiological disorders result strictly from non-living (abiotic) factors and require preventive and adaptive management strategies such as adjusting sowing time, optimizing irrigation schedules, improving nutrient delivery, and adopting heat- or cold-tolerant varieties. The visual symptoms of physiological disorders can sometimes be misleading and easily confused with biotic diseases. For example, leaf chlorosis caused by iron deficiency may resemble fungal leaf spot symptoms but lacks diagnostic features such as pathogen sporulation or concentric lesions. Moreover, physiological disorders often follow specific patterns linked to environmental exposure—for instance, heat stress damage is more pronounced in sun-exposed parts of the canopy. In contrast, nitrogen deficiency typically begins in older leaves and progresses upward. Conversely, biotic diseases often exhibit a random or patchy distribution in the field due to vector transmission or pathogen spread. Accurate diagnosis is therefore essential, as misdiagnosis can lead to inappropriate management decisions, resulting in economic losses and ineffective mitigation. Under global climate change scenarios, the frequency of climatic extremes is expected to increase, thereby intensifying the incidence and severity of physiological disorders in wheat (Hasanuzzaman et al. 2013). A comprehensive understanding of the causes, symptoms, and preventive measures of these disorders will enable researchers and farmers to make informed decisions, minimize yield losses, and ensure stable wheat production and food security.

### 6.1.3 Impact of Physiological Disorders on Growth, Grain Quality, and Yield

All growth stages of wheat are highly vulnerable to physiological disorders, which can reduce tillering, impair spike development, lower grain weight, and ultimately decrease crop yield. For example, zinc or boron deficiencies may cause deformed spikes, floret sterility, and poor grain set. Similarly, terminal heat stress during grain filling reduces starch accumulation and accelerates leaf senescence, leading to yield losses of up to 30–50% (Talukder et al. 2014). Wheat quality is also significantly affected by physiological disorders. For instance, waterlogging caused by poor drainage or flooding can induce root hypoxia, reduce nutrient uptake, and cause nitrogen leaching, ultimately resulting in pale leaves, reduced photosynthesis, and lower grain protein levels (Setter and Waters 2003). Temperature fluctuations during reproductive growth stages further compromise grain quality, as high temperatures during grain development reduce gluten strength and protein content, thereby affecting flour quality for baking and processing (Reynolds and Ortiz 2010). India, along with other major wheat-growing nations, faces increasingly unpredictable rainfall patterns and temperature extremes, making physiological disorders an escalating challenge to achieving stable and high yields, even under well-managed cultivation.

### 6.1.4 Importance of Accurate Diagnosis and Timely Mitigation in Sustaining Wheat Production

Sustainable wheat production under rising environmental stresses caused by climate change requires the accurate identification and timely management of physiological disorders. Early diagnosis enables the implementation of corrective measures such as using heat- or drought-tolerant varieties, applying balanced fertilization based on soil testing, adjusting sowing dates to avoid terminal heat, applying foliar nutrients during critical growth stages, and improving field drainage and irrigation scheduling (Barbedo 2025). Advanced technologies, including machine learning models, spectral imaging, remote sensing, and chlorophyll meters, provide non-destructive, real-time data, greatly enhancing the ability to monitor stress indicators in wheat fields. Future efforts must focus on climate-resilient agronomic practices, integration of genomic tools, and development of early warning systems to safeguard wheat crops against the increasing threat of physiological stress. Ensuring sustainable wheat production underscores the importance of systematic approaches for early diagnosis of physiological disorders and building capacity to implement cost-effective mitigation practices (Sivakumar et al. 2023).

## 6.2 Historical Perspectives on Diagnosis and Mitigation of Wheat Physiological Disorders

Understanding wheat disorders has a long history, evolving from early observations in farming communities to systematic scientific approaches in the modern era. Ancient farmers primarily relied on visible symptoms to assess crop health, although their interpretations were often influenced by cultural beliefs rather than scientific reasoning. Over time, these early observations, combined with empirical practices, laid the foundation for the development of plant pathology and physiology as structured disciplines.

### 6.2.1 Early Observations and Documentation of Abnormal Wheat Growth

The earliest records of wheat problems were largely descriptive, as farmers and observers noticed unusual symptoms such as yellowing, stunted growth, or shriveled grains, but could not separate environmental stress from infectious diseases. In ancient Rome, for instance, wheat rust was considered such a serious threat that a deity, Robigus, was worshipped in the hope of warding it off (Boyd 2005), illustrating how symptom recognition was present but explanations were rooted in superstition. Similarly, outbreaks of ergot during the Middle Ages caused mass poisonings, yet the fungal origin of the problem remained unknown at the time. A recurring challenge was the inability to distinguish between stresses of different origins, as chlorosis, purpling, or dwarfing of plants—symptoms we now link to nutrient deficiencies such as lack of nitrogen, phosphorus, or sulfur—would have been indistinguishable from fungal or viral infections to early farmers (Chen et al. 2018). As a result, their "diagnoses" tended to group all deviations from normal growth under the broad label of "disease," with mitigation strategies equally unspecific, often relying on ritual or generalized farming practices.

A more scientific direction began to emerge in the eighteenth century. In 1755, Tillet demonstrated that wheat smut was caused by a transmissible agent, breaking with the older idea of spontaneous generation (Murray 2024). Later, Anton de Bary provided crucial evidence that fungi such as *Phytophthora infestans* were independent organisms responsible for plant disease, rather than byproducts of unhealthy plants

**Table 6.1** The historical conflation of wheat stress symptoms

| S. No. | Historical symptom/ observation | Historical explanation | Modern diagnosis |
|---|---|---|---|
| 1. | Black, sooty heads ("smut"), stinking smut | "Corruption" of the grain | Fungal diseases (*Ustilago tritici, Tilletia tritici*) |
| 2. | Rust | Divine anger from the rust god, Robigus | Fungal disease (*Puccinia* spp.) |
| 3. | Blasting, stunted growth, shriveled grain | Spontaneous generation (autogenesis), lack of water | Heat stress, drought stress, nutrient deficiencies (e.g., N, P, K) |
| 4. | Yellowing of leaves, poor tillering | Curses, poor soil | Nutrient deficiencies (N, S, Mg), insect damage (e.g., Russian wheat aphid) |

(Kutschera and Hossfeld 2012). These findings shifted thinking away from "autogenesis" and toward germ theory, allowing researchers to separate true infections from physiological stresses like drought or nutrient deficiency (Abbas et al. 2021). From an evolutionary perspective, wheat itself contributed to its resilience, as the development of hexaploid bread wheat (*Triticum aestivum*) through natural hybridization provided a broader genetic base for stress tolerance (El Baidouri et al. 2017). Traits inherited from ancestral species, such as improved germination and better sodium retention, gave wheat a built-in capacity to cope with harsh environments, enabling practices such as dryland farming to succeed long before formal breeding programs began. Table 6.1 highlights how symptoms of stress were often misinterpreted in historical times, contrasting past explanations with modern diagnoses.

### 6.2.2 Traditional Agricultural Practices and Farmer Knowledge

Long before the rise of modern science, farmers developed strategies to reduce crop losses from environmental and nutritional stresses, and many of these practices, refined through trial and error, anticipated principles later confirmed by agronomy. In the Fertile Crescent, where wheat was first domesticated, dryland farming practices emerged to cope with moisture scarcity (Jaradat 2013). Summer fallowing was used to store soil water across seasons, planting dates were adjusted to match rainfall, and crop residues were left on the soil surface to limit evaporation. Irrigation systems in Mesopotamia and Egypt represent some of the earliest deliberate interventions against drought (Nayee 2024). Many of these methods align closely with modern water conservation principles; for instance, residue retention is now known to improve infiltration and organic matter, while experiments have shown that "dust mulching" by tillage, once thought to conserve moisture, actually accelerates water loss (Gill et al. 1977; Du et al. 2022). Crop rotation was another long-standing practice, with systems such as the Norfolk rotation—including nitrogen-fixing legumes—reflecting an early understanding of nutrient cycling (Kjaergaard 2003). By alternating crops, farmers not only restored soil fertility but also disrupted disease and pest cycles, and the

**Table 6.2** Comparison of traditional farming practices and their historical rationale with modern agronomic principles

| S. No. | Traditional practice | Historical rationale (Farmer's View) | Modern agronomic principle |
|---|---|---|---|
| 1. | Summer fallowing | "Resting" the land, allowing it to recover nutrients | Water conservation, nutrient mineralization, Pest/disease cycle disruption |
| 2. | Crop residue retention (no-till) | Keeping the soil "cool" or "moist" and preventing wind erosion | Increased water infiltration and soil organic matter |
| 3. | Roman rotation (food, feed, fallow) | A way to grow different crops and allow the land to recover | Improved soil health and fertility, nutrient cycling |
| 4. | Planting maslins (mixed grains) | To ensure a stable crop even in a bad year | Yield stability and risk mitigation, built-in climate resilience |
| 5. | Use of legumes in rotation | To make the next crop grow better without adding inputs | Biological nitrogen fixation and improved soil fertility |

"rotation effect," a consistent yield boost when crops are alternated, was recognized long before its scientific explanation. Legumes in particular reduced the risk of nitrogen deficiency in subsequent wheat crops, effectively mitigating a major physiological stress. An older but less widely remembered strategy was the use of "maslins," or mixed cereal plantings, in which wheat was often grown together with barley or other grains, ensuring that if one crop failed due to drought or another stress, the companion crop could compensate. This practice served as a buffer against climate variability and gradually adapted seed stocks to local conditions. Farmers may not have used terms such as "genetic diversity" or "resilience," but their mixed plantings achieved both. Table 6.2 synthesizes these traditional farming practices with their modern agronomic principles, highlighting the enduring wisdom of early agriculture.

### 6.2.3 Limitations of Early Knowledge

For much of history, diagnosing wheat problems was limited by the lack of scientific tools. Without microscopes or chemical analyses, early descriptions relied on visible signs such as yellowing or stunted growth, which were often non-specific. For example, a pale wheat plant could indicate nitrogen deficiency, sulfur shortage, or even molybdenum limitation (Watanabe et al. 2022). Since autogenesis—the belief that pathogens arose spontaneously—was widely accepted until the mid-nineteenth century, it was nearly impossible to distinguish non-living stressors from infections. Consequently, management strategies remained broad, relying largely on observation and tradition rather than targeted solutions.

### 6.2.4 Role of Early Researchers in Identifying Nutrient Deficiencies and Abiotic Stress

Despite these constraints, the eighteenth and nineteenth centuries saw significant progress. Researchers began to recognize patterns in symptom appearance and linked them to nutrient mobility. For example, deficiencies in mobile nutrients such as nitrogen, phosphorus, and potassium first manifested in older leaves, whereas shortages of immobile nutrients such as calcium or boron appeared in younger tissues (Zhao et al. 2024). Other characteristic signs were documented, including interveinal chlorosis in magnesium or iron deficiency and leaf purpling in phosphorus-deficient plants (Shenker and Chen 2005; Hermans and Verbruggen 2005). Simple field experiments, such as strip plots where fertilizers were applied to one area but withheld from another, helped confirm the role of specific nutrients (Kyveryga et al. 2018). These early efforts laid the groundwork for more systematic approaches to plant nutrition and stress diagnosis.

### 6.2.5 The Green Revolution and Its Impact on Physiological Disorders

The Green Revolution of the mid-twentieth century marked a turning point in wheat production. The development of semi-dwarf, high-yielding varieties was particularly significant, as it reduced the risk of lodging, which had previously limited the use of high-nitrogen inputs in traditional tall varieties (Solanki et al. 2024). With shorter, stronger stems, these new cultivars could support heavier grain loads when combined with fertilizers.

Alongside improved varieties came a broader package of technologies, including chemical fertilizers, irrigation, and pesticides, which together enabled unprecedented yield levels. However, these advances also transformed the management of physiological stresses, as crops became more dependent on precise nutrient and water supply, and productivity became directly linked to access to external inputs. While the gains were undeniable—helping to avert widespread famine in many regions—new challenges emerged. Expanding irrigation placed pressure on groundwater reserves, heavy fertilizer use altered soil health over time, and the focus on uniform, input-intensive varieties often overlooked the diverse growing conditions faced by smallholder farmers. Consequently, although the Green Revolution alleviated immediate yield constraints, it also highlighted longer-term concerns regarding sustainability, soil fertility, and water management.

## 6.3 Scientific Advancements and Principles Developed Over the Past Centuries

The transition from observation-based farming knowledge to modern plant science has been gradual but transformative. As plant physiology developed as a discipline, researchers began employing experiments and diagnostic tools to distinguish infectious diseases from environmental or nutritional problems. This shift enabled the development of more targeted strategies for managing wheat disorders.

### 6.3.1 Evolution of Plant Physiology and Its Relevance to Wheat

The study of plant function emerged as a scientific field in the eighteenth century. Stephen Hales, often called the father of plant physiology, conducted early experiments on the movement of water and gases through plants, providing the first mechanistic insights into their internal processes (Volkov and Schwenke 2021). Later, German scientists Julius von Sachs and Wilhelm Knop advanced the study of mineral nutrition. By growing plants in nutrient solutions rather than soil, they demonstrated that specific elements are essential for growth (Treitel 2020). Von Sachs's hydroponic experiments helped establish the principle that plant growth depends on distinct mineral requirements (Zhaksylykkelini 2025). These studies laid the groundwork for understanding wheat responses to stress, as recognizing that deficiencies in single nutrients could limit growth provided a basis for addressing physiological disorders through precise nutrient management.

### 6.3.2 Development of Diagnostic Methods

As plant physiology developed, diagnosis became increasingly systematic. One of the earliest insights was that symptoms varied depending on whether a nutrient was mobile or immobile. For example, deficiencies of nitrogen or potassium first appeared in older leaves, whereas calcium deficiency manifested in younger tissues. This knowledge improved field diagnosis, but more reliable methods soon followed. By the early twentieth century, soil testing had been introduced to estimate nutrient availability and guide fertilizer application (Beegle 2005). Initial tests focused on simple measures such as soil acidity or phosphorus content, but later laboratory methods could assess a wider range of nutrients along with pH and organic matter. Around the same time, tissue testing emerged as a complementary tool. Unlike soil tests, tissue analysis revealed the nutrients actually absorbed by the crop, enabling confirmation or refutation of visual symptoms. When soil and tissue test results were used together, they allowed for more precise fertilizer recommendations and helped prevent physiological stress.

### 6.3.3 Advances in Stress Physiology

Better diagnostic tools opened the door to understanding how wheat copes with environmental stress. Drought, heat, and salinity are now known to affect photosynthesis directly by limiting carbon fixation and damaging chloroplasts. These stresses also trigger the production of reactive oxygen species (ROS), which can harm membranes and proteins. To protect itself, wheat activates antioxidant enzymes such as superoxide dismutase (SOD) and catalase (CAT), which detoxify ROS before they cause damage. Other adaptive responses include osmotic adjustment, in which compounds like proline accumulate to help cells maintain turgor, and hormonal regulation. Abscisic acid (ABA), for example, plays a central role in closing stomata and coordinating stress responses throughout the plant. Understanding these mechanisms has been critical for guiding both breeding programs and crop management strategies to reduce yield losses under stress.

### 6.3.4 Fertilization, Irrigation, and Agronomic Adjustments

Scientific progress has also transformed field crop management. Liebig's "law of the minimum" remains a guiding principle: plant growth is limited by the essential nutrient in shortest supply (Goulding et al. 2008). This understanding has led to balanced fertilization strategies that supply all nutrients in the correct proportions, rather than focusing solely on nitrogen or phosphorus. Irrigation practices have likewise advanced. Instead of responding only to visible water stress, modern systems use soil moisture sensors, evapotranspiration estimates, and field monitoring to apply water more precisely. This approach reduces both drought stress and problems such as nutrient leaching or waterlogging that can result from excessive irrigation. Some traditional methods have also been reevaluated; for instance, "dust mulching" was once believed to conserve water but is now known to often increase evaporation (Du et al. 2022). In contrast, residue retention and no-till practices have been proven to conserve soil moisture, improve soil structure, and protect against heat and drought. Additional innovations include the adoption of stress-tolerant cultivars and the use of precision agriculture tools, such as NDVI sensors, for site-specific nutrient management (Bouskour et al. 2023).

## 6.4 Recent Developments in Diagnosis and Mitigation of Wheat Physiological Disorders

The management of crop disorders has traditionally relied primarily on broad-spectrum agronomic practices. However, there has been a gradual shift toward more precise and targeted interventions, driven by advancements in crop physiology, digital agriculture, and artificial intelligence (Fig. 6.2). The integration of soil nutrition science, rapid diagnostic tools, regular monitoring, and high-resolution sensor technologies has accelerated the detection and mitigation of various physiological

**Fig. 6.2** Diagnosis and mitigation strategies of wheat physiological disorders

disorders in crops. Several phenotypic traits—such as color, shape, temperature, moisture level, and stomatal characteristics—can now be monitored using sensors, enabling early identification of disorders.

Once detected, mitigation strategies can be implemented promptly using anti-transpirants, micronutrient sprays, soil management, water management, and other interventions. These strategies can be applied through both genetic and agronomic approaches, supported by machinery and AI-based tools. To manage physiological disorders effectively, it is essential to understand their causes, consequences, diagnostic techniques, and appropriate mitigation strategies.

### 6.4.1 Use of Micronutrient Sprays, Growth Regulators, and Anti-transpirants

Nearly 99% of the water taken up by plants is lost to the environment as water vapor. Plants regulate this loss through stomatal control and leaf area modulation, which is especially crucial under high temperature and drought conditions. Water conservation can be enhanced by applying anti-transpirants that close stomata, reduce water loss, and improve plant water status. These are broadly categorized as

**Table 6.3** Management of various abiotic stresses through different chemicals and nutrients, along with their modes of action

| Stress | Chemical/Nutrient | Action | References |
|---|---|---|---|
| Drought | Antitranspirants | | |
| | Pinolene (Di-1-p-menthene) | Creates a semi-permeable, physical film on the leaf surface that reduces water vapor diffusion from the plant | Blum (2017) |
| | Waxes/Silicones | Forms a thin, physical coating on the leaf epidermis, acting as a barrier to reduce transpiration | Rehman et al. (2021) |
| | Abscisic Acid (ABA) | A natural plant hormone that triggers stomatal closure by affecting guard cell turgor, reducing water loss | Davies (2010) |
| | Chitosan | A biopolymer that enhances drought tolerance by inducing stomatal closure and improving plant antioxidant defenses | Zeng and Luo (2012) |
| | Salicylic Acid (SA) | A plant signaling molecule that can regulate stomatal opening and closure, contributing to a plant's drought response | Decsi et al. (2025) |
| | Kaolin Clay | Forms a white, reflective film that lowers leaf temperature by increasing solar radiation reflection, thus reducing transpiration | Glenn et al. (2010) |
| | Calcium Carbonate ($CaCO_3$) | Applied as a fine powder, it acts as a reflective agent to reduce heat load and leaf temperature | Ahmed et al. (2011) |
| | Chlormequat Chloride (Cycocel) | Reduces internode elongation and promotes root growth, leading to a more compact plant with a better root-to-shoot ratio and improved water use efficiency | PirastehAnosheh et al. (2016) |
| | Micronutrients | | |
| | Fe | Enhances plant height, RWC, proline, and protein content | Jalilvand et al. (2014) |
| | Se | Enhances RWC, leaf area | Teimouri et al. (2014) |
| | K | Enhances plant height, spike length, and yield | Aown et al. (2012) |
| Salinity | Fe, Zn, Mn | Enhances growth, nutrient uptake | El-Fouly et al. (2011) |
| Heat | Se | Enhances yield, photosynthetic pigments, antioxidants, biomass, and antioxidants | Iqbal et al. (2015) |
| | Zn, Fe | Enhances seed yield and quality | Habib (2012) |

film-forming agents (reducing transpiration without hindering gas exchange), stomatal regulators (controlling stomatal opening and closure), and reflective compounds (increasing leaf albedo by reflecting solar radiation).

Micronutrient deficiency, often termed hidden hunger because it is not easily visible, severely affects plant growth and productivity. Since the gap between deficiency and toxicity is narrow—for example, boron sufficiency is 10–200 ppm, but toxicity may appear at 50–200 ppm—precise management is essential. Most disorders show distinct symptoms such as yellow patches (S), interveinal chlorosis (Mg, Mn, Fe), and deformed leaves (B, Zn, Mo). Timely soil or foliar application can alleviate these deficiencies; for instance, foliar Fe, Zn, and Mn improve growth and uptake (El-Fouly et al. 2011), Fe enhances height, RWC, proline, and protein under drought (Jalilvand et al. 2014), and Se increases RWC and leaf area (Teimouri et al. 2014). Table 6.3 summarizes anti-transpirants, micronutrients, and chemicals used for disorder management. Recently, nanoparticle-based micronutrient formulations have emerged as cost-effective foliar options that also minimize environmental pollution.

### 6.4.2 Use of Soil and Tissue Nutrient Testing Kits for Real-Time Diagnosis

Atomic absorption spectrophotometer (AAS), CNS analyzer, and inductively coupled plasma emission spectrometer (ICP-emission) are widely used for estimating K, Mg, Ca, Fe, Mn, Cu, Ni, Zn, and Mo, while ICP instruments can also measure P, B, and S. Colorimetric methods are primarily used for estimating phosphate, ammonium-, nitrate-, and nitrite-nitrogen, as well as B, whereas ion chromatography is employed for Cl, sulfate, nitrate-nitrogen, and nitrite-nitrogen. Although reliable, these methods are often laborious, expensive, and time-consuming. To address these limitations, several countries have developed rapid nutrient testing kits that are portable, provide quick results, are relatively inexpensive, and require minimal training to operate. In Africa, such kits are widely used, including the Garden Kit for N, P, K, and pH; Model EL for Fe, Mn, Cl, N, P, K, Ca, Mg, S, pH, and organic matter; LaMotte AST-5 for NO3-N, P, K, Ca, Mg, Cl, and Cu, along with pH and humus; and mobile labs for N, P, K, Ca, and Mg (Dimkpa et al. 2017).

### 6.4.3 Enhanced Understanding of Interactions Between Environmental Factors and Physiological Responses

Relatively little is known about genotype × environment (G × E) interactions for physiological traits in most crops, including wheat. Several studies have attempted to understand the genetics, physiology, and their interaction with environmental factors, particularly water deficits prior to anthesis (Baker 1996; Haji and Hunt 1999). High-temperature stress accelerates senescence and reduces photosynthetic efficiency in wheat, with stay-green types identified as a key adaptive mechanism (Joshi et al. 2007). Similarly, drought triggers stomatal closure and induces the accumulation of osmolytes such as proline to support survival (Verslues and Sharma 2010). Exposure to combined drought and heat stress alters transcriptional programming in a manner distinct from either stress alone (Rizhsky et al. 2004). Advances in molecular biology have further highlighted the importance of hormonal crosstalk, particularly the interaction of ABA and ethylene in regulating stomatal dynamics under drought stress (Wilkinson and Davies 2010).

In rice, the *Sub1A* gene for submergence tolerance illustrates the integration of physiological responses with environmental stress perception (Xu et al. 2006). Wheat shows a sharp decline in grain filling under terminal heat stress, as high temperatures disrupt photosystem II and destabilize membranes. In response, heat-shock proteins are produced to support stability and recovery, while stay-green genotypes maintain better tolerance (Joshi et al. 2007). Nutrient availability also influences physiological responses; nitrogen (N) deficiency reduces chlorophyll content, photosynthesis, and biomass, which plants counter through nutrient remobilization, root architecture adjustments, and symbiosis with N-solubilizing microbes (Mickelbart et al. 2015). These examples highlight the direct link between environmental stresses and physiological processes, which translate into breeding targets. Traits such as stay-green in wheat, osmotic adjustment under drought, and *Sub1A*-mediated submergence tolerance in rice exemplify the successful translation of physiological understandings into breeding. Furthermore, advances in omics technologies, crop modeling, and high-throughput phenotyping are enabling precise dissection of stress responses at cellular, organ, and whole-plant levels.

### 6.4.4 Development of Stress-Resilient Agronomic Packages Tailored to Specific Agro-climatic Zones

Stress-resilient agronomic packages are designed and recommended by integrating genetics, soil and nutrient management, water management, and integrated pest and disease management for specific agro-climatic zones. These packages aim not only to protect crops from environmental stresses but also to improve resource-use efficiency and safeguard yields. Recent advancements in field testing, remote sensing, and modeling have accelerated the development of location-specific approaches, ensuring that agronomic interventions align with local climate, soil conditions, and socioeconomic contexts.

#### 6.4.4.1 Drought-Prone Zones

In arid and semi-arid areas, the agronomic practices emphasize the conservation of soil moisture and the incorporation of tolerant varieties. Practices such as mulching, ridge-furrow planting, conservation tillage, and rainwater harvesting are promoted. Irrigation through micro and drip irrigation is highly suitable. Foliar sprays of osmoprotectants such as proline and $KNO_3$, during mid-season droughts, have also been shown to reduce losses.

#### 6.4.4.2 Heat Stress Zones

Terminal heat stress is a major challenge in wheat-growing regions of South Asia. Agronomic strategies such as modified crop geometry, the use of heat-tolerant cultivars, and optimization of planting dates have been suggested (Joshi et al. 2007). In the hot zones of Rajasthan, early sowing of pearl millet is recommended (Yadav and

Kumar 2012). Zero-tillage wheat following rice harvest has also been promoted in the Indo-Gangetic Plains, enabling timely sowing and reducing heat exposure during grain filling (Lobell et al. 2012). Additionally, foliar sprays of growth regulators such as salicylic acid or glycine betaine under high temperatures have shown promise in mitigating oxidative damage.

#### 6.4.4.3 Nutrient-Stressed and Degraded Zones

In rainfed zones with nutrient deficiencies, balanced and site-specific nutrient management is recommended. Foliar application of micronutrients and split doses of nitrogen are also advised. The application of zinc (Zn) and boron (B) has been shown to improve stress tolerance and enhance grain quality in cereals and oilseeds.

#### 6.4.4.4 Salinity and Sodicity Affected Zones

Salinity and sodicity are major challenges, particularly in coastal areas. Recommended practices include the use of salt-tolerant crop varieties, soil amendments, and improved water management. Gypsum has been widely applied to improve soil structure, enhance nutrient availability, and support better crop development (Qadir et al. 2007). In rice-based systems, alternate wetting and drying irrigation cycles are encouraged to minimize salt buildup and conserve water. More recently, biofertilizers such as halotolerant plant growth-promoting rhizobacteria (PGPR) have been incorporated into management packages to enhance nutrient uptake and alleviate salt-induced physiological stress.

## 6.5 Evolution in the Contemporary Hi-Tech Era

Agriculture has evolved from traditional practices to the integration of digital tools, biotechnology, genomics-based methodologies, and data-driven approaches. Many crops, including wheat, have benefited from these advancements. Remote sensing, drones, machine learning (ML), artificial intelligence (AI), and IoT-based sensors and devices are enabling timely and accurate assessments of crop health, soil status, and environmental conditions, allowing researchers and farmers to make rapid decisions that enhance crop stands and productivity (Kamilaris and Prenafeta-Boldú 2018). For example, real-time soil moisture mapping combined with satellite imagery has optimized irrigation in rice–wheat systems in India, while ML models are being trained to predict wheat rust outbreaks with high accuracy (Pantazi et al. 2019). The convergence of these technologies reflects agriculture's ongoing transformation into a smart, resilient, and climate-adaptive system.

### 6.5.1 Application of Remote Sensing, Thermal Imaging, and Multispectral Analysis for Stress Detection

Remote sensing technologies, such as multispectral and thermal imaging, are increasingly being used to detect physiological stress in wheat. Multispectral sensors measure light reflectance across several bands, enabling the early detection of chlorophyll content, water stress, and nutrient deficiencies. The normalized difference vegetation index (NDVI) has been successfully applied to evaluate drought stress and canopy vigor in wheat fields. Thermal imaging, meanwhile, generates canopy temperature maps that help detect water deficits by identifying areas with elevated leaf temperatures and has also been used to assess terminal heat stress. Satellite-based remote sensing from platforms such as Sentinel-2 and Landsat, when combined with ground data, has further enabled the identification of areas prone to terminal heat during critical grain-filling stages, allowing timely irrigation and management interventions.

### 6.5.2 AI and ML Models for Early Prediction and Classification of Disorders

Machine learning models have been trained on large datasets of wheat leaf images to detect and distinguish among leaf rust, stripe rust, stem rust, and spot blotch at early infection stages (Kamilaris and Prenafeta-Boldú 2018). For instance, convolutional neural networks (CNNs) have achieved over 90% accuracy in differentiating wheat leaf rust lesions from other foliar symptoms (Pantazi et al. 2019). Mobile applications integrated with AI systems have been deployed for field-level disease diagnosis in wheat-growing regions of South Asia. Additionally, AI-based weather forecasting models are being used to predict rust pathogen outbreaks, enabling farmers to take timely preventive measures.

### 6.5.3 Use of Drones and Satellite Imagery for Large-Scale Crop Monitoring

Drones and satellite imagery offer valuable opportunities for large-scale wheat crop monitoring. Drones equipped with high-resolution cameras and multispectral sensors can collect real-time data on crop canopy, plant density, and disease spread at the micro-field level (Singh et al. 2023). For instance, unmanned aerial vehicles (UAVs) have been used to map yellow rust infestations in wheat fields across Punjab, enabling early containment. Satellite imagery, such as MODIS and Sentinel-2, facilitates regional-scale crop monitoring, yield forecasting, and stress detection (Pan et al. 2021). The integration of drone-based data with satellite imagery allows for both high-resolution local monitoring and large-area assessment. These technologies are particularly suitable for India, given its large agricultural acreage and diverse cropping systems.

### 6.5.4 Integration of IoT-Based Sensors for Real-Time Soil and Plant Health Monitoring

The Internet of Things (IoT) is transforming wheat farming by enabling continuous, real-time monitoring of soil and plant health parameters. For example, soil moisture sensors connected to automated irrigation systems have improved water-use efficiency in wheat under semi-arid conditions. IoT-based leaf wetness sensors are being used to predict the onset of fungal diseases such as rust and powdery mildew (Patle et al. 2021). Integration of these sensors with cloud-based platforms allows farmers to receive alerts and recommendations directly on their smartphones. Biosensors for NPK measurement have advanced through several innovative approaches. Miniature multi-ion sensors integrated with artificial neural networks and ion-selective field-effect transistors provide precise nutrient detection. Flexible laser-induced graphene (LIG) on polyimide substrates enables low-cost electrochemical sensing of plant-available nitrogen (NH4+ and NO3-). Color sensors, such as TCS3200, are employed in Smart Soil Quality Predictors to assess N, P, and K, along with pH, salinity, and moisture, by comparing absorption wavelengths. An IoT-based nitrogen detection model uses soil solutions mixed with chemicals, in which color changes are measured via RGB light sensors and processed by microcontrollers to classify nitrogen levels as high, moderate, or low. Similarly, fiber-optic sensor systems employ laser light through optical fibers for colorimetric analysis to determine nutrient concentrations. Collectively, these biosensor technologies provide rapid, low-cost, and real-time soil fertility assessments, enhancing precision agriculture and promoting sustainable crop management (Dimkpa et al. 2017).

### 6.5.5 Precision Farming Approaches for Mitigating Specific Disorders Via Targeted Inputs

Precision farming employs advanced technologies to deliver inputs such as water, fertilizers, and pesticides at the optimal time, place, and quantity, thereby reducing waste and improving crop health. For instance, the GreenSeeker handheld sensor has been successfully deployed in India's wheat belts to guide in-season nitrogen application, minimizing wastage while maintaining yields. Rust outbreaks have been managed with minimal chemical input through site-specific fungicide applications guided by disease risk models (Newlands 2018). Similarly, precision irrigation using drip and sprinkler systems has mitigated the effects of terminal heat stress during grain filling. These practices not only enhance yield and grain quality but also reduce the environmental footprint of wheat production.

## 6.6 Statistical Insights in Diagnosis and Mitigation of Wheat Physiological Disorders in India

Statistical analyses of field and experimental data reveal significant correlations between environmental stresses, nutrient imbalances, and the occurrence of wheat physiological disorders. These findings support prognostic modeling and the development of targeted mitigation strategies to maintain yield and grain quality across diverse agro-climatic conditions worldwide.

### 6.6.1 Regional Distribution and Frequency of Major Physiological Disorders

Terminal heat stress is a major concern in the North-West Indo-Gangetic Plains (NWIGP) of India, particularly in Punjab, Haryana, and western Uttar Pradesh, where it significantly affects grain filling and causes grain shriveling due to early warm spells in February and March. Episodes of heat stress in these plains are increasing in magnitude, frequency, and intensity, with the largest projected rise in Punjab, resulting in yield losses of approximately 25% (Dubey et al. 2020; Zachariah et al. 2021). Terminal heat during March and April affects around 13.5 Mha of wheat crops in these regions (Kumar et al. 2021). Other zones, including the Northeastern Plains (eastern UP, Bihar, Jharkhand, West Bengal, and Assam), the Central zone (Madhya Pradesh and eastern Rajasthan), and the Peninsular zone (Maharashtra, Karnataka, and Tamil Nadu), are also impacted by terminal heat (Kumar et al. 2021; Nesar et al. 2022). Another major cause of physiological disorders in wheat is lodging. Unexpectedly severe rains or hailstorms in the northwestern plains and sporadically in central and peninsular regions can lead to significant yield and quality losses (Khobra et al. 2019). Wheat rust also contributes to physiological stress. Stripe rust, previously confined to cooler hill and northwestern regions, is now spreading south and west due to climate change. Leaf and stem rust show regional variation, with stem rust predominantly occurring in central and peninsular regions. Between 2019 and 2022, approximately 56% of the 963 leaf-rust samples that were pathotyped exhibited a dominant pathotype, indicating continued diversity and circulation of the pathogen (Bhardwaj et al. 2019).

### 6.6.2 Correlation of Climatic Events (Heat Waves, Cold Spells) with Disorder Incidence

Terminal heat stress, defined as high-temperature exposure during critical reproductive stages such as flowering and grain filling, is strongly associated with reductions in membrane integrity, antioxidant activity, chlorophyll content, photosynthesis, pollen viability, grain size, and grain weight, all of which negatively impact total yield (Lal et al. 2022; Farhad et al. 2023). Temperatures exceeding 34 °C accelerate wheat senescence, shortening the grain-filling period and reducing yields in northern India (Lobell et al. 2012). Factors such as crop variety, planting density, nitrogen management, and extreme weather events are closely correlated with lodging (Khobra et al. 2019). Similarly, early cold spells (frost or low temperatures) during critical stages can delay development, cause tiller loss, stunt early growth, and even induce sterility during reproductive phases (Hassan et al. 2021).

### 6.6.3 Trends in Yield Loss Attributed to Specific Physiological Issues over the Decades

Since the 1960s, damaging episodes of terminal heat stress in the Indo-Gangetic Plains have been increasing in frequency, intensity, and spatial extent (Dubey et al. 2020; Zachariah et al. 2021). Lodging and field losses have also been reported due to unseasonal rain and wind events, with the extent of loss influenced by wind and rain intensity, nitrogen fertilizer usage, and wheat variety (Khan et al. 2023). Drought, characterized by persistent and regionally variable soil moisture stress, affects both vegetative and reproductive stages. Year-to-year variability in yield losses due to drought is strongly correlated with irrigation practices, groundwater availability, and the frequency of monsoon and winter rains (Zhang et al. 2017).

### 6.6.4 Adoption Rates of Corrective Practices (Foliar Nutrition, Irrigation Systems)

Foliar sprays of potassium chloride (KCl), salicylic acid, thiourea, and boron are increasingly adopted, as these nutrients enhance yield potential, improve stress tolerance, increase leaf area index and relative water content, and help mitigate heat stress (Solanki et al. 2020). Regarding irrigation, micro-irrigation systems, including drip and sprinkler irrigation, are gaining popularity in India; however, coverage remains limited to approximately 17.4% at the time of reporting. Other modern irrigation technologies are also underutilized, emphasizing the need for targeted programs to survey and promote efficient irrigation practices among wheat growers nationwide (Palanisami and Nagothu 2024).

### 6.6.5 Assessment of Farmer Awareness and Institutional Support in Managing Physiological Disorders

A survey conducted in Haryana found that approximately 61.1% of farmers had medium-level knowledge and 56.9% had adopted modern scientific wheat cultivation practices (Singh et al. 2014). Another survey in the Jamui district of Bihar reported that 56.7% of wheat growers had medium-level knowledge, 25% had low knowledge, and 18.3% had high-level knowledge. While farmers were generally aware of optimal sowing times and irrigation management, they lacked knowledge about plant growth regulators and intercropping systems. Farmer knowledge was positively correlated with age, education, social media exposure, and agricultural land size (Kumar et al. 2024). Numerous similar surveys have been conducted across India, and continued assessments are recommended to evaluate farmers' awareness of modern agricultural tools and practices.

ICAR–Indian Institute of Wheat and Barley Research (ICAR–IIWBR) is the leading center for wheat research and development in India and coordinates the All India Coordinated Research Project on Wheat & Barley (AICRP–W&B). The institute releases wheat varieties for farmers and conducts research on a range of wheat diseases, including physiological stresses such as heat, drought, and salinity. It also conducts agronomic trials, water- and nutrient-management studies, and stress-tolerance screenings across the country, providing guidance to farmers based on local conditions. Similarly, other ICAR institutes, agricultural universities, and Krishi Vigyan Kendras (KVKs) actively support farmers in adopting improved wheat cultivation practices.

## 6.7 Future Projections and Global Challenges

### 6.7.1 Anticipated Impact of Climate Change on Frequency and Severity of Physiological Disorders

Major climatic changes are projected by 2025 and mid-century. Elevated levels of carbon dioxide ($CO_2$) are expected to increase total biomass in wheat but reduce grain concentrations of protein, zinc, and iron (Myers et al. 2014). Heat waves and frequent terminal heat episodes are also projected to become more common, particularly in March and April. Terminal heat during anthesis and grain filling is the primary cause of increased spikelet sterility and shriveled grains, and its incidence is expected to rise in the future (Asseng et al. 2015). Simulations in India indicate that terminal heat can reduce yields by double digits, thereby increasing the frequency and severity of heat-related physiological issues such as shortened grain filling and floret/pollen sterility (Dubey et al. 2020). Frequent drought events combined with heat episodes are projected to cause prolonged dry spells in major wheat-growing regions. The combined effects of heat and drought are expected to exacerbate physiological disorders, including inhibited root growth, early senescence, and year-to-year yield reductions (Daloz et al. 2021).

### 6.7.2 Increasing Challenge of Multiple Concurrent Stresses

The Indian wheat cropping system is facing major challenges due to overlapping or concurrent stresses such as salinity, micronutrient deficiencies, drought, and terminal heat stress. The combined effects of heat and drought are increasingly frequent in Northwest and Central India, accelerating senescence, reducing root growth and water uptake, and ultimately lowering wheat grain yields (Dubey et al. 2020). The simultaneous impact of salinity, sodicity, heat, and drought further reduces tiller survival and root function, while causing physiological disorders such as impaired photosynthesis, weakened antioxidative defense systems, and nutrient and water deficiencies. Field studies and remote sensing data indicate that salinity, along with other stress factors, is continuously increasing in the Indo-Gangetic Plain (Kalambukattu et al. 2023). Wheat crops exposed to heat stress on soils deficient in nutrients such as zinc and iron exhibit physiological symptoms including leaf yellowing and poor grain filling, which further degrade grain quality (Khokhar et al. 2024).

### 6.7.3 Emerging Concerns with Intensive Farming Practices on Plant Physiological Health

Extensive irrigation practices combined with inadequate drainage systems are exacerbating salinity, sodicity, and waterlogging. These stresses lead to osmotic damage, ionic toxicity, reduced tillering, and impaired nutrient uptake, immediately affecting grain filling and other physiological functions. Key wheat cultivars in the Indo-Gangetic Plain are experiencing yield penalties, and salinity/sodicity hotspots are expanding (Sheoran et al. 2021; Kalambukattu et al. 2023). Additionally, the monoculture rice–wheat system and infrequent crop rotations are depleting soil organic matter, altering microbial ecosystems, concentrating pests and pathogens, and causing gradual nutritional imbalances. These factors collectively impair crop resilience, root health, and water and nutrient uptake (Yang et al. 2024).

### 6.7.4 Gaps in Stress-Mitigation Breeding and Genotype–Environment Interaction Studies

The inability to accurately and efficiently phenotype drought-resilient traits, such as root architecture, poses a major barrier to trait-based selection under stress conditions (Bapela et al. 2022). The polygenic nature, weak heritability, and high environmental sensitivity of stress-tolerance traits make their selection particularly challenging. Identifying and maintaining stable QTLs across diverse environments is difficult using traditional breeding techniques, especially under abrupt environmental changes (Bapela et al. 2022; Gahlaut et al. 2023). Linkage drag and integration challenges with elite germplasm have limited the utilization of valuable stress-resilience genes present in wheat landraces and wild relatives. Furthermore, the lack of comprehensive germplasm-bridging strategies and pre-breeding efforts has hindered the effective mobilization of these diverse genetic resources into breeding programs, resulting in the underutilization of genetically varied wheat landraces and wild relatives (Babalola et al. 2025).

### 6.7.5 Importance of Global Collaboration in Abiotic Stress Tolerance Research and Technological Transfer

Environments are rapidly changing due to climate change, and no single nation or organization can independently conduct coordinated, large-scale testing across multiple climates or perform futuristic scenario modeling to breed heat- and drought-resistant varieties. Therefore, global collaboration in abiotic stress tolerance research and technology transfer is essential (Xiong et al. 2021). Pre-breeding initiatives and shared germplasm resources help reduce research duplication while providing breeders access to elite varieties, landraces, and wild relatives, thereby accelerating the incorporation of desirable traits. Moreover, open access to datasets and standardized procedures enables the development of genomic selection strategies and environmentally relevant models (Xu et al. 2023). Genotype–environment (G × E) interactions, which are difficult to detect in single-location studies, can be identified through coordinated multi-environment trial networks that include sites in heat- and drought-prone areas (Xiong et al. 2021).

### 6.7.6 Policy and Extension Priorities for Widespread Adoption of Diagnostic and Mitigation Tools

Extension capacity should be strengthened by providing grants for extensive training, including the use of diagnostic tools, and by educating agricultural traders to deliver decision-support systems that facilitate the adoption of fungicides, improved agronomic practices, and high-performing varieties (Ma and Rahut 2024). Priority should be given to affordable, portable, and real-time diagnostic tools such as the Mobile and Real-time PLant disEase (MARPLE) platform. The point-of-care (POC) MARPLE diagnostic system identifies specific strains of wheat stem rust (*Puccinia graminis* f. sp. *tritici*; Pgt) and wheat yellow rust (*Puccinia striiformis* f. sp. *tritici*; Pst) pathogens using amplicon re-sequencing on the portable MinION sequencer. The technique targets genes that are highly variable between Pst and Pgt strains. Following amplicon re-sequencing, the gene sequences can be used for phylogenetic analysis, comparing new samples with a global set of reference isolates. The MARPLE diagnostics platform is user-friendly, requiring minimal equipment and training, making it particularly suitable for resource-limited environments (Vinopal et al. 2025).

## 6.8 Conclusion

Under climate change and resource constraints, physiological disorders in wheat present a growing challenge to sustaining stable yields and grain quality. These non-infectious disorders have far-reaching impacts on crop physiology, affecting grain filling, causing sterility, and reducing nutritional quality. Historical reliance on traditional practices has gradually shifted toward scientific diagnosis and mitigation strategies, culminating in the adoption of precision agronomy and digital agriculture. However, the increasing frequency of concurrent stresses such as salinity, heat, drought, and nutrient deficiencies demands more integrated solutions. Future research must focus on unraveling genotype–environment interactions, identifying stress-resilient germplasm, and translating physiological insights into breeding programs. Equally important is the dissemination of diagnostic tools and mitigation packages to farmers through robust extension systems and supportive policies. Global collaborations, data sharing, and advanced modeling will be indispensable for developing climate-resilient wheat production systems. Strengthening the diagnosis and management of physiological disorders will not only secure wheat productivity but also contribute to global food and nutritional security.

## References


Abbas S, Amna, Javed MT et al (2021) Nutrient deficiency stress and relation with plant growth and development. In: Fahad S, Sonmez O, Saud S et al (eds) Engineering tolerance in crop plants against abiotic stress, 1st edn. CRC Press, Boca Raton

Ahmed FF, Mokhtar MS, Ahmed MKA (2011) Protecting crimson seedless grapevines growing in hot climates from sunburn. Res J Agric Biol Sci 7:135–141

Aown M, Raza S, Saleem MF et al (2012) Foliar application of potassium under water deficit conditions improved the growth and yield of wheat (*Triticum aestivum* L.). J Anim Plant Sci 22(2):431–437

Asseng S, Ewert F, Martre P et al (2015) Rising temperatures reduce global wheat production. Nat Clim Chang 5:143–147. https://doi.org/10.1038/nclimate2470

Atta K, Singh AP, Dash SSS et al (2023) Physiological and biochemical approaches for mitigating the effect of abiotic stresses in wheat. In: Khan MK, Pandey A, Hamurcu M et al (eds) Abiotic stresses in wheat. Academic

Babalola KO, Monacelli N, Gozzi M et al (2025) Genetic diversity and climate change adaptation in wheat: a systematic review of landraces, composite cross populations, and evolutionary populations. Front Sustain Food Syst 9:1504922. https://doi.org/10.3389/fsufs.2025.1504922

Baker RJ (1996) Oslo and Biggar spring wheats respond differently to controlled temperature and moisture stress. Can J Plant Sci 76(3):413–416. https://doi.org/10.4141/cjps96-073

Bapela T, Shimelis H, Tsilo TJ et al (2022) Genetic improvement of wheat for drought tolerance: Progress, challenges and opportunities. Plants 11(10):1331. https://doi.org/10.3390/plants11101331

Barbedo JGA (2025) A review of artificial intelligence techniques for wheat crop monitoring and management. Agronomy 15(5):1157. https://doi.org/10.3390/agronomy15051157

Beegle D (2005) Assessing soil phosphorus for crop production by soil testing. Phosphorus Agric Environ 46:123–143. https://doi.org/10.2134/agronmonogr46.c5

Bhardwaj SC, Singh GP, Gangwar OP et al (2019) Status of wheat rust research and progress in rust management—Indian context. Agronomy 9(12):892. https://doi.org/10.3390/agronomy9120892

Blum A (2017) Osmotic adjustment is a prime drought stress adaptive engine in support of plant production. Plant Cell Environ 40(1):4–10. https://doi.org/10.1111/pce.12800

Bouskour S, Bahatti L, Zaggaf M (2023) The use of NDVI to improve cereals agriculture: a review. In: 2023 3rd international conference on Innovative Research in Applied Science, Engineering and Technology (IRASET), Mohammedia, 1–7. https://doi.org/10.1109/IRASET57153.2023.10153054

Boyd LA (2005) Can Robigus defeat an old enemy?—yellow rust of wheat. J Agric Sci 143(4):233–243. https://doi.org/10.1017/S0021859605005095

Chen CT, Lee CL, Yeh DM (2018) Effects of nitrogen, phosphorus, potassium, calcium, or magnesium deficiency on growth and photosynthesis of Eustoma. HortScience 53(6):795–798. https://doi.org/10.21273/HORTSCI12947-18

Daloz AS, Rydsaa JH, Hodnebrog Ø et al (2021) Direct and indirect impacts of climate change on wheat yield in the Indo-Gangetic plain in India. J Agric Food Res 4:100132. https://doi.org/10.1016/j.jafr.2021.100132

Davies P (2010) Plant hormones: biosynthesis, signal transduction, action! Springer, Dordrecht. https://doi.org/10.1007/978-1-4020-2686-7

Decsi K, Ahmed M, Abdul-Hamid D et al (2025) The role of salicylic acid in activating plant stress responses—results of the past decade and future perspectives. Int J Mol Sci 26:4447. https://doi.org/10.3390/ijms26094447

Dimkpa C, McLean J, Gatere L et al (2017) Methods for rapid testing of plant and soil nutrients. Sustain Agric Rev 1:1–43. https://doi.org/10.1007/978-3-319-58679-3_1

Du C, Li L, Effah Z (2022) Effects of straw mulching and reduced tillage on crop production and environment: a review. Water 14(16):2471. https://doi.org/10.3390/w14162471

Dubey R, Pathak H, Chakrabarti B et al (2020) Impact of terminal heat stress on wheat yield in India and options for adaptation. Agric Syst 181:102826. https://doi.org/10.1016/j.agsy.2020.102826

El Baidouri M, Murat F, Veyssiere M et al (2017) Reconciling the evolutionary origin of bread wheat (*Triticum aestivum*). New Phytol 213:1477–1486. https://doi.org/10.1111/nph.14113

El-Fouly MM, Mobarak ZM, Salama ZA (2011) Micronutrients (Fe, Mn, Zn) foliar spray for increasing salinity tolerance in wheat (*Triticum aestivum* L.). Afr. J Plant Sci 5(5):314–322

FAOSTAT (2024) FAOSTAT statistical database. Food and Agriculture Organization of the United Nations. https://www.fao.org/faostat/en/

Farhad M, Kumar U, Tomar V et al (2023) Heat stress in wheat: a global challenge to feed billions in the current era of the changing climate. Front Sustain Food Syst 7:1203721. https://doi.org/10.3389/fsufs.2023.1203721

Gahlaut V, Gautam T, Wani SH (2023) Abiotic stress tolerance in wheat (*Triticum aestivum* L.): molecular breeding perspectives. In: Wani SH, Wang D, Singh GP (eds) QTL mapping in crop improvement. Academic

Gill KS, Jalota SK, Prihar SS et al (1977) Water conservation by soil mulch in relation to soil type, time of tillage, tilth and evaporativity. J Indian Soc Soil Sci 25(4):360–366

Glenn DM, Cooley N, Walker R et al (2010) Impact of kaolin particle film and water deficit on wine grape water use efficiency and plant water relations. HortScience 45:1178–1187

Goulding K, Jarvis S, Whitmore A (2008) Optimizing nutrient management for farm systems. Philos Trans R Soc Lond Ser B Biol Sci 363(1491):667–680. https://doi.org/10.1098/rstb.2007.2177

Habib M (2012) Effect of supplementary nutrition with Fe, Zn chelates and urea on wheat quality and quantity. Afr J Biotechnol 11(11):2661–2665. https://doi.org/10.5897/AJB11.1762

Haji HM, Hunt LA (1999) Genotype × environment interactions and underlying environmental factors for winter wheat in Ontario. Can J Plant Sci 79(4):497–505. https://doi.org/10.4141/P98-075

Hasanuzzaman M, Nahar K, Alam MM et al (2013) Physiological, biochemical, and molecular mechanisms of heat stress tolerance in plants. Int J Mol Sci 14(5):9643–9684. https://doi.org/10.3390/ijms14059643

Hassan MA, Xiang C, Farooq M et al (2021) Cold stress in wheat: plant acclimation responses and management strategies. Front Plant Sci 12:676884. https://doi.org/10.3389/fpls.2021.676884

Hermans C, Verbruggen N (2005) Physiological characterization of Mg deficiency in *Arabidopsis thaliana*. J Exp Bot 56(418):2153–2161. https://doi.org/10.1093/jxb/eri215

ICAR-IIWBR (2025) In: Tiwari R, Tyagi BS, Sareen S et al (eds) Director's report of AICRP on wheat and barley 2024–25. ICAR-Indian Institute of Wheat and Barley Research, Karnal

IPCC (2012) Managing the risks of extreme events and disasters to advance climate change adaptation. In: Field CB, Barros V, Stocker TF et al (eds) A special report of working groups I and II of the intergovernmental panel on climate change. Cambridge University Press, Cambridge/New York

Iqbal M, Hussain I, Liaqat H et al (2015) Exogenously applied selenium reduces oxidative stress and induces heat tolerance in spring wheat. Plant Physiol Biochem 94:95–103. https://doi.org/10.1016/j.plaphy.2015.05.012

Jalilvand S, Roozbahani A, Hasanpour J (2014) Effect of foliar application of iron on morpho-physiological traits of wheat under drought stress. Bull Environ Pharmacol Life Sci 3:167–177

Jaradat AA (2013) Wheat landraces: a mini review. Emir J Food Agric 25(1):20–29. https://doi.org/10.9755/ejfa.v25i1.15376

Joshi AK, Kumari M, Singh VP et al (2007) Stay green trait: variation, inheritance and its association with spot blotch resistance in spring wheat (*Triticum aestivum* L.). Euphytica 153:59–71. https://doi.org/10.1007/s10681-006-9235-z

Kalambukattu JG, Johns B, Kumar S et al (2023) Temporal remote sensing-based soil salinity mapping in Indo-Gangetic plain employing machine-learning techniques. Proc Indian Natl Sci Acad 89:290–305. https://doi.org/10.1007/s43538-023-00157-x

Kamilaris A, Prenafeta-Boldú F (2018) Deep learning in agriculture: a survey. Comput Electron Agric 147:70–90. https://doi.org/10.1016/j.compag.2018.02.016

Khan H, Mamrutha HM, Mishra CN et al (2023) Harnessing high yield potential in wheat (*Triticum aestivum* L.) under climate change scenario. Plants 12(6):1271. https://doi.org/10.3390/plants12061271

Khobra R, Sareen S, Meena BK et al (2019) Exploring the traits for lodging tolerance in wheat genotypes: a review. Physiol Mol Biol Plants 25(3):589–600. https://doi.org/10.1007/s12298-018-0629-x

Khokhar JS, Broadley MR, Ander EL (2024) Soil zinc surveillance frameworks can inform human nutrition studies: opportunities in India. Front Soil Sci 4:1421652. https://doi.org/10.3389/fsoil.2024.1421652

Kjaergaard T (2003) A plant that changed the world: the rise and fall of clover 1000–2000. Landsc Res 28(1):41–49. https://doi.org/10.1080/01426390306531

Kumar P, Gupta V, Singh G et al (2021) Assessment of terminal heat tolerance based on agro-morphological and stress selection indices in wheat. Cereal Res Commun 49:217–226. https://doi.org/10.1007/s42976-020-00112-2

Kumar S, Meena N, Doharey RK et al (2024) Knowledge of wheat growers regarding improved wheat production technology in Jamui district of Bihar, India. J Exp Agric Int 46(8):42–48. https://doi.org/10.9734/jeai/2024/v46i82678

Kutschera U, Hossfeld U (2012) Physiological phytopathology: origin and evolution of a scientific discipline. J Appl Bot Food Qual 85(1):1–5

Kyveryga PM, Mueller TA, Mueller DS (2018) On-farm replicated strip trials. In: Shannon DK, Clay DE, Kitchen NR (eds) Precision agriculture basics. American Society of Agronomy, crop science Society of America, and soil science Society of America, Madison

Lal MK, Tiwari RK, Gahlaut V et al (2022) Physiological and molecular insights on wheat responses to heat stress. Plant Cell Rep 41(3):501–518. https://doi.org/10.1007/s00299-021-02784-4

Lata C, Kumar A, Sharma SK et al (2017) Tolerance to combined boron and salt stress in wheat varieties: biochemical and molecular characterization. Indian J Exp Biol 55:321–328

Lata C, Prasad P, Gangwar OP et al (2022) Temporal behavior of wheat–*Puccinia striiformis* interaction prompted defense-responsive genes. J Plant Interact 17(1):674–684. https://doi.org/10.1080/17429145.2022.2082570

Le Gall H, Philippe F, Domon JM et al (2015) Cell wall metabolism in response to abiotic stress. Plants 4:112–166. https://doi.org/10.3390/plants4010112

Levitt J (1980) Responses of plants to environmental stress. Volume 1: chilling, freezing, and high temperature stresses. Academic, Cambridge

Lobell D, Sibley A, Ortiz-Monasterio JI (2012) Extreme heat effects on wheat senescence in India. Nat Clim Chang 2:186–189. https://doi.org/10.1038/nclimate1356

Ma W, Rahut DB (2024) Climate-smart agriculture: adoption, impacts, and implications for sustainable development. Mitig Adapt Strateg Glob Change 29:44. https://doi.org/10.1007/s11027-024-10139-z

Mickelbart M, Hasegawa P, Bailey-Serres J (2015) Genetic mechanisms of abiotic stress tolerance that translate to crop yield stability. Nat Rev Genet 16:237–251. https://doi.org/10.1038/nrg3901

Murray GM (2024) GP Darnell-smith and the introduction of copper carbonate 'dry pickling' of wheat seed. Hist Rec Aust Sci 35(2):98–104. https://doi.org/10.1071/HR23027

Myers S, Zanobetti A, Kloog I et al (2014) Increasing $CO_2$ threatens human nutrition. Nature 510:139–142. https://doi.org/10.1038/nature13179

Nayee HV (2024) Impact of ancient water management systems on agricultural productivity and sustainability. Res Rev J Indian Knowl Syst 1(1):11–16

Nesar NA, Hameedi A, Attri M et al (2022) Terminal heat stress and its mitigation options through agronomic interventions in wheat crop: a review. Int J Environ Climate Change 12(11):131–139. https://doi.org/10.9734/IJECC/2022/v12i1130955

Newlands NK (2018) Model-based forecasting of agricultural crop disease risk at the regional scale, integrating airborne inoculum, environmental, and satellite-based monitoring data. Front Environ Sci 6:63. https://doi.org/10.3389/fenvs.2018.00063

Palanisami K, Nagothu US (2024) India's water future in a changing climate. Springer, Singapore. https://doi.org/10.1007/978-981-97-1785-9

Pan Q, Gao M, Wu P et al (2021) A deep-learning-based approach for wheat yellow rust disease recognition from unmanned aerial vehicle images. Sensors 21:6540. https://doi.org/10.3390/s21196540

Pantazi XE, Moshou D, Tamouridou AA (2019) Automated leaf disease detection in different crop species through image features analysis and one-class classifiers. Comput Electron Agric 156:96–104. https://doi.org/10.1016/j.compag.2018.11.005

Patle KS, Saini R, Kumar A et al (2021) IoT enabled leaf wetness sensor on flexible substrates for in-situ plant disease management. IEEE Sensors J 21(17):19481–19491. https://doi.org/10.1109/JSEN.2021.3089722

PirastehAnosheh H, Emam Y, Khaliq A (2016) Response of cereals to cycocel application. Iran Agric Res 35:1–12

Qadir M, Oster JD, Schubert S et al (2007) Phytoremediation of sodic and saline-sodic soils. Adv Agron 96:197–247. https://doi.org/10.1016/S0065-2113(07)96006-X

Rehman MU, Ilahi H, Adnan M et al (2021) Application of silicon: a useful way to mitigate drought stress: an overview. Curr Res Agric Farming 2:9–17. https://doi.org/10.18782/2582-7146.134

Reynolds MP, Ortiz R (2010) Adapting crops to climate change: a summary. In: Reynolds MP (ed) Climate change and crop production. CABI, Wallingford

Rizhsky L, Liang H, Shuman J et al (2004) When defense pathways collide: the response of Arabidopsis to a combination of drought and heat stress. Plant Physiol 134(4):1683–1696. https://doi.org/10.1104/pp.103.033431

Setter T, Waters I (2003) Review of prospects for germplasm improvement for waterlogging tolerance in wheat, barley and oats. Plant Soil 253:1–34. https://doi.org/10.1023/A:1024573305997

Shenker M, Chen Y (2005) Increasing iron availability to crops: fertilizers, organo-fertilizers, and biological approaches. Soil Sci Plant Nutr 51(1):1–17. https://doi.org/10.1111/j.1747-0765.2005.tb00001.x

Sheoran P, Kumar A, Sharma R et al (2021) Quantitative dissection of salt tolerance for sustainable wheat production in sodic agro-ecosystems through farmers' participatory approach: an Indian experience. Sustainability 13(6):3378. https://doi.org/10.3390/su13063378

Singh R, Hansra BS, Singh R et al (2014) Knowledge and adoption level of farmers of Haryana about scientific wheat cultivation practices. Indian J Agric Res 48(1):52–56. https://doi.org/10.5958/j.0976-058X.48.1.009

Singh SK, Saxena R, Porwal AN et al (2017) Assessment of hailstorm damage in wheat crop using remote sensing. Curr Sci 112:2095–2100. https://doi.org/10.18520/cs/v112/i10/2095-2100

Singh H, Roy A, Setia R et al (2023) Classification of yellow rust of wheat from Sentinel-2 satellite imagery using deep learning artificial neural network. Arab J Geosci 16:626. https://doi.org/10.1007/s12517-023-11750-1

Sivakumar R, Kalarani MK, Karthikeyan R et al (2023) Physiological disorders and their corrective measures. Nat Farming Sustain Agric 77

Solanki RL, Swami P, Nagar K et al (2020) Effect of fertilizers and foliar application of nutrient on wheat (*Triticum aestivum* L.) grain yield through conduct on farm trials at farmers' fields. Int J Curr Microbiol Appl Sci 9(9):1551–1556. https://doi.org/10.20546/ijcmas.2020.907.179

Solanki H, Thakker R, Ochieng LA (2024) Green revolution and its impacts on the environment: a review. Towards Excellence 16(4)

Talukder ASMH, McDonald GK, Gill GS (2014) Effect of short-term heat stress prior to flowering and early grain set on the grain yield of wheat. Field Crop Res 160:54–63. https://doi.org/10.1016/j.fcr.2014.01.013

Teimouri S, Hasanpour J, Tajali AA (2014) Effect of selenium spraying on yield and growth indices of wheat (*Triticum aestivum* L.) under drought stress condition. Int J Adv Biol Biom Res 2(6):2091–2103

Tiwari V, Mamrutha HM, Sareen S et al (2017) Managing abiotic stresses in wheat. In: Minhas PS (ed) Abiotic stress management for resilient agriculture. Springer, Singapore

Treitel C (2020) Nutritional modernity: the German case. Osiris 35(1):183–203

Verslues PE, Sharma S (2010) Proline metabolism and its implications for plant-environment interaction. Arabidopsis Book 8:e0140

Vinopal D, Bryan A, Hodson DP et al (2025) MARPLE: Mobile and real-time plant disease diagnostics for enhancing wheat rust surveillance. Methods Mol Biol 2898:93–111. https://doi.org/10.1007/978-1-0716-4378-5_6

Volkov V, Schwenke H (2021) A quest for mechanisms of plant root exudation brings new results and models, 300 years after hales. Plants 10(1):38. https://doi.org/10.3390/plants10010038

Watanabe T, Okada R, Tokunaga S et al (2022) Nitrogen deficiency-induced molybdenum accumulation in wheat. J Plant Nutr 45(9):1413–1424. https://doi.org/10.1080/01904167.2021.2020838

Wilkinson S, Davies WJ (2010) Drought, ozone, ABA and ethylene: new insights from cell to plant to community. Plant Cell Environ 33(4):510–525. https://doi.org/10.1111/j.1365-3040.2009.02052.x

Xiong W, Reynolds MP, Crossa J et al (2021) Increased ranking change in wheat breeding under climate change. Nat Plants 7:1207–1212. https://doi.org/10.1038/s41477-021-00988-w

Xu K, Xu X, Fukao T et al (2006) *Sub1A* is an ethylene-response-factor-like gene that confers submergence tolerance to rice. Nature 442:705–708. https://doi.org/10.1038/nature04920

Xu Y, Qin F, Chu C et al (2023) Abiotic stress tolerance: genetics, genomics, and breeding. Crop J 11(4):969–974. https://doi.org/10.1016/j.cj.2023.07.002

Yadav OP, Kumar A (2012) Long-term shifts in genetic diversity and yield stability of pearl millet in stress-prone environments. Exp Agric 48(3):334–346. https://doi.org/10.1017/S0014479711001084

Yang R, Shen Y, Kong X et al (2024) Effects of diverse crop rotation sequences on rice growth, yield, and soil properties: a field study in Gewu station. Plants 13:3273. https://doi.org/10.3390/plants13233273

Zachariah M, Mondal A, AghaKouchak A (2021) Probabilistic assessment of extreme heat stress on Indian wheat yields under climate change. Geophys Res Lett 48(20):e2021GL094702. https://doi.org/10.1029/2021GL094702

Zeng D, Luo X (2012) Physiological effects of chitosan coating on wheat growth and activities of protective enzyme with drought tolerance. Open J Soil Sci 2:282–288. https://doi.org/10.4236/ojss.2012.23034

Zhaksylykkelini U (2025) Everything can be grown at home using the hydroponics method. Eurasian Sci Rev Int Peer-Rev Multidiscip J 1(3):2574–2586

Zhang X, Obringer R, Wei C et al (2017) Droughts in India from 1981 to 2013 and implications to wheat production. Sci Rep 7:44552. https://doi.org/10.1038/srep44552

Zhao S, Huq ME, Fahad S et al (2024) Boron toxicity in plants: understanding mechanisms and developing coping strategies; a review. Plant Cell Rep 43:238. https://doi.org/10.1007/s00299-024-03317-5

# Advances in Understanding and Enhancing Drought and Heat Tolerance in Wheat

7

Rinki, H. M. Mamrutha, V. Rajeswari, Preety Rani, Vikram Singh, Zeenat Wadhwa, Kiran, Vanita Pandey, and Jagadish Rane

**Abstract**

Wheat (*Triticum aestivum* L.) productivity is increasingly threatened by climate change-induced drought and heat stress, which frequently co-occur and severely disrupt physiological, biochemical, and reproductive processes, especially during anthesis and grain filling. These stresses diminish photosynthetic capacity, destabilize cellular membranes and enzymes, accelerate senescence, and cause oxidative damage, resulting in significant yield and quality losses. Wheat plants employ complex adaptive mechanisms, including abscisic acid-mediated signaling, activation of heat shock transcription factors and proteins, osmotic adjustment, and enhanced antioxidant defense systems. However, the differing physiological requirements for drought and heat tolerance complicate the development of cultivars with combined stress resilience. Recent advances in genomics, quantitative trait loci (QTL) mapping, genome-wide association studies (GWAS), and multi-omics approaches have identified key genomic regions and candidate genes linked to traits such as the stay-green phenotype, canopy temperature depression, and stress-resilient root architecture. The integration of marker-assisted and genomic selection, introgression of beneficial alleles from wild relatives, and the use of high-throughput phenotyping and predictive breeding tools are accelerating the development of climate-resilient wheat cultivars. These strategies offer promising avenues to sustain wheat productivity under increasingly variable and extreme

Rinki · V. Rajeswari · P. Rani · V. Singh · Z. Wadhwa · Kiran · V. Pandey
ICAR-Indian Institute of Wheat and Barley Research, Karnal, Haryana, India

H. M. Mamrutha (✉)
ICAR-National Institute of Seed Science and Technology, Regional Station, Bengaluru, Karnataka, India
e-mail: mamrutha.m@icar.gov.in

J. Rane
ICAR-Central Institute for Arid Horticulture, Bikaner, Rajasthan, India



R. Tiwari et al. (eds.), *Wheat Stress Management*,
https://doi.org/10.1007/978-981-92-0902-6_7

environmental conditions. This chapter reviews historical and recent advances aimed at improving heat and drought tolerance in wheat.

**Keywords**

Wheat · Drought and heat stress tolerance · High-throughput phenotyping · Genomic selection and QTL mapping · Genome editing (CRISPR/Cas) · Climate-resilient wheat breeding

## 7.1 Introduction

Wheat (*Triticum aestivum* L.) is a staple crop globally, providing nearly 20% of the calories and protein consumed by the world population. Its productivity and yield stability are increasingly threatened by climate change-induced abiotic stresses, particularly drought and heat. The frequency, intensity, and duration of high-temperature events and water scarcity episodes have increased markedly in recent decades and are expected to intensify further, posing significant risks to wheat-based production systems. Yield losses from drought and heat stress are most pronounced when these stresses coincide with sensitive developmental stages such as flowering and grain filling. Notably, drought and heat frequently co-occur in major wheat-growing regions, and their combined effects are generally more damaging than either stress alone, resulting in accelerated senescence, impaired reproductive development, and significant reductions in grain number and weight (Zhao et al. 2017). At the physiological level, drought stress primarily limits plant growth by reducing soil moisture availability, which leads to decreased leaf water potential, stomatal closure, reduced $CO_2$ assimilation, and diminished photosynthetic efficiency. In contrast, heat stress directly compromises membrane stability, enzyme activity, and protein structure, thereby disrupting metabolic processes such as photosynthesis and respiration. Elevated temperatures during anthesis can impair pollen viability and fertilization, while heat stress during grain filling shortens this phase, reduces starch accumulation, and ultimately decreases 1000-kernel weight. Both stresses increase the production of reactive oxygen species (ROSs), causing oxidative damage to lipids, proteins, and nucleic acids if not effectively neutralized by antioxidant systems. The complexity of combined drought and heat stress responses is partly due to their contrasting effects on stomatal behavior: drought promotes closure to conserve water, whereas heat often requires partial opening for transpirational cooling. This interplay complicates the physiological adaptation strategies necessary for tolerance (Zandalinas et al. 2018).

Advances in molecular biology and genomics have substantially improved our understanding of the mechanisms underlying stress perception and tolerance in wheat. Stress signaling pathways involve intricate networks of phytohormones, transcription factors, and downstream protective genes. Abscisic acid (ABA) plays a central role in mediating drought responses, regulating stomatal conductance, and activating stress-responsive gene expression. In contrast, heat stress induces the expression of heat shock transcription factors (HSFs) and heat shock proteins (HSPs), which function as molecular chaperones to stabilize proteins and maintain cellular homeostasis under elevated temperatures (Bita and Gerats 2013). Additionally, osmotic adjustment through the accumulation of compatible solutes such as proline and soluble sugars, maintenance of membrane integrity, and activation of antioxidant defense systems are critical adaptive mechanisms contributing to stress resilience. The genetic architecture underlying drought and heat tolerance in wheat is complex and polygenic, involving numerous quantitative trait loci (QTLs) and gene networks. The availability of high-quality wheat genome assemblies and high-throughput genotyping platforms has enabled genome-wide association studies (GWAS) and QTL mapping to identify genomic regions linked to traits such as canopy temperature depression, the stay-green phenotype, root architecture, and grain-filling duration under stress. Recent research has identified candidate genes, including members of the *WRKY*, *DREB*, and *SnRK* families, that enhance tolerance by regulating stress-responsive pathways (Gupta et al. 2020). The integration of marker-assisted selection (MAS) and genomic selection into breeding programs has accelerated the development of climate-resilient cultivars by enabling early, precise selection of desirable alleles. Furthermore, introgression of beneficial alleles from wild relatives and landraces has expanded the genetic base of modern wheat, improving adaptation to high-temperature and water-limited environments without significant yield penalties under favorable conditions (Reynolds et al. 2021).

Technological advancements in phenotyping have also transformed research on drought and heat tolerance. High-throughput phenotyping platforms using remote sensing, thermal imaging, and spectral reflectance analysis enable rapid, nondestructive evaluation of large breeding populations under field conditions. Traits such as canopy temperature, chlorophyll fluorescence, normalized difference vegetation index (NDVI), and stay-green characteristics have become valuable proxies for stress tolerance. Coupled with artificial intelligence and predictive modeling, these approaches enhance the efficiency of selection for complex adaptive traits and support precision breeding strategies. Recent advances in physiology, molecular genetics, genomics, and phenomics have substantially deepened understanding of wheat responses to drought and heat stress. Nevertheless, the dynamic and multifactorial nature of climate change necessitates integrated approaches that combine multi-omics analyses, advanced phenotyping, and predictive breeding models to develop wheat cultivars with durable tolerance to multiple abiotic stresses. Strengthening these interdisciplinary efforts is essential to safeguard global wheat production and ensure food security amid increasingly variable and extreme climatic conditions. Various abiotic stresses affecting wheat production in India, along with the adaptive strategies employed by the wheat crop, are listed in Table 7.1.

**Table 7.1** Different abiotic stresses and their adaptive strategies in wheat

| Abiotic stress | Stress effects in wheat | Plant response mechanisms | Research and breeding tools | Expected outcomes |
|---|---|---|---|---|
| High temperature | Membrane instability | Heat shock proteins (HSPs) | QTL mapping and GWAS | Consistent and stable grain development |
| Heat stress during flowering (anthesis) | Pollen sterility | Antioxidant defense (SOD, CAT, POD) | Transcriptomics profiling | Enhanced reproductive success under high temperature |
| Water deficit | Reduced turgor & stomatal closure | Accumulation of osmoprotectants (proline, glycine betaine) | Metabolomics profiling | Improved water-use efficiency |
| Terminal heat stress | Reduced grain filling duration | Expression of stay-green characteristics | Genomic prediction and selection | Sustained yield under late heat |
| Prolonged drought | Root growth limitation | Deep and prolific root systems | High-throughput phenotyping and root system imaging | Enhanced soil moisture extraction |
| Combined heat + drought | ROS accumulation | Enhanced ROS detoxification systems | Integrative systems biology approaches | Multi-stress resilience |
| Irregular rainfall | Poor crop establishment, lodging susceptibility | Early vigor and phenological plasticity | High-throughput phenotyping (UAV, thermal imaging) | Improved crop establishment |

## 7.2 Historical Perspectives on Understanding and Enhancing Drought and Heat Tolerance in Wheat

Early identification of weather-induced crop failures highlighted varietal differences in stress tolerance, guiding traditional farmer-led selection of locally adapted Triticum aestivum landraces. Early physiological investigations documented visible symptoms of drought and heat injury, including wilting, delayed flowering, sterility, and leaf scorching, and breeders selected genotypes based on visual performance and yield stability under adverse conditions. During the Green Revolution, however, breeding priorities shifted toward maximizing yield under optimal, high-input systems, and stress adaptation regained prominence only as climate-related risks intensified. Climate change-induced water scarcity and rising temperatures now pose serious threats to wheat production worldwide, particularly when drought and heat coincide during sensitive developmental stages such as flowering and anthesis. Even brief episodes of terminal stress can cause irreversible fertility loss, reduced grain number, kernel shrinkage, and, in extreme cases, total crop failure. Early physiological studies demonstrated that heat stress accelerates development and disrupts reproductive processes through pollen abortion, anther indehiscence, impaired pollen tube growth, and reduced seed reserve accumulation, while also damaging photosynthesis, respiration, chlorophyll stability, and photosystem II integrity. Drought similarly restricts growth by reducing leaf area, assimilate mobilization, amyloplast formation, and grain set. When both stresses occur simultaneously, their combined effects intensify through stomatal closure, elevated leaf temperature, suppressed photosynthesis, and altered respiration, often in genotype-specific synergistic or antagonistic patterns.

Recognition that drought tolerance is a complex, polygenic trait frequently associated with trade-offs in yield potential led to deeper investigations into the relationship between yield stability and stress resilience. Initial phenotypic observations of survival and productivity under stress evolved into systematic physiological studies identifying adaptive traits such as altered phenology, improved water-use efficiency, photosynthetic stability, reproductive resilience, canopy temperature depression, stay-green behavior, and osmotic adjustment. Advances in molecular biology and genomics further elucidated the regulatory networks governing stress perception and tolerance, shifting breeding strategies from purely phenotypic selection to the identification of genes, QTLs, and GWAS linked to drought and heat tolerance. The intrinsic polyploidy of wheat contributes significantly to stress adaptation by enhancing genetic buffering and functional diversity; comparative studies show that although diploid, tetraploid, and hexaploid wheats all experience reductions in biomass and yield under drought, polyploid forms maintain higher grain yield and water-use efficiency through improved biomass partitioning and physiological regulation. Contemporary breeding programs now integrate physiological, molecular, and genomic knowledge to develop climate-resilient cultivars. Tools such as MAS, genomic selection, high-throughput phenotyping, systems biology, and CRISPR/Cas genome editing enable precise manipulation of stress-responsive genes and accelerated identification of resilient genotypes. This transition—from breeding for optimal environments to breeding for stability under combined drought and heat stress—reflects a strategic shift toward developing widely adaptable wheat varieties capable of sustaining productivity under increasingly variable climatic conditions.

## 7.3 Scientific Advancements and Principles Developed Over the Past Centuries

Plant stress evaluation now encompasses a multidimensional strategy that integrates physiological, biochemical, and molecular indicators. Physiological stress is commonly assessed by measuring plant water status, including leaf water potential, relative water content, and transpiration, as well as gas exchange traits such as stomatal conductance, water-use efficiency, and intracellular $CO_2$, along with thermal responses. Biochemical stress responses are characterized by the accumulation of osmolytes and hormones, including proline, soluble sugars, and ABA, and by the activation of antioxidant defense enzymes such as ATPase, catalase (CAT), glutathione reductase, and superoxide dismutase (SOD). The understanding of drought and heat stress has advanced from initial descriptive observations to mechanistic insights and, more recently, to the development of applied genetic and breeding strategies. Foundational theories have been refined into practical tools for crop improvement through the application of modern technologies.

### 7.3.1 Development of Physiological Markers

*Relative water content*: It reflects the balance between water uptake and transpiration and serves as a reliable indicator of plant water status under drought conditions. Studies have confirmed that drought-tolerant genotypes retain more water than sensitive genotypes.

*Canopy temperature*: It is a rapid, nondestructive indicator of transpiration efficiency and plant water relations. Genotypes that maintain lower canopy temperature under drought and heat stress demonstrate superior stomatal conductance and yield stability. As a result, canopy temperature has been widely validated as a reliable selection criterion for drought-resilient wheat.

*Chlorophyll fluorescence*: It provides noninvasive insights into photosynthetic efficiency and stress-induced damage to photosystem II. Higher chlorophyll content and improved quantum efficiency are associated with stress tolerance and serve as physiological markers for identifying stress-tolerant genotypes.

*Root architecture*: It is fundamental to drought and heat tolerance because deeper and denser root systems enhance water uptake and hydraulic conductance. Hormonal regulation, particularly through ABA, mediates stress-induced plasticity in root traits, facilitating access to deeper soil moisture and thereby improving yield stability.

*Stomatal regulation*: It mediates the balance between $CO_2$ assimilation, water conservation, and thermal control. Dynamic stomatal responses enable genotypes to sustain photosynthesis and regulate leaf temperature under variable drought and heat conditions.

*Stay-green trait*: It is characterized by delayed senescence and maintains photosynthetic activity during grain filling under terminal stress. Functional stay-green genotypes exhibit enhanced assimilate supply, increased antioxidant protection, and greater yield stability, establishing this trait as a critical target for climate-resilient wheat breeding (Kamal et al. 2019).

### 7.3.2 Development of Biochemical Markers

*Osmotic adjustment*: Osmotic adjustment allows plants to maintain cellular turgor during stress by accumulating compatible solutes such as proline and soluble sugars (Qaseem et al. 2019). By reducing cellular osmotic potential, osmotic adjustment facilitates water balance and sustains metabolic activity under drought and salinity conditions. Additionally, osmotic adjustment preserves cellular hydration and enzyme function through compatible solute accumulation and ABA-mediated regulation, and operates synergistically with antioxidant systems during concurrent drought and heat stress.

*Antioxidant enzyme activity*: Drought and heat stress generate oxidative stress by increasing reactive oxygen species (ROS) production, which damages cellular structures. Stress-tolerant genotypes demonstrate elevated activities of antioxidant enzymes, including SOD, CAT, and ascorbate peroxidase (APX). These enzymes detoxify ROS and maintain cellular integrity, establishing them as reliable biochemical markers of stress resilience.

### 7.3.3 Establishment of Screening Methods for Stress Breeding

Reliable evaluation of drought and heat tolerance in wheat requires integrated screening across controlled environments and field conditions. Controlled screening enables precise manipulation of stress intensity and timing, allowing repeatable assessment of traits such as seedling vigor, chlorophyll fluorescence, and canopy temperature, while high-throughput phenotyping platforms enhance efficiency through rapid, nondestructive measurements. However, field validation in stress-prone regions is essential to capture the complexity of natural environments. Managed stress trials, modified sowing dates, and multilocation testing exposed genotypes to natural drought and heat during sensitive developmental stages, ensuring realistic evaluation. Best practices, therefore, combine controlled precision with field realism to identify genotypes with stable performance under climate-induced stress, supported by specialized facilities developed for both environments.

Stress breeding aims to develop cultivars that maintain high yield potential under optimal conditions while ensuring resilience under drought and heat stress, emphasizing the simultaneous improvement of productivity and stability. Selection across contrasting environments prevents the release of narrowly adapted cultivars and identifies genotypes with broad adaptability. Secondary physiological traits—such as canopy temperature, chlorophyll stability, and water-use efficiency—serve as reliable early indicators of stress resilience. The integration of these physiological indices with molecular breeding tools, including MAS and genomic selection, enhances the improvement of complex polygenic traits by capturing both small-effect QTLs and genome-wide contributions. Quantitative genetics has been fundamental in deciphering the inheritance of drought and heat tolerance, with QTL mapping and GWAS identifying genomic regions and candidate genes associated with yield, root architecture, and water-use efficiency under stress. Moderate to high heritability estimates for key traits support genomic-assisted breeding, although strong genotype × environment (G×E) interactions necessitate multilocation and multiyear testing. Stress indices such as the Stress Susceptibility Index (SSI), Drought Tolerance Index (DTI), and Heat Tolerance Index (HTI) further refine selection by quantifying tolerance while preserving yield potential. Recent advances that combine high-throughput phenotyping, spectral and thermal imaging, genomic prediction, and machine learning-based models provide a comprehensive framework to enhance selection accuracy, accelerate breeding cycles, and ensure the development of wheat genotypes that perform competitively under both stress and optimal conditions.

### 7.3.4 High-Throughput Phenotyping for Stress Tolerance in Wheat

HTP enables rapid, nondestructive assessment of physiological and morphological traits in large populations of wheat, overcoming the limitations of conventional visual scoring. Thermal imaging is widely used to quantify canopy temperature depression as a proxy for transpiration efficiency and drought tolerance through ground-based and unmanned aerial vehicle (UAV) platforms. Multispectral and hyperspectral imaging techniques, including indices such as the NDVI and Photochemical Reflectance Index (PRI), facilitate monitoring of biomass, chlorophyll content, and early water stress signals, while root phenotyping approaches such as shovelomics and imaging-based systems characterize root architecture associated with deep water capture. Under heat stress conditions, HTP tools—including thermal imaging, chlorophyll fluorescence (Fv/Fm), red–green–blue (RGB)-based senescence tracking, and hyperspectral indices—enable the identification of genotypes with thermostable photosynthesis and functional stay-green traits. The integration of HTP with genomics and machine-learning-based prediction models strengthens genotype–phenotype associations and accelerates validation of genetically edited and stress-resilient lines, thereby enhancing breeding efficiency for climate adaptation.

### 7.3.5 Crop Simulation Models and Use of AI in Enhancing Stress Tolerance

Crop simulation models, including DSSAT-CERES, APSIM-Wheat, and STICS, integrate crop physiology, soil processes, and weather data to predict wheat growth and yield under varying climatic conditions (Jones et al. 2002). When combined with weather forecasts, these models facilitate proactive decision-making regarding drought, heat, and frost stress. These models simulate soil water balance, photosynthesis, biomass partitioning, and grain development under drought conditions, enabling the optimization of irrigation strategies, sowing dates, and phenological development to minimize yield losses (Teixeira et al. 2013; Zheng et al. 2014). Large-scale modeling studies have demonstrated that each 1 °C increase in temperature can reduce wheat yield by approximately 6% (Asseng et al. 2015). Decision-support systems, including DSSAT-based advisories and Yield Prophet, integrate seasonal forecasts with crop models to deliver field-specific recommendations

(Hochman et al. 2007). Nevertheless, these systems face persistent challenges, including forecast uncertainty, substantial calibration requirements, and inadequate representation of extreme events. Recent developments in remote sensing, Machine learning, Internet of Things (IoT) sensors, and hybrid, process-based AI modeling have enhanced yield prediction accuracy and real-time stress monitoring (Crane-Droesch 2018). As climate variability increases, integrated modeling and AI-driven decision-support systems are expected to become essential for sustainable wheat production and global food security.

## 7.4 Transcriptomic, Proteomic, and Metabolomic Responses Under Heat and Drought Stress

Omics technologies—including transcriptomics, proteomics, and metabolomics—have substantially advanced understanding of stress adaptation in wheat by revealing multilayered molecular regulatory networks underlying heat and drought responses. Integrated metabolomic and transcriptomic analyses of flag leaves during reproductive-stage heat stress have shown pronounced alterations in lipids, amino acids, phenolics, and alkaloids, along with enrichment of endoplasmic reticulum protein-processing pathways (including HSP20 chaperones) and ATP-binding cassette (ABC) transporters at anthesis and post-anthesis stages, with the latter exhibiting extensive differential gene expression. Transcriptome studies have identified key transcription factor families—MYB, WRKY, NAC, AP2/ERF, bZIP, and HSF—implicated in photosynthesis, oxidative stress regulation, and stress-responsive gene expression, highlighting strong candidates for functional validation. Complementary metabolomic analyses have documented stress-induced shifts in amino acids, antioxidants, and secondary metabolites, some of which correlate positively or negatively with agronomic traits and may serve as potential biomarkers, while enhanced activity of the ascorbate–glutathione cycle highlights its role in ROS detoxification. Proteomic investigations further reveal differential protein abundance associated with photosynthesis, redox homeostasis, energy metabolism, and protein quality control in tolerant genotypes, including RuBisCO activase and PEP carboxylase.

## 7.5 Evolution in the Contemporary Hi-Tech Era

### 7.5.1 Introgression of Donor Alleles from Landraces and Wild Relatives for Stress Tolerance

Heat and drought tolerance in wheat are complex, polygenic traits strongly influenced by GE interactions and its hexaploid genome, necessitating continuous broadening of the genetic base through the incorporation of underutilized landraces, synthetic wheat, and wild relatives. Synthetic hexaploid wheat, developed by crossing tetraploid wheat with *Aegilops tauschii*, introduces novel variation for drought, salinity, and heat tolerance from the secondary gene pool. Similarly, wild emmer wheat (*Triticum dicoccoides*) represents a rich reservoir of stress-adaptive alleles, exhibiting superior water-use efficiency and productivity under water-limited conditions. Synthetic derivatives from interspecific crosses enhance root biomass allocation, improve water-use efficiency, and stabilize grain yield under stress; however, introgression from tertiary gene pools remains constrained by polygenic inheritance, meiotic instability, and linkage drag associated with undesirable alleles.

GWAS exploiting wild emmer diversity have identified significant marker–trait associations for chlorophyll retention and grain yield under heat stress, while evaluations of durum wheat backcross populations confirm that traits such as heading time and 1000-kernel weight contribute substantially to yield stability under combined drought and heat stress. MAS enables efficient tracking of introgressed alleles, and genomic selection (GS) facilitates prediction of breeding values for complex stress-related traits. Together, synthetic wheat development and advanced genomic tools provide a sustainable, strategic framework for enhancing wheat resilience to climate change.

### 7.5.2 Genome Editing Tools to Modify Stress-Responsive Genes

Genome editing technologies, including zinc finger nucleases (ZFNs), transcription activator-like effector nucleases (TALENs), and CRISPR/Cas systems, have revolutionized wheat improvement by enabling precise modification of stress-responsive genes. CRISPR/Cas9 demonstrates greater efficiency due to its simplicity and ability to generate targeted edits without integrating foreign DNA (Kumar et al. 2019). Although wheat hexaploidy introduces complexity, gene redundancy facilitates flexible manipulation of homoeologs (Shan et al. 2013). CRISPR-based approaches in wheat encompass gene knockouts (such as *TaERF3*), promoter editing (*TaGS3*), multiplex targeting of ABA signaling genes (*TaPYL*, *TaPP2C*, *TaSnRK2*, *TaABF*), and homoeolog-specific editing of transcription factors like TaNAC2. These strategies have resulted in enhanced stress tolerance with minimal yield penalties (Kim et al. 2018). Genome edits targeting stomatal regulation (*TaEPFL9*, *TaOST1*), root architecture (*DRO1*), aquaporins (*TaPIP1*), and heat-response regulators (*TaHSFA6f*, *TaFAD7*, *TaCIPK14*) have improved water-use efficiency, thermal stability, pollen viability, and yield under stress conditions (Osakabe et al. 2016). Future advancements are expected to depend on the integration of CRISPR with speed breeding, genomic selection, and high-throughput phenotyping, which could reduce variety development timelines from over a decade to 3–5 years.

## 7.6 Research and Development for Enhancing Drought and Heat Tolerance in Wheat in India

Long-term statistical analyses of climatic trends, yield data, and varietal adoption indicate an increasing impact of heat and drought stress on Indian wheat productivity. Intensified terminal heat and moisture stress during critical growth stages has exacerbated yield losses across major wheat-producing regions, emphasizing regional vulnerability (Dubey et al. 2020). Conversely, yield responses in regions adopting stress-tolerant varieties demonstrate the effectiveness of genetic improvement and advanced agronomic practices in mitigating climate-induced losses. National programs, supported by international collaborations, multilocation trials, and impact assessments, have been instrumental in enhancing wheat resilience (Reynolds et al. 2021). These findings highlight the need for evidence-based breeding, policy support, and efficient dissemination of technologies to maintain wheat productivity amid climate change.

### 7.6.1 Trends in Drought and Heat Events During Critical Wheat Growth Stages in Major Agro-Climatic Zones

Indian wheat is cultivated during the rabi season (October–April), with flowering and grain filling identified as the most climate-sensitive stages that determine final yield (Carrera et al. 2024). Long-term climatic analyses demonstrate a consistent increase in temperature extremes and drought frequency during these critical phases. Maximum temperatures have risen significantly during late winter and early spring, especially from March to April (Kamal et al. 2019). The northwestern and central wheat belts, including Punjab, Haryana, western Uttar Pradesh, Rajasthan, and Madhya Pradesh, exhibit the strongest warming trends and frequently experience terminal heat stress when temperatures surpass the optimal threshold of approximately 30 °C during grain filling. These conditions shorten grain filling duration, limit starch accumulation, and ultimately reduce yields.

Rainfall variability during the rabi season has intensified, leading to increased moisture stress during early reproductive stages. Although wheat cultivation in the Indo-Gangetic Plains is predominantly irrigated, declining groundwater levels and unpredictable winter rainfall have heightened vulnerability in both irrigated and rainfed systems (Kar et al. 2024). Central and eastern India are experiencing more frequent droughts during the tillering and booting stages, which disrupt canopy development and sink formation. Statistical analyses indicate a growing frequency and severity of compound heat and drought events, where diminished transpirational cooling further elevates canopy temperatures (Zhou et al. 2024). Late-sown wheat, which is common after rice harvest, is especially susceptible because grain filling coincides with peak temperatures (Upadhyaya and Bhandari 2022).

### 7.6.2 Analysis of Yield Losses and Area Affected by Terminal Heat and Water Stress Over the Decades

Over the past three decades, statistical analyses have identified terminal heat stress and water scarcity as the primary factors contributing to yield variability in Indian wheat. Studies that integrate historical yield data, meteorological records, and crop simulation models consistently demonstrate significant yield reductions during the reproductive and grain-filling stages due to elevated temperatures and moisture deficits (Fan et al. 2025). An increase of 1 °C in mean temperature during grain filling can decrease wheat yield by 0.6–8.9%, depending on genotype, sowing date, and water availability (Lobell and Field 2007). The north-western Indo-Gangetic Plains, a key wheat production region, have frequently experienced terminal heat events, resulting in early crop maturity and reduced grain size (Kuriachen et al. 2022).

Water stress exacerbates yield losses, particularly in eastern and central India, where groundwater depletion and irregular rainfall reduce irrigation reliability. Rainfed and partially irrigated systems are especially vulnerable to drought during tillering and flowering, resulting in decreased spike number and grain set (Chai et al. 2016). Empirical evidence consistently demonstrates that the combination of heat and drought stress leads to greater yield reductions than either stress alone. The expansion of late sowing following rice harvest has increased the area exposed to terminal heat stress. Simulation studies indicate that climate change will continue to reduce wheat yields, with the magnitude of impact varying across agro-ecological zones, irrigation access, and varietal selection (Yadav et al. 2025). As a result, wheat research priorities have shifted from maximizing yield under optimal conditions to stabilizing production under multiple stress factors (Hossain et al. 2021).

### 7.6.3 Adoption Rates of Stress-Tolerant Wheat Varieties by Farmers

The adoption of stress-tolerant wheat varieties has been the most effective on-farm strategy to mitigate heat and drought stress in India. Over the past two decades, national breeding programs have released several cultivars that combine high yield potential with tolerance to terminal heat and moisture stress (Mohammadi 2018). Varieties such as HD 2967, HD 3086, DBW 187, and DBW 222 currently dominate wheat acreage in the Indo-Gangetic Plains (Singh et al. 2023). HD 2967 achieved rapid adoption in Punjab, Haryana, western Uttar Pradesh, and Rajasthan due to its yield stability across variable climates. In contrast, HD 3086 expanded cultivation in favorable zones because of its superior performance. DBW 187 and DBW 222 have demonstrated particular effectiveness under late-sown and terminal heat conditions (Press Information Bureau of India 2022).

Multilocation trials conducted under the All India Coordinated Research Project (AICRP) on Wheat and Barley have confirmed the superiority of these varieties under stress environments. This success has been further supported by seed multiplication and dissemination efforts through Indian Council of Agricultural Research

(ICAR) institutes and state seed corporations (Mishra et al. 2025). Indirect indicators, including seed distribution data and farmer surveys, indicate that stress-tolerant varieties now occupy 60–80% of wheat acreage in several northern states. Adoption decisions are primarily influenced by extension support, minimum support price (MSP) assurance, grain quality, and disease resistance (Masere and Worth 2022). In summary, genetic replacement with stress-tolerant varieties is India's most successful climate-adaptation strategy for wheat (Yadav et al. 2022).

### 7.6.4 Regional Yield Gains Attributed to Stress Mitigation Through Genetic Improvement and Improved Agronomy

Recent regional yield gains in Indian wheat are primarily attributable to effective stress mitigation rather than to yield maximization under optimal conditions. Genetic improvement programs have prioritized traits such as heat tolerance during grain filling, early vigor, deep rooting, and efficient assimilate partitioning (Jha et al. 2014). Results from multilocation trials consistently indicate that newer stress-tolerant genotypes outperform older varieties in heat- and drought-prone environments, particularly in late-sown systems (Liu et al. 2023). In the north-western Indo-Gangetic Plains, yield stability has increased despite heightened climatic variability (Roy et al. 2022), while eastern regions have experienced gains through the integration of stress-tolerant cultivars with improved management practices.

Agronomic interventions such as timely sowing, optimized irrigation, balanced fertilization, and residue retention have further improved productivity and resilience (Shah et al. 2025). Early planting enables grain filling to occur before the onset of extreme heat, while strategic irrigation helps regulate canopy temperature and mitigate moisture stress. The combination of advanced agronomic practices and stress-resilient varieties has reduced yield variability and promoted sustainability across diverse regions, including central and peninsular India (Pehlivan et al. 2025). Regional yield data demonstrate that integrated genetic and agronomic strategies have substantially reduced climate-induced production losses (Yuan et al. 2024), supporting continued investment in climate-smart wheat improvement.

### 7.6.5 Role of National Programs and International Collaborations in Advancing Stress Tolerance Research

National programs and international partnerships have played a critical role in advancing research on drought and heat tolerance in Indian wheat. The AICRP on Wheat and Barley evaluates breeding lines across diverse agro-climatic zones, using both early and late-sowing, to identify genotypes resilient to stress. The zonal testing system facilitates region-specific varietal recommendations and enhances adoption efficiency. The Indian Institute of Wheat and Barley Research (IIWBR) functions as the national center for wheat improvement, integrating conventional breeding with physiological and molecular approaches. International collaboration with the International Maize and Wheat Improvement Center (CIMMYT) has substantially accelerated progress by providing heat- and drought-tolerant germplasm and advanced breeding methodologies, such as shuttle breeding, high-throughput phenotyping, and genomic selection (Listman 2023). Materials derived from CIMMYT have been extensively utilized in Indian breeding programs to develop locally adapted, climate-resilient varieties. These partnerships have enhanced research infrastructure, facilitated data sharing, and improved climate risk assessment, thereby aligning national objectives with global adaptation goals (The Wheat Initiative 2024).

## 7.7 Conclusion

Recent advances in understanding drought and heat tolerance in wheat have evolved from traditional selection of adapted landraces to an integrated, technology-driven approach that combines physiological insights, quantitative genetics, omics tools, and digital innovation. The identification of key adaptive traits and stress-responsive genes, together with marker-assisted and genomic selection, genome editing, high-throughput phenotyping, and artificial-intelligence-based predictive modeling, has substantially enhanced the capacity to breed climate-resilient cultivars. Although national and international programs have achieved measurable progress in mitigating the impacts of terminal heat and drought, the increasing prevalence of multi-stress environments due to climate change necessitates more rapid deployment of next-generation breeding tools, expanded germplasm exchange, and closer integration of agronomy, policy support, and digital advisory systems. Sustaining wheat productivity and ensuring food system security amid escalating climatic challenges will require a coordinated, multidisciplinary effort.

## References

Asseng S, Ewert F, Martre P et al (2015) Rising temperatures reduce global wheat production. Nat Clim Chang 5:143–147. https://doi.org/10.1038/nclimate2470

Bita CE, Gerats T (2013) Plant tolerance to high temperature in a changing environment: scientific fundamentals and production of heat stress-tolerant crops. Front Plant Sci 4:273. https://doi.org/10.3389/fpls.2013.00273

Carrera CS, Savin R, Slafer GA (2024) Critical period for yield determination across grain crops. Trends Plant Sci 29(3):329–342. https://doi.org/10.1016/j.tplants.2023.08.012

Chai Q, Gan Y, Zhao C et al (2016) Regulated deficit irrigation for crop production under drought stress: a review. Agron Sustain Dev 36:3. https://doi.org/10.1007/s13593-015-0338-6

Crane-Droesch A (2018) Machine learning methods for crop yield prediction and climate change impact assessment in agriculture. Environ Res Lett 13:114003. https://doi.org/10.1088/1748-9326/aae159

Dubey R, Pathak H, Chakrabarti B et al (2020) Impact of terminal heat stress on wheat yield in India and options for adaptation. Agric Syst 181:102826. https://doi.org/10.1016/j.agsy.2020.102826

Fan YC, Yuan YQ, Yuan YC et al (2025) Research progress on the impact of climate change on wheat production in China. PeerJ 13:e18569. https://doi.org/10.7717/peerj.18569

Gupta A, Rico-Medina A, Caño-Delgado AI (2020) The physiology of plant responses to drought. Science 368(6488):266–269. https://doi.org/10.1126/science.aaz7614

Hochman Z, Dang YP, Schwenke GD et al (2007) Simulating the effects of saline and sodic subsoils on wheat crops growing on Vertosols. Aust J Agric Res 58(8):802–810. https://doi.org/10.1071/AR06365

Hossain A, Skalicky M, Brestic M et al (2021) Consequences and mitigation strategies of abiotic stresses in wheat (*Triticum aestivum* L.) under the changing climate. Agronomy 11(2):241. https://doi.org/10.3390/agronomy11020241

Jha UC, Bohra A, Singh NP (2014) Heat stress in crop plants: its nature, impacts and integrated breeding strategies to improve heat tolerance. Plant Breed 133(6):679–701. https://doi.org/10.1111/pbr.12217

Jones JW, Hoogenboom G, Porter CH et al (2002) The DSSAT cropping system model. Eur J Agron 18(3–4):235–265. https://doi.org/10.1016/S1161-0301(02)00107-7

Kamal NM, Alnor Gorafi YS, Abdelrahman M et al (2019) Stay-green trait: a prospective approach for yield potential, and drought and heat stress adaptation in globally important cereals. Int J Mol Sci 20(23):5837. https://doi.org/10.3390/ijms20235837

Kar G, Singh AK, Datta D (2024) Potential impact of climate change on water productivity and water footprints of rice and wheat in the Indo-Gangetic plains of India. Environ Monit Assess 196:1101. https://doi.org/10.1007/s10661-024-13287-6

Kim D, Alptekin B, Budak H (2018) CRISPR/Cas9 genome editing in wheat. Funct Integr Genomics 18(1):31–41. https://doi.org/10.1007/s10142-017-0572-x

Kumar R, Kaur A, Pandey A (2019) CRISPR-based genome editing in wheat: a comprehensive review and future prospects. Mol Biol Rep 46:3557–3569. https://doi.org/10.1007/s11033-019-04761-3

Kuriachen P, Devi A, Sam AS (2022) Wheat yield responses to rising temperature: insights from northern plains of India. Theor Appl Climatol 150:1157–1172. https://doi.org/10.1007/s00704-022-04218-x

Listman M (2023) Hot, dry climates call for resilient, high-performing wheat varieties. International Maize and Wheat Improvement Center (CIMMYT). https://www.cimmyt.org/news/hot-dry-climates-call-for-resilient-high-performing-wheat-varieties/

Liu J, He Q, Zhou G et al (2023) Effects of sowing date variation on winter wheat yield: conclusions for suitable sowing dates for high and stable yield. Agronomy 13(4):991. https://doi.org/10.3390/agronomy13040991

Lobell DB, Field CB (2007) Global-scale climate–crop yield relationships and the impacts of recent warming. Environ Res Lett 2:014002. https://doi.org/10.1088/1748-9326/2/1/014002

Masere TP, Worth SH (2022) Factors influencing adoption, innovation of new technology and decision-making by small-scale resource constrained farmers: the perspective of farmers in lower Gweru, Zimbabwe. Afr J Food Agric Nutr Dev 22(3):20013–20035. https://doi.org/10.18697/ajfand.108.20960

Mishra GP, Dadlani M, Vijay D et al (2025) Development of seed sector in India. In: Dadlani M, Das Gupta PR, Dadlani NK (eds) Indian seed sector. Springer, Singapore. https://doi.org/10.1007/978-981-96-0714-3_2

Mohammadi R (2018) Breeding for increased drought tolerance in wheat: a review. Crop Pasture Sci 69(3):223–241. https://doi.org/10.1071/CP17387

Osakabe Y, Watanabe T, Sugano SS et al (2016) Optimization of CRISPR/Cas9 genome editing to modify abiotic stress responses in plants. Sci Rep 6:26685. https://doi.org/10.1038/srep26685

Pehlivan N, Altaf MT, Emamverdian A et al (2025) Beyond the lab: future-proofing agriculture for climate resilience and stress management. Front Plant Sci 16:1565850. https://doi.org/10.3389/fpls.2025.1565850

Press Information Bureau (2022) Promoting a heat-resistant variety of wheat (press release ID: 1882246). Ministry of Agriculture and Farmers Welfare, Government of India. https://www.pib.gov.in/PressReleasePage.aspx?PRID=1882246

Qaseem MF, Qureshi R, Shaheen H (2019) Effects of pre-anthesis drought, heat and their combination on the growth, yield and physiology of diverse wheat (*Triticum aestivum* L.) genotypes varying in sensitivity to heat and drought stress. Sci Rep 9:6955. https://doi.org/10.1038/s41598-019-43477-z

Reynolds MP, Lewis JM, Ammar K et al (2021) Harnessing translational research in wheat for climate resilience. J Exp Bot 72(14):5134–5157. https://doi.org/10.1093/jxb/erab256

Roy T, Kalambukattu JG, Biswas SS et al (2022) Agro-climatic variability in climate change scenario: adaptive approach and sustainability. In: Chatterjee U, Akanwa AO, Kumar S et al (eds) Ecological footprints of climate change. Springer Climate, Cham. https://doi.org/10.1007/978-3-031-15501-7_12

Shah SS, van Dam J, Singh A (2025) Impact of irrigation, fertilizer, and pesticide management practices on groundwater and soil health in the rice–wheat cropping system—a comparison of conventional, resource conservation technologies and conservation agriculture. Environ Sci Pollut Res 32:533–558. https://doi.org/10.1007/s11356-024-35661-0

Shan Q, Wang Y, Li J (2013) Targeted genome modification of crop plants using a CRISPR-Cas system. Nat Biotechnol 31:686–688. https://doi.org/10.1038/nbt.2650

Singh SK, Kumar S, Kashyap PL et al (2023) Wheat. In: Ghosh PK, Das A, Saxena R et al (eds) Trajectory of 75 years of Indian agriculture after independence. Springer Nature Singapore

Teixeira EI, Fischer G, Van Velthuizen H et al (2013) Global hot-spots of heat stress on agricultural crops due to climate change. Agric For Meteorol 170:206–215. https://doi.org/10.1016/j.agrformet.2011.09.002

The Wheat Initiative (2024) The Wheat Initiative reinforced its commitment to supporting global food security during the G20 meeting of the agricultural chief scientists. https://www.wheatinitiative.org/press-releases/the-wheat-initiative-reinforced-its-commitment-to-supporting-global-food-security-during-the-g20-meeting-of-the-agricultural-chief-scientists/

Upadhyaya N, Bhandari K (2022) Assessment of different genotypes of wheat under late sowing condition. Heliyon 8(1):e08726. https://doi.org/10.1016/j.heliyon.2022.e08726

Yadav MR, Choudhary M, Singh J et al (2022) Impacts, tolerance, adaptation, and mitigation of heat stress on wheat under changing climates. Int J Mol Sci 23(5):2838. https://doi.org/10.3390/ijms23052838

Yadav PK, Bhujel P, Bhandari N (2025) Screening tomato genotypes for early-stage drought tolerance using polyethylene glycol-induced osmotic stress. BMC Plant Biol 25:1476. https://doi.org/10.1186/s12870-025-07508-4

Yuan X, Li S, Chen J et al (2024) Impacts of global climate change on agricultural production: a comprehensive review. Agronomy 14(7):1360. https://doi.org/10.3390/agronomy14071360

Zandalinas SI, Mittler R, Balfagón D et al (2018) Plant adaptations to the combination of drought and high temperatures. Physiol Plant 162(1):2–12. https://doi.org/10.1111/ppl.12540

Zhao C, Liu B, Piao S et al (2017) Temperature increase reduces global yields of major crops in four independent estimates. Proc Natl Acad Sci USA 114(35):9326–9331. https://doi.org/10.1073/pnas.1701762114

Zheng B, Chapman S, Christopher J et al (2014) Frost trends and their estimated impact on yield in the Australian wheatbelt. Procedia Environ Sci 29:171–172. https://doi.org/10.1016/j.proenv.2015.07.244

Zhou S, Wu S, Gao J et al (2024) Increased stress from compound drought and heat events on vegetation. Sci Total Environ 949:175113. https://doi.org/10.1016/j.scitotenv.2024.175113

# Advances in Enhancing Wheat Resilience to Salt Stress

# 8

Neeraj Kulshreshtha

**Abstract**

Soil salinity and sodicity are major abiotic stresses that severely constrain wheat productivity, particularly in arid and semi-arid regions. Elevated salt concentrations disrupt plant physiological processes through osmotic stress, ion toxicity, nutrient imbalance, and oxidative damage, ultimately causing poor germination, stunted growth, and significant yield losses. With salinization intensifying due to unsustainable irrigation practices, groundwater depletion, climate change, and seawater intrusion, the threat to wheat (*Triticum aestivum* L.) cultivation is expected to escalate further. This chapter synthesizes both historical progress and recent advances in enhancing wheat resilience to salinity stress, with an emphasis on physiological and molecular mechanisms, plant breeding, genomics, transgenics, genome editing, and omics-based approaches. It also highlights the transformative potential of digital agriculture, high-throughput phenotyping, and big data analytics in accelerating research and deployment. Statistical insights into salinity-affected areas, yield impacts, and adoption trends are discussed, alongside future projections and policy frameworks. The integration of traditional breeding, advanced technologies, and farmer-centric strategies is crucial to achieving sustainable wheat production and ensuring global food security under escalating salinity stress.

**Keywords**

Wheat · Salinity stress · Sodicity · Abiotic stress tolerance · Genome editing · Omics · Marker-assisted selection · Sustainable agriculture

N. Kulshreshtha (✉)
ICAR-Central Soil Salinity Research Institute, Karnal, Haryana, India



R. Tiwari et al. (eds.), *Wheat Stress Management*,
https://doi.org/10.1007/978-981-92-0902-6_8

## 8.1 Introduction

Soil salinization and sodicity represent an escalating global threat to wheat production. This crisis is driven by a complex interplay of natural and anthropogenic factors that form a self-reinforcing cycle, undermining agricultural sustainability. Unsustainable irrigation practices, excessive groundwater extraction, and the intensifying impacts of climate change are major contributors to the accumulation of salts in the soil. Although the current scale of the problem is unprecedented, the phenomenon itself is not new. Historical accounts from ancient Mesopotamia, often regarded as the cradle of civilization, offer a sobering precedent, as agricultural collapse in this region was closely linked to soil degradation, highlighting the vulnerability of staple crops to salinity stress.

Jacobsen and Adams (1958), in their seminal work *Salt and Silt in Ancient Mesopotamian Agriculture*, documented a catastrophic period of human-induced salinization between 2400 and 1700 BC. Initially, wheat and barley were cultivated in roughly equal proportions around 3500 BC. However, centuries of large-scale irrigation have gradually led to salt accumulation in the soil. By 2100 BC, wheat cultivation had declined to a mere 2%, and by 1700 BC, it had disappeared entirely. In response, farmers turned to the more salt-tolerant barley, but even barley yields eventually declined. This agricultural collapse contributed to the downfall of the Sumerian civilization after nearly five millennia of irrigated farming. The Mesopotamian experience serves as a stark reminder that agricultural systems, if mismanaged, can become engines of their own destruction.

Currently, salt-affected soils cover more than 1.38 billion hectares globally, accounting for over 10% of the Earth's land surface (Abrol et al. 1988). Asia, Africa, and Australia are among the most severely impacted regions. Such soils arise from both natural processes, such as rock weathering and saline groundwater, and human activities, including poor irrigation management, inadequate drainage, and excessive fertilizer use. Irrigated lands are particularly vulnerable, with more than 80 million hectares affected by salinity. This figure continues to rise due to climate change, unsustainable farming practices, and seawater intrusion (Qadir and Oster 2004).

The physiological impact of salinity on wheat is profound and multifaceted. During germination, elevated salt concentrations hinder water uptake, delaying or preventing sprouting. After germination, plants experience osmotic stress that restricts water absorption, coupled with ionic stress that disrupts cellular functions. High sodium ($Na^+$) concentrations interfere with the uptake of essential nutrients such as potassium ($K^+$), leading to nutrient imbalances and impaired metabolic activity (Munns and Tester 2008; Roy et al. 2014). These combined stresses reduce biomass accumulation, compromise photosynthetic efficiency, and ultimately diminish yield.

Addressing this challenge requires more than scientific innovation; it demands a holistic, multi-stakeholder approach. Enhancing wheat resilience to salinity, especially on marginal lands, involves integrating advanced research with farmer-centric policies and socio-economic strategies. Bridging the gap between laboratory breakthroughs and field-level implementation is essential to ensure that scientific progress translates into tangible benefits. Without coordinated efforts across disciplines, institutions, and sectors, even the most promising solutions will fall short of achieving lasting and sustainable impact.

## 8.2 Historical Perspectives on Enhancing Wheat Resilience to Salt Stress

The challenges of cultivating wheat in saline soils are not new; early agricultural observations documented poor crop performance in salt-affected fields. For centuries, farmers in salt-prone regions relied on traditional salt-tolerant landraces and their own selection practices to sustain productivity. These farmer-led selections, rooted in generations of practical experience, represented the earliest efforts toward breeding for salinity tolerance. Initial scientific interventions to address salinity focused primarily on soil reclamation and mapping of saline areas. Early soil scientists investigated the spatial distribution of salts and experimented with management practices such as leaching, improved drainage, and crop rotation to mitigate salinity (Qadir and Oster 2004). These foundational studies established the basis for understanding the physical and chemical dynamics of saline environments. In parallel, early plant breeders began identifying and crossing wheat varieties exhibiting natural resilience to salinity, marking the first steps toward systematic genetic improvement. Over time, this understanding evolved from an exclusive focus on soil management to enhancing the inherent tolerance of plants themselves. This shift represented a major paradigm change, from soil-centric solutions to integrated approaches, combining genetic, physiological, and agronomic improvements. While soil reclamation remains an essential component of salinity management, emphasis has increasingly shifted toward developing wheat varieties capable of maintaining growth and yield under high salt concentrations through improved water relations and ion homeostasis (Munns and Tester 2008). This historical transition from empirical observation to scientific innovation has been instrumental in shaping modern strategies for improving wheat resilience to salinity stress.

## 8.3 Scientific Advancements and Principles Developed over the Past Centuries

When wheat plants are cultivated in salt-affected soils, they activate a range of complex physiological and biochemical mechanisms to sustain growth and survival. These adaptive strategies are crucial for mitigating the three major effects of salinity stress: reduced water uptake due to osmotic imbalance, toxic ion accumulation, and consequent cellular damage (Munns and Tester 2008).

### 8.3.1 Identification of Physiological Mechanisms: Osmotic Adjustment, Ion Homeostasis ($Na^+$ Exclusion and $K^+$ Retention), and Antioxidant Defense Systems

A critical strategy for wheat under saline conditions is the maintenance of ion homeostasis, which involves regulating the concentration of salts within plant tissues. Salt-tolerant varieties possess an enhanced ability to restrict the entry of harmful sodium ($Na^+$) ions into their roots, a process known as ion exclusion (Flowers and Colmer 2008). When $Na^+$ ions do enter the plant, they are actively transported and sequestered into the central vacuole. This compartmentalization isolates toxic salts from the cytoplasm, thereby protecting essential metabolic machinery and preserving the health of newly developing leaves (Munns and Tester 2008). By confining toxic ions, the plant maintains cellular integrity and continues its vital physiological functions with minimal disruption (Fig. 8.1). High salt concentrations in the soil lower the water potential, making it difficult for wheat roots to absorb water and leading to osmotic stress. To counter this, wheat plants perform osmotic adjustment, an adaptive mechanism that maintains turgor pressure and water balance (Chen and Murata 2011). This process involves the synthesis and accumulation of compatible solutes, such as proline, glycine betaine, and various sugars (Ashraf and Foolad 2007). These solutes reduce the osmotic potential within cells, enabling water uptake from the surrounding saline soil. Beyond facilitating water absorption, these molecules serve dual functions: as osmolytes, they help regulate water flow and maintain cell turgor, and as osmoprotectants, they stabilize proteins and membranes, preventing damage from dehydration and ionic toxicity (Anwar et al. 2022). Because these compounds are metabolically inert, they can accumulate to high concentrations without interfering with cellular metabolism. In addition to osmotic adjustment, wheat plants adopt other strategies to optimize water relations

**Fig. 8.1** Diagrammatic representation of the defense mechanisms in wheat plants against salt stress

```
1. Ion Homeostasis
├─ Na⁺ Exclusion (roots)
├─ Na⁺ Sequestration (vacuoles)
└─ K⁺ Retention
2. Osmotic Adjustment
├─ Compatible solutes:
│  ├─ Proline
│  ├─ Glycine betaine
│  └─ Sugars
└─ Maintains turgor & protects enzymes
3. Antioxidant Defense
├─ Enzymatic:
│  ├─ SOD
│  └─ CAT
└─ Non-Enzymatic
├─ Ascorbic acid
└─ Glutathione
```

(Zou et al. 2022). Some genotypes modify their root architecture, developing deeper roots to access less saline water or expanding laterally to exploit moisture in the topsoil. Others regulate stomatal conductance by partially closing stomata to reduce transpiration losses. However, this also limits carbon dioxide uptake and may reduce photosynthesis and growth. Consequently, osmotic adjustment remains one of the most efficient mechanisms for maintaining water balance while sustaining physiological activity under salt stress. Salt stress also induces the excessive generation of reactive oxygen species (ROS), including superoxide ($O_2^-$) and hydrogen peroxide ($H_2O_2$), which are toxic byproducts of metabolism. These molecules can cause severe oxidative damage to lipids, proteins, and nucleic acids. To mitigate this, wheat plants activate a comprehensive antioxidant defense system (Gill and Tuteja 2010; Rao et al. 2025). This system includes enzymatic components such as superoxide dismutase (SOD), which converts superoxide radicals into hydrogen peroxide, and catalase (CAT), which subsequently breaks down hydrogen peroxide into water and oxygen. Complementing these are non-enzymatic antioxidants like ascorbic acid (vitamin C) and glutathione, which directly neutralize ROS. Together, these enzymatic and non-enzymatic defenses form a coordinated network that mitigates oxidative stress, preserving cellular homeostasis and enabling sustained growth under saline conditions.

### 8.3.2 Role of Root System Architecture and Membrane Transporters

The ability of a wheat plant to tolerate salinity is strongly influenced by its root system architecture, membrane transporters, and the synthesis of compatible solutes. A well-developed root system provides a key physical adaptation to salt stress, either by penetrating deeper soil layers to access less saline water or by expanding laterally to increase water uptake from a larger soil volume (Zou et al. 2022). At the cellular level, maintaining ion balance is critical. Membrane transporters encoded by the HKT (High-affinity $K^+$ transporter) and SOS (Salt Overly Sensitive) gene families play essential roles in this process. HKT genes facilitate the removal of excess sodium ($Na^+$) from the xylem, thereby preventing its accumulation in leaves, while SOS genes regulate $Na^+$ exclusion from root cells (Ji et al. 2013). Simultaneously, NHX ($Na^+/H^+$ exchanger) genes contribute to salinity tolerance by sequestering $Na^+$ into vacuoles, effectively isolating it from sensitive cytoplasmic machinery (Munns and Tester 2008). To mitigate osmotic stress, wheat plants also synthesize and accumulate compatible solutes such as proline and glycine betaine. These organic osmolytes maintain cellular water balance and protect enzymes and proteins from denaturation, thereby sustaining physiological function under saline conditions (Ashraf and Foolad 2007).

### 8.3.3 Breeding Strategies for Salt Tolerance: Backcrossing, Selection in Saline Environments, and Polygenic Trait Improvement

To develop wheat cultivars capable of thriving in saline environments, breeders employ various strategies ranging from traditional selection to advanced biotechnological interventions. The most established approach is conventional breeding, which involves crossing salt-tolerant local landraces or wild relatives of wheat with high-yielding commercial cultivars. This strategy exploits the natural genetic diversity available within germplasm collections. Institutions such as the ICAR–Central Soil Salinity Research Institute (ICAR-CSSRI), Karnal, have successfully applied this approach to develop several salt-tolerant wheat varieties (Sharma et al. 2012). Its major advantages lie in its proven effectiveness and low cost. However, conventional breeding for salt tolerance remains challenging due to the polygenic nature of the trait, which complicates accurate screening and selection. Consequently, the process is often time-consuming and labor-intensive. Molecular breeding offers greater precision and efficiency by integrating genetic and genomic tools into the selection process. Marker-assisted selection (MAS) is a key component of this approach, allowing breeders to identify and select plants carrying alleles linked to salt tolerance using DNA markers (Flowers et al. 1997). This enables selection at the seedling stage, thereby accelerating the breeding cycle and improving selection accuracy. Despite its advantages, molecular breeding is constrained by the limited availability of well-characterized and stable quantitative trait loci (QTLs) associated with salt tolerance in wheat. The complex, multigenic control of this trait continues to hinder the precise mapping and utilization of all relevant genomic regions. Genetic engineering, or transgenic technology, provides a more direct and targeted avenue for improving salt tolerance. By introducing specific genes, such as those regulating ion transport or the biosynthesis of compatible solutes, directly into the wheat genome, researchers can achieve substantial enhancement of tolerance within a single generation (Flowers and Colmer 2008). Nonetheless, this approach faces significant challenges, including stringent regulatory frameworks, high development costs, and widespread public resistance to genetically modified (GM) crops. These barriers currently limit the large-scale deployment of transgenic salt-tolerant wheat varieties.

#### 8.3.3.1 Major Problems in Salinity Breeding

Breeding for salt tolerance is fraught with challenges that complicate the effectiveness of all the aforementioned approaches.

##### 8.3.3.1.1 Polygenic Nature of the Trait

The most fundamental challenge is that salt tolerance is a polygenic trait, governed by multiple genes, each exerting a relatively small effect. This complexity makes it exceedingly difficult to isolate and transfer all the requisite genes from a tolerant parent into a new cultivar. Moreover, salinity tolerance in plants arises not from a single mechanism but from a coordinated network of physiological responses, including ion exclusion, osmotic adjustment, and antioxidant defense. Consequently, breeding programs must simultaneously manage and integrate numerous genetic factors, posing significant logistical and scientific challenges.

#### 8.3.3.1.2 Environmental and Developmental Variability

A plant's response to salt stress is highly dynamic rather than static. Its tolerance level can vary depending on the type of salt present (e.g., sodium chloride versus sodium sulfate), the plant's developmental stage, and the presence of additional environmental stressors such as drought or heat (Munns and Tester 2008). This environmental and developmental variability greatly complicates accurate screening, as a genotype that performs well under one set of conditions may exhibit poor tolerance under another. Consequently, breeding trials often yield inconsistent results, posing a major challenge to the reliable identification and selection of salt-tolerant lines.

#### 8.3.3.1.3 Challenges in Screening and Phenotyping

Accurate and reliable screening for salt tolerance remains a critical bottleneck in breeding programs. Field-based screening is often unreliable because salinity levels can fluctuate considerably across a field, creating localized "hotspots" that confound data interpretation. Moreover, field environments typically expose plants to multiple concurrent stresses, making it difficult to isolate the specific effects of salinity (Fig. 8.2). In contrast, controlled-environment screening in laboratories or greenhouses provides greater uniformity but is costly and may not accurately simulate the dynamic and multifactorial conditions encountered in actual agricultural fields (Roy et al. 2014). This disconnect between controlled-environment performance and field outcomes often leads to the selection of genotypes that perform well under experimental conditions but fail to sustain their tolerance in commercial cultivation.

### 8.3.3.2 Methods to Increase Selection Efficiency for Tolerance in Wheat

Selecting wheat genotypes for salt tolerance is a complex process that demands a multifaceted approach. The primary objective is to enhance selection efficiency by addressing challenges such as soil heterogeneity and by identifying genotypes that

**Fig. 8.2** Screening of wheat genotypes in sodic soils

**Fig. 8.3** Selection in microplots provides uniform salt-affected conditions (pH 2: 9.3)

combine salinity tolerance with high yield potential. Strategies to achieve this include the use of controlled-environment experiments and the integration of advanced physiological and molecular screening techniques.

#### 8.3.3.2.1 Overcoming Environmental Challenges

One of the major challenges in breeding for salt tolerance is the substantial variability of salt-affected soils. Although natural field conditions are ideal for evaluating performance in target environments, they often exhibit inconsistent salt levels, resulting in unreliable outcomes. To address this issue, researchers have developed methods to establish controlled, artificial environments. At institutions such as ICAR–CSSRI, Karnal, scientists use microplots and pots to generate uniform saline or alkaline conditions with defined electrical conductivity (EC) or pH (Fig. 8.3). This approach enables the consistent and reliable screening of a large number of genotypes under precisely controlled stress levels (Kesh et al. 2022). For example, alkali soils of a specific pH or exchangeable sodium percentage (ESP) can be prepared by calculating and adding the required amount of a salt such as $NaHCO_3$ to achieve the desired level. This ensures that all plants in the experiment experience uniform stress, thereby improving the comparability and reliability of the results.

#### 8.3.3.2.2 Using Tolerance Indices and Physiological Traits

Effective selection for salt tolerance is not merely about plant survival; it focuses on identifying genotypes that can maintain high yields under stress. This requires the establishment of specific threshold stress levels and the use of quantitative tolerance indices. In wheat, the threshold salinity level is approximately 6.0 dS $m^{-1}$, where noticeable yield reduction begins, while a 50% yield reduction is typically observed at around 14.0 dS $m^{-1}$. Understanding these thresholds is essential for designing reliable screening experiments. Moreover, screening should be conducted at multiple growth stages, germination, seedling, and maturity, since a genotype's tolerance may vary throughout its life cycle.

Beyond simple yield measurements, breeders employ a range of indices to gain a more comprehensive understanding of salt tolerance. The Germination Rate Index

(GRI), modified from Maguire (1962), for instance, helps distinguish genotypes with similar final germination percentages but differing germination speeds (Bai et al. 2018), with faster germination considered advantageous. For assessing yield performance, indices such as Geometric Mean Productivity (GMP) and the Stress Tolerance Index (STI) are particularly useful, as they evaluate a genotype's performance under both stressed and non-stressed conditions, enabling the identification of lines that are both tolerant and high-yielding (Akram et al. 2025). Additionally, the AMMI (Additive Main Effects and Multiplicative Interaction) model is a powerful tool for evaluating genotype stability across environments (Prasad et al. 2016). This model can be applied not only to yield data but also to key physiological parameters such as the $K^+/Na^+$ ratio, an important indicator of a plant's capacity to exclude toxic sodium ions.

#### 8.3.3.2.3 Integrating Traditional and Modern Breeding Approaches

Combining traditional breeding approaches with modern molecular and phenotyping techniques can substantially accelerate the development of salt-tolerant wheat varieties. The pedigree selection (PS) method, which relies on rigorous visual selection in each generation, has proven effective but is often time-consuming. An alternative, the modified single seed descent (MSSD) approach, involves limited selection during the main growing season while advancing generations without selection in the off-season. Studies have shown that both PS and MSSD can be equally effective for yield improvement in salt-affected soils (Kulshreshtha et al. 2019). While morphological traits such as plant height and days to heading generally exhibit high heritability, yield-related traits tend to have low to intermediate heritability, emphasizing the importance of selection based on multiple indices. Modern breeding programs increasingly incorporate high-throughput phenotyping, which employs non-destructive technologies such as hyperspectral imaging to rapidly collect large-scale physiological and morphological data. This enables the early identification of tolerant genotypes, often before visible stress symptoms appear. On the molecular front, marker-assisted selection (MAS) utilizes DNA markers linked to salt-tolerance genes, allowing for rapid and accurate screening of seedlings under laboratory conditions, thereby reducing dependence on lengthy field evaluations. This approach is supported by quantitative trait loci (QTL) mapping, which identifies specific genomic regions associated with tolerance. Advancing further, genomic selection (GS) uses genome-wide marker information to predict the overall genetic potential of individual plants for complex traits, thereby improving selection accuracy and efficiency. Ultimately, the most effective breeding strategy integrates the precision of molecular and phenomic tools with the robustness of conventional field-based evaluations. Such an approach ensures that the resulting wheat cultivars are not only genetically resilient to salinity but also agronomically stable and high-yielding under real-world salt-affected environments.

#### 8.3.3.2.4 Integration of Plant Physiology and Agronomy in Early Screening Methodologies

Early screening methods for salt tolerance in wheat represented an important integration of plant physiology, agronomy, and breeding. Agronomists played a key role in identifying fields with varying salinity levels and systematically screening a diverse range of germplasm, from landraces to advanced breeding lines. Concurrently, plant physiologists evaluated critical physiological traits such as the $K^+/Na^+$ ratio and the accumulation of compatible solutes such as proline (Munns and Tester 2008). This multidisciplinary approach, combining field-level observations with detailed physiological assessments, proved fundamental in understanding salinity tolerance. It enabled breeders to establish clear correlations between field performance and underlying physiological mechanisms, thereby enhancing the efficiency and precision of selecting promising genotypes for further breeding.

### 8.3.4 Soil and Water Management Techniques: Gypsum Application, Organic Amendments, and Efficient Drainage Systems

Agronomic management strategies are essential for the successful cultivation of wheat in salt-affected soils. These practices focus on improving soil conditions and managing water and nutrient availability. For saline soils—characterized by high concentrations of soluble salts, leaching is a key practice. This involves applying large volumes of good-quality water to flush excess salts below the plant's root zone, often supported by subsurface drainage systems to prevent waterlogging and further salt accumulation. Efficient irrigation methods such as drip or furrow irrigation, combined with smaller and more frequent water applications, help maintain a consistently low salt concentration around the roots. When only poor-quality water is available, blending it with higher-quality water can mitigate its adverse effects (Sharma et al. 2018). For sodic soils, those with high sodium ($Na^+$) content, the use of soil amendments is critical. The application of gypsum is a widely adopted and effective practice, as calcium ($Ca^{2+}$) from gypsum replaces exchangeable sodium, thereby improving soil structure, porosity, and drainage (Abrol et al. 1988). Incorporating organic matter such as farmyard manure or compost further enhances soil structure and water-holding capacity, while providing a buffering effect against salt stress. Laser land leveling, although not directly correcting sodicity, ensures a uniform soil surface, thereby improving the efficiency of amendments like gypsum and minimizing water stagnation. Nutrient management also plays a vital role in improving crop resilience. Increasing phosphorus (P) and potassium (K) fertilization helps wheat plants counteract the toxic effects of sodium and maintain ionic balance (Hussain et al. 2020). Additionally, selecting salt-tolerant wheat varieties adapted to local conditions and planting on raised beds or ridges can protect seeds and young seedlings from high surface-salt concentrations. Using a slightly higher seed rate can also compensate for potential germination losses under saline or sodic conditions.

## 8.4 Recent Developments in Enhancing Wheat Resilience to Salinity

As a vital staple for global food security, wheat production is increasingly threatened by rising soil salinity (Traye et al. 2025). Salinity stress severely reduces crop yield by inducing osmotic and ionic imbalances, along with oxidative damage at the cellular level. However, recent advancements in plant science and breeding have opened new avenues for developing resilient wheat cultivars. Researchers are now integrating traditional breeding approaches with advanced molecular and genomic tools to effectively address this challenge (Wang and Li 2024).

### 8.4.1 Discovery of Salinity-Tolerance QTLs and Candidate Genes

A breakthrough in developing salt-tolerant wheat has been the discovery and characterization of quantitative trait loci (QTLs) and their associated candidate genes. QTLs are genomic regions that harbor genes contributing to quantitative traits, such as salinity tolerance. Advances in high-throughput genotyping and mapping approaches, including genome-wide association studies (GWAS) and bi-parental mapping, have enabled researchers to precisely identify key genomic regions linked to salt tolerance. For instance, QTLs on wheat chromosomes 2A, 2D, and 5A have been associated with critical tolerance mechanisms such as maintaining shoot growth, regulating ion homeostasis, and controlling the $Na^+/K^+$ ratio (Asif et al. 2021; Yacoubi et al. 2022). The identification and functional characterization of specific genes within these regions, including ion transporters and stress-inducible genes, provide valuable molecular markers for use in breeding programs.

### 8.4.2 Marker-Assisted Selection (MAS) and Gene Pyramiding

Once QTLs are identified, plant breeders can exploit them through marker-assisted selection (MAS). This technique uses DNA markers tightly linked to desirable traits, allowing the identification of superior genotypes within breeding populations without the need for extensive field evaluations (Ashraf et al. 2012; Wang and Li 2024). MAS significantly accelerates the breeding cycle, enhancing both precision and efficiency compared to conventional selection methods. An additional, highly effective approach is gene pyramiding, which involves combining multiple salinity-resistance loci into a single elite wheat genotype. By stacking several favorable alleles, breeders can develop cultivars with a more durable and comprehensive defense against salt stress. This strategy is particularly advantageous for complex, polygenic traits such as salinity tolerance.

### 8.4.3 Transgenic and RNAi Approaches

Modern molecular biology tools provide powerful avenues for enhancing wheat resilience to salinity. Transgenic approaches involve introducing genes from other species, such as halophytes, that confer superior salt tolerance (Paul and Roychoudhury 2018). These genes typically encode proteins involved in ion transport, antioxidant defense, or osmolyte biosynthesis, all of which are essential for maintaining cellular homeostasis under saline conditions. For example, the expression of a gene encoding a $Na^+/H^+$ antiporter facilitates the sequestration of excess sodium ions into vacuoles, thereby preventing cytoplasmic toxicity and maintaining ion balance (Paul and Roychoudhury 2018). Similarly, RNA interference (RNAi) can be employed to silence genes that negatively affect stress tolerance. By downregulating genes responsible for excessive ion accumulation or reactive oxygen species (ROS) generation, researchers can enhance the plant's ability to withstand salt stress. These targeted molecular interventions offer a precise and effective means of engineering improved stress tolerance in wheat.

### 8.4.4 Insights from Multi-omics Technologies

To unravel the complex mechanisms underlying wheat's response to salt stress, researchers are increasingly employing integrated *omics* approaches. Transcriptomics provides insight into gene expression dynamics, revealing which genes are activated or suppressed in response to stress. Proteomics examines the resulting protein profiles, while metabolomics characterizes the small molecules involved in metabolic adjustments (Wang et al. 2023). Integrating data from these complementary disciplines offers a comprehensive, systems-level understanding of the plant's adaptive responses. Such an approach enables the identification of key regulatory pathways and molecular networks essential for stress resilience, thereby revealing novel targets for breeding and genetic engineering. For example, metabolomic studies have shown that salt-tolerant wheat genotypes maintain elevated levels of specific amino acids and secondary metabolites that function as osmolytes and antioxidants, contributing to improved tolerance (Wang et al. 2023).

### 8.4.5 Development of Salinity-Tolerant Cultivars

Wheat improvement programs across the world have successfully developed several salt-tolerant varieties and genetic materials, substantially improving the livelihoods of farmers in salt-affected regions. These varieties not only ensure stable yields on previously unproductive lands but also contribute to economic growth, land reclamation, and enhanced food security.

#### 8.4.5.1 Salt-Tolerant Wheat Varieties and Genetic Materials

Significant progress has been made in developing wheat germplasm for salt tolerance, with major contributions from the ICAR–CSSRI, Karnal, which systematically screens both Indian and exotic wheat lines. Through selection, hybridization, mutation, and shuttle breeding, numerous improved derivatives of the salt-tolerant landrace *Kharchia* have been developed. Although some of these derivatives were not released as commercial varieties, they were registered as valuable genetic stocks with ICAR–National Bureau of Plant Genetic Resources (NBPGR), New Delhi, and have served as crucial resources for future breeding programs. As a result of these efforts, several salt-tolerant varieties have been released in India. The first was KRL 1–4 in 1990, followed by KRL 19 in 2000, both showing strong tolerance to saline soils with an electrical conductivity (ECe) up to 7.3 dS $m^{-1}$ and sodic soils with a $pH_2$ up to 9.3. Subsequent releases, KRL 213 and KRL 210 in 2011 and 2012, respectively, offered enhanced yield potential, improved disease resistance, and greater adaptability to stress conditions. Notably, KRL 210 also exhibits additional tolerance to waterlogging. These varieties marked a significant improvement over earlier cultivars and contributed to the biological amelioration of salt-affected soils. More recently, KRL 283, released in 2018, has demonstrated superior tolerance to salinity ($EC_e$ up to 7.5 dS $m^{-1}$) and sodicity ($pH_2$ up to 9.35), along with remarkable waterlogging tolerance. Its high yield stability under both normal and stress conditions represents a major advancement. In addition to these released varieties, CSSRI has developed and registered several important genetic stocks, such as KRL 35, KRL 99, and KRL 3–4, with ICAR–NBPGR, New Delhi. These genotypes are widely used in breeding programs as strong donors of salt and waterlogging tolerance, characterized by low sodium uptake and robust stress adaptation.

Globally, other countries have also made significant strides in breeding for salinity tolerance. In Australia, wheat improvement programs at the University of Adelaide and CSIRO have developed lines such as MW293 and improved durum selections with strong sodium exclusion traits for saline soils (Deinlein et al. 2014). In Egypt, varieties such as Sakha 8, Sakha 93, Sids 1, and Giza 168 are widely cultivated in the Nile Delta, combining high tolerance with yield stability (Elhag 2023). In Bangladesh, selections such as BARI Gom-25 and BARI Gom-26 are promoted for cultivation in the saline coastal belt. Similarly, in Pakistan, varieties including Sarsabz, Bhittai, and Kiran-95 are grown in the salt-affected areas of Sindh (Kesh et al. 2022), while Pasban-90 serves as a benchmark cultivar due to its ability to maintain a high potassium-to-sodium ratio ($K^+/Na^+$) in plant tissues (Rehman et al. 2025). At the global level, CIMMYT has played a pivotal role in identifying and distributing salinity-tolerant germplasm through international nurseries. Modern breeding programs now integrate conventional selection with physiological screening, marker-assisted selection, and genomics-based tools to accelerate the development of resilient, high-yielding wheat cultivars for saline environments.

#### 8.4.5.2 Impact on Farmers' Livelihoods

The adoption of salt-tolerant wheat varieties has had a profound and positive impact on the livelihoods of farmers in salt-affected regions, transforming previously unproductive land into productive and valuable assets.

##### 8.4.5.2.1 Increased and Stable Yields

The primary advantage of these varieties lies in their ability to produce reasonable yields under conditions where conventional cultivars would fail completely. This ensures farmers a more stable and predictable income, substantially reducing the risk of crop failure. The introduction of varieties such as KRL 210 and KRL 213 has provided a practical and economically viable solution for cultivating saline and sodic soils that were once considered barren. The widespread distribution of breeder and truthfully labeled seeds by ICAR–CSSRI, Karnal, and collaborating agencies has facilitated large-scale adoption in Indian states such as Haryana, Punjab, Uttar Pradesh, and Rajasthan. KRL 210 has been adopted on more than 1.18 lakh hectares, while KRL 213 covers over 1.23 lakh hectares (Kulshreshtha et al. 2022). Farmer-to-farmer seed exchange has further accelerated their dissemination, likely expanding the actual area under these varieties beyond formal estimates.

##### 8.4.5.2.2 Economic Benefits

The higher and more stable yields achieved by these varieties directly translate into increased farm income, often resulting in a higher benefit–cost ratio. It is estimated that these cultivars have generated wheat grain worth over ₹1513 crore from lands that were previously barren or low-yielding. Such economic gains enable farmers to break the cycle of poverty, reinvest in their farms, and build greater financial stability and resilience against climatic variability. Varieties like KRL 210 have gained widespread popularity due to their superior adaptability, reduced input requirements, and desirable traits such as early maturity and excellent chapati-making quality, making them highly suitable for diverse saline ecosystems.

##### 8.4.5.2.3 Land Reclamation

By enabling the cultivation of productive crops on previously underutilized saline and sodic lands, these varieties help farmers effectively reclaim degraded areas for agriculture. This expansion of the cultivable land area represents a major advantage in regions with limited resources and high population density.

##### 8.4.5.2.4 Food and Nutritional Security

Ensuring a consistent food supply from these reclaimed lands contributes significantly to food and nutritional security for farming families and the wider community. It helps meet the food demands of a growing population in regions where soil salinity remains a major constraint to agricultural productivity. The widespread adoption of these salt-tolerant varieties has not only increased farm incomes but also strengthened food security and promoted ecological sustainability, highlighting the critical role of crop breeding in addressing site-specific agricultural challenges.

## 8.5 Evolution in the Contemporary Hi-Tech Era

Contemporary agriculture is being transformed by high-tech innovations that are fundamentally redefining how crop resilience is enhanced. In wheat, this shift represents a move from traditional breeding to a precise, data-driven, and proactive approach for addressing soil salinity. This technological evolution is accelerating the development of salt-tolerant cultivars, offering renewed hope for sustainable food production in an era of shrinking arable land.

### 8.5.1 Genome-Editing (CRISPR/Cas9) for Precise Trait Manipulation

Genome editing, particularly through the CRISPR/Cas9 system, has emerged as a powerful tool for precisely manipulating genes associated with salt tolerance. Unlike conventional transgenic methods that involve inserting foreign genes, CRISPR/Cas9 functions as a molecular "scissor," enabling targeted modifications to the plant's own DNA. This technology allows scientists to knock out genes that contribute to salt sensitivity or activate genes that enhance tolerance, such as those involved in ion transport or stress signaling (Yacoubi et al. 2022; Ma et al. 2024). By precisely editing one or multiple genes simultaneously—a process known as multiplexing—researchers can develop new wheat varieties with enhanced resilience without introducing foreign DNA, thereby easing potential regulatory and public acceptance concerns (Smedley et al. 2021).

### 8.5.2 High-Throughput Phenotyping and Remote Sensing

Phenotyping has long been a major bottleneck in breeding programs. Today, high-throughput phenotyping platforms integrate sensors, robotics, and imaging systems to rapidly and non-destructively measure plant growth, chlorophyll content, and other physiological traits under controlled saline conditions (Wu et al. 2015). This enables the real-time screening of thousands of genotypes, greatly accelerating the identification of promising breeding lines. Complementing these systems are remote sensing tools such as unmanned aerial vehicles (UAVs) and satellite imagery, which provide powerful means to map and monitor salt-affected areas on a large scale. By analyzing spectral data—reflectance patterns across different wavelengths—scientists and farmers can accurately delineate saline zones and evaluate crop performance in the field (Chen et al. 2024). These technologies support targeted management practices and supply valuable data for breeders to assess new cultivars under real-world conditions.

### 8.5.3 Machine Learning and Big Data Analytics

The vast amount of data generated by modern genomic and phenotyping platforms can be overwhelming. This is where machine learning and big data analytics play a transformative role. Machine learning algorithms can process complex datasets to uncover subtle relationships between a plant's genetic makeup (genotype) and its performance in a specific environment (phenotype). This capability is particularly valuable for understanding Genotype × Environment (G × E) interactions, which are inherently complex and difficult to predict using traditional analytical approaches (Roorkiwal et al. 2018). By training predictive models on data from multi-environment trials, breeders can accurately forecast which wheat genotypes are most likely to perform well in particular saline conditions, thereby reducing the need for costly and time-consuming field evaluations (Fernandes et al. 2024). These data-driven models have become a cornerstone of modern breeding, enabling more efficient selection and accelerating the development of salt-tolerant wheat varieties.

### 8.5.4 Digital Decision Support Systems and Farmer Advisory Apps

The final piece of this high-tech puzzle is ensuring that information reaches those who need it most: the farmers. Digital decision-support systems and farmer advisory applications serve as vital links between scientific research and on-farm implementation. These platforms integrate data from multiple sources, including remote sensing, weather forecasts, and soil sensors, to provide farmers with actionable, real-time recommendations. For instance, an app could suggest the most suitable salt-tolerant wheat cultivar, determine the optimal planting window, or recommend specific land management practices based on the field's salinity profile (Leng and Yimit 2008). By empowering farmers with data-driven insights, these tools help translate advanced research into tangible outcomes, enhancing productivity, sustainability, and resilience in salt-affected agroecosystems.

## 8.6 Statistical Insight into Enhancing Wheat Resilience to Salinity

Contemporary statistical analyses and data-driven insights are essential for understanding and mitigating the effects of soil salinity on wheat production. This approach transcends isolated experiments, offering a broader and more accurate understanding of the problem and the effectiveness of potential solutions. By quantitatively assessing the extent of salinity stress and evaluating the outcomes of various interventions, researchers can formulate more targeted and effective strategies to enhance wheat resilience.

### 8.6.1 Extent and Distribution of Salt-Affected Soils

Globally, soil salinity poses a major threat to agricultural productivity, with estimates indicating that nearly 1 billion hectares are affected by salt, and projections suggesting that up to 50% of the world's arable land could be impacted by 2050. In India, approximately 6.74 million hectares are affected by salinity and alkalinity. These problem soils are unevenly distributed, with concentrations in key wheat-growing regions—particularly the arid alluvial tracts of Punjab, Haryana, and Rajasthan in the north, as well as parts of the Indo-Gangetic Plains. The distribution is often patchy and variable, highlighting the need for large-scale statistical mapping and continuous monitoring.

### 8.6.2 Estimated Yield Losses and Their Impact

The economic and food security implications of soil salinity are profound. Wheat, though moderately salt-tolerant, can experience yield reductions of up to 50% under moderate salinity levels. Globally, salinity-induced wheat production losses are estimated at 10–15 million metric tons per year (Traye et al. 2025). In India, annual crop production losses due to salinity and sodicity are estimated at approximately 17 million tons, translating to a monetary loss exceeding ₹23,000 crores (Kaledhonkar and Meena 2022). These statistics highlight the urgent need to develop and implement resilient farming strategies to safeguard food and nutritional security.

### 8.6.3 Trends in the Adoption of Salt-Tolerant Varieties and Impact Assessment

The development and adoption of salt-tolerant wheat varieties are key to mitigating yield losses due to soil salinity. Adoption trends indicate that genetic and agronomic innovations are steadily narrowing these productivity gaps. Salt-tolerant wheat varieties bred for sodic and saline soils, such as KRL 1–4, KRL 19, KRL 210, KRL 213, and KRL 283, are now widely recommended across the North-Western and Eastern zones of India. Participatory demonstrations have consistently shown 20–50% yield gains over local checks in saline-prone districts (Meena and Singh 2020). In one study, frontline demonstrations (FLDs) conducted on saline-affected lands in Rajasthan revealed that salt-tolerant varieties provided an average yield advantage of 11.1 quintals per hectare compared to local varieties. The higher benefit–cost ratio (2.4), relative to 1.8 for traditional varieties, further demonstrates their economic viability, encouraging wider adoption among farmers (Meena and Singh 2020). Comparative on-farm analyses have shown that KRL 210 sustains significantly higher yields than popular non-tolerant varieties as soil pH and residual alkalinity increase (Sheoran et al. 2021). Complementary agronomic practices, such as balanced nitrogen management with higher economic optima on sodic soils, gypsum or press-mud amendments, improved drainage, and conjunctive use of better-quality irrigation water, have been widely disseminated through Krishi Vigyan Kendras (KVKs) and state programs. These interventions have enhanced input-use efficiency and stabilized farm returns (Kumar and Sharma 2020; Sheoran et al. 2021). The wheat variety KRL 210, developed by the ICAR-CSSRI, Karnal, represents a significant technological innovation addressing the twin challenges of soil and water salinity. Its impact assessment reveals a multi-dimensional positive influence on agricultural productivity, farmer livelihoods, and environmental sustainability, particularly in salt-affected regions of India. The primary impact of KRL 210 lies in its ability to deliver high and stable yields under stress conditions where conventional wheat varieties perform poorly. Field studies and on-farm demonstrations consistently show that KRL 210 provides a marked yield advantage. For example, in areas with high soil sodicity (pH >8.7) and high residual alkalinity in irrigation water, KRL 210 exhibits a lower yield reduction (15.7%) compared to popular but susceptible varieties such as HD 2967 (21.9%) (Sheoran et al. 2022). While its performance is most remarkable under stress, KRL 210 also demonstrates strong yield potential, producing up to 7.0 t $ha^{-1}$ in normal soils and 3.0–5.0 t $ha^{-1}$ in sodic soils. This dual adaptability makes it a versatile choice for farmers managing heterogeneous land conditions. Its tolerance to sodicity (up to pH $\approx$ 9.3) and salinity (up to 7.0 dS $m^{-1}$) allows it to maintain key physiological functions, such as leaf relative water content and photosynthetic rate, under saline stress. This physiological resilience translates into stable and reliable harvests, buffering farmers against the unpredictability of salt-induced yield losses. The adoption of KRL 210 has delivered tangible economic benefits, particularly for marginal and smallholder farmers. The higher yields translate into significant increases in net farm income. For instance, in the Ghaghar basin of Haryana, economic analyses revealed substantial gains in regions with high soil pH and residual alkalinity in groundwater (Sheoran et al. 2022). A success story from ICAR-CSSRI (2017) also documents a farmer achieving a markedly higher benefit–cost ratio through KRL 210 adoption. By transforming previously low-yielding or uncultivable saline lands into productive assets, KRL 210 strengthens household food security and enhances farmers' socio-economic resilience (Meena and Singh 2020). Beyond direct economic returns, the large-scale adoption of KRL 210 offers broader environmental and social benefits. Its ability to thrive in salt-affected soils promotes the productive utilization of degraded land, thereby supporting the sustainability of agricultural systems while reducing dependence on chemical soil amendments. Moreover, its tolerance to poor-quality irrigation water enables efficient use of limited water resources, a critical advantage in water-scarce regions. The successful dissemination of KRL 210 through farmer-participatory approaches and knowledge-cum-seed networks has empowered farmers to make informed decisions and improve their livelihoods (Meena and Singh 2020). This participatory model has facilitated the rapid spread of the variety across Punjab, Haryana, Uttar Pradesh, and Rajasthan, underscoring its adaptability and impact.

### 8.6.4 Impact Assessments of Salinity Mitigation Projects

Rigorous impact assessments are essential for evaluating the effectiveness of large-scale agricultural interventions. The CSSRI-led initiatives on alkali land reclamation and subsurface drainage have demonstrated remarkable success in restoring soil productivity and improving livelihoods. Through the adoption of CSSRI technologies, approximately 1.8 million hectares of alkali land have been reclaimed, contributing an additional 12 million tonnes of rice–wheat grain annually to India's food basket (Kaledhonkar and Meena 2022). At the national scale, gypsum-based sodic soil reclamation has been implemented on about 2.07 million hectares, with reclaimed lands contributing an estimated 16.6 million tonnes of food grains each year. Typical productivity trajectories show wheat yields increasing from ~2.0 t $ha^{-1}$ in the first year to ~3.0 t $ha^{-1}$ by the third year after reclamation. These yield improvements translate into substantial farm income gains and employment generation. Overall, the programs have enhanced national food security while creating approximately 250 million man-days of rural employment (Kaledhonkar and Meena 2022). The success of these initiatives provides robust statistical evidence supporting the effectiveness of scientific approaches in managing saline and sodic soils and improving farmer livelihoods. Case studies from the Indo-Gangetic Plains corroborate these outcomes, linking land reclamation efforts to improved soil health indicators, higher net returns, and enhanced livelihood security for resource-poor households (Mandal et al. 2022).

### 8.6.5 Role of Government Schemes and Policy Incentives

Government policies play a critical role in promoting salinity-resilient agriculture. Schemes such as the National Mission for Sustainable Agriculture (NMSA), which encompasses the Soil Health Card (SHC) and the Pradhan Mantri Krishi Sinchayee Yojana (PMKSY), indirectly support farmers in managing saline and sodic conditions (PIB 2025). The SHC scheme provides detailed information on soil nutrient status, enabling more informed nutrient management decisions that help mitigate the adverse effects of salinity. The PMKSY, with its strong emphasis on micro-irrigation, promotes efficient water use, a critical factor in preventing secondary salinization (PIB 2025). Its "Per Drop More Crop" component accelerates the adoption of micro-irrigation technologies, thereby reducing the risk of salt accumulation through improved water-use efficiency (PMKSY 2023). Importantly, the Reclamation of Problem Soils (RPS) sub-scheme under the Rastriya Krishi Vikas Yojana (RKVY) provides targeted funding for gypsum-based soil reclamation and related interventions through a project-based, state-led model. Parliamentary and ministerial reports reaffirm the RPS objective—to restore degraded saline and sodic lands to productive cultivation and curb the further spread of soil degradation (MoA&FW 2016). By offering a robust institutional framework and financial incentives, these government initiatives collectively empower farmers to adopt best practices and technologies, such as salt-tolerant crop varieties, balanced nutrient management, and improved land and water-use strategies, thereby strengthening agricultural resilience at the national level.

## 8.7 Future Projections and Global Challenges

Future projections for enhancing wheat resilience to salinity face several complex challenges, primarily driven by climate change and a need for greater global cooperation. While scientific advancements have been significant, scaling these solutions and ensuring their effectiveness across diverse environments remains a formidable task.

### 8.7.1 Anticipated Expansion of Salt-Affected Soils

The future of agriculture will be increasingly shaped by the growing crisis of soil salinization. Climate change, through its associated shifts in rainfall patterns and increased evapotranspiration, is expected to intensify salinity problems, particularly in arid and semi-arid regions (Zahid et al. 2025). This challenge is further exacerbated by rising sea levels, which promote saltwater intrusion into coastal aquifers and river deltas, leading to the contamination of both groundwater and topsoil (FAO 2024).

In addition to climatic drivers, poor irrigation management, including inadequate drainage and the use of saline groundwater, continues to contribute to secondary salinization (Ullah et al. 2021). According to the Food and Agriculture Organization of the United Nations (FAO), nearly 1.4 billion hectares of land worldwide are currently affected by salinity, and this figure is projected to increase substantially in the coming decades (FAO 2024).

### 8.7.2 Combined Stresses and Genetic Bottlenecks

A major challenge for future breeding programs is the development of wheat varieties capable of withstanding multiple stresses, not only salinity but also high temperatures and drought. Salinity stress frequently co-occurs with other environmental constraints, and the plant's physiological responses are complex and interconnected (Ullah et al. 2021). Breeding for a single trait, such as salt tolerance, may inadvertently compromise a variety's resilience to heat or water scarcity. This highlights the difficulty faced by plant breeders, who must now select for multiple, often interacting and competing traits simultaneously. A further bottleneck lies in the limited genetic variation for high-level salinity tolerance within the domesticated wheat gene pool (Ahmad et al. 2013). Although certain landraces exhibit moderate tolerance, the existing genetic resources available to breeders are insufficient to achieve the resilience required under future climate projections. This limitation highlights the need to explore non-conventional genetic resources, including wild relatives, synthetic derivatives, and molecular approaches, to broaden the adaptive base for breeding salinity-resilient wheat.

### 8.7.3 Global Collaboration and Novel Genetic Resources

To overcome these challenges, a new paradigm of global collaboration is essential. The establishment of open-access databases and gene banks is critical for the exchange of germplasm and genetic information related to salinity tolerance (Crop Trust 2020). International organizations such as the International Center for Biosaline Agriculture (ICBA) and the Global Crop Diversity Trust are actively engaged in conserving and disseminating diverse plant genetic resources, including those originating from marginal and saline environments. This collaborative model ensures that breeders and researchers worldwide can access and utilize the most promising genetic materials, thereby accelerating the development of salinity-resilient crops.

One of the most promising strategies for broadening the genetic base of wheat is to explore its wild relatives and halophytes. Halophytes, naturally adapted to high-salinity environments, possess distinctive physiological and genetic mechanisms for salt tolerance, such as efficient ion exclusion, osmotic adjustment, and the formation of salt glands. Studying these species enables scientists to identify novel genes and adaptive mechanisms that can be incorporated into wheat through advanced genetic engineering or interspecific hybridization, effectively overcoming the limitations of the domesticated gene pool (Grigore and Vicente 2023).

### 8.7.4 Policy Focuses on Sustainable Agriculture

The success of scientific and collaborative efforts ultimately depends on supportive government policies and effective farmer training at the grassroots level. Policies that promote sustainable saline agriculture, incentivize the adoption of salt-tolerant varieties, and build farmer capacity in water- and resource-efficient cropping systems are essential. When coupled with farmer advisory applications and digital decision-support tools, these policy measures form the critical final link between research and real-world impact, ensuring that laboratory innovations translate into enhanced productivity, resilience, and livelihood security for the farmers most affected by salinization.

## 8.8 Conclusion

The escalating crisis of soil salinization poses a major threat to global food security, demanding an urgent and integrated response to strengthen wheat resilience. Advances in understanding wheat's physiological defense mechanisms, such as ion homeostasis ($Na^+$ exclusion and $K^+$ retention), osmotic adjustment, and antioxidant systems, have shifted the focus from soil-centric management to a more effective plant-centric strategy. The convergence of genomics, molecular breeding, and advanced biotechnologies such as transgenics and CRISPR–Cas9 genome editing offers unprecedented precision in developing salt-tolerant wheat genotypes. Complementarily, digital agriculture and multi-omics integration through high-throughput phenotyping, big data analytics, and systems biology approaches are revolutionizing the accuracy and speed of identifying resilient lines. Ensuring sustainable wheat production in salt-affected environments, however, requires a unified approach that blends molecular precision with robust field breeding and sound agronomic management practices. Ultimately, the success of these scientific innovations depends on enabling policy frameworks and farmer-oriented implementation, ensuring that cutting-edge research translates into tangible gains for global food security in the face of the growing challenge of soil salinity.

## References

Abrol IP, Yadav JSP, Massoud FI (1988) Salt-affected soils and their management. FAO Soils Bull 39. Food and Agriculture Organization of the United Nations (FAO), Rome

Ahmad M, Shahzad A, Iqbal M et al (2013) Morphological and molecular genetic variation in wheat for salinity tolerance at germination and early seedling stage. Aust J Crop Sci 7(1):66–74

Akram M, Ali Q, Ali S et al (2025) Selection of salt-tolerant wheat genotypes for better yield considering physio-biochemical attributes and antioxidative defence potential: agronomic traits and stress tolerance indices. Acta Agric Scand Sect B Soil Plant Sci 75:1. https://doi.org/10.1080/09064710.2025.2466433

Anwar A, Zhang S, He L et al (2022) Understanding the physiological and molecular mechanism of salinity stress tolerance in plants. Not Bot Horti Agrobo 50(4):12959. https://doi.org/10.15835/nbha50312959

Ashraf M, Foolad MR (2007) Roles of glycine betaine and proline in improving plant abiotic stress tolerance. Environ Exp Bot 59(2):206–216. https://doi.org/10.1016/j.envexpbot.2005.12.006

Ashraf M, Akram NA, Mehboob-Ur-Rahman et al (2012) Marker-assisted selection in plant breeding for salinity tolerance. Methods Mol Biol 913:305–333. https://doi.org/10.1007/978-1-61779-986-0_21

Asif MA, Garcia M, Tilbrook J et al (2021) Identification of salt tolerance QTL in a wheat RIL mapping population using destructive and non-destructive phenotyping. Funct Plant Biol 48(2):131–140. https://doi.org/10.1071/FP20167

Bai J, Yan W, Wang Y et al (2018) Screening oat genotypes for tolerance to salinity and alkalinity. Front Plant Sci 9:1302. https://doi.org/10.3389/fpls.2018.01302

Chen TH, Murata N (2011) Glycine betaine protects plants against abiotic stress: mechanisms and biotechnological applications. Plant Cell Environ 34(1):1–20. https://doi.org/10.1111/j.1365-3040.2010.02232.x

Chen H, Wu J, Xu C (2024) Monitoring soil salinity classes through remote sensing-based ensemble learning concept: considering scale effects. Remote Sens 16(4):642. https://doi.org/10.3390/rs16040642

Crop Trust (2020) ICBA and Crop Trust join forces to conserve plant genetic resources for marginal environments. ICBA News, Dubai

Deinlein U, Stephan AB, Horie T et al (2014) Plant salt-tolerance mechanisms. Trends Plant Sci 19(6):371–379. https://doi.org/10.1016/j.tplants.2014.02.001

Elhag D (2023) Performance of some Egyptian bread wheat cultivars under saline soil conditions at North Delta of Egypt. Egypt J Agron 45(3):271–286. https://doi.org/10.21608/agro.2024.270930.1418

FAO (2024) FAO launches first major global assessment of salt-affected soils in 50 years. Food and Agriculture Organization of the United Nations, Rome

Fernandes IK, Vieira CC, KOG D et al (2024) Using machine learning to combine genetic and environmental data for maize grain yield predictions across multi-environment trials. Theor Appl Genet 137:189. https://doi.org/10.1007/s00122-024-04687-w

Flowers TJ, Colmer TD (2008) Salinity tolerance in halophytes. New Phytol 179(4):945–963. https://doi.org/10.1111/j.1469-8137.2008.02531.x

Flowers TJ, Garcia A, Koyama M et al (1997) Breeding for salt tolerance in crop plants—the role of molecular biology. Acta Physiol Plant 19:427–433. https://doi.org/10.1007/s11738-997-0039-0

Gill SS, Tuteja N (2010) Reactive oxygen species and antioxidant machinery in abiotic stress tolerance in crop plants. Plant Physiol Biochem 48(12):909–930. https://doi.org/10.1016/j.plaphy.2010.08.016

Grigore MN, Vicente O (2023) Wild halophytes: tools for understanding salt tolerance mechanisms of plants and for adapting agriculture to climate change. Plants (Basel) 12(2):221. https://doi.org/10.3390/plants12020221

Hussain Z, Khattak RA, An P et al (2020) Wheat growth and nutrient uptake after phosphorus and potassium applications in saline-sodic field of semi-arid region. Biosci J 36(6):1867–1878. https://doi.org/10.14393/BJ-v36n6a2020-42494

ICAR-CSSRI (2017) Farmer strikes gold with salt-tolerant wheat variety: KRL 210. ICAR-Central Soil Salinity Research Institute, Karnal

Jacobsen T, Adams RM (1958) Salt and silt in ancient Mesopotamian agriculture: progressive changes in soil salinity and sedimentation contributed to the breakup of past civilizations. Science 128(3334):1251–1258. https://doi.org/10.1126/science.128.3334.1251

Ji H, Pardo JM, Batelli G et al (2013) The salt overly sensitive (SOS) pathway: established and emerging roles. Mol Plant 6(2):275–286. https://doi.org/10.1093/mp/sst017

Kaledhonkar M, Meena BL (2022) Significant technologies along with impact based on 50 years of salinity research under ICAR-AICRP on management of salt affected soils & use of saline water in agriculture. https://doi.org/10.13140/RG.2.2.19344.79360

Kesh H, Devi S, Kumar N et al (2022) Insights into physiological, biochemical and molecular responses in wheat under salt stress. Intech Open. https://doi.org/10.5772/intechopen.102740

Kulshreshtha N, Sharma SK, Yaduvanshi NPS et al (2019) Genetic improvement of wheat (*Triticum aestivum* L.) for waterlogged sodic soils through modified single-seed descent and pedigree selection approaches. J Soil Sal Water Qual 11(2):250–262

Kulshreshtha N, Kumar A, Kumar A et al (2022) Genetic interventions to improve salt and micro-element toxicity tolerance in wheat. In: Kashyap PL et al (eds) New horizons in wheat and barley research. Springer, Singapore. https://doi.org/10.1007/978-981-16-4449-8_18

Kumar P, Sharma PK (2020) Soil salinity and food security in India. Front Sustain Food Syst 4:533781. https://doi.org/10.3389/fsufs.2020.533781

Leng ZX, Yimit H (2008) Environmental decision support system development for soil salinization in the arid area oasis. In: International seminar on business and information management. IEEE, Wuhan, pp 449–452. https://doi.org/10.1109/ISBIM.2008.242

Ma YX, Yang S, Lang SP (2024) Application of CRISPR/Cas9 in wheat genetic improvement. Biosci Methods 15(6):315–326. https://doi.org/10.5376/bm.2024.15.0031

Maguire JD (1962) Speed of germination—aid in selection and evaluation for seedling emergence and vigor. Crop Sci 2(2):176–177. https://doi.org/10.2135/cropsci1962.0011183X000200020033x

Mandal S, Mishra VK, Verma CL et al (2022) Managing waterlogged sodic soil through land modification in canal irrigated Indo-Gangetic plain of India—a socio-economic evaluation. Indian J Agric Econ 77(4):575–591. https://doi.org/10.63040/25827510.2022.04.001

Meena ML, Singh D (2020) Dissemination of salt-tolerant wheat varieties through frontline demonstrations approach for sustainable wheat production in Pali district of Rajasthan. J Cereal Res 12(1):61–68. https://doi.org/10.25174/2582-2675/2020/87064

MoA&FW (2016) Guidelines for reclamation of problem soils (RPS)—a sub-scheme of RKVY. Department of Agriculture & Farmers Welfare, Ministry of Agriculture & Farmers Welfare, Government of India, New Delhi

Munns R, Tester M (2008) Mechanisms of salinity tolerance. Annu Rev Plant Biol 59:651–681. https://doi.org/10.1146/annurev.arplant.59.032607.092911

Paul S, Roychoudhury A (2018) Transgenic plants for improved salinity and drought tolerance. In: Gosal S, Wani S (eds) Biotechnologies of crop improvement, vol 2. Springer. https://doi.org/10.1007/978-3-319-90650-8_7

PIB (2025) Promotion of climate-resilient and sustainable farming practices. Press Information Bureau, Government of India, New Delhi

PMKSY (2023) Operational guidelines of per drop more crop. Pradhan Mantri Krishi Sinchayee Yojana, Department of Agriculture & Farmers Welfare, Ministry of Agriculture & Farmers Welfare, Government of India, New Delhi

Prasad I, Kulshreshtha N, Chinchmalatpure AR et al (2016) Genetic evaluation and AMMI analysis for salinity tolerance in diverse wheat germplasm. Cereal Res Commun 44:217–228. https://doi.org/10.1556/0806.43.2015.053

Qadir M, Oster JD (2004) Crop and irrigation management strategies for saline-sodic soils and waters aimed at environmentally sustainable agriculture. Sci Total Environ 323(1–3):1–19. https://doi.org/10.1016/j.scitotenv.2003.10.012

Rao MJ, Duan M, Zhou C et al (2025) Antioxidant defense system in plants: reactive oxygen species production, signaling, and scavenging during abiotic stress-induced oxidative damage. Horticulturae 11(5):477. https://doi.org/10.3390/horticulturae11050477

Rehman S, Yang J, Zhang J et al (2025) Salt stress in wheat: a physiological and genetic perspective. Plant Stress 16:100832. https://doi.org/10.1016/j.stress.2025.100832

Roorkiwal M, Jarquin D, Singh MK et al (2018) Genomic-enabled prediction models using multi-environment trials to estimate the effect of genotype × environment interaction on prediction accuracy in chickpea. Sci Rep 8:11701. https://doi.org/10.1038/s41598-018-30027-2

Roy SJ, Negrão S, Tester M (2014) Salt tolerance and crop production in the 21st century. J Exp Bot 65(4):843–855. https://doi.org/10.1093/jxb/ert303

Sharma PC, Pandey KC, Kumar S (2012) Recent advances in research on saline and sodic soils of India: a review. Indian J Soil Conserv 40(1):1–15

Sharma DK, Singh R, Kumar V (2018) Agronomic management of salt-affected soils for sustainable crop production. Indian J Agric Sci 88(4):11–20

Sheoran P, Kumar A, Sharma R et al (2021) Quantitative dissection of salt tolerance for sustainable wheat production in sodic agro-ecosystems through farmers' participatory approach: an Indian experience. Sustainability 13:3378. https://doi.org/10.3390/su13063378

Sheoran P, Kamboj P, Kumar A et al (2022) Varietal-led ecological resilience through KRL 210 in salt-stressed wheat. Indian Farming 72(8):47–49

Smedley MA, Hayta S, Clarke M et al (2021) CRISPR-Cas9-based genome editing in wheat. Curr Protoc 1:e65. https://doi.org/10.1002/cpz1.65

Traye ID, Oli NM, Weng X et al (2025) Salinity tolerance in wheat: mechanisms and breeding approaches. Plants (Basel) 14(11):1641. https://doi.org/10.3390/plants14111641

Ullah A, Bano A, Khan N (2021) Climate change and salinity effects on crops and chemical communication between plants and plant growth-promoting microorganisms under stress. Front Sustain Food Syst 5:618092. https://doi.org/10.3389/fsufs.2021.618092

Wang HP, Li HM (2024) Application of molecular marker-assisted selection in wheat stress resistance breeding. Triticeae Genomics Genet 15:1–9. https://doi.org/10.5376/tgg.2024.15.0001

Wang W, Cao J, Huang S et al (2023) Integrated transcriptomics and metabolomics analyses provide insights into salt-stress response in germination and seedling stage of wheat (*Triticum aestivum* L.). Curr Plant Biol 33:100274. https://doi.org/10.1016/j.cpb.2023.100274

Wu H, Shabala L, Zhou M et al (2015) Developing and validating a high-throughput assay for salinity tissue tolerance in wheat and barley. Planta 242(4):847–857. https://doi.org/10.1007/s00425-015-2317-1

Yacoubi I, Gadaleta A, Mathlouthi N et al (2022) Abscisic acid-stress-ripening genes involved in plant response to high salinity and water deficit in durum and common wheat. Front Plant Sci 13:789701. https://doi.org/10.3389/fpls.2022.789701

Zahid A, Raza MY, Ahmed M et al (2025) Soil salinity and climate change: strategies for mitigation of salt stress to sustainable agriculture. In: Ahmed M (ed) Climate resilient and sustainable agriculture, volume 1: advances in global change research, vol 79. Springer, Cham. https://doi.org/10.1007/978-3-032-00190-0_6

Zou Y, Zhang Y, Testerink C (2022) Root dynamic growth strategies in response to salinity. Plant Cell Environ 45(3):695–704. https://doi.org/10.1111/pce.14205

# Innovations in Wheat Breeding for Enhanced Lodging Resistance

9

Rinki, Kiran, Nidhi, Tanya Singh, Thokchom Malemnganbi Devi, Sonia Sheoran, and V. Rajeswari

**Abstract**

The increasing frequency of extreme weather events has exacerbated lodging-related yield losses in wheat, presenting a significant barrier to crop stability, harvest efficiency, and global food security. Lodging, particularly when it occurs near physiological maturity, substantially reduces both grain yield and quality, highlighting the critical need to improve lodging resistance in modern wheat cultivars. Susceptibility to lodging results from complex interactions among stem morphological, anatomical, and biochemical characteristics, in addition to environmental and management variables. Elucidating these interactions and identifying the key traits distinguishing lodging-resistant from susceptible genotypes remain essential research objectives. Current mitigation approaches combine chemical regulation, agronomic optimization, and genetic enhancement. Recent advances in quantitative trait loci mapping, genome-wide association studies, genomic selection, and genome editing have significantly increased the precision and efficiency of identifying and manipulating loci associated with lodging resistance. Simultaneously, developments in plant physiology and phenotyping, including high-throughput phenomics, remote sensing, and artificial intelligence–driven predictive modelling, have enabled large-scale, environmentally responsive assessment of lodging-related traits. These methodologies support the integration of complementary resistance mechanisms and the development of ideotypes adapted to diverse agro-climatic environments. Nevertheless, substantial gaps remain in understanding genotype–phenotype–

Rinki (✉) · Kiran · Nidhi · S. Sheoran · V. Rajeswari
ICAR-Indian Institute of Wheat and Barley Research, Karnal, Haryana, India
e-mail: rinki85@icar.org.in

T. Singh · T. M. Devi
ICAR-Indian Institute of Wheat and Barley Research, Karnal, Haryana, India

Karnal Hub of IARI Mega University, Karnal, Haryana, India



R. Tiwari et al. (eds.), *Wheat Stress Management*,
https://doi.org/10.1007/978-981-92-0902-6_9

environment interactions and in translating genetic improvements into reliable field performance. This chapter reviews recent scientific progress, emerging technologies, and ongoing challenges in lodging resistance research and identifies future priorities for integrated breeding and crop management strategies to develop high-yielding, lodging-resilient wheat cultivars for a changing climate.

**Keywords**

Wheat · Lodging · Stem architectural traits · Dwarfing gene · Lignin

## 9.1 Introduction

The rapid increase in global population necessitates the development of high-yielding, resilient crop varieties to ensure long-term food security. Wheat (*Triticum aestivum* L.), as one of the three principal cereal crops worldwide, plays a critical role in human nutrition and is cultivated across diverse agroclimatic regions. Sustained yield improvement in wheat depends on enhancing photosynthetic efficiency, increasing total biomass production, and optimizing assimilate partitioning to grains. Additionally, maintaining adequate structural stability to prevent lodging is also important.

Lodging is the permanent displacement of a plant from its upright position. It can be stem-lodging, where culms bend or break, or root lodging, when roots fail to anchor the plant. Usually, lodging occurs when mechanical support is weak and harsh weather, such as heavy rain, strong winds, or hail, is present. As a result, cereals such as wheat and rice can suffer significant yield and quality losses (Mulsanti et al. 2018). Lodging also disrupts canopy structure, reduces light capture, inhibits photosynthesis and nutrient movement, and affects grain filling. Yield losses from lodging can reach 80%, depending on the angle, crop growth stage, and severity (Shah et al. 2017). Lodging during flowering or early grain filling is particularly harmful because this is a key time for kernel growth. Fischer and Stapper (1987) found that each day of lodging at this stage can reduce yield by about 0.5% and worsen bread-making quality. This occurs due to smaller kernels, more shrivelled grain, and lower flour extraction, resulting in a lower market value. Lodging later in crop growth can cause greater quality loss by inducing pre-harvest sprouting under humid conditions, particularly in varieties lacking seed dormancy (Luo et al. 2022). Lodging also makes machine harvesting more difficult, increases post-harvest disease, raises costs, and results in economic losses.

Lodging arises from complex interactions among genetic, physiological, and environmental factors. Wheat exhibits resistance to lodging through a combination of traits. Aboveground characteristics include plant height, internode length, stem diameter, wall thickness, and stem solidity. Belowground attributes, such as root system architecture, soil structure, and anchorage strength, as well as key biochemical properties, also play significant roles. Shorter plants generally exhibit greater lodging resistance by lowering the centre of gravity and reducing the lever arm. In contrast, taller genotypes, despite often producing higher yields, are more susceptible to mechanical stress. Stem strength is primarily determined by structural features: thicker, lower internodes with increased wall thickness and dry matter content confer greater rigidity. Enhanced supportive tissues and densely packed vascular bundles further reinforce stem strength and facilitate efficient water and nutrient transport (Khobra et al. 2019). Solid or partially filled stems are more resistant to lodging than hollow stems because they bend more slowly. Deeper and wider root systems, particularly those with robust coronal roots, enhance the soil–root interface and improve plant anchorage, thereby reducing the likelihood of lodging. In addition to these structural features, biochemical components such as lignin, cellulose, silica, and sugars contribute to lodging resistance by increasing stem stiffness, toughness, and flexibility. Silica deposition enhances stem hardness and flexibility, while stored sugars maintain stem strength and delay senescence under stress. Agronomic practices also influence lodging risk; excessive nitrogen application and high planting density can promote excessive stem elongation and weaken structural integrity, increasing susceptibility to lodging (Hu et al. 2025). Environmental factors, including strong winds, heavy rainfall, and hail, can induce lodging, while soil properties such as type, moisture content, slope, and overall crop health modulate lodging severity (Berry et al. 2004).

A significant variation in lodging resistance exists among wheat cultivars with diverse plant heights and stem architectures. The introduction of semi-dwarfing genes (*Rht1* and *Rht2*) has enhanced lodging resistance by reducing plant height; however, further reductions in height may adversely affect biomass accumulation and photosynthetic capacity, thereby limiting yield potential. Consequently, future breeding efforts should focus on optimizing structural and physiological traits to improve lodging resistance without compromising grain yield. Key strategies include efficient biomass partitioning, reinforcement of stem and root structures, and enhanced nitrogen-use efficiency. While high nitrogen inputs can increase yield, they often exacerbate lodging risk by promoting rapid stem elongation and weakening, highlighting the necessity for balanced, genotype-specific nitrogen management.

Lodging continues to pose a significant challenge for yield stability and grain quality in wheat production, particularly under increasingly variable climatic conditions. Effective mitigation requires a comprehensive understanding of plant morphology, anatomy, physiology, and biomechanics in relation to environmental stressors. Breeding for lodging resistance, combined with optimized agronomic practices and climate-resilient crop management, offers a sustainable pathway to maintaining wheat productivity. Recent advancements in molecular breeding, quantitative trait locus (QTL) mapping, high-throughput phenotyping, and precision agriculture provide valuable tools for addressing lodging. In particular, aerial imaging and real-time lodging detection facilitate the elucidation of the complex genetic basis of lodging resistance. This chapter, therefore, reviews recent progress in elucidating the mechanisms underlying lodging and presents multidimensional strategies for developing and managing lodging-resistant wheat cultivars suited to future environmental challenges.

## 9.2 Historical Perspectives on Wheat Breeding for Enhanced Lodging Resistance

India's shift from chronic wheat deficit to national self-sufficiency demonstrates the significant impact of genetic improvement and agronomic intensification on food security. At independence, wheat production was approximately 6.46 million tonnes (Dronamraju 2018), primarily due to the dominance of traditional tall landraces that were low-yielding and highly susceptible to lodging in irrigated, nitrogen-intensive systems. These cultivars exhibited elongated culms, reduced basal internode wall thickness, and lower lignin accumulation, resulting in inferior stem mechanical strength compared to modern semi-dwarf varieties (Zhang et al. 2016). Under high-nitrogen conditions, tall genotypes experienced substantial yield losses, primarily due to increased lodging susceptibility (Piñera-Chavez et al. 2016a). Their excessive height and high leaf area index also increased vulnerability to foliar diseases by creating dense, humid canopies favourable for pathogen proliferation (Zetzsche et al. 2020).

To address these limitations, wheat breeding programs from the mid-twentieth century onward focused on modifying plant architecture and developing lodging-resistant ideotypes. The introduction of semi-dwarfing alleles, particularly the gibberellin (GA)-insensitive *Rht-B1b* (*Rht1*) and *Rht-D1b* (*Rht2*) genes, marked a significant advance in global wheat improvement. These alleles reduce sensitivity to endogenous gibberellins, limiting cell elongation by decreasing cell-wall extensibility and resulting in shorter, sturdier culms (Keyes et al. 1989; Hoogendoorn et al. 1990). The widespread adoption of *Rht* genes significantly reduced lodging and improved harvest index by reallocating assimilates from vegetative to reproductive organs. However, these advantages were accompanied by pleiotropic effects, such as reduced coleoptile length and restricted sub-crown internode elongation, which can hinder seedling emergence under deep sowing, residue retention, or moisture-limited conditions (Allan et al. 1962; Acuña et al. 2001).

In contrast, GA-sensitive height-reducing genes such as *Rht8* provide moderate reductions in plant height while maintaining coleoptile length and early seedling vigour (Ellis et al. 2004; Rebetzke et al. 2012). These characteristics make *Rht8* and related alleles especially valuable for conservation agriculture and sustainable production systems, which often involve residue retention and variable sowing depths. As a result, the strategic use of both GA-insensitive and GA-sensitive dwarfing genes has broadened the options available to breeders for optimizing plant architecture across diverse agro-ecological environments.

Although semi-dwarfism has greatly improved lodging resistance, modern high-yielding wheat cultivars remain susceptible to lodging under excessive nitrogen fertilization, delayed sowing, and extreme weather events such as strong winds or heavy rainfall. This ongoing vulnerability has led to the complementary use of plant growth regulators (PGRs), which suppress excessive internode elongation, increase stem wall thickness, and enhance culm mechanical strength (Peng et al. 2023). PGR application is particularly critical in humid, wind-prone agroecologies, in systems aiming for high grain protein content through elevated nitrogen inputs (Wu et al. 2021a, b), and under late-sown conditions where rapid stem elongation coincides with adverse environmental stressors (Peake et al. 2020).

## 9.3 Scientific Advancements and Principles Developed Over the Past Centuries

Progress in plant biology, genetics, and agronomy over several centuries has advanced the understanding of lodging tolerance in wheat and guided the development of structurally resilient cultivars. In India, where wheat cultivation spans nearly 30 million hectares, sustained productivity gains have relied on cultivars that perform consistently under high-input production systems (ICAR 2020). Traditional tall wheat cultivars, although suited to low-input conditions, were highly susceptible to lodging in fertile, irrigated environments, thereby limiting yield improvement. The introduction of semi-dwarf, strong-straw cultivars overcame this constraint and stands as a key achievement of the Green Revolution (Borlaug 1968; Hedden 2003).

A significant advance in lodging research was the identification of plant height as a key factor influencing lodging susceptibility (Pinthus 1974). Subsequent genetic and physiological investigations established the central role of gibberellins (GAs) in controlling stem elongation and internode growth (Sasaki et al. 2002). This understanding facilitated the targeted deployment of GA-insensitive semi-dwarfing genes, such as *Rht-B1b* and *Rht-D1b*, which revolutionized global wheat breeding (Peng et al. 1999). These alleles limit cell elongation by reducing cell-wall extensibility, thereby producing shorter, mechanically robust culms that can support higher nitrogen inputs and achieve increased yield potential (Flintham et al. 1997). The discovery of GA-sensitive dwarfing genes, including *Rht8*, further diversified breeding strategies by enabling moderate height reduction without compromising coleoptile length or early seedling vigour, traits essential for conservation agriculture and residue-retained cropping systems (Ellis et al. 2004).

Advancements in plant biomechanics have improved lodging research by enabling quantitative evaluation of stem and root mechanical properties, such as bending strength, flexural stiffness, Young's modulus, and anchorage strength (Kokubo et al. 1989; Berry et al. 2004). Molecular investigations have identified key structural biosynthesis genes, including *PAL*, *COMT*, *4CL3*, *CAD2/7*, *CCR20*, and *C4H*, which regulate the deposition of lignin, cellulose, and hemicellulose in the culm. These cell-wall components are critical for stem strength and lodging resistance. Coordinated regulation of these genes reinforces the mechanical stability observed in modern wheat cultivars (Ma 2010; Liu et al. 2018).

Simultaneously, agronomic practices have been refined to support genetic advances in lodging tolerance. Optimizing sowing time aligns crop development with climatic conditions, reducing exposure to high winds and rainfall during sensitive stages such as heading and grain filling (Reynolds et al. 2007). Modifying planting density decreases excessive interplant competition, which can otherwise lead to internode elongation and structural weakness (Calderini and Reynolds 2000). Conservation agriculture practices, especially crop residue retention, have improved soil physical properties, increased moisture conservation, and strengthened root anchorage, thereby lowering lodging risk (Jat et al. 2014). Balanced nutrient management, including adequate potassium, silicon, and essential micronutrients, has further enhanced cell wall integrity and stem mechanical strength, thereby improving lodging resistance (Ma and Yamaji 2006; Hossain et al. 2021).

Recent methodological innovations, such as quantitative trait locus (QTL) mapping, genome-wide association studies (GWAS), transcriptomics, and high-throughput phenotyping, have significantly advanced the understanding of lodging as a complex, polygenic trait (Spielmeyer et al. 2002; Zanke et al. 2015). Researchers have identified genetic loci and pathways related to root system architecture, tiller anchorage, stem anatomy, and hormone regulation. These discoveries highlight the intricate genetic and physiological networks that control lodging tolerance.

## 9.4 Advances in Wheat Breeding for Enhanced Lodging Resistance

### 9.4.1 QTL Mapping and Genome-Wide Association Studies (GWAS) for Lodging Resistance in Wheat

Phenotypic evaluation of lodging resistance in wheat presents significant challenges due to pronounced environmental influences and genotype-by-environment interactions. This complexity necessitates the use of advanced genetic approaches, including QTL mapping and GWAS. GWAS, facilitated by next-generation sequencing technologies, leverages historical recombination and linkage disequilibrium within diverse germplasm panels to identify genotype–phenotype associations at higher mapping resolution than traditional biparental populations. This approach is particularly effective for dissecting complex, polygenic traits such as lodging resistance (Zhu et al. 2008; Wang et al. 2020).

Early bi-parental QTL analyses in wheat identified multiple genomic regions associated with lodging-related traits. For instance, Keller et al. (1999) reported nine QTLs, while Hai et al. (2005) identified loci controlling stem wall thickness and culm diameter. Piñera-Chavez et al. (2021) mapped 16 major QTLs. Subsequent studies expanded the genomic landscape of lodging resistance, revealing additional QTLs on chromosomes 2D, 4D, 7B, and 7D (Dreccer et al. 2020). Stem-associated loci were also identified on chromosomes 1B, 3B, 4A, 4B, and 5A (Daetwyler et al. 2007). Despite these advances, many early studies relied primarily on plant height as a proxy for lodging resistance, often overlooking critical biomechanical traits such as stem bending strength, flexural stiffness, and composite lodging indices. These biomechanical traits more directly reflect mechanical strength under field conditions (Berry et al. 2007). Environmental factors, particularly nitrogen availability, further influence lodging susceptibility by interacting with genetic determinants of stem structure. Elevated nitrogen inputs promote excessive vegetative growth and internode elongation, often reducing culm mechanical integrity and increasing the risk of lodging. Under these conditions, stem biochemical composition, especially lignin concentration, and associated mechanical properties become critical determinants of lodging tolerance (Dreccer et al. 2020). Concurrently, large-scale genomic analyses of yield-related traits have identified approximately 145 meta-QTLs and several candidate genes, which are currently undergoing functional validation and cloning. These findings highlight the extensive genetic overlap among yield formation, biomass partitioning, and lodging resistance (Yang et al. 2021a, b).

### 9.4.2 Discovery of New Dwarfing and Semi-Dwarfing Genes with Minimal Yield Penalty in Wheat

Plant height serves as a critical determinant of lodging resistance, canopy architecture, and yield stability in wheat. The identification of the semi-dwarfing genes *Rht1* (*Rht-B1b*) and *Rht2* (*Rht-D1b*) during the 1960s transformed wheat breeding and was instrumental to the Green Revolution by enabling reduced plant height, enhanced lodging resistance, and increased harvest index. These alleles, located on chromosomes 4B and 4D, encode gibberellin (GA)-insensitive, truncated DELLA proteins that constitutively repress gibberellin signalling. This repression reduces height by approximately 20% and 24%, respectively, and exerts pleiotropic effects on heading date, spike fertility, and kernel number (Peng et al. 1999; Hedden 2003; Miralles and Slafer 1995). Additional GA-insensitive dwarfing genes, such as *Rht3* (retrotransposon insertion), *Rht10* (duplication of *Rht-D1b*), and allelic variants including *Rht11* and *Rht17*, also restrict stem elongation. However, these variants are frequently associated with reduced coleoptile length and diminished early vigour under stress-prone conditions (Flintham and Gale 1983; Divashuk et al. 2013).

Recent advances in wheat genomics have broadened the range of height-reducing loci by facilitating the identification of several gibberellin-sensitive dwarfing genes, including *Rht8* (2D), *Rht12* (5A), *Rht13* (7B), *Rht18* (6A), and *Rht24* (6A). These genes modulate plant stature through distinct molecular mechanisms involving RNase H-like proteins, GA2-oxidases, or NB-LRR genes. Generally, these loci confer more favourable agronomic trade-offs, as they are associated with enhanced photosynthetic performance, improved nitrogen-use efficiency, and better seedling establishment (Borrill et al. 2017; Buss et al. 2020; Chai et al. 2021; Tian et al. 2022). Although extensive genetic mapping has been conducted, many dwarfing genes remain insufficiently characterized at the molecular level. These include *Rht4* (2B deletion), *Rht5* (3BS), *Rht6* (4D), *Rht7*, *Rht21* (2AS), *Rht22* (7AS), and additional loci such as *Rht14*, *Rht16*, and *Rht25* on 6AS (Ellis et al. 2004; Wu et al. 2021a, b). Further functional analysis and targeted deployment of these underexplored loci could significantly enhance the precision of plant height regulation, lodging resistance, and yield stability in future wheat breeding programs.

### 9.4.3 Breeding for Improved Stem Anatomy and Carbohydrate Remobilization in Relation to Lodging Resistance

Stem architectural and biochemical traits are primary determinants of lodging resistance in wheat. Plant height, internode length, stem diameter, and cell wall composition collectively govern mechanical stability (Duan et al. 2012; Xiang et al. 2019). Lodging-resistant genotypes of wheat, rice, and maize typically exhibit thicker culms, shorter, stronger basal internodes, and increased accumulation of structural carbohydrates, including starch, sucrose, cellulose, and lignin. These features enhance stem rigidity and the ability to withstand wind and mechanical stress (Berry et al. 2004; Kashiwagi et al. 2008; Wang et al. 2014; Piñera-Chavez et al. 2016b). Stem strength is closely associated with the development of robust secondary cell walls. Experimental evidence from Arabidopsis mutants indicates that defects in secondary wall formation significantly reduce mechanical strength and lead to severe lodging susceptibility (Li et al. 2022). While stems serve as transient carbohydrate reservoirs supporting grain filling, excessive remobilization of these reserves during reproductive growth can weaken culm strength and increase susceptibility to lodging.

At the biochemical level, lignin and cellulose are key contributors to stem mechanical strength, both individually and in combination, particularly in basal internodes that experience the maximum bending stress (Kong et al. 2013). Elevated levels of lignin, cellulose, pectin, and structural proteins, along with increased activities of lignin biosynthetic enzymes such as phenylalanine ammonia-lyase (PAL), tyrosine ammonia-lyase (TAL), and peroxidases (POD), are consistently linked to enhanced lodging resistance (Berry et al. 2004; Nguyen et al. 2016). Although reduced lignin content generally weakens stem integrity, exceptions exist. For instance, the rice cultivar *Leaf Star* demonstrates strong lodging resistance despite relatively low lignin levels, due to thick cortical fibres and multiple reinforcing anatomical features. This example highlights that lodging resistance is a polygenic and multifactorial trait that integrates stem anatomy, wall composition, and carbohydrate dynamics (Wang et al. 2020).

### 9.4.4 Screening of Global Wheat Germplasm for Alleles Contributing to Stem Strength and Root System Development

Lodging, which can be categorized as root lodging (failure of crown roots) or stem lodging (culm bending or breakage), remains a significant constraint on wheat productivity and yield stability worldwide (Pinthus 1974). The integration of physiological and biomechanical models that account for stem, root, and shoot traits has advanced the understanding of lodging mechanisms and established a rational framework for selection in both winter and spring wheat (Baker et al. 1998; Berry et al. 2007; Piñera-Chavez et al. 2016a). Long-term breeding efforts have produced shorter plants with thicker stem walls and higher lignin content, as observed in modern Chinese and UK cultivars, thereby enhancing stem strength and reducing lodging risk (Berry et al. 2000; Zhang et al. 2016). Nevertheless, recent yield improvements achieved without further reductions in plant height have raised renewed concerns regarding lodging susceptibility, highlighting the limitations of relying solely on height-based selection.

Research from CIMMYT indicates that high biomass accumulation and grain yield are attainable in semi-dwarf wheat when robust stem and root systems are present, suggesting that lodging resistance can be enhanced largely independently of plant height (Aisawi et al. 2015). Lodging susceptibility varies significantly across different environments and management practices, and is influenced by interacting traits such as spike mass, basal stem mechanical strength, and root plate spread (Berry et al. 2007; Mirabella et al. 2019). Despite this complexity, comprehensive global screening of germplasm for stem and root mechanical properties remains limited, particularly in regions such as Australia. Broadening phenotypic and genetic evaluation of diverse germplasm for lodging-related traits is essential to identify favourable alleles and to develop high-yielding wheat cultivars with improved structural resilience across diverse production environments.

## 9.5 Evolution in the Contemporary Hi-Tech Era

Rapid advances in agricultural biotechnology and computational sciences have fundamentally reshaped crop improvement strategies, enabling more precise and efficient breeding for complex traits such as lodging resistance. Genomic innovations, including genomic selection and CRISPR–Cas–based genome editing, facilitate the targeted manipulation of plant architectural and biomechanical traits while preserving or enhancing yield potential. Genomic selection accelerates genetic gain by capturing the cumulative effects of many small-effect loci underlying lodging resistance. In contrast, genome editing allows functional validation and precise modification of key genes controlling stem strength, root architecture, and hormone signalling.

Concurrently, high-throughput phenotyping (HTP) platforms—integrating imaging, sensor technologies, and unmanned aerial systems—combined with machine learning and artificial intelligence approaches, enable robust, large-scale quantification of lodging-related traits across diverse environments and management conditions. These data-rich frameworks support the development of multi-trait selection indices that simultaneously optimize yield, grain quality, and structural resilience, thereby advancing ideotype breeding.

### 9.5.1 Application of Genomic Selection to Predict Lodging-Resistance Traits with High Accuracy

Over the past decade, breeding strategies for improving lodging resistance in wheat have shifted from conventional phenotype-based selection and major QTL discovery to genomic selection, which exploits genome-wide marker information to predict complex, polygenic traits. Earlier approaches focused on identifying a limited number of large-effect loci; however, numerous small-effect genes govern lodging resistance and the coordinated expression of multiple component traits, including stem mechanical strength, internode anatomy, plant height, and root anchorage. By simultaneously capturing the effects of both major and minor loci, genomic selection enables earlier and more accurate selection, reducing reliance on costly, environment-sensitive field lodging assessments (Daetwyler et al. 2007; Tessema et al. 2020; Dreisigacker et al. 2021).

Recent methodological advances, particularly multi-trait and secondary-trait genomic selection models, have substantially improved prediction accuracy for lodging-related traits. High-throughput phenotyping–derived proxies such as normalized difference vegetation index (NDVI), canopy temperature, stem diameter, and internode structural parameters show strong genetic correlations with lodging and, when incorporated into genomic selection frameworks, markedly enhance predictive performance, analogous to gains reported for yield prediction (Shahi et al. 2022). Further improvements have been achieved by integrating prior biological knowledge from GWAS and QTL mapping, for example, through marker weighting or fixation of validated lodging-related loci, especially within multi-kernel and hybrid genomic selection models (Leonova and Ageeva 2022; Rabieyan et al. 2024).

Operational implementation of GS for lodging resistance, exemplified by CIMMYT's global wheat breeding program, highlights the importance of high-quality phenotyping, well-designed training populations, and robust modelling of genotype-by-environment (G×E) interactions (Dreisigacker et al. 2021). While moderate to high prediction accuracies have been consistently reported for highly heritable component traits such as stem thickness and internode strength, direct lodging indices remain more environmentally dependent, reinforcing the strategic value of selecting for component traits with higher heritability (Rabieyan et al. 2024). A standard genomic selection pipeline involves assembling a diverse training population with comprehensive lodging and proxy trait data, followed by genome-wide genotyping using GBS, SNP arrays, or low-coverage sequencing, and model implementation using GBLUP, Bayesian, kernel-based, or deep-learning approaches that accommodate multi-trait and multi-environment data structures (Tessema et al. 2020; Shahi et al. 2022; Wang et al. 2023). Despite substantial progress, challenges remain for traits with low heritability, limited training populations, or strong environmental sensitivity. Addressing these constraints will require expanded multi-environment datasets, improved G×E modelling, routine integration of high-throughput phenotyping, and the incorporation of functional biological knowledge—such as cell wall biosynthesis and mineral nutrition—into informed marker-weighting and prediction frameworks.

### 9.5.2 CRISPR–Cas Genome Editing for Targeted Modification of Height and Culm Architecture Genes

Simultaneously reducing plant height and enhancing culm mechanical strength constitutes the most effective genetic approach for mitigating lodging in wheat. Historically, the widespread adoption of semi-dwarfing *Rht-B1b* and *Rht-D1b* alleles, which are gain-of-function mutations in DELLA proteins that confer gibberellin insensitivity, has significantly improved lodging resistance. However, these alleles have also resulted in reduced grain size, shortened coleoptiles, and diminished nitrogen-use efficiency (NUE) (Song et al. 2023). The development of CRISPR–Cas genome editing now enables precise manipulation of height-regulatory pathways beyond gibberellin insensitivity, facilitating coordinated modification of plant architecture without compromising yield components. For example, disruption of the zinc finger E3 ligase gene ZnF-B in the brassinosteroid (BR) signalling pathway stabilizes TaBKI1, reduces BR perception, and generates compact, semi-dwarf plants without the yield penalties associated with *Rht-B1b and Rht-D1b*. Functional analyses of znf-bb, eama-bb, and rht1-bb mutants indicate that loss of *ZnF-B* confers semi-dwarfism with minimal pleiotropic effects. At the same time, deletion of *Rht-B1* increases plant height and thousand-grain weight. Notably, the double mutant achieves a balance between reduced height and improved spike architecture and grain traits, underscoring BR signalling as a CRISPR-compatible pathway for developing lodging-resistant ideotypes (Song et al. 2023).

Refining plant stature through minor-effect height loci further enhances the precision of architectural optimization. CRISPR-mediated analysis of *Rht8* demonstrated that two natural *Rht8-B1* variants reduced plant height by 3–6% in mapping populations, while edited $T_2$ lines exhibited an additional reduction of approximately 5.6%. Although this effect is less pronounced than that of *Rht8-D1*, it is advantageous in elite genetic backgrounds where excessive dwarfism can hinder establishment (Zhou et al. 2023). Other promising architectural targets include the circadian clock gene *TaPRR95*, whose editing influences both plant height and flowering time (Fu et al. 2024), and the strigolactone biosynthesis gene *TaD27*, where knockout results in a doubling of tiller number and increased grain number per plant without compromising structural integrity (Awan et al. 2024). Optimizing tiller density can enhance stand uniformity and redistribute mechanical load, thereby indirectly reducing the risk of lodging.

Culm rigidity, a key factor in stem lodging resistance, depends on the coordinated deposition of lignin, cellulose, and hemicellulose in secondary cell walls (Shah et al. 2019; Li et al. 2022). Although targeted CRISPR edits of these pathways in wheat remain in early stages, studies of barley COMT knockouts and rice lignin-pathway mutants confirm the essential role of lignin biosynthesis in culm strength (Guha et al. 2024). Wheat homologs such as *TaCOMT*, *TaCCR*, *TaCAD*, and *TaCesA* are therefore promising candidates for enhancing stem mechanical properties while maintaining optimal semi-dwarf stature. CRISPR editing in wheat has become routine, with protoplast-based gRNA screening increasing efficiency (Arndell et al. 2019) and advanced modular Cas9 and Cas12a systems, including

GRF–GIF enhancers, enabling efficient single and multiplex genome edits (Lawrenson et al. 2024).

### 9.5.3 High-Throughput Phenotyping

High-throughput phenotyping (HTP) enables scalable, objective, and repeatable quantification of wheat lodging by converting field-scale imagery into biologically meaningful structural traits. Unmanned aerial vehicles (UAVs) equipped with RGB sensors employ photogrammetric reconstruction to generate dense three-dimensional point clouds and canopy height models (CHMs) from overlapping images. Deviations from an upright canopy surface provide robust metrics of lodging incidence, extent, and severity at both plot and plant scales (Wilke et al. 2019; Volpato et al. 2021). UAV–RGB platforms have been validated for time-series monitoring of plant height and canopy deformation across thousands of breeding plots, demonstrating their practicality for large-scale selection. Threshold-based analyses of CHMs derived from nadir flights reliably quantify lodging percentage and severity by capturing changes in canopy surface roughness and collapse dynamics. Furthermore, integration of UAV imagery with deep learning enables spatially explicit lodging mapping at regional scales for breeding and damage assessment (Wilke et al. 2019; Volpato et al. 2021; Zhang et al. 2023). In breeding nurseries, the combined use of spectral and structural UAV-derived traits with machine learning consistently outperforms single-trait approaches for lodging prediction (Rabieyan et al. 2023).

In addition to canopy-level metrics, digital analysis of stem inclination offers finer resolution for assessing lodging severity. The crop angle of inclination (CAI), defined as the angular deviation of stems from the vertical axis (0° upright to 90° prostrate), enables quantitative discrimination among lodging classes (Rabieyan et al. 2024). Recent point-cloud–based pipelines convert UAV-derived three-dimensional data into two-dimensional representations and apply convolutional neural networks (CNNs), such as MobileNet, achieving classification accuracies exceeding 90% during grain filling and maturity (Li et al. 2022). Phenotyping precision is strongly influenced by flight parameters such as altitude, image overlap, viewing geometry, and multi-angle perspectives, which determine point-cloud density and reconstruction fidelity (Sangjan et al. 2022). Emerging deep-learning frameworks that jointly segment lodged regions and classify severity further enhance robustness and repeatability (Zang et al. 2024). An effective HTP workflow integrates UAV–RGB or LiDAR acquisition, CHM and CAI derivation, threshold- or learning-based lodging assessment, and machine learning-driven genotype ranking. These standardized lodging phenotypes can be readily incorporated into genomic selection pipelines and are particularly valuable for evaluating architectural innovations such as semi-dwarf, lodging-resistant ideotypes (Volpato et al. 2021; Rabieyan et al. 2023).

### 9.5.4 Integration of Machine Learning and AI to Model Lodging Risks Under Different Environments and Genotypes

Machine learning (ML) and artificial intelligence (AI) have significantly improved lodging assessment in wheat by offering accurate, efficient, and scalable alternatives to subjective manual scoring. Current high-throughput workflows utilize UAV-based RGB and multispectral imagery, digital surface models (DSM), and agronomic and environmental covariates to detect lodged areas and predict lodging risk across diverse genotype × environment (G×E) scenarios (Rabieyan et al. 2023; Zhang et al. 2023). Standard ML/AI pipelines employ high-resolution imagery enhanced with spectral indices such as NDVI, canopy height models, texture descriptors, and time-series phenological features. These data are analyzed using algorithms ranging from Random Forest, which is effective for plot-level lodging estimation, to deep-learning segmentation architectures such as U-Net and its derivatives that enable pixel-level classification of lodged canopies (Yang et al. 2021a, b; Ali et al. 2023). Data fusion strategies, including integrating RGB imagery with DSMs or multispectral data with temporal features, consistently enhance robustness and accuracy across developmental stages (Yang et al. 2021a, b).

The methodological diversity observed across studies reflects the multi-scale complexity of lodging assessment. Random Forest-based models have been utilized for genotype evaluation in breeding trials (Rabieyan et al. 2023), while multi-branch convolutional neural network (CNN) architectures that incorporate canopy structural features have improved UAV-based classification (Zhang et al. 2023). Additionally, Sentinel-2 time-series data combined with ML techniques have enabled regional-scale lodging monitoring. Height-informed Mobile U-Net models further improve segmentation accuracy and severity estimation, particularly when structural data are explicitly included (Yang et al. 2021a, b; Ali et al. 2023). A persistent challenge remains in translating detection outputs into genotype-specific lodging risk predictions. Addressing this challenge requires integrating trial metadata, including varietal identity, management practices such as sowing density and nitrogen inputs, and morphological traits associated with stem and root strength, and using multi-environment datasets to adequately model G×E interactions. Models trained in a single environment often demonstrate limited generalizability, highlighting the need for transfer learning and domain adaptation strategies (Togninalli et al. 2023). Emerging multimodal deep-learning frameworks that jointly utilize genomic, phenomic, and environmental data represent a promising direction, though comprehensive datasets linking lodging phenotypes to genetic information remain limited. Future advancements will rely on the development of interpretable, lightweight, and scalable models that can identify key physiological and environmental drivers of lodging and provide reliable predictions across genotypes, environments, and sensor platforms (Zhang et al. 2023; Rabieyan et al. 2023).

### 9.5.5 Development of Multi-trait Selection Indices Combining Yield Quality and Lodging Resistance for Ideotype Breeding

Developing wheat ideotypes that achieve high yield, superior grain quality, and strong lodging resistance concurrently remains challenging because of genetic and physiological antagonism among these traits. Multi-trait selection indices provide a comprehensive framework for addressing this complexity by evaluating genotypes based on aggregate performance rather than individual traits. While the classical Smith–Hazel index (Smith 1936; Hazel 1943) has been widely adopted for this purpose, its effectiveness is limited by multicollinearity among correlated traits and the inherent subjectivity and difficulty in assigning reliable economic weights (Olivoto and Nardino 2020).

Recent developments, such as the Multi-Trait Genotype–Ideotype Distance Index (MGIDI), address these challenges by employing factor analysis to reduce dimensionality and eliminate multicollinearity prior to estimating each genotype's Euclidean distance from a breeder-defined ideotype (Olivoto and Nardino 2020). This methodology facilitates objective and biologically relevant selection for balanced trait enhancement. In wheat, MGIDI has demonstrated effectiveness in identifying genotypes that simultaneously improve yield, grain quality, and lodging resistance across diverse environments (Djenadi et al. 2025; Rahmati and Karimizadeh 2025). The integration of the Multi-Trait Stability Index (MTSI), which combines mean performance and stability into a single metric (Zuffo et al. 2020), further enhances this ideotype-based approach by prioritizing genotypes that sustain structural integrity and consistent performance under variable and stress-prone conditions.

## 9.6 Statistical Insight into Wheat Breeding for Lodging Resistance

Lodging resistance in wheat is a complex, quantitatively inherited trait governed by interactions among plant architecture, environmental conditions, and crop management practices. This multifactorial nature necessitates robust statistical analyses to disentangle genetic effects from environmental noise and to quantify lodging-associated yield penalties reliably. Multi-environment trial datasets are particularly critical, as they enable precise estimation of genotype × environment interactions and facilitate the assessment of lodging-induced yield losses across contrasting agro-climatic conditions. Long-term analyses further demonstrate that the widespread adoption of semi-dwarf germplasm has been a major contributor to reduced lodging risk, primarily through shortened plant stature and improved stem strength. Systematic evaluation of breeding cycles, varietal release histories, and trait interrelationships provides empirical evidence of genetic gains in lodging resistance and clarifies trade-offs with yield and its components. Correlation and trend analyses reveal how selection for yield, plant height, stem traits, and biomass allocation has reshaped wheat ideotypes over time, often balancing productivity with structural stability.

### 9.6.1 Trends in Lodging Incidence and Associated Yield Losses in Major Wheat-Growing Zones of India and Globally

Lodging continues to constrain wheat productivity and harvest efficiency, resulting from the interplay of genetic susceptibility, intensive crop management, and increasing climatic volatility. High-input production systems, particularly those with elevated nitrogen fertilization and dense plant populations, promote excessive biomass accumulation, elevate the canopy centre of gravity, and reduce culm mechanical strength, thereby increasing the risk of lodging (Wang et al. 2023). Field- and regional-scale studies consistently report yield penalties of approximately 10–50%, depending on the timing and severity of lodging, with greater losses observed when lodging occurs during grain filling (Zang et al. 2024; Chen et al. 2024). Persistent lodging risk is most pronounced in intensively managed temperate regions of China and Europe. In contrast, subtropical irrigated systems, such as the Indo-Gangetic Plains, experience less frequent but often severe lodging, typically triggered by extreme rainfall and storm events occurring near or after anthesis (Wang et al. 2023; Feng et al. 2024).

Global and India-specific analyses indicate that lodging-induced yield losses are strongly influenced by cultivar age, sowing time, and management practices, although losses generally remain within the 10–50% range. While genetic improvements, particularly the adoption of semi-dwarf ideotypes and enhanced stem strength, have reduced baseline lodging susceptibility in modern cultivars, these benefits are increasingly counteracted by agronomic intensification and climatic instability (Wang et al. 2023; Peng et al. 2025). Recent advances in UAV-based remote sensing and deep-learning methods enable near-real-time detection and spatial quantification of lodging, thereby improving loss estimation and supporting targeted management interventions (Zang et al. 2024).

### 9.6.2 Impact of Adoption of Semi-Dwarf Varieties on Lodging Reduction and Yield Stability

The introduction of semi-dwarf wheat varieties containing reduced height (*Rht*) genes during the Green Revolution significantly reduced lodging by decreasing plant stature and shifting the canopy centre of gravity. This improvement enhanced mechanical stability under high-input conditions (Wang et al. 2024). The commonly used alleles *Rht-B1b* and *Rht-D1b* confer gibberellin insensitivity, restrict stem elongation, and reduce bending moments caused by wind and rainfall. These effects contribute to the superior yield stability of semi-dwarf cultivars in irrigated and fertilizer-intensive systems (Chai et al. 2021; Peng et al. 2025). However, these classical *Rht* alleles can also shorten coleoptile length and reduce early seedling vigour, thereby limiting establishment under deep sowing or moisture-stressed conditions and constraining adaptability (Xu et al. 2023).

To overcome these limitations, breeding programs have increasingly targeted alternative height-reducing sources, such as allelic variation in *TaERF-A1* and novel dwarfing mutations, such as *Rht-yz*. These alternatives achieve moderate stature reduction without negatively affecting coleoptile length or early growth (Hao et al. 2024). Results from multi-environment trials demonstrate that semi-dwarf genotypes combining optimized plant height with enhanced stem mechanical properties, including increased culm wall thickness and lignin deposition, as well as improved root anchorage, provide the highest lodging resistance and yield stability across diverse agro-climatic conditions. Given the rise in climatic extremes, integrative, multi-trait breeding strategies are necessary to maintain and expand the benefits of dwarfing, supporting both high average productivity and resilience across seasons (Wang et al. 2023; Peng et al. 2025).

### 9.6.3 Genetic Gain in Lodging Resistance Over Breeding Cycles and Varietal Releases

Wheat breeding has produced significant genetic gains in lodging resistance, particularly through the introduction of semi-dwarfing *Rht* alleles in the mid-twentieth century. This intervention sharply reduced plant height and substantially improved standability, constituting the most significant advance in lodging resistance to date (Muszynska et al. 2021). Subsequent progress has been incremental and focused on specific traits, such as enhanced stem mechanical properties, improved tissue composition, and increased root anchorage, rather than further reductions in plant height (Muszynska et al. 2021; Peng et al. 2025). Longitudinal analyses of varietal releases indicate that modern cultivars are both shorter and structurally stronger. However, intensified management practices, including higher nitrogen inputs and denser planting, along with increasing climatic variability, have either sustained or increased lodging risk in some regions. This highlights the context dependence of realized genetic gain (Wang et al. 2023; Peng et al. 2025).

Recent advances in genomics have accelerated progress by enabling analysis of lodging as a complex trait controlled by loci associated with stem anatomy, root architecture, and phenological development. Genome-wide association studies and genomic prediction models now provide sufficient accuracy to support marker-assisted and genomic selection, thereby increasing selection intensity, reducing reliance on extensive phenotyping, and shortening breeding cycles (Rabieyan et al. 2024). Modern breeding strategies combine multi-environment lodging evaluations, high-throughput phenotyping of biomechanical and root traits, and genomic selection frameworks. However, concerns persist regarding bias in estimates of genetic gain arising from evolving agronomic practices, changing test environments, and genotype-by-environment interactions. To address these challenges, several authors recommend standardized, long-term trials and systematic use of archived check cultivars to enable unbiased, multi-decadal assessment of genetic progress in lodging resistance. Ongoing improvement will require integration of mechanistic trait understanding, advanced phenotyping platforms, and robust genomic selection within rigorously controlled experimental designs (Raymond et al. 2023; Peng et al. 2025).

### 9.6.4 Correlation Between Lodging Resistance and Yield Components in Different Agro-climatic Zones

The relationships between lodging resistance and yield components are highly context-dependent, varying across agro-climatic zones due to strong genotype-by-environment-by-management interactions. In temperate, high-input irrigated systems such as the North China Plain, agronomic interventions can enhance both lodging resistance and yield. For instance, a high–low plot sowing configuration increased grain yield by approximately 25%, primarily through higher spike density. At the same time, reduced plant height and a lowered centre of gravity improved structural stability (Wu et al. 2021a, b). These findings demonstrate that, under favourable conditions, increases in sink capacity (spikes per square meter) can be achieved without increasing lodging risk, provided plant architecture is effectively managed.

In more diverse production systems, including mixed-cropping and cultivar-mixture approaches, positive relationships between lodging resistance and yield have also been observed. Research by Kong et al. (2024) shows that combining wheat genotypes within a field can reduce lodging through complementary stem and root traits, while maintaining or improving key yield components such as grain number per spike and thousand-kernel weight through enhanced canopy structure and root anchorage. In contrast, low-input, rainfed subtropical systems typically experience lodging at later developmental stages, at which correlations with grain weight loss are weaker. However, in irrigated or stress-prone environments, early lodging events result in severe reductions in spike fertility and grain filling. Overall, these findings suggest that favourable trade-offs between lodging resistance and yield are possible when spatial arrangement, varietal diversity, and local environmental conditions are explicitly considered. Therefore, both breeding and agronomic strategies should be designed for specific agro-climatic scenarios, integrating optimized plant architecture, genotype combinations, and sowing patterns to sustain yield while minimizing lodging risk (Wu et al. 2021a, b; Kong et al. 2024).

### 9.6.5 National and International Breeding Programmes Focused on Lodging Mitigation

Wheat breeding programs at both global and national levels have increasingly prioritized lodging resistance as a central breeding objective, shifting from a focus on single traits such as reduced plant height to a more holistic, system-based approach. Internationally, CIMMYT prioritizes accelerated breeding cycles, genomic selection, and comprehensive multi-location testing to identify genotypes with improved stem mechanical strength and yield stability in lodging-prone environments (CIMMYT 2023). In India, the Indian Council of Agricultural Research (ICAR), through the Indian Institute of Wheat and Barley Research (IIWBR) and the All India Coordinated Research Project (AICRP) on Wheat, systematically evaluates advanced breeding lines for culm and root architectural traits, internode thickness, and tolerance to late-season rainfall under high-input management. These evaluations are subsequently translated into region-specific agronomic recommendations for farmers (ICAR-IIWBR 2022).

Simultaneously, breeding strategies have expanded to broaden the genetic base of lodging resistance beyond traditional *Rht* dwarfing alleles. The use of elite multiparent populations, nested association mapping panels, and multi-parent advanced generation inter-cross (MAGIC) populations has enabled the identification and recombination of quantitative trait loci associated with culm diameter, stem wall thickness, and root anchorage strength. This approach facilitates the accumulation of complementary alleles that improve standability without reducing yield potential (Dreccer et al. 2022; Rabieyan et al. 2024). These genetic advancements are increasingly supported by trait-focused pre-breeding and high-throughput phenotyping platforms, such as near-infrared reflectance for stem strength, NDVI-derived canopy vigour indices, and canopy temperature measurements. Additionally, the adoption of alternative Rht-like alleles and trait pyramiding strategies aims to mitigate the yield penalties sometimes associated with classical semi-dwarfism. Overall, current evidence suggests that the most significant reductions in lodging losses result from integrating genetic improvements with optimized agronomic practices, including refined sowing dates, planting densities, and careful application of growth regulators. This highlights the importance of coordinated genotype and management strategies in contemporary wheat improvement (Rabieyan et al. 2024).

## 9.7 Future Projections and Global Challenges

Wheat resilience to lodging is expected to encounter increasing challenges in the coming decades due to climatic changes, agronomic trade-offs, genetic limitations, and the need for sustainable and climate-adaptive varieties. Addressing these challenges requires an integrated approach that incorporates predictive modelling, breeding advancements, physiological research, and supportive policy measures to ensure yield stability under heightened lodging risk.

### 9.7.1 Anticipated Increase in Lodging Risk Due to Climate Change, High Rainfall Events, and High-Input Cultivation

Climate projections consistently indicate that both the frequency and intensity of extreme weather events, particularly intense rainfall, storms, and concurrent heat stress, will increase under future climate scenarios. This trend is expected to amplify lodging risk in major wheat-growing regions. Under high-emission pathways, up to

83% of global wheat-growing areas are projected to experience extreme heat during reproductive stages by mid-century, thereby increasing the likelihood of compound climatic shocks across key breadbasket regions such as South Asia and Eastern Europe (Biess et al. 2024). Elevated temperatures accelerate crop phenology and shorten the grain-filling period. When combined with high nitrogen-driven biomass accumulation and insufficient stem strength, these factors substantially increase susceptibility to lodging.

Simultaneously, climate change is expected to intensify extreme rainfall and storm events, which directly impose mechanical forces that induce stem bending and root failure. The interaction of these stresses, including thermal acceleration of development, excessive vegetative growth under intensive management, and increased precipitation- and wind-induced loads, creates a convergence of risk factors that threaten the stability of modern wheat ideotypes. As a result, mitigating lodging under future climates has become an urgent priority, necessitating coordinated advances in breeding for structural resilience and the development of agronomic strategies tailored to increasingly volatile weather regimes.

### 9.7.2 Challenges in Balancing Lodging Resistance with Traits Like Biomass, Grain Size, and Harvest Index

Physiological constraints and trade-offs in resource allocation, particularly for yield-related traits, limit efforts to enhance lodging resistance in wheat. In China, long-term breeding programs have achieved notable genetic gains in lodging resistance, estimated at approximately 15% from the 1950s to the 1980s and 172.8% from the 1990s to the 2020s. However, these advances are often diminished under high nitrogen conditions, which induce excessive stem elongation and decrease cellulose concentration, ultimately reducing culm mechanical strength (Wang et al. 2023). This persistent challenge highlights the difficulty of maintaining structural stability while increasing yield through intensive input use.

Lodging resistance is also influenced by the partitioning of biomass among stems, leaves, and reproductive organs, which generates trade-offs with grain size, quality, and overall yield potential. Morphological traits, such as increased culm wall thickness and higher cellulose content, are consistently associated with improved lodging tolerance. However, these characteristics often compete with the allocation of assimilates to grain development (Khobra et al. 2019). Therefore, the identification and combination of genotypes that achieve both high yield and strong lodging resistance remain significant challenges, emphasizing the importance of integrative breeding strategies that balance structural reinforcement with efficient biomass use.

### 9.7.3 Limited Genetic Diversity in Current Dwarfing Genes and the Need for Novel Sources of Resistance

Most lodging-resistant wheat varieties depend on Green Revolution semi-dwarfing genes, such as *Rht-B1* and *Rht-D1*. However, the limited genetic base of these alleles constrains future adaptive potential. Broader genetic sources, including landraces and wild relatives, remain insufficiently explored. Although not specific to lodging resistance, research in drought tolerance demonstrates that landraces, such as Norin 10 and Aka Komugi, can provide valuable alleles, including *Rht8c*, which may be repurposed for structural resilience (Bapela et al. 2022). Expanding genetic diversity for lodging resistance, particularly by incorporating root-anchorage traits and lignin biosynthesis pathways, is essential for future breeding success.

### 9.7.4 Importance of Interdisciplinary Collaboration Among Breeders, Physiologists, and Data Scientists

Addressing lodging physiology and resistance development necessitates multidisciplinary collaboration. As noted in the review, strategies for mitigating lodging in wheat and rice, investigations of physiological traits such as lignin content and stem mechanics, and agronomic management practices, including growth regulators and root anchorage, require further detailed study and remain insufficiently characterized (Shah et al. 2019). The complexity of lodging, which is influenced by plant morphology, biochemistry, structural mechanics, and environmental interactions, underscores the need for cooperation among breeders, mechanistic physiologists, and data scientists. The integration of advanced phenotyping, modelling approaches such as lodging mechanics and Bayesian yield projections, and machine learning platforms may accelerate the elucidation of lodging-resistance mechanisms and support predictive breeding efforts.

### 9.7.5 Policy Support for Resilient Agriculture and Investment in Cutting-Edge Breeding Technologies

Breeding pipelines alone are inadequate without robust institutional and policy support. Climate-resilient agriculture (CRA) frameworks, which integrate varietal innovation, agronomic interventions, and adaptive extension services, are increasingly prioritized in tropical regions, including India. Despite lodging being a major yield-limiting factor, it has not received adequate policy attention. Investment in infrastructure, including controlled stress phenotyping nurseries, growth regulator trials, and data platforms, as well as research utilizing these advanced techniques, remains underfunded. Furthermore, the pace of lodging-resistance innovation will depend on regulatory clarity regarding emerging technologies such as gene editing, for example, the use of CRISPR to modify stem architecture.

### 9.7.6 Emphasis on Breeding Climate-Smart Wheat Varieties Tailored for Different Agro-ecologies

Varietal strategies for lodging resistance must be closely aligned with regional agro-ecological conditions and changing climate regimes. In subtropical wheat production systems, factors such as limited sowing windows, high interannual rainfall variability, and occasional late-season storms substantially affect crop architecture and lodging risk. Empirical evidence indicates that modifications in sowing time (for example, early versus late November), combined with optimized agronomic practices such as split nitrogen application and precise irrigation scheduling, can significantly influence lodging indices and related yield losses by affecting crop phenology, biomass accumulation, and stem strength (Jat et al. 2014).

These findings indicate that genetic improvement alone is inadequate without corresponding management adaptation. The development of climate-smart wheat varieties that integrate moderate plant height, enhanced stem structure, and phenological flexibility tailored to local sowing dates and stress conditions is therefore critical. Incorporating these ideotypes within region-specific climate-resilient agriculture (CRA) frameworks, in which varietal deployment is coordinated with targeted agronomic practices, represents a practical approach to reducing lodging risk and maintaining yield stability in increasingly variable subtropical environments.

## 9.8 Conclusion and Way Forward

Lodging resistance in wheat has progressed from a focus on reduced plant height to an integrated breeding and management approach that incorporates advances in genetics, physiology, and agronomy. The use of semi-dwarfing genes, the enhancement of stem and root mechanical strength, and the optimization of nutrient and crop management have significantly reduced lodging incidence. Nevertheless, lodging remains a challenge under conditions of climatic variability and intensive agricultural inputs. Notable genetic improvements have resulted from enhanced stem architecture and root anchorage, facilitated by QTL mapping and genomic selection, thereby supporting the balance between high yield potential and structural stability. The integration of traditional breeding with molecular tools, high-throughput and field-based phenotyping, remote sensing, artificial intelligence-driven analytics, and genomic prediction is advancing wheat improvement by enabling more precise identification and selection of lodging-related traits. Additionally, emerging technologies, such as genome editing, offer targeted opportunities to modify key aspects of plant architecture and biomechanics. Ongoing, collaborative research is necessary to achieve durable lodging resistance without reducing productivity. This requires combining complementary above- and below-ground traits and developing ideotypes suited to diverse agro-ecological environments. Future directions include identifying and utilizing new genetic resources, refining environment-specific phenotyping to better capture genotype-by-environment interactions, and engaging with farmers to ensure practical application and adoption. The overarching goal is to deliver climate-resilient, lodging-resistant wheat cultivars that maintain high yields and support global food security in increasingly complex production systems.

## References

Acuña T, Rebetzke GJ, Condon AG et al (2001) The effect of *Rht* genotype and temperature on coleoptile growth and dry matter partitioning in young wheat seedlings. Funct Plant Biol 28:417–423. https://doi.org/10.1071/PP01010

Aisawi KAB, Reynolds MP, Singh RP et al (2015) The physiological basis of the genetic progress in yield potential of CIMMYT spring wheat cultivars from 1966 to 2009. Crop Sci 55(4):1749–1764. https://doi.org/10.2135/cropsci2014.09.0601

Ali N, Mohammed A, Bais A et al (2023) LodgeNet: an automated framework for precise detection and classification of wheat lodging severity levels in precision farming. Front Plant Sci 14:1255961. https://doi.org/10.3389/fpls.2023.1255961

Allan RE, Vogel OA, Peterson CJ Jr (1962) Seedling emergence rate of fall-sown wheat and its association with plant height and coleoptile length. Agron J 54(4):347–350. https://doi.org/10.2134/agronj1962.00021962005400040022x

Arndell T, Sharma N, Langridge P et al (2019) gRNA validation for wheat genome editing with the CRISPR-Cas9 system. BMC Biotechnol 19(1):71. https://doi.org/10.1186/s12896-019-0565-z

Awan MJA, Amin I, Rasheed A et al (2024) Knockout mutation in *TaD27* enhances number of productive tillers in hexaploid wheat. Front Genome Ed 6:1455761. https://doi.org/10.3389/fgeed.2024.1455761

Baker CJ, Berry PM, Spink JH et al (1998) A method for the assessment of the risk of wheat lodging. J Theor Biol 194(4):587–603. https://doi.org/10.1006/jtbi.1998.0778

Bapela T, Shimelis H, Tsilo TJ et al (2022) Genetic improvement of wheat for drought tolerance: progress, challenges and opportunities. Plants 11(10):1331. https://doi.org/10.3390/plants11101331

Berry PM, Griffin JM, Sylvester-Bradley R et al (2000) Controlling plant form through husbandry to minimise lodging in wheat. Field Crop Res 67(1):59–81. https://doi.org/10.1016/S0378-4290(00)00084-8

Berry PM, Sterling M, Spink JH et al (2004) Understanding and reducing lodging in cereals. Adv Agron 84:217–271. https://doi.org/10.1016/S0065-2113(04)84005-7

Berry PM, Sylvester-Bradley R, Berry S (2007) Ideotype design for lodging-resistant wheat. Euphytica 154:165–179. https://doi.org/10.1007/s10681-006-9284-3

Biess B, Gudmundsson L, Windisch MG et al (2024) Future changes in spatially compounding hot, wet or dry events and their implications for the world's breadbasket regions. Environ Res Lett 19(6):064011. https://doi.org/10.1088/1748-9326/ad4619

Borlaug NE (1968) Wheat breeding and its impact on world food supply. Aust Acad Sci, Canberra, pp 1–36

Borrill P, Harrington SA, Uauy C (2017) Genome-wide sequence and expression analysis of the NAC transcription factor family in polyploid wheat. G3 (Bethesda) 7(9):3019–3029. https://doi.org/10.1534/g3.117.043679

Buss W, Ford BA, Foo E et al (2020) Overgrowth mutants determine the causal role of gibberellin GA2oxidaseA13 in *Rht12* dwarfism of wheat. J Exp Bot 71:7171–7178. https://doi.org/10.1093/jxb/eraa443

Calderini DF, Reynolds MP (2000) Changes in grain weight as a consequence of de-graining treatments at pre- and post-anthesis in synthetic hexaploid lines of wheat (*Triticum durum* × *T. tauschii*). Aust J Plant Physiol 27:183–191

Chai S, Yao Q, Zhang X et al (2021) The semi-dwarfing gene *Rht-dp* from dwarf Polish wheat (*Triticum polonicum* L.) is the "Green Revolution" gene *Rht-B1b*. BMC Genomics 22:63. https://doi.org/10.1186/s12864-021-07367-x

Chen X, Liu Q, Zheng B et al (2024) Optimized phosphorus application enhances wheat stem lodging resistance under spring low-temperature stress. Plants 13:2980. https://doi.org/10.3390/plants13212980

CIMMYT (2023) Wheat breeding strategy and progress report. Int Maize Wheat Improv Cent, Mexico

Daetwyler HD, Villanueva B, Bijma P et al (2007) Inbreeding in genome-wide selection. J Anim Breed Genet 124:369–376. https://doi.org/10.1111/j.1439-0388.2007.00693.x

Divashuk MG, Bespalova LA, Vasilyev AV et al (2013) Reduced height genes and their importance in winter wheat cultivars grown in southern Russia. Euphytica 190:137–144. https://doi.org/10.1007/s10681-012-0789-7

Djenadi C, Yahiaoui S, Abdelguerfi-Laouar M et al (2025) Genetic analysis and marker-assisted selection using KASP markers in a doubled haploid bread wheat population. Euphytica 221:126. https://doi.org/10.1007/s10681-025-03554-z

Dreccer MF, Condon AG, Macdonald B et al (2020) Genotypic variation for lodging tolerance in spring wheat: wider and deeper root plates, a feature of low lodging, high-yielding germplasm. Field Crop Res 258:107942. https://doi.org/10.1016/j.fcr.2020.107942

Dreccer MF, Macdonald B, Farnsworth CA et al (2022) Multi-donor × elite-based populations reveal QTL for low-lodging wheat. Theor Appl Genet 135:1685–1703. https://doi.org/10.1007/s00122-022-04063-6

Dreisigacker S, Crossa J, Pérez-Rodríguez P et al (2021) Implementation of genomic selection in the CIMMYT global wheat program: findings from the past 10 years. Crop Breed Genet Genom 3:e210005. https://doi.org/10.20900/cbgg20210005

Dronamraju K (2018) A century of geneticists: mutation to medicine. CRC Press, Boca Raton

Duan J, Wu J, Liu Y et al (2012) New cis-regulatory elements in the *Rht-D1b* locus region of wheat. Funct Integr Genomics 12:489–500. https://doi.org/10.1007/s10142-012-0283-2

Ellis MH, Rebetzke GJ, Chandler P et al (2004) The effect of different height reducing genes on the early growth of wheat. Funct Plant Biol 31:583–589. https://doi.org/10.1071/FP03207

Feng S, Shi C, Wang P et al (2024) Stem characteristics and yield of wheat are regulated to improve planting efficiency and reduce lodging risk by fertilizer rate and irrigation stage. Agric Water Manag 306:109192. https://doi.org/10.1016/j.agwat.2024.109192

Fischer RA, Stapper M (1987) Lodging effects on high-yielding crops of irrigated semidwarf wheat. Field Crop Res 17:245–258. https://doi.org/10.1016/0378-4290(87)90038-4

Flintham JE, Gale MD (1983) The Tom Thumb dwarfing gene *Rht3* in wheat. II. Effects on height, yield and grain quality. Theor Appl Genet 66:249–256. https://doi.org/10.1007/BF00251155

Flintham JE, Börner A, Worland AJ et al (1997) Optimizing wheat grain yield: effects of *Rht* (gibberellin-insensitive) dwarfing genes. J Agric Sci 128:11–25. https://doi.org/10.1017/S0021859696003942

Fu M, Liu S, Che Y et al (2024) Genome-editing of a circadian clock gene *TaPRR95* facilitates wheat peduncle growth and heading date. J Genet Genomics 51:1101–1110. https://doi.org/10.1016/j.jgg.2024.05.011

Guha PK, Magar ND, Kommana M et al (2024) Strong culm: a crucial trait for developing next-generation climate-resilient rice lines. Physiol Mol Biol Plants 30:665–686. https://doi.org/10.1007/s12298-024-01445-6

Hai L, Guo H, Xiao S et al (2005) Quantitative trait loci (QTL) of stem strength and related traits in a doubled-haploid population of wheat (*Triticum aestivum* L.). Euphytica 141:1–9. https://doi.org/10.1007/s10681-005-4713-2

Hao J, Zhao Z, Fu X et al (2024) Effect of a novel dwarfing mutant site on chromosome 4B on agronomic traits in common wheat. Front Plant Sci 15:1338425. https://doi.org/10.3389/fpls.2024.1338425

Hazel LN (1943) The genetic basis for constructing selection indexes. Genetics 28:476–490. https://doi.org/10.1093/genetics/28.6.476

Hedden P (2003) The genes of the Green Revolution. Trends Genet 19:5–9. https://doi.org/10.1016/S0168-9525(02)00009-4
Hoogendoorn J, Rickson JM, Gale MD (1990) Differences in leaf and stem anatomy related to plant height of tall and dwarf wheat (*Triticum aestivum* L.). J Plant Physiol 136:72–77. https://doi.org/10.1016/S0176-1617(11)81618-4
Hossain A, Skalicky M, Brestic M et al (2021) Consequences and mitigation strategies of abiotic stresses in wheat (*Triticum aestivum* L.) under the changing climate. Agronomy 11:241. https://doi.org/10.3390/agronomy11020241
Hu X, Tian P, Fu W et al (2025) Effects of nitrogen fertilizer application on the lodging resistance traits, yield, and quality of two gluten types of wheat. Agriculture 15:637. https://doi.org/10.3390/agriculture15060637
ICAR (2020) Wheat production statistics of India. Indian Council of Agricultural Research, New Delhi
ICAR-IIWBR (2022) Vision 2050. ICAR-Indian Institute of Wheat and Barley Research, Karnal
Jat RK, Sapkota TB, Singh RG et al (2014) Seven years of conservation agriculture in a rice–wheat rotation of Eastern Gangetic Plains of South Asia: yield trends and economic profitability. Field Crop Res 164:199–210. https://doi.org/10.1016/j.fcr.2014.04.015
Kashiwagi T, Togawa E, Hirotsu N et al (2008) Improvement of lodging resistance with QTLs for stem diameter in rice (*Oryza sativa* L.). Theor Appl Genet 117:749–757. https://doi.org/10.1007/s00122-008-0816-1
Keller M, Karutz C, Schmid J et al (1999) Quantitative trait loci for lodging resistance in a segregating wheat × spelt population. Theor Appl Genet 98:1171–1182. https://doi.org/10.1007/s001220051182
Keyes GJ, Paolillo DJ, Sorrells ME (1989) The effects of dwarfing genes *Rht1* and *Rht2* on cellular dimensions and rate of leaf elongation in wheat. Ann Bot 64:683–690. https://doi.org/10.1093/oxfordjournals.aob.a087894
Khobra R, Sareen S, Meena BK et al (2019) Exploring the traits for lodging tolerance in wheat genotypes: a review. Physiol Mol Biol Plants 25:589–600. https://doi.org/10.1007/s12298-018-0629-x
Kokubo A, Kuraishi S, Sakurai N (1989) Culm strength of barley: correlation among maximum bending stress, cell wall dimensions, and cellulose content. Plant Physiol 91:876–882. https://doi.org/10.1104/pp.91.3.876
Kong E, Liu D, Guo X et al (2013) Anatomical and chemical characteristics associated with lodging resistance in wheat. Crop J 1:43–49. https://doi.org/10.1016/j.cj.2013.07.012
Kong X, Zhao G, Sun X et al (2024) Coordinating lodging incidence and grain yield through wheat genetic diversity. Field Crop Res 315:109468. https://doi.org/10.1016/j.fcr.2024.109468
Lawrenson T, Clarke M, Kirby R et al (2024) An optimised CRISPR Cas9 and Cas12a mutagenesis toolkit for barley and wheat. Plant Methods 20:123. https://doi.org/10.1186/s13007-024-01234-y
Leonova IN, Ageeva EV (2022) Localization of the quantitative trait loci related to lodging resistance in spring bread wheat (*Triticum aestivum* L.). Vavilovskii Zh Genet Selektsii 26:675–683. https://doi.org/10.18699/VJGB-22-82
Li Y, Yang B, Zhou S et al (2022) Identification of lodging degree of wheat using point cloud data and convolutional neural network. Front Plant Sci 13:968479. https://doi.org/10.3389/fpls.2022.968479
Liu Q, Luo L, Zheng L (2018) Lignins: biosynthesis and biological functions in plants. Int J Mol Sci 19:335. https://doi.org/10.3390/ijms19020335
Luo X, Wu Z, Fu L et al (2022) Evaluation of lodging resistance in rice based on an optimized parameter from lodging index. Crop Sci 62:1–12. https://doi.org/10.1002/csc2.20712
Ma QH (2010) Functional analysis of a cinnamyl alcohol dehydrogenase involved in lignin biosynthesis in wheat. J Exp Bot 61:2735–2744. https://doi.org/10.1093/jxb/erq107
Ma JF, Yamaji N (2006) Silicon uptake and accumulation in higher plants. Trends Plant Sci 11:392–397. https://doi.org/10.1016/j.tplants.2006.06.007

Mirabella NE, Abbate PE, Alonso MP et al (2019) Identifying traits at crop maturity and models for estimation of lodging susceptibility in bread wheat. Crop Pasture Sci 70:95–106. https://doi.org/10.1071/CP17347

Miralles DJ, Slafer GA (1995) Yield, biomass and yield components in dwarf, semi-dwarf and tall isogenic lines of spring wheat under recommended and late sowing dates. Plant Breed 114:392–396. https://doi.org/10.1111/j.1439-0523.1995.tb00818.x

Mulsanti IW, Yamamoto T, Ueda T et al (2018) Finding the superior allele of japonica-type for increasing stem lodging resistance in indica rice varieties using chromosome segment substitution lines. Rice 11:25. https://doi.org/10.1186/s12284-018-0216-3

Muszynska A, Guendel A, Melzer M et al (2021) A mechanistic view on lodging resistance in rye and wheat: a multiscale comparative study. Plant Biotechnol J 19:2646–2661. https://doi.org/10.1111/pbi.13689

Nguyen TN, Son S, Jordan MC et al (2016) Lignin biosynthesis in wheat (*Triticum aestivum* L.): its response to waterlogging and association with hormonal levels. BMC Plant Biol 16:28. https://doi.org/10.1186/s12870-016-0717-4

Olivoto T, Nardino M (2020) MGIDI: a novel multi-trait index for genotype selection in plant breeding. bioRxiv. https://doi.org/10.1101/2020.07

Peake AS, Bell KL, Fischer RA et al (2020) Cultivar × management interaction to reduce lodging and improve grain yield of irrigated spring wheat: optimising plant growth regulator use, N application timing, row spacing and sowing date. Front Plant Sci 11:401. https://doi.org/10.3389/fpls.2020.00401

Peng J, Richards D, Hartley N et al (1999) 'Green Revolution' genes encode mutant gibberellin response modulators. Nature 400:256–261. https://doi.org/10.1038/22307

Peng C, Guo Y, Huang G et al (2023) Effects of a novel plant growth regulator B2 on stalk quality and grain yield of winter wheat in North China. Plant Soil 487:437–453. https://doi.org/10.1007/s11104-023-05941-8

Peng D, Xu H, Guo Z et al (2025) Characteristics of lodging resistance of wheat cultivars from different breeding decades as affected by the application of paclobutrazol under shading stress. Agronomy 15:1848. https://doi.org/10.3390/agronomy15081848

Piñera-Chavez FJ, Berry PM, Foulkes MJ et al (2016a) Avoiding lodging in irrigated spring wheat. I. Stem and root structural requirements. Field Crop Res 196:325–336. https://doi.org/10.1016/j.fcr.2016.06.009

Piñera-Chavez FJ, Berry PM, Foulkes MJ et al (2016b) Avoiding lodging in irrigated spring wheat. II. Genetic variation of stem and root structural properties. Field Crop Res 196:64–74. https://doi.org/10.1016/j.fcr.2016.06.007

Piñera-Chavez FJ, Berry PM, Foulkes MJ et al (2021) Identifying quantitative trait loci for lodging-associated traits in the wheat doubled-haploid population Avalon × Cadenza. Crop Sci 61:2371–2386. https://doi.org/10.1002/csc2.20485

Pinthus MJ (1974) Lodging in wheat, barley, and oats: the phenomenon, its causes, and preventive measures. Adv Agron 25:209–263. https://doi.org/10.1016/S0065-2113(08)60782-8

Rabieyan E, Darvishzadeh R, Alipour H (2023) Identification and estimation of lodging in bread wheat genotypes using machine learning predictive algorithms. Plant Methods 19:109. https://doi.org/10.1186/s13007-023-01088-w

Rabieyan E, Darvishzadeh R, Alipour H (2024) Genetic analyses and prediction for lodging-related traits in a diverse Iranian hexaploid wheat collection. Sci Rep 14:275. https://doi.org/10.1038/s41598-023-49927-z

Rahmati M, Karimizadeh R (2025) Selection of bread wheat genotypes based on multi-trait selection index (MGIDI and SIIG). Plant Prod 48:189–205. https://doi.org/10.22055/ppd.2025.48230.2221

Raymond J, Mackay I, Penfield S et al (2023) Continuing genetic improvement and biases in genetic gain estimates revealed in historical UK variety trials data. Field Crop Res 302:109086. https://doi.org/10.1016/j.fcr.2023.109086

Rebetzke GJ, Ellis MH, Bonnett DG et al (2012) Height reduction and agronomic performance for selected gibberellin-responsive dwarfing genes in bread wheat (*Triticum aestivum* L.). Field Crop Res 126:87–96. https://doi.org/10.1016/j.fcr.2011.09.022

Reynolds MP, Pierre CS, Saad AS et al (2007) Evaluating potential genetic gains in wheat associated with stress-adaptive trait expression in elite genetic resources under drought and heat stress. Crop Sci 47:S172–S189. https://doi.org/10.2135/cropsci2007.10.0022IPBS

Sangjan W, McGee RJ, Sankaran S (2022) Optimization of UAV-based imaging and image processing orthomosaic and point cloud approaches for estimating biomass in a forage crop. Remote Sens 14:2396. https://doi.org/10.3390/rs14102396

Sasaki A, Ashikari M, Ueguchi-Tanaka M et al (2002) A mutant gibberellin-synthesis gene in rice. Nature 416:701–702. https://doi.org/10.1038/416701a

Shah AN, Tanveer M, Rehman AU et al (2017) Lodging stress in cereal – effects and management: an overview. Environ Sci Pollut Res 24:5222–5237. https://doi.org/10.1007/s11356-016-8237-1

Shah L, Yahya M, Shah SMA et al (2019) Improving lodging resistance: using wheat and rice as classical examples. Int J Mol Sci 20:4211. https://doi.org/10.3390/ijms20174211

Shahi D, Guo J, Pradhan S et al (2022) Multi-trait genomic prediction using in-season physiological parameters increases prediction accuracy of complex traits in US wheat. BMC Genomics 23:298. https://doi.org/10.1186/s12864-022-08487-8

Smith HF (1936) A discriminant function for plant selection. Ann Eugenics 7:240–250. https://doi.org/10.1111/j.1469-1809.1936.tb02143.x

Song L, Liu J, Cao B et al (2023) Reducing brassinosteroid signalling enhances grain yield in semi-dwarf wheat. Nature 617:118–124. https://doi.org/10.1038/s41586-023-06023-6

Spielmeyer W, Ellis MH, Chandler PM (2002) Semidwarf (*sd-1*), "Green Revolution" rice, contains a defective gibberellin 20-oxidase gene. Proc Natl Acad Sci USA 99:9043–9048. https://doi.org/10.1073/pnas.132266399

Tessema BB, Liu H, Sørensen AC et al (2020) Strategies using genomic selection to increase genetic gain in breeding programs for wheat. Front Genet 11:578123. https://doi.org/10.3389/fgene.2020.578123

Tian X, Xia X, Xu D et al (2022) *Rht24b*, an ancient variation of *TaGA2ox-A9*, reduces plant height without yield penalty in wheat. New Phytol 233:738–750. https://doi.org/10.1111/nph.17808

Togninalli M, Wang X, Kucera T et al (2023) Multi-modal deep learning improves grain yield prediction in wheat breeding by fusing genomics and phenomics. Bioinformatics 39:btad336. https://doi.org/10.1093/bioinformatics/btad336

Volpato L, Pinto F, González-Pérez L et al (2021) High-throughput field phenotyping for plant height using UAV-based RGB imagery in wheat breeding lines. Front Plant Sci 12:591587. https://doi.org/10.3389/fpls.2021.591587

Wang C, Ruan R, Yuan X et al (2014) Relationship between lignin metabolism and lodging resistance of culm in buckwheat. J Agric Sci 6:29. https://doi.org/10.5539/jas.v6n9p29

Wang X, Mace E, Tao Y et al (2020) Large-scale genome-wide association study reveals drought-induced lodging in grain sorghum is associated with plant height and carbon remobilisation traits. Theor Appl Genet 133:3201–3215. https://doi.org/10.1007/s00122-020-03665-2

Wang H, Bernardo A, St Amand P et al (2023) Skim exome capture genotyping in wheat. Plant Genome 16:e20381. https://doi.org/10.1002/tpg2.20381

Wang D, Guo M, Liu S et al (2024) Spatiotemporal evolution of winter wheat planting area and meteorology-driven effects on yield under climate change in Henan Province, China. Plants 13:2109. https://doi.org/10.3390/plants13152109

Wilke N, Siegmann B, Klingbeil L et al (2019) Quantifying lodging percentage and lodging severity using a UAV-based canopy height model. Remote Sens 11:515. https://doi.org/10.3390/rs11050515

Wu L, Han X, Islam S et al (2021a) Effects of sowing mode on lodging resistance and grain yield in winter wheat. Agronomy 11:1378. https://doi.org/10.3390/agronomy11071378

Wu Q, Chen Y, Xie J et al (2021b) A 36 Mb terminal deletion of chromosome 2BL is responsible for a wheat semi-dwarf mutation. Crop J 9:873–881. https://doi.org/10.1016/j.cj.2020.06.015

Xiang D, Song Y, Wu Q et al (2019) Relationship between stem characteristics and lodging resistance of Tartary buckwheat (*Fagopyrum tataricum*). Plant Prod Sci 22:202–210. https://doi.org/10.1080/1343943X.2019.1577143

Xu D, Hao Q, Yang T et al (2023) Impact of "Green Revolution" gene *Rht-B1b* on coleoptile length of wheat. Front Plant Sci 14:1147019. https://doi.org/10.3389/fpls.2023.1147019

Yang B, Zhu Y, Zhou S (2021a) Accurate wheat lodging extraction from multi-channel UAV images using a lightweight network model. Sensors 21:6826. https://doi.org/10.3390/s21206826

Yang Y, Amo A, Wei D et al (2021b) Large-scale integration of meta-QTL and genome-wide association study discovers genomic regions for yield-related traits in bread wheat. Theor Appl Genet 134:3083–3109. https://doi.org/10.1007/s00122-021-03881-4

Zang H, Su X, Wang Y et al (2024) Automatic grading evaluation of winter wheat lodging based on deep learning. Front Plant Sci 15:1284861. https://doi.org/10.3389/fpls.2024.1284861

Zanke CD, Ling J, Plieske J et al (2015) Analysis of main-effect QTL for thousand grain weight in European winter wheat by genome-wide association mapping. Front Plant Sci 6:644. https://doi.org/10.3389/fpls.2015.00644

Zetzsche H, Friedt W, Ordon F (2020) Breeding progress for pathogen resistance is a second major driver for yield increase in German winter wheat. Sci Rep 10:20374. https://doi.org/10.1038/s41598-020-77200-0

Zhang Y, Xu W, Wang H et al (2016) Progress in improving stem lodging resistance of Chinese wheat cultivars. Euphytica 212:275–286. https://doi.org/10.1007/s10681-016-1768-1

Zhang K, Zhang R, Yang Z et al (2023) Efficient wheat lodging detection using UAV remote sensing images and a multi-branch classification framework. Remote Sens 15:4572. https://doi.org/10.3390/rs15184572

Zhou Y, Wang J, Liu X et al (2023) Fine-tuning wheat plant height through CRISPR/Cas editing of *Rht8* alleles. Plant Biotechnol J 21:1764–1776. https://doi.org/10.1111/pbi.14083

Zhu C, Gore M, Buckler ES et al (2008) Status and prospects of association mapping in plants. Plant Genome 1:1–10. https://doi.org/10.3835/plantgenome2008.02.0089

Zuffo AM, Steiner F, Aguilera JG et al (2020) Multi-trait stability index: a tool for simultaneous selection of soybean genotypes under drought and saline stress. J Agron Crop Sci 206:815–822. https://doi.org/10.1111/jac.12409

# Advances in Mitigating Waterlogging Stress in Wheat

# 10

Priti Duhan, Meenu Dua, Charu Lata, Arpita Sharma,
Naresh Kumar, Poonam Bansal, and Vikas Joon

**Abstract**

Wheat production is increasingly constrained by waterlogging, particularly in regions characterized by heavy soils, unpredictable rainfall, and insufficient drainage, all of which impair normal root function. This chapter provides a comprehensive overview of scientific and technological advancements that have transformed strategies for managing waterlogging stress in wheat. It reviews progress from traditional field practices, such as raised beds and surface drainage, to recent discoveries regarding physiological responses, including root aerenchyma formation, metabolic adaptation to hypoxia, and enhanced antioxidant defences. Advances in breeding, facilitated by transcriptomic profiling, genomic technologies, and molecular markers, have improved the identification of tolerant genotypes and the incorporation of key traits. Innovative approaches, including CRISPR-based genome editing, high-throughput phenotyping, remote

P. Duhan (✉) · M. Dua
K.L. Mehta Dayanand College for Women, Faridabad, Haryana, India
e-mail: pduhanbiotech@kuk.ac.in

C. Lata
ICAR-Indian Institute of Wheat and Barley Research, Karnal, Haryana, India

A. Sharma
ICAR-Indian Institute of Wheat and Barley Research, Karnal, Haryana, India

Thapar Institute of Engineering and Technology, Patiala, Punjab, India

N. Kumar
Department of Chemistry and Biochemistry, Eternal University, Baru Sahib, HP, India

P. Bansal
MMDU University, Ambala, Haryana, India

V. Joon
ICAR-Central Plantation Crops Research Institute, Kasaragod, Kerala, India



R. Tiwari et al. (eds.), *Wheat Stress Management*,
https://doi.org/10.1007/978-981-92-0902-6_10

sensing, and artificial intelligence-driven advisory systems, further support early detection and adaptive management. Biological interventions, such as microbial inoculants and seed priming, confer additional resilience during critical growth stages. However, significant challenges persist due to limited genetic diversity, climate variability, and socioeconomic barriers to technology adoption. This chapter emphasizes the strategic actions required to develop climate-resilient wheat production systems capable of withstanding future waterlogging stress.

**Keywords**

Wheat · Waterlogging · Tolerance · Genomics · Agronomy

## 10.1 Introduction

Waterlogging stress significantly limits wheat productivity across diverse agro-ecosystems. Prolonged root zone saturation typically results from excessive rainfall or irrigation, especially in regions characterized by heavy soils, low-lying topography, or insufficient drainage (Sharma and Kumar 2023). These conditions impede soil aeration and create an environment unfavourable for wheat growth and development. The increasing frequency and severity of waterlogging, driven by climate variability and unpredictable rainfall, place considerable pressure on farmers and wheat-based cropping systems (Chhabra et al. 2025). Physiologically, waterlogging induces a range of adverse effects primarily due to oxygen deficiency in the rhizosphere. When soil pores become saturated, hypoxic, or anoxic conditions develop around the roots, severely limiting gas exchange. Oxygen deprivation disrupts root metabolism, inhibits aerobic respiration, and reduces ATP synthesis (Shahid and Ali 2025). As root activity declines, plants experience reduced uptake of essential nutrients, including nitrogen, iron, and micronutrients. The accumulation of toxic metabolites, hormonal imbalances, and altered water relations further intensify plant stress (Lee et al. 2025). These physiological impairments also heighten susceptibility to soil-borne pathogens, accelerating root decay and reducing overall plant vigour.

These physiological disruptions have significant agronomic consequences. Persistent waterlogging in wheat often leads to poor tillering, reduced leaf area, diminished photosynthetic efficiency, and delayed phenological development (Pais et al. 2022). Ultimately, these effects can cause substantial yield losses, frequently exceeding 20–50%, depending on the duration and timing of the stress. Grain quality also declines, as evidenced by reductions in grain weight, protein content, and milling performance. In severe cases, prolonged submersion may result in complete crop failure. Because waterlogging stress involves complex interactions amongst soil physics, plant physiology, microbial ecology, and environmental variability, effective mitigation requires integrated, multifaceted approaches (Fig. 10.1; Tyagi et al. 2023). Conventional breeding has produced tolerant genotypes, whilst agronomic practices such as improved drainage, raised-bed planting, and optimized

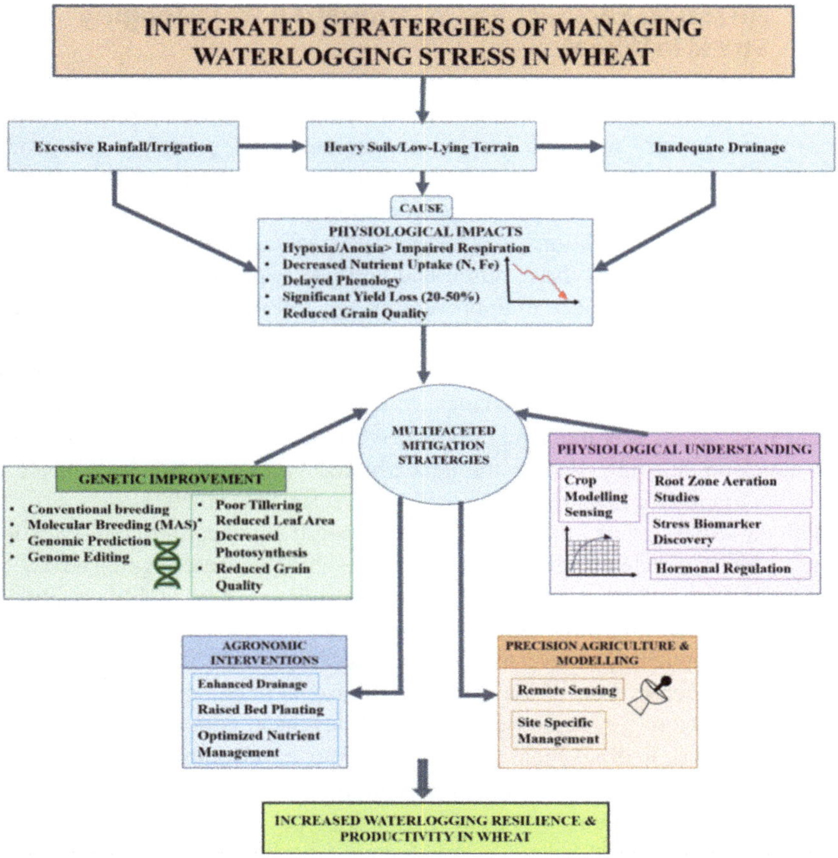


**Fig. 10.1** Integrated strategies of managing waterlogging stress in wheat

nutrient management offer additional solutions. Advances in biotechnology, including molecular breeding, marker-assisted selection, genomic prediction, and genome editing, have expanded the range of tools available to researchers and breeders. Furthermore, crop modelling, precision agriculture, and remote sensing technologies provide new opportunities for early detection and site-specific management of waterlogging (Sharma et al. 2025). This chapter provides an overview of the scientific understanding, historical context, and recent advances in managing waterlogging stress in wheat. It presents a comprehensive synthesis of innovative technologies, recent experimental findings, and foundational research that collectively enhance waterlogging resilience. By highlighting key discoveries, existing knowledge gaps, and future directions, this chapter aims to support researchers, breeders, agronomists, and policymakers in developing effective strategies to improve wheat productivity under waterlogged conditions.

## 10.2 Historical Perspectives on Mitigating Waterlogging Stress in Wheat

Efforts to reduce waterlogging stress in wheat have been shaped by practical farming experience, early scientific inquiry, and advances in agronomic and genetic technologies over centuries. In flood-prone plains and river basins, traditional farming communities developed context-specific strategies to enhance wheat survival under saturated soil conditions well before systematic research on abiotic stress tolerance began. Amongst these, raised-bed farming, which elevates planting rows above the natural soil surface to improve drainage and reduce the risk of standing water around the root zone, was a widely adopted indigenous technique (Chhabra et al. 2025). To divert excess runoff from crops, farmers also dug drainage trenches between plots or along field boundaries. In several regions of South Asia and the Middle East, wheat was frequently cultivated on elevated bunds or ridges formed after monsoon flooding, demonstrating early recognition of the importance of micro-topography for managing excess moisture (Qamar et al. 2025).

With the expansion of formal agricultural research in the early to mid-twentieth century, scientific understanding of wheat's vulnerability to waterlogging advanced. Early studies revealed that traditional tall wheat varieties were highly sensitive to saturated soil conditions due to specific structural and physiological characteristics, despite their adaptability to diverse environments (Zahra et al. 2025). Extended vegetative cycles often exposed these varieties to seasonal flooding, whilst their large, vertically oriented root systems were prone to decay under poor aeration (Woolway et al. 2025). These findings highlighted the need for improved wheat varieties that can tolerate intermittent or prolonged soil saturation. They laid the groundwork for understanding the diverse responses of wheat genotypes to waterlogging.

Historical patterns of irrigation development further complicated efforts to address waterlogging. The expansion of canal systems and insufficient drainage infrastructure in many irrigated wheat-growing regions led to rising groundwater tables, inadequate land levelling, and unintended water accumulation in low-lying fields (Minhas and Qadir 2024). Over-irrigation, driven by low water prices, unpredictable rainfall, or limited awareness, exacerbated these issues. As a result, waterlogging evolved from a naturally occurring, rainfall-dependent event to a persistent management problem in intensively cultivated areas (Aggarwal et al. 2024). By the latter half of the twentieth century, scientists and agricultural planners increasingly recognized the importance of water and land management in sustaining wheat productivity in vulnerable regions. Large-scale initiatives focused on land levelling, reshaping field bunds, and improving drainage became integral to agricultural development programs. Simultaneously, agronomists refined recommendations for sowing dates, irrigation scheduling, and soil structure management to reduce the risk of soil saturation during critical stages of wheat growth (Cheng et al. 2024).

Breeding programs began to incorporate strategies to avoid or escape waterlogging stress. The introduction of early maturing cultivars enabled crops to complete sensitive vegetative and reproductive stages before peak rainfall or irrigation-induced saturation. Additionally, shallower-rooted genotypes reduced exposure to anaerobic soil layers, offering a partial solution for fields with persistently high-water tables. Although these early varietal improvements did not fully resolve the complexity of waterlogging, they represented a significant shift from reactive field practices to proactive biological mitigation (Cheng et al. 2024). Overall, the historical progression of efforts to reduce waterlogging stress in wheat illustrates the convergence of traditional knowledge, scientific insight, and technological innovation. This foundation has enabled the development of modern, integrated approaches that combine genetic tolerance, advanced agronomy, and environmental monitoring, which are now essential for sustaining wheat productivity in water-affected agroecosystems.

## 10.3 Scientific Advancements and Principles Developed over the Past Centuries

Over the past few centuries, research on wheat responses to waterlogging has progressed from descriptive observations to specialized physiological, genetic, and engineering-based approaches. Foundational studies clarified how wheat plants perceive and adapt to oxygen-deficient environments, whilst recent advances have translated this understanding into targeted breeding and management strategies. A key finding is the identification of physiological and anatomical responses that enable specific wheat genotypes to tolerate saturated soils (Nirmalaruban et al. 2025). For instance, wheat roots frequently develop aerenchyma, or air-filled spaces within cortical tissues, under hypoxic or anoxic conditions, which facilitates oxygen transport from shoots to submerged roots.

This adaptation, initially documented in wetland species, was subsequently recognized as a significant mechanism for partial tolerance in upland crops such as wheat. Concurrently, research revealed the importance of antioxidant defence systems in mitigating oxidative stress during reoxygenation events. Increased activity of enzymes such as superoxide dismutase, peroxidases, and catalases has been shown to limit cellular injury and support recovery following water recession (Basit et al. 2024). Additionally, ethylene, a stress-responsive plant hormone, has been identified as a central signalling molecule regulating adaptive responses, including aerenchyma formation, adventitious root initiation, leaf senescence, and shoot elongation under waterlogged conditions.

Advancements in plant genetics over the past century have enabled the identification of natural variability in waterlogging tolerance within global wheat germplasm collections. Studies of landraces, wild relatives, and region-specific cultivars have revealed substantial phenotypic diversity in traits such as root porosity, chlorosis tolerance, biomass maintenance, and survival under periodic flooding. Genetic mapping and quantitative trait locus (QTL) analyses have demonstrated the polygenic basis of waterlogging tolerance, supporting the incorporation of diverse tolerance sources into modern breeding programs (Ali et al. 2024). To facilitate these efforts, researchers have established systematic screening procedures for precise and repeatable evaluation of wheat genotypes under controlled or semi-controlled wet conditions. These protocols simulate varying durations and intensities of water saturation, allowing assessment of plant responses during both vegetative and reproductive stages. Standardized screening environments, such as water-filled tanks, lysimeter systems, and artificially elevated water tables, are now essential for comparing genotypic performance and identifying promising lines for further improvement (McCauley et al. 2021).

Several morphological and physiological traits have been identified as effective markers of tolerance within these screening approaches. Survival in hypoxic environments is often closely associated with root-related traits such as adventitious root formation, root porosity, and maintenance of root biomass (Devi et al. 2025). Additional reliable selection criteria include above-ground indicators such as the extent of leaf chlorosis, shoot elongation during partial inundation, tiller survival, and retention of green leaf area. These quantifiable traits have enhanced the efficiency of selecting tolerant breeding materials and improved the repeatability of screening programs.

Substantial progress has been made in soil and water engineering solutions to reduce waterlogging at both field and regional scales, complementing advances in biology and genetics. Research in soil physics and hydrology has guided the development of surface and subsurface drainage systems that efficiently remove excess water and maintain adequate aeration in the root zone (Ucar and Miseckaite 2025). Many wheat-producing regions have implemented innovations such as mole drains, tile drainage, controlled drainage structures, and improved field levelling. These engineering interventions, often integrated with agronomic practices such as bed planting and residue management, have significantly reduced the frequency and severity of waterlogging.

## 10.4 Recent Developments in Mitigating Waterlogging Stress in Wheat

Recent advancements in plant breeding, biotechnology, soil–plant–microbe interactions, and precision agronomy have substantially accelerated efforts to mitigate waterlogging stress in wheat. These innovations have shifted traditional methods towards integrated, resilient, and scientifically robust strategies that address both the physiological limitations of wheat and the ecological challenges of water-affected agricultural systems. Figure 10.2 illustrates the revolution in waterlogging-related research and management approaches for wheat in the contemporary hi-tech era.

### 10.4.1 Breeding of Tolerant Varieties with Improved Root Architecture and Metabolic Efficiency

The development of wheat varieties capable of sustaining production under partial to continual waterlogging has become a central objective in both traditional and modern breeding programs. Recent breeding initiatives prioritize root architectural

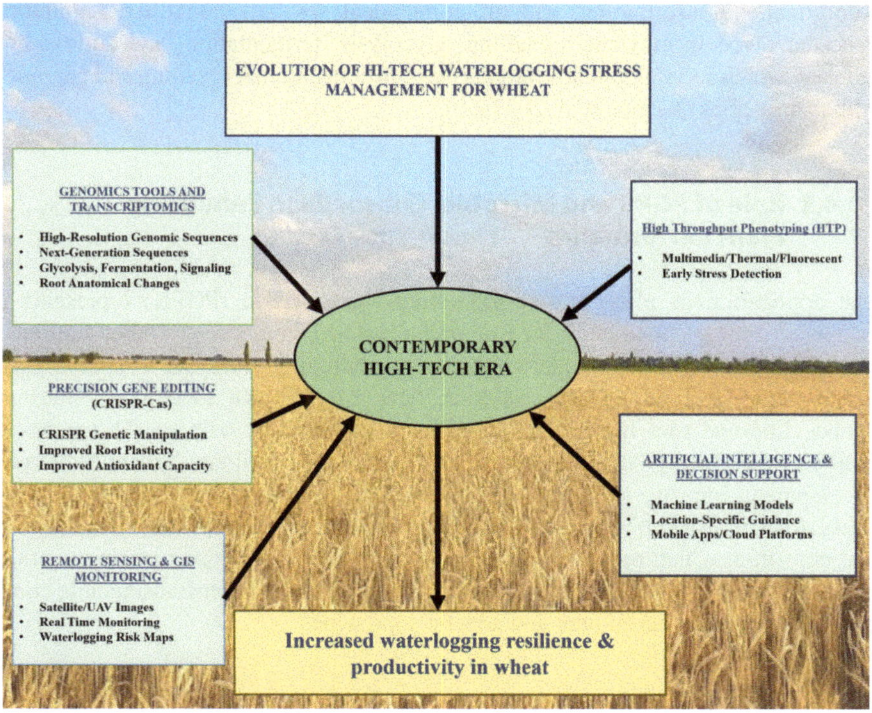


**Fig. 10.2** Schematic representation of the revolutionization of waterlogging-related research and management approaches in wheat in the contemporary hi-tech era

traits such as the formation of adventitious roots, increased root porosity, and shallow yet horizontally extensive root systems that maintain function in oxygen-limited soils (Bettcher 2024). Metabolic adaptations, including enhanced anaerobic respiration, reduced accumulation of toxic byproducts, and improved energy-use efficiency in hypoxic conditions, are also emphasized. The integration of these traits, together with improved shoot vigour and stable phenology, has led to the release of several promising waterlogging-tolerant wheat cultivars suitable for diverse agroclimatic regions (Kumar et al. 2021a, b).

## 10.4.2 Molecular Markers and Genomic Tools for Waterlogging Tolerance

Key traits, including root porosity, aerenchyma formation, and tolerance-associated quantitative trait loci (QTLs), can now be rapidly incorporated through marker-assisted selection (MAS) and marker-assisted backcrossing (MABC) (Raj and Nadarajah 2022). The use of high-throughput genotyping platforms has accelerated the identification of single-nucleotide polymorphisms (SNPs) associated with waterlogging resilience, allowing breeders to screen large populations efficiently.

Additionally, transcriptomic and proteomic analyses have identified regulatory genes involved in ethylene signalling, glycolysis, fermentation, and antioxidant defence, which serve as valuable targets for both conventional breeding and genome-editing strategies (Kumar et al. 2021a, b).

### 10.4.3 Role of PGPR and Microbial Consortia in Enhancing Plant Performance

The application of plant growth-promoting rhizobacteria (PGPR) represents a promising biological strategy for mitigating waterlogging stress. Under saturated soil conditions, beneficial microbial strains, including *Bacillus*, *Pseudomonas*, *Azospirillum*, and *Trichoderma* species, have been shown to enhance nutrient uptake, improve root health, and regulate plant hormone balance (Hasan et al. 2024). By producing ACC deaminase, these microbes facilitate adventitious root development, improve oxidative stress management, and reduce ethylene-induced stress. Recent research has highlighted the synergistic effects of multi-strain microbial consortia, which provide wheat plants with greater support than single-strain inoculants (Wu et al. 2023). These biological interventions offer a sustainable complement to existing agronomic and genetic approaches.

### 10.4.4 Seed Priming and Hormonal Treatments for Early-Stage Resilience

Recent advances have established seed priming as an effective strategy to enhance early vigour and resistance to waterlogging (Zhang et al. 2025). In moisture-stressed environments, approaches such as osmopriming, hydropriming, and nutrient-based priming improve seedling establishment, root emergence, and germination uniformity. Gibberellin treatments promote coleoptile elongation and seedling growth under anaerobic conditions, whilst ethylene inhibitors mitigate premature leaf senescence (Setiawan et al. 2024). Hormonal treatments have shown significant potential. These interventions are particularly effective during early growth stages, when seedlings are highly susceptible to temporary waterlogging.

### 10.4.5 Integrated Agronomic Practices for Holistic Waterlogging Management

Recent developments in agronomic practices have emphasized integrated management systems that include nutrient optimization, strategic crop placement, and soil modification. Raised-bed planting is increasingly adopted for its ability to enhance root-zone aeration, reduce soil compaction, and improve drainage. Adjustments to crop scheduling, such as early sowing or selecting crop phenologies that avoid periods of peak rainfall, further reduce the risk of saturation events (Minoli et al. 2022).

Nutrient management strategies, including split nitrogen application and the use of slow-release fertilizers, have been shown to maintain nutrient availability under variable moisture conditions (Rashid et al. 2021). When combined with conservation agriculture practices such as residue retention and minimal tillage, these approaches contribute to more resilient soil structures and greater long-term crop stability.

## 10.5 Evolution in the Contemporary Hi-Tech Era

The emergence of advanced technologies has fundamentally transformed research and management strategies for waterlogging stress in wheat. The integration of genomics, digital agriculture, remote sensing, artificial intelligence, and high-throughput phenotyping has enhanced both the understanding of stress responses and the development of practical interventions for breeders, agronomists, and farmers. These technological advancements provide exceptional precision, efficiency, and scalability in addressing persistent challenges within wheat production systems.

### 10.5.1 Genomic Tools and Transcriptomics for Identifying Candidate Genes

The identification of genes and regulatory networks associated with waterlogging tolerance has advanced with the availability of high-resolution genomic sequences and next-generation sequencing platforms (Tyagi et al. 2023). Unique gene sets activated under hypoxic stress have been identified through transcriptomic profiling, utilizing RNA-seq, microarrays, and single-cell expression technologies (Ahmed et al. 2024). These genes regulate glycolysis, fermentation, ethylene signalling, detoxification of reactive oxygen species (ROS), root anatomical modifications, and hormonal interactions (Chandra-Kuntal 2022). Comparative transcriptomics between tolerant and susceptible genotypes has revealed candidate genes underlying adaptive traits, including adventitious root initiation, aerenchyma formation, and cellular energy maintenance under oxygen-deficient conditions. These findings provide valuable molecular targets for genetic engineering and breeding programs.

### 10.5.2 Precision Gene Editing Using CRISPR–Cas Systems

Genome editing technologies, particularly CRISPR–Cas9 and its recent advancements such as Cas12a and base editors, now enable precise manipulation of genes associated with waterlogging tolerance (Choudry et al. 2024). Unlike conventional breeding, genome editing facilitates targeted, minor genetic modifications that enhance specific stress-related traits without introducing undesirable genetic linkages. Edited wheat lines have demonstrated improvements in root plasticity, ethylene perception, carbohydrate mobilization, and antioxidant capacity (Alrajhi et al. 2024). Ongoing research focuses on removing negative regulators of stress tolerance and modifying promoters of hypoxia-responsive genes to optimize their expression. As regulatory frameworks evolve, genome-edited wheat varieties are expected to provide farmers with high-performing cultivars resilient under wet conditions.

### 10.5.3 Remote Sensing and GIS-Based Monitoring Systems

Recent advancements in satellite imaging, unmanned aerial vehicles (UAVs), and ground-based sensors have enabled real-time monitoring of waterlogging across extensive agricultural landscapes (Kumar et al. 2025). Remote sensing instruments that use multispectral and hyperspectral imagery support early detection of surface drainage patterns, soil moisture accumulation, and plant canopy health. Geographic Information Systems (GIS) platforms integrate spatial datasets with elevation models, rainfall records, and hydrological flow patterns to generate comprehensive waterlogging risk maps (Hossan et al. 2025). By using these digital tools, farm managers and policymakers can efficiently identify high-risk areas, prioritize drainage interventions, adjust sowing schedules, or introduce crop varieties with tolerance.

### 10.5.4 Artificial Intelligence and Decision Support Systems

The integration of artificial intelligence (AI) and machine learning into agricultural decision-making has significantly enhanced the accuracy of waterlogging mitigation techniques (Rastogi et al. 2024). AI-driven models predict the probability of waterlogging and recommend adaptive strategies by analysing historical climate data, soil characteristics, and crop growth information (Selvakumar et al. 2025). Farmers receive location-specific recommendations for drainage management, irrigation scheduling, field levelling, and cultivar selection via mobile applications and cloud-based platforms. These decision support systems translate complex agronomic data into accessible recommendations, thereby facilitating the application of advanced research at the farm level.

### 10.5.5 High-Throughput Phenotyping for Accelerated Screening

High-throughput phenotyping (HTP) is a practical approach for identifying wheat genotypes tolerant of waterlogging stress, providing a faster, more objective alternative to traditional visual assessments (Kumari et al. 2024). Through the integration of imaging technologies, including multispectral, hyperspectral, thermal, and chlorophyll fluorescence sensors, HTP platforms enable the detection of early physiological changes such as alterations in canopy temperature, leaf pigmentation, and photosynthetic activity prior to the appearance of visible symptoms (He et al. 2024).

Drone-based imaging facilitates rapid evaluation across extensive breeding trials, whilst automated greenhouse and conveyor systems provide continuous data to monitor plant responses throughout the stress period (Al Najadi 2024). The adoption of advanced root phenotyping tools, such as X-ray computed tomography, three-dimensional imaging, and field-based shovelomics, has enhanced the characterization of critical traits, including aerenchyma formation, root porosity, and adventitious rooting, all of which support oxygen diffusion in saturated soils. When combined with genomic datasets, HTP outputs support marker discovery, refine predictive models in genomic selection, and expedite the identification of superior genotypes (Sheikh et al. 2024). Collectively, HTP reduces bottlenecks in trait evaluation and strengthens breeding pipelines to develop wheat varieties that are resilient to waterlogging.

## 10.6 Statistical Insight into Mitigating Waterlogging Stress in Wheat

Quantitative assessments of waterlogging stress and associated mitigation strategies are essential for identifying patterns of vulnerability, informing research priorities, and evaluating the effectiveness of technological interventions in wheat-based production systems. The following subsections present key statistical trends and insights relevant to India and the broader South Asian context, where waterlogging is a persistent challenge in low-lying, monsoon-dependent agroecological zones. These findings are further detailed in Table 10.1.

### 10.6.1 Trends in Waterlogging-Affected Wheat Areas in India and South Asia

Waterlogging has emerged as a significant constraint in wheat-growing ecosystems across India and several regions of South Asia, primarily due to altered rainfall patterns, increased irrigation intensity, and inadequate surface drainage. Agricultural survey data reveal a steady expansion of flood-prone and persistently saturated fields, especially in the eastern Indo-Gangetic Plains, where clay-rich soils and shallow water tables increase the risk of prolonged inundation (Sharma et al. 2022). States including Bihar, eastern Uttar Pradesh, and parts of West Bengal frequently experience episodic waterlogging during critical wheat growth stages, leading to inconsistent yield patterns across years (Bundela and Singh 2021). Comparable trends are evident in Nepal's Terai region and Pakistan's Punjab province, where unregulated canal irrigation and monsoon variability further exacerbate waterlogging. These developments emphasize the urgent need for region-specific management approaches and the development of climate-resilient wheat varieties.

**Table 10.1** Statistical insights on waterlogging stress in wheat

| S. no. | Sections | Key content | References |
|---|---|---|---|
| 1. | Trends in affected areas | Waterlogging is increasing in India and South Asia due to monsoon shifts, intensive irrigation, and poor drainage; major hotspots include Bihar, Eastern UP, West Bengal, the Nepal Terai, and the Punjab in Pakistan | Sharma et al. (2022), Bundela and Singh (2021) |
| 2. | Yield losses and production decline | Wheat yield declines 15–50% under moderate–severe waterlogging; linked to reduced oxygen, nutrient uptake, and high-rainfall episodes | Qian et al. (2025), Rupngam and Messiga (2024) |
| 3. | Agronomic practices impact | Raised beds, furrow irrigation, residue retention, and organic matter additions improve drainage and yield by 8–25% and enhance soil health | Smith et al. (2024) |
| 4. | Adoption of tolerant cultivars and drainage | Adoption of tolerant varieties is increasing; drainage methods are slower due to cost; uptake is higher with training and demonstrations | Akter and Mazumder (2025), Amin et al. (2024) |
| 5. | Breeding program performance | Marker-assisted selection and phenotyping improve gains; focus on root traits and vigour; collaborations accelerate variety release; greater investment is needed | Belliappa et al. (2024), Khedr and Khedr (2025), Thingujam et al. (2025) |

### 10.6.2 Yield Losses and Production Decline Due to Waterlogging

Yield penalties resulting from waterlogging exhibit considerable variation across environments; however, statistical models consistently indicate substantial losses, particularly when flooding coincides with tillering or grain filling. Multi-year trial data indicate that wheat yields may decline by 15–50% under moderate to severe waterlogging, contingent on soil type, genotype, and stress duration (Qian et al. 2025). Extended periods of soil saturation reduce oxygen availability in the root zone, thereby impairing nutrient uptake, limiting biomass accumulation, and accelerating premature leaf senescence. These factors collectively contribute to a decline in yield. At the regional level, production variability is closely associated with high-rainfall events and poorly drained landscapes, establishing waterlogging as a significant but often overlooked yield constraint in rainfed and irrigated lowlands of South Asia (Rupngam and Messiga 2024). These findings highlight the economic importance of timely drainage interventions and the deployment of resilient cultivars.

### 10.6.3 Impact of Improved Agronomic Practices on Yield and Soil Health

Agronomic interventions are essential for mitigating the adverse effects of waterlogging, as demonstrated by increasing evidence from on-farm evaluations. Raised-bed planting and furrow-irrigated systems consistently improve yields by facilitating rapid water removal and minimizing root-zone anoxia (Smith et al. 2024). The addition of organic matter, retention of crop residues, and implementation of controlled traffic farming enhance soil aggregation and infiltration, enabling fields to recover more efficiently following heavy rainfall. Statistical analyses of demonstration trials report an 8–25% increase in yield when these improved practices are adopted, compared with conventional flat sowing (Maredia et al. 2025). In addition to yield improvements, these practices support long-term soil health by increasing microbial activity, enhancing porosity, and stabilizing soil structure. These factors collectively reduce the severity of waterlogging in subsequent seasons.

### 10.6.4 Adoption Rates of Tolerant Cultivars and Drainage Interventions

The adoption of waterlogging-tolerant wheat cultivars has increased gradually in regions with persistent flooding risk; however, uptake rates vary according to seed availability, market linkages, and farmer awareness. Extension program surveys indicate that farmers in chronically affected districts are more likely to experiment with tolerant varieties, with adoption rates often surpassing those of improved agronomic practices (Akter and Mazumder 2025). In contrast, drainage-related interventions such as field levelling, surface drains, and subsurface channels exhibit slower but consistent adoption, primarily due to higher labour and installation costs. Cluster-level analyses demonstrate that communities receiving targeted training and access to demonstration plots achieve significantly higher adoption rates, highlighting the importance of participatory extension approaches for widespread dissemination (Amin et al. 2024). Although progress has been made, considerable potential remains for scaling up both varietal and drainage solutions through enhanced policy support and strengthened supply chains.

### 10.6.5 Comparative Performance of Breeding Programs Targeting Waterlogging Tolerance

Breeding programs in India and South Asia have achieved significant advancements in developing wheat lines with enhanced tolerance to waterlogging. However, the effectiveness and reach of these initiatives differ amongst institutions (Belliappa et al. 2024). Programs that incorporate marker-assisted selection, high-throughput phenotyping, and multilocation testing consistently report the most significant genetic gains, as these approaches more accurately capture trait expression across a range of stress environments (Khedr and Khedr 2025). Comparative analyses indicate that breeding pipelines focusing on root traits, early vigour, and efficient carbohydrate metabolism achieve higher selection accuracy than those that rely exclusively on visual assessment of stress symptoms. Collaborative networks that connect national agricultural research systems with international centres have also facilitated faster release rates and broader adaptability of tolerant lines. Nevertheless, sustained investment in phenotyping infrastructure, genomic technologies, and long-term field trials remains critical to maintain breeding progress and ensure stable varietal performance under future climate variability (Thingujam et al. 2025).

## 10.7 Future Projections and Global Challenges

Rapid climate shifts and evolving agronomic pressures are transforming wheat production, with waterlogging emerging as a significant global constraint. Intensifying extreme precipitation events and increasing vulnerability of farming systems have exposed the limitations of existing germplasm, infrastructure, and policy frameworks. Addressing these challenges requires a proactive strategy that integrates climate prediction, advanced breeding technologies, and socioeconomic considerations.

### 10.7.1 Climate-Driven Increases in Rainfall Variability and Extreme Precipitation

Climate projections for major wheat-growing regions indicate that both the frequency and duration of extreme rainfall events are expected to increase substantially in the coming decades. Simulations from multiple global circulation models suggest that regions such as the Indo-Gangetic Plains, the Australian wheat belt, and parts of Eastern Europe will experience increased frequency of unpredictable storm clusters and prolonged wet spells during the early vegetative phase of wheat (Kingra and Kukal 2024). These events create conditions in which soil saturation persists for extended periods, intensifying the risk of hypoxia-induced metabolic disruption, root decay, and impaired nutrient uptake. As rainfall patterns become increasingly erratic, alternating between brief droughts and sudden oversaturation, wheat cultivation and agronomic management must adapt to a future in which waterlogging is both more widespread and more temporally variable (Langan et al. 2022). Consequently, predictive modelling and climate-responsive cropping systems are essential for sustainable wheat production.

### 10.7.2 Genetic Bottleneck and Limited Diversity for Waterlogging Tolerance in Elite Germplasm

Although significant progress has been made in improving yield and disease resistance, most elite wheat cultivars continue to exhibit a relatively narrow genetic base for waterlogging tolerance (Pais 2024). Physiological traits such as rapid aerenchyma formation, maintenance of ATP production under hypoxic conditions, adventitious root development, and regulated ethylene signalling are found in landraces and wild relatives, including *Triticum dicoccoides* and *Aegilops* species. However, these genetic resources remain underutilized. Contemporary breeding programs often prioritize uniformity and grain quality, inadvertently eliminating alleles that confer stress tolerance but may cause minor yield reductions under non-stress conditions. This genetic bottleneck is further exacerbated by limited phenotyping capabilities for below-ground traits, making it challenging to identify tolerant genotypes at early developmental stages (Bian et al. 2022). Addressing these challenges will require broader exploration of germplasm, pangenome-based allele mining, and targeted introgression through genomic selection and gene editing to enhance the adaptive capacity of future wheat varieties.

### 10.7.3 Breeding Imperatives for Multi-stress-Tolerant Wheat Varieties

Waterlogging in agricultural fields frequently coincides with additional environmental stresses, resulting in compounded physiological challenges for wheat. For instance, standing water in coastal deltas often causes salinity accumulation, whilst late-season drainage failures may expose recovering plants to sudden heat waves. These combined stresses intensify cellular damage by simultaneously disrupting ion homeostasis, oxidative balance, and membrane stability. Existing waterlogging-tolerant cultivars often lose effectiveness under multiple stressors because their underlying tolerance mechanisms function independently rather than synergistically. Therefore, future wheat ideotypes should integrate traits such as enhanced root porosity, salt exclusion, heat-responsive antioxidant pathways, and rapid post-flood recovery metabolism. Multi-trait genomic prediction models and CRISPR-enabled gene pyramiding are promising approaches for developing varieties capable of withstanding the complex stress matrices anticipated under future climate scenarios.

### 10.7.4 Socio-Economic Barriers to the Adoption of Drainage Technology and Improved Practices

Although agronomic strategies for managing waterlogging have been scientifically validated, their adoption at the farm level remains inconsistent due to socio-economic barriers. Smallholder farmers frequently lack the capital necessary to install surface drains, subsurface tiles, or laser-assisted land levelling systems. In flood-prone areas, communal land ownership and fragmented fields hinder the implementation of coordinated drainage networks (Effiong et al. 2024). Furthermore, limited awareness constrains the adoption of raised-bed planting, controlled irrigation, and short-duration tolerant crop varieties. Many farmers perceive new technologies as requiring high initial investment and offering uncertain returns, particularly in regions with weak or inconsistent extension services (Alam et al. 2024). Unless these structural barriers are addressed, the advantages of scientific progress in waterlogging mitigation will likely remain confined to mechanized or commercially oriented farming systems.

### 10.7.5 Need for International Collaboration in Waterlogging Stress Research and Resource Sharing

Waterlogging stress poses a significant agricultural challenge; however, global research efforts remain fragmented across institutions and exhibit limited collaboration (Schubert and Qadir 2024). Numerous promising quantitative trait loci, transcriptome datasets, and phenotyping protocols are restricted to regional programs, which impedes progress in breeding initiatives (Upadhyay et al. 2024). Enhanced international collaboration through shared germplasm repositories, distributed phenotyping networks, and collaborative mapping populations could expedite the identification of robust tolerance mechanisms (Mondal et al. 2023). Collaborative research amongst countries susceptible to monsoon variability or coastal flooding would facilitate experimental standardization and enable more accurate quantification of genotype-by-environment interactions. Additionally, coordinated training programs and open-access data platforms would provide emerging wheat-growing regions, particularly in Africa and Southeast Asia, with equitable access to essential tools for developing waterlogging-resilient cultivars.

### 10.7.6 Policy Priorities for Climate-Resilient Infrastructure, Research Investment, and Farmer Capacity Building

Long-term mitigation of waterlogging stress requires policy frameworks that integrate scientific advances with accessible, farmer-oriented solutions. Governments should prioritize investments in regional drainage infrastructure, flood-monitoring systems, and early warning communication platforms (Bukhari et al. 2025). Strengthening national breeding programs through consistent funding, multilocation testing networks, and support for advanced genomic technologies is essential for developing climate-resilient cultivars. Additionally, farmer training initiatives should incorporate waterlogging management into broader climate-smart agriculture strategies, ensuring that producers understand both the technological and ecological dimensions of stress mitigation. Adoption incentives, including subsidized drainage equipment, seed distribution programs, and community-based extension services, can facilitate the translation of research findings into practical field applications. Robust policy interventions are critical for maintaining wheat productivity amid intensifying climate change pressures.

## 10.8 Conclusion

The challenge of waterlogging stress in wheat has driven a wide range of scientific, technological, and agronomic innovations over the past several decades. This chapter has summarized the significant advances that have shaped current understanding of how wheat plants respond to hypoxia, the genetic and physiological traits that contribute to tolerance, and the ways integrated management strategies can reduce crop losses. Advances in root anatomy research, stress-induced metabolic pathways, precision breeding, and field engineering have contributed to a multidimensional and interdisciplinary approach to mitigating waterlogging stress. Recent breakthroughs in genomics, phenomics, and digital agriculture have significantly enhanced wheat resilience. Genomic tools and transcriptomic analyses have expanded the catalogue of candidate genes for waterlogging tolerance, enabling more targeted breeding strategies. Gene-editing technologies, such as CRISPR–Cas systems, enable precise modification of regulatory pathways involved in root plasticity, ethylene signalling, and oxidative stress management. In addition to genetic tools, advancements in remote sensing, GIS-based monitoring, and AI-driven advisory platforms enable farmers and researchers to detect, manage, and anticipate waterlogging at both field and regional scales. These innovations, combined with improved agronomic practices such as raised-bed planting, optimized drainage, and microbial bioinoculants, constitute a robust, integrated approach to mitigation. Despite these advances, significant research gaps and capacity constraints continue to limit the widespread implementation of waterlogging mitigation strategies. Key challenges include inadequate drainage infrastructure, insufficient farmer training, and limited access to improved cultivars. Furthermore, the genetic diversity for waterlogging tolerance in elite breeding pools remains narrow, highlighting the need for sustained breeding efforts and the incorporation of diverse germplasm sources. Strengthening research capacity through coordinated international collaboration, multilocation testing networks, and investments in high-throughput phenotyping will be essential for accelerating the development and dissemination of resilient wheat varieties.

## References

Aggarwal S, Negi A, Barthwal R et al (2024) Mitigating drought stress in various grain crops: strategies for alleviation. In: Singh N, Babu SA (eds) Climate crisis and sustainable solutions. Springer, Singapore. https://doi.org/10.1007/978-981-97-7110-3_5

Ahmed YB, Ababneh OE, Al-Khalili AA et al (2024) Identification of hypoxia prognostic signature in glioblastoma multiforme based on bulk and single-cell RNA-seq. Cancer 16(3):633. https://doi.org/10.3390/cancers16030633

Akter R, Mazumder MSU (2025) Adaptation strategies to waterlogging among coastal farmers in Bangladesh: practices, determinants, and implications for resilient agriculture. Clim Serv 39:100598. https://doi.org/10.1016/j.cliser.2025.100598

Al Najadi RR (2024) Effectiveness of using drone-based thermal/infrared camera in enhancing the performance of controlled environment agriculture. MSc thesis, Sultan Qaboos University, Oman

Alam MJ, Sarma PK, Begum IA et al (2024) Agricultural extension service, technology adoption, and production risk nexus: evidence from Bangladesh. Heliyon 10(14):e34226. https://doi.org/10.1016/j.heliyon.2024.e34226

Ali L, Khan MB, Ahmed F et al (2024) Influence of abiotic stresses on the production and sustainability of rice crop. Agrobiol Rec 16:99–108. https://doi.org/10.47278/journal.abr/2024.017

Alrajhi A, Alharbi S, Beecham S et al (2024) Regulation of root growth and elongation in wheat. Front Plant Sci 15:1397337. https://doi.org/10.3389/fpls.2024.1397337

Amin AA, Mehrez A, Bichni H (2024) Enhancing agricultural extension programs: a practical guide to designing and implementing randomized controlled trials. An-Najah Univ J Res A 39(2):101–116. https://doi.org/10.35552/anujr.a.39.2.2319

Basit F, Khalid M, El-Keblawy A et al (2024) Hypoxia stress: plant sensing, responses, and tolerance mechanisms. Environ Sci Pollut Res 31(55):63458–63472. https://doi.org/10.1007/s11356-024-35439-4

Belliappa SH, Bomma N, Pranati J et al (2024) Breeding for waterlogging tolerance in pigeonpea: current status and future prospects. CABI Agric Biosci 5:98. https://doi.org/10.1186/s43170-024-00299-y

Bettcher M (2024) The hydrology, soils, and vegetation of six southeastern geographically isolated depressional wetlands in the context of restoration. MSc thesis, University of Georgia

Bian L, Zhang H, Ge Y et al (2022) Closing the gap between phenotyping and genotyping: review of advanced image-based phenotyping technologies in forestry. Ann For Sci 79:22. https://doi.org/10.1186/s13595-022-01143-x

Bukhari SAS, Shafi I, Ahmad J et al (2025) Review of flood monitoring and prevention approaches: a data analytic perspective. Nat Hazards 121(5):5103–5128. https://doi.org/10.1007/s11069-024-07050-w

Bundela DS, Singh OP (2021) Waterlogging and salinity problems in irrigation commands. Sustain Agric 74:1–15

Chandra-Kuntal K (2022) Ethylene and ROS crosstalk in plant developmental processes. In: Singh S, Husain T, Singh VP et al (eds) Ethylene in plant biology. Wiley, pp 125–177. https://doi.org/10.1002/9781119744719.ch7

Cheng L, Chang X, Li J (2024) Effects of recommended fertilizer application strategies based on yield goal and nutrient requirements on drip-irrigated spring wheat yield and nutrient uptake. Agronomy 14(11):2491. https://doi.org/10.3390/agronomy14112491

Chhabra V, Sreethu S, Kaur G et al (2025) Response of wheat crop to water-logged conditions under different land configurations and nutrient management. Sci Rep 15(1):7168. https://doi.org/10.1038/s41598-024-83752-2

Choudry MW, Riaz R, Nawaz P et al (2024) CRISPR-Cas9 mediated understanding of plants' abiotic stress-responsive genes to combat changing climatic patterns. Funct Integr Genomics 24(4):132. https://doi.org/10.1007/s10142-024-01405-z

Devi EL, Rangappa K, Devi AG et al (2025) Harnessing root phenotyping and root growth plasticity of landraces of maize (*Zea mays* L.) for enhancing waterlogging tolerance in fragile agro-ecologies of Eastern Himalaya. Heliyon 11(3):e42340. https://doi.org/10.1016/j.heliyon.2025.e42340

Effiong C, Ngang E, Ekott I (2024) Land use planning and climate change adaptation in river-dependent communities in Nigeria. Environ Dev 49:100970. https://doi.org/10.1016/j.envdev.2024.100970

Hasan A, Tabassum B, Hashim M et al (2024) Role of plant growth-promoting rhizobacteria as a plant growth enhancer for sustainable agriculture. Bacteria 3(2):59–75. https://doi.org/10.3390/bacteria3020005

He S, Li X, Chen M et al (2024) Crop HTP technologies: applications and prospects. Agriculture 14(5):723. https://doi.org/10.3390/agriculture14050723

Hossan MR, Hossain MI, Shaon ZH (2025) Flood and waterlogging risk assessment of Jashore district leveraging analytical hierarchy process through spatial analysis. Res Sq. https://doi.org/10.21203/rs.3.rs-7106425/v1

Khedr EH, Khedr N (2025) Genetic gains in crop yield through genomic selection under drought stress. In: Custom-Des Crop Breed, 2nd edn, pp 627–650. https://doi.org/10.1002/9781394192595.ch21

Kingra PK, Kukal SS (2024) Managing agricultural water productivity in a changing climate scenario in Indo-Gangetic plains. In: Behnassi M, Al-Shaikh AA, Gurib-Fakim A et al (eds) The water, climate, and food nexus. Springer, Cham. https://doi.org/10.1007/978-3-031-50962-9_13

Kumar A, Anju T, Kumar S et al (2021a) Integrating omics and gene editing tools for rapid improvement of traditional food plants for diversified and sustainable food security. Int J Mol Sci 22(15):8093. https://doi.org/10.3390/ijms22158093

Kumar S, Mahapatra S, Chakraborty S et al (2021b) Effect of abiotic stresses and mitigation strategy associated with their tolerance in wheat. J Cereal Res 13(1):16–37. https://doi.org/10.25174/2582-2675/2021/111211

Kumar R, Chaurasiya DK, Singh S (2025) Advanced technological innovations for mitigating biotic stresses in horticulture: from remote sensing to AI-driven solutions. In: Rai AC, Raj VP, Kumar A et al (eds) Biotic stress tolerance in horticultural crops. Woodhead Publishers, pp 231–242. https://doi.org/10.1016/B978-0-443-27324-7.00014-8

Kumari P, Gangwar H, Kumar V et al (2024) Crop phenomics and high-throughput phenotyping. In: Priyadarshan PM, Jain SM, Penna S et al (eds) Digital agriculture. Springer, Cham. https://doi.org/10.1007/978-3-031-43548-5_13

Langan P, Bernád V, Walsh J et al (2022) Phenotyping for waterlogging tolerance in crops: current trends and future prospects. J Exp Bot 73(15):5149–5169. https://doi.org/10.1093/jxb/erac243

Lee BR, Park SH, Muchlas M et al (2025) Biochar application alleviates drought-induced oxidative stress by activating the salicylic acid-mediated glutathione synthesis pathway in *Brassica napus*. BMC Plant Biol 25(1):1582. https://doi.org/10.1186/s12870-025-07575-7

Maredia MK, Farris J, Mason NM et al (2025) Does providing free trial packs with demonstration plots increase the adoption of agricultural technologies? Cost-effectiveness evidence from Tanzania. Agric Econ. https://doi.org/10.1111/agec.70033

McCauley D, Levin A, Nackley L (2021) Reviewing mini-lysimeter controlled irrigation in container crop systems. HortTechnology 31(6):634–641. https://doi.org/10.21273/HORTTECH04826-21

Minhas PS, Qadir M (2024) Managing saline water for irrigating agricultural crops. In: Irrigation sustainability with saline and alkali waters. Springer, Singapore. https://doi.org/10.1007/978-981-97-4102-1_4

Minoli S, Jägermeyr J, Asseng S et al (2022) Global crop yields can be lifted by timely adaptation of growing periods to climate change. Nat Commun 13(1):7079. https://doi.org/10.1038/s41467-022-34411-5

Mondal R, Kumar A, Gnanesh BN (2023) Crop germplasm: current challenges, physiological-molecular perspective, and advanced strategies towards development of climate-resilient crops. Heliyon 9(1):e12973. https://doi.org/10.1016/j.heliyon.2023.e12973

Nirmalaruban R, Yadav R, Sugumar S et al (2025) Root traits: a key for breeding climate-smart wheat (*Triticum aestivum*). Plant Breed 144(3):310–334. https://doi.org/10.1111/pbr.13248

Pais IMPR (2024) Exploring bread wheat diversity in waterlogging responses: root system, plant development, ecophysiological traits and yield implications. PhD thesis, Universidade NOVA de Lisboa, Portugal

Pais IP, Moreira R, Semedo JN et al (2022) Wheat crop under waterlogging: potential soil and plant effects. Plants 12(1):149. https://doi.org/10.3390/plants12010149

Qamar R, Ashraf S, Abbas FT et al (2025) Terrace farming for better water management in agriculture. In: Mubeen M, Jatoi WN, Hashmi MZ et al (eds) Innovations in agricultural water management. Springer, Cham. https://doi.org/10.1007/978-3-031-91883-4_7

Qian L, Luo Y, Duan K (2025) A simple drainage-friendly approach for estimating waterlogging impacts on cotton yields under high temperature conditions. Sustainability 17(2):20474. https://doi.org/10.3390/su17020474

Raj SRG, Nadarajah K (2022) QTL and candidate genes: techniques and advancement in abiotic stress resistance breeding of major cereals. Int J Mol Sci 24(1):6. https://doi.org/10.3390/ijms24010006

Rashid M, Hussain Q, Khan KS et al (2021) Carbon-based slow-release fertilizers for efficient nutrient management: synthesis, applications, and future research needs. J Soil Sci Plant Nutr 21(2):1144–1169. https://doi.org/10.1007/s42729-021-00429-9

Rastogi M, Kolur SM, Burud A et al (2024) Advancing water conservation techniques in agriculture for sustainable resource management: a review. J Geogr Environ Earth Sci Int 28(3):41–53. https://doi.org/10.9734/jgeesi/2024/v28i3755

Rupngam T, Messiga AJ (2024) Unraveling the interactions between flooding dynamics and agricultural productivity in a changing climate. Sustainability 16(14):6141. https://doi.org/10.3390/su16146141

Schubert S, Qadir M (2024) Soil salinity and salt resistance of crop plants. Springer, Cham. https://doi.org/10.1007/978-3-031-73250-8

Selvakumar P, Usha MS, Dhokpande SR et al (2025) Water conservation: AI applications in sustainable water use. In: Tariq MU, Serigo RP (eds) Cases on AI-driven solutions to environmental challenges. IGI Glob Science Publication, pp 269–296. https://doi.org/10.4018/979-8-3693-7483-2

Setiawan E, Budiarto R, Hamdani JS et al (2024) An overview of gibberellin inhibitors for regulating vegetable growth and development. Kultivasi 23(3):375–385. https://doi.org/10.24198/kultivasi.v23i3.58797

Shahid M, Ali S (2025) Chlorpyrifos and imidacloprid phytotoxicity triggers oxidative stress and hinders lettuce growth, cellular, and metabolic functions. Physiol Plant 177(6):e70620. https://doi.org/10.1111/ppl.70620

Sharma PK, Kumar S (2023) Soil physical environment and plant growth. Springer, Berlin, Heidelberg. https://doi.org/10.1007/978-3-031-28057-3

Sharma VK, Azad RK, Chowdary VM et al (2022) Delineation of frequently flooded areas using remote sensing: a case study in part of Indo-Gangetic basin. In: Pandey A, Chowdary VM, Behera MD et al (eds) Geospatial technologies for land and water resources management, Water science and technology library, vol 103. Springer, Cham. https://doi.org/10.1007/978-3-030-90479-1_27

Sharma A, Prakash A, Bhambota S et al (2025) Investigations of precision agriculture technologies with application to developing countries. Environ Dev Sustain 27(7):15135–15171. https://doi.org/10.1007/s10668-024-04572-y

Sheikh M, Iqra F, Ambreen H et al (2024) Integrating artificial intelligence and high-throughput phenotyping for crop improvement. J Integr Agric 23(6):1787–1802. https://doi.org/10.1016/j.jia.2023.10.019

Smith LJ, Scheikowski L, Kafle D (2024) Distribution of reniform nematode (*Rotylenchulus reniformis*) in cotton fields in Central Queensland and population dynamics in response to cropping regime. Pathogens 13(10):888. https://doi.org/10.3390/pathogens13100888

Thingujam D, Gouli S, Cooray SP et al (2025) Climate-resilient crops: integrating AI, multi-omics, and advanced phenotyping to address global agricultural and societal challenges. Plants 14(17):2699. https://doi.org/10.3390/plants14172699

Tyagi A, Ali S, Park S et al (2023) Exploring the potential of multiomics and other integrative approaches for improving waterlogging tolerance in plants. Plants 12(7):1544. https://doi.org/10.3390/plants12071544

Ucar Y, Miseckaite O (2025) Sustainable irrigation and irrigation indices: a theoretical review. In: Bahadir M, Haarstrick A, Güneş K et al (eds) Agricultural water management, water and wastewater management. Springer, Cham. https://doi.org/10.1007/978-3-031-96134-2_19

Upadhyay VR, Ramesh V, Kumar H et al (2024) Phenomics in livestock research: bottlenecks and promises of digital phenotyping and other quantification techniques on a global scale. OMICS 28(8):380–393. https://doi.org/10.1089/omi.2024.0109

Woolway RI, Zhang Y, Jennings E et al (2025) Extreme and compound events in lakes. Nat Rev Earth Environ 6:1–19. https://doi.org/10.1038/s43017-025-00710-w

Wu D, Wang W, Yao Y et al (2023) Microbial interactions within beneficial consortia promote soil health. Sci Total Environ 900:165801. https://doi.org/10.1016/j.scitotenv.2023.165801

Zahra A, Ahmad F, Asghar A et al (2025) Phenotypic and functional trait variation in *Malvastrum coromandelianum* under different tree plantations in Pakistan. Sci Nat 112(4):52. https://doi.org/10.1007/s00114-025-02003-0

Zhang Y, Wang X, Malko MM et al (2025) Unravelling the long-term beneficial effects of stress priming-induced tolerance to waterlogging stress in wheat. Plant Growth Regul 105:1–18. https://doi.org/10.1007/s10725-025-01371-x

# 11 Genetic, Physiological, and Digital Innovations for Enhancing Waterlogging Resilience in Wheat

Vikas Gupta, Charan Singh, Rakesh Kumar Bairwa, Jitendra Kumar Meena, and Gyanendra Singh

**Abstract**

Wheat crops face various biotic and abiotic stresses, with waterlogging being a major abiotic factor negatively impacting productivity worldwide. The severity of waterlogging has increased due to unexpected and drastic climate changes, especially the rise in unseasonal heavy rainfall over the past two decades in irrigated and high-rainfall areas. Waterlogging causes multiple problems for wheat growth and yield, including restricted gas exchange, nutrient deficiencies, and microelement toxicities. Its effects vary depending on depth, duration, plant growth stage, and weather conditions. Waterlogging reduces tiller number, spike number, grains per spike, and grain weight, leading to significant yield loss. The situation worsens when waterlogging occurs during crop lodging, particularly around heading or later stages. Wheat plants develop morphological, physiological, and metabolic adaptations to survive waterlogged conditions. The negative effects of waterlogging can be mitigated through suitable agronomic, breeding, and biotechnological strategies. Practices such as raised-bed cultivation, laser land levelling, and nutrient management have proven effective, along with breeding approaches. Many studies have identified promising genotypes, QTLs, and traits associated with waterlogging tolerance, enabling rapid development of tolerant cultivars through high-throughput phenotyping. This chapter reviews recent progress in agronomic, breeding, and genomic methods to address waterlogging stress, aiming to sustain annual productivity and improve resilience in future wheat varieties amid ongoing climate challenges.

**Keywords**

Wheat · Waterlogging · Climate resilience · Genomic tools

V. Gupta · C. Singh · R. K. Bairwa · G. Singh (✉)
ICAR-Indian Institute of Wheat and Barley Research, Karnal, Haryana, India

J. K. Meena
ICAR-Central Research Institute for Jute and Allied Fibres, Barrackpore, West Bengal, India



R. Tiwari et al. (eds.), *Wheat Stress Management*,
https://doi.org/10.1007/978-981-92-0902-6_11

## 11.1 Introduction

Wheat (*Triticum aestivum* L.) is the staple food for about 2.5 billion people in more than 89 countries. It is cultivated on approximately 220.76 million hectares worldwide, producing 770.88 million tons. Wheat supplies roughly 19% of global dietary calories and 21% of protein intake (Shiferaw et al. 2013). In the face of climate change, meeting the nutritional needs of a projected population exceeding 9 billion by 2050 poses a major challenge. Wheat yields are affected by various abiotic stresses, including heat, drought, salinity, and waterlogging. Amongst these, waterlogging is a primary constraint on wheat productivity globally. An estimated 10–15 million hectares of wheat are impacted by waterlogging, resulting in yield losses of 20%–50% (Hossain and Uddin 2011). In India, waterlogging is especially problematic in the eastern, central, and peninsular regions (Singh et al. 2018a). The frequency of flooding has increased recently due to intense and unpredictable rainfall linked to climate change. Global warming has likely increased the frequency of waterlogging (Feyen et al. 2012). Waterlogging stress is widespread, particularly in irrigated and high-rainfall areas. It occurs when water remains on the soil surface for long periods or when soil moisture exceeds field capacity by at least 20% (Aggarwal et al. 2006). Although water is vital for plant growth, excessive soil moisture reduces oxygen supply and hampers diffusion, leading to hypoxic or anoxic conditions (Sasidharan et al. 2017). These conditions limit root growth and can cause root death, impairing water and nutrient uptake and transport to the shoots, thereby harming overall plant growth and development. During waterlogging, plants face multiple stresses, including restricted gas exchange, mineral nutrient deficiencies, and microelement toxicities. Common soil toxicities include Mn, Fe, Na, Al, and B (Setter et al. 2009). The effects of waterlogging depend on factors such as water depth and duration, plant growth stage, and weather conditions (Herzog et al. 2016). Waterlogging reduces the number of spikes, grains per spike, and grain weight, leading to significant yield losses (Singh et al. 2017). It also interferes with photosynthesis, respiration, and transpiration in wheat. Post-anthesis stem elongation is especially vulnerable to waterlogging stress. Therefore, breeding for improved tolerance to excess moisture is crucial for increasing production and stabilizing yields across different environments. As climate change is expected to increase precipitation, enhancing waterlogging stress tolerance is a priority for wheat breeders, especially in regions at risk of soil waterlogging.

The effects of waterlogging stress in agricultural fields can be mitigated through the timely implementation of agronomic and engineering interventions. Agronomic practices, such as zero tillage, raised-bed cultivation, laser-assisted land levelling, early planting, and balanced nutrient management, including the enhancement of soil organic matter, have demonstrated effectiveness in reducing waterlogging stress in multiple studies (Ali et al. 2018; Du et al. 2021; Rautaray 2021). Effective water management strategies encompassing adequate supply and efficient drainage are essential. Additionally, the use of molecular tools to monitor and direct the flow of alleles is critical for addressing this stress. Several QTLs associated with waterlogging tolerance have been identified. The development of waterlogging-tolerant varieties can be accelerated by integrating QTL mapping, functional genomics, and high-throughput sequencing. This chapter reviews recent advances in agronomy, breeding, and biotechnology to address waterlogging stress.

## 11.2 Historical Perspectives on Mitigating Waterlogging Stress in Wheat

Before the scientific understanding of how waterlogging affects wheat, farmers in flood-prone areas developed skilled, lasting practices to protect their crops. These methods, passed down through generations, demonstrated detailed knowledge of how crops respond to local conditions and soil types. Raised-bed planting, also called the ridge-and-furrow system, has been one of the most successful and widely used techniques (Govaerts et al. 2007). In this method, wheat is grown on raised beds, with the furrows between acting as drainage channels, improving aeration and reducing stress. Recent research has confirmed that this old practice remains effective even in waterlogged conditions (Du et al. 2021). Another common approach is to create drainage trenches or ditches around fields to remove excess surface water during heavy rains and floods. These traditional methods were simple to set up and relied on local labour, making them accessible to both small- and large-scale farmers (Rautaray 2021). As advanced cultivation methods developed, researchers began systematically studying how waterlogging affects crops. Even though there is no direct evidence that tall wheat varieties are more susceptible to waterlogging, there is a clear negative link between plant height and grain yield per plot (Singh et al. 2014). Historically, most farmers relied on flood irrigation, which often led to temporary waterlogging in uneven fields. Over the last 50–60 years, India has greatly expanded its canal network. However, poor management of these systems has led to waterlogged root zones and excessive moisture during critical growth stages, reducing yields. This relationship between irrigation methods and poor land management highlights the importance of both supplying adequate water and ensuring proper drainage. Recognizing waterlogging as a systemic issue prompted a major change in farming approaches, shifting from simply providing water to comprehensively managing the water–soil environment. As a result, land and water management now play vital roles in wheat production in water-prone areas. Important measures include developing better drainage systems, using laser-guided land levelling to achieve uniform fields, and adopting water-saving irrigation methods. The main goal of these practices is to prevent excess water from building up around the roots. Additionally, planting wheat earlier in the season helps reduce waterlogging stress, as early sowing helps wheat avoid damage during late flowering and grain filling (Ali et al. 2018).

Alongside advancements in land and water management, plant breeders have been developing new crop varieties that tolerate moderate waterlogging. Waterlogging-tolerant genotypes can sustain high growth rates, accumulate significant biomass, and produce higher grain yields under stress conditions (Setter and Waters 2003). Several studies have sought to identify traits associated with waterlogging tolerance. For example, dry weight and leaf area have been identified as key selection criteria for waterlogging-tolerant cultivars (Jiang et al. 2022). Additionally, varieties with high dry weight and longer root length have shown greater tolerance to waterlogging stress (Ghobadi et al. 2017). The number of tillers has also been used as an effective selection criterion when waterlogging stress occurs during the vegetative stage in saline soils (Singh et al. 2018a; b). Statistical analyses suggest that climate change has delayed peak waterlogging events by approximately 1.1–3.7 days per century. Adjusting sowing dates can help farmers ensure that sensitive crop growth stages avoid periods of waterlogging stress. Under these circumstances, early maturing wheat varieties may have an advantage due to their escape mechanism (Cheng et al. 2024; Deng et al. 2025). The transition from basic field techniques to advanced land management and breeding for tolerant cultivars highlights the evolution of agricultural science in addressing persistent waterlogging challenges.

## 11.3 Scientific Advancements and Principles Developed over the Past Centuries

Over recent decades, advancements in scientific research have significantly deepened our understanding of wheat responses to waterlogging stress. The field has evolved from early qualitative observations of flooding damage to a comprehensive knowledge of the physiological, anatomical, and molecular mechanisms behind wheat survival. Wheat withstands waterlogged conditions through various morphological, physiological, and metabolic adaptations. Root growth under waterlogged conditions serves as a reliable indicator for selecting tolerant genotypes. In such environments, wheat plants develop seminal roots and numerous nodal roots with aerenchyma, a common adaptation. Genotypic responses to waterlogging stress differ, with tolerant genotypes generally exhibiting superior root growth and higher root biomass (Huang et al. 1994). The performance of these genotypes can be evaluated by their ability to recover from temporary hypoxic conditions. Ethylene production has also been linked to waterlogging tolerance in several studies. Under oxygen-deficient conditions, plant roots show increased 1-aminocyclopropane-1-carboxylic acid (ACC) synthase activity, leading to the accumulation of ACC in roots. Waterlogging promotes the transport of ACC from roots to shoots, significantly increasing ethylene production, as ACC is a direct precursor of ethylene in plants. Applying external ACC further raises ethylene levels, promoting aerenchyma formation by upregulating RBOH gene expression (Yamauchi et al. 2014). Aerenchyma tissues in adventitious roots facilitate aerobic respiration and root growth (Jackson 2007). Ethylene helps resist waterlogging stress by managing morphological changes, hypoxia-responsive gene expression, and the reactive oxygen species (ROS) scavenging system (Chen et al. 2024). Wetland plants usually have a structural barrier in the root epidermis that prevents oxygen loss from the basal root zone—a process called radial oxygen loss (ROL)—which is an effective adaptation to wetland environments. Most wetland plants possess this ROL barrier, whilst non-wetland plants such as wheat do not. The barrier also enhances oxygen transfer from roots to the apex. Root growth in anaerobic, waterlogged conditions depends on oxygen supplied by diffusion from above-ground parts through root aerenchyma (Colmer 2002). Aerenchyma provides a pathway for oxygen diffusion in roots. Wetland species have constitutive aerenchyma, with its volume increasing further under oxygen deficiency in the root zone. In contrast, dryland species like wheat develop aerenchyma at lower volumes compared to wetland species such as rice and *Hordeum marinum*. Adventitious roots in *H. marinum* develop a strong ROL barrier. The superior root aeration traits of *H. marinum* have been transferred to wheat through wide hybridization (Malik et al. 2011). In waterlogged conditions, some oxygen is lost from root tips into the soil, where it oxidizes, alleviating the toxic effects of ferrous and sulfate ions.

Trait diversity is essential for developing waterlogging-tolerant genotypes. Numerous studies have been conducted under both field and greenhouse conditions to identify promising genotypes for waterlogging tolerance. Table 11.1 provides an overview of these studies and the genotypes identified.

Screening germplasm under abiotic stress involves creating artificial stress conditions using various methods. In the field, waterlogging can be induced by either unusually high natural rainfall or artificial irrigation. This method is straightforward and cost-effective, but outcomes are heavily affected by environmental variability.

**Table 11.1** Genetic variability for enhancing waterlogging tolerance in wheat

| S. no. | No. of genotypes evaluated | Promising genotypes identified | Reference |
|---|---|---|---|
| 1. | 15 diverse genotypes | Terral LA 422, Shelby, Pioneer 2691 | Collaku and Harrison (2002) |
| 2. | 5 diverse genotypes | Nishikazekomugi and Iwainodaichi | Hayashi et al. (2013) |
| 3. | 100 elite diverse genotypes | PBW 621, DBW 52, HD 3027, KRL 213, RW 3684 | Singh et al. (2014) |
| 4. | 28 soft red winter wheat genotypes | 26R22, AR01167-3-1, Magnolia, USG 3555 | Arguello et al. (2016) |
| 5. | 109 Indian and exotic lines | BH 1146, DBW39, DBW 52, NW 1014, NW 1067, NW 4081, PBW 621, PBW 631, PBW 590, RW 3684, HD 2967, HD 2997, NW 4083 | Singh et al. (2017) |
| 6. | 149 elite Indian and Australian germplasm lines | Cunderin, Ducula 4, HD 3086, KRL 105, RW 3684 | Singh et al. (2018b) |
| 7. | 181 spring wheat genotypes | Altar84/*Ae. tauschii* (219)//2 × Seri, CETA/*Ae. Tauschii*, Bjarne, Kukri, Mirakel, Zebra | Sundgren et al. (2018) |
| 8. | 60 wheat genotypes | SSDC3-264, SSDC3-347, NW 5054, PBW 550 | Gupta et al. (2022) |
| 9. | 23 bread wheat genotypes | PL-1, PL-5, GR-1, GR-3, AdvL-2, Austrl-2, Austrl-4, and Austrl-5 | Pais et al. (2024) |
| 10. | 100 wheat genotypes | Kzyl-Sark, Misr4, Hofed, BAW-1157, and Corack | Elhabashy et al. (2025) |

For reliable, repeatable screening, experiments must be consistent, though they can be labour-intensive and time-consuming. Alternatively, pot experiments involve growing plants in pots with drainage holes inside watertight containers, with water added to induce waterlogged conditions. This approach allows for precise water control, but may not fully replicate field conditions due to other confounding factors. Conducting these experiments in environmentally controlled greenhouses can enhance reliability. Pot experiments are particularly useful for screening germplasm for waterlogging tolerance at the seedling stage, as plant growth is limited at later stages (Hossain and Uddin 2011). Recently, high-throughput phenotyping platforms using imaging systems, high-precision sensors, and automated robotics have been employed for screening. These systems produce highly reliable automated data, though their setup costs are high.

Engineering developments in soil and water management, combined with breeding efforts, have significantly reduced waterlogging at the field level. Precision land levelling, raised-bed planting, and shallow surface drainage canals are examples of modern surface drainage techniques that effectively control waterlogging. Together, advances in physiology, breeding, agronomy, and engineering have created a strong, layered system for managing waterlogging stress in wheat. These integrated strategies have greatly improved the resilience of wheat production systems to rising climate variability.

## 11.4 Recent Developments in Mitigating Waterlogging Stress in Wheat

Waterlogging is a major abiotic stress limiting wheat production. Under these conditions, hypoxia and anoxia significantly reduce yield by directly inhibiting aerobic respiration and nutrient uptake, reducing photosynthesis, and increasing plant vulnerability to diseases and pathogen attacks, ultimately leading to substantial yield losses (Setter and Waters 2003). This problem is especially severe in irrigated and high-moisture ecosystems, where water stagnation continues. Currently, wheat cultivation faces a global challenge posed by waterlogging stress, worsened by the rising frequency of extreme rainfall events linked to climate change. Recent advances in physiology, genetics, biotechnology, microbial intervention, and agronomy have helped mitigate the adverse effects of waterlogging on wheat growth and yield.

### 11.4.1 Adaptations and Molecular Breeding Interventions

Recent research has examined the physiological and biochemical responses of wheat to waterlogging stress. Polyamines such as spermidine (Spd) and spermine (Spm) have been shown to confer protection by increasing ethanol production and alcohol dehydrogenase (ADH) activity, whilst inhibiting lactate dehydrogenase (LDH) activity and lactate accumulation. Du et al. (2018) reported that these changes reduce the cytotoxic effects of lactic acid buildup and enhance the efficiency of anaerobic respiration. Wheat genotypes display variation in antioxidant responses; tolerant genotypes exhibit higher activities of enzymes, including superoxide dismutase (SOD), catalase (CAT), and peroxidase (POD), which mitigate oxidative damage from ROS during hypoxia and reaeration (Alizadeh-Vaskasi et al. 2018). Proline accumulation also contributes to stress tolerance by stabilizing proteins and membranes and facilitating osmotic adjustment as an osmolyte. Metabolic adaptations further support wheat survival in low-oxygen soils. Under hypoxic conditions, wheat shifts from aerobic respiration to fermentative pathways. Adequate soluble carbohydrate reserves are critical for maintaining glycolysis and fermentation when mitochondrial respiration is impaired (Sairam et al. 2008). Anatomical and morphological adaptations, particularly root characteristics and metabolic adjustments, are essential for survival under waterlogging. Hypoxia reduces the survival of seminal roots and inhibits adventitious root extension, resulting in lower root-to-shoot ratios and compromised nutrient uptake. During waterlogging, the development of adventitious roots and root cortical aerenchyma forms internal aeration channels that facilitate oxygen transport from shoots to roots (Herzog et al. 2016). Genetic variation in root porosity and oxygen transport efficiency is significant for breeding programs.

Advancements in genetic and molecular research have greatly improved the understanding of how waterlogging tolerance is inherited in wheat. Early studies showed that a few major genes, such as *Wt1*, are consistently found in tolerant genotypes (Boru et al. 2001). Combining multiple tolerance genes has been shown to enhance resilience, highlighting the value of pyramiding strategies in breeding programs. Many QTLs have recently been identified at different growth stages. Genome-wide association studies (GWAS) have identified QTLs associated with adventitious root and aerenchyma formation, particularly on chromosome 4B (Xu et al. 2022). Yu and Chen (2013) discovered 36 QTLs related to traits such as root and shoot dry weight, whilst Ding et al. (2023) identified stable QTLs, including QWTC.caas-4AL, which account for a large part of the variation in tolerance and are linked to KASP markers for marker-assisted selection (MAS). Pang et al. (2022) also found key QTLs (*qWlg5A*, *qWlg7B*, and *qWlg2D*) during germination. Breeding strategies focused on yield stability, biomass growth, and efficient metabolic adaptation under low-oxygen conditions are crucial for creating resilient wheat varieties. Transcriptomic studies have revealed important molecular pathways, emphasizing the roles of steroid and brassinosteroid biosynthesis, ethylene signalling, and ROS detoxification (Yang et al. 2022).

Future wheat waterlogging stress management should integrate diverse control mechanisms, including high-throughput phenotyping technologies for precise screening of tolerance. Multi-omics approaches, such as transcriptomics, metabolomics, and genomics, can elucidate the complex physiological mechanisms underlying tolerance. The selection of climate-resilient varieties will depend on specific local contexts and will require farmer-driven innovations and participatory breeding strategies. Ultimately, coordinated efforts across genetics, biotechnology, biochemistry, microbiology, and sustainable farming systems will be essential to maintain or enhance wheat productivity amid unpredictable climate conditions.

### 11.4.2 Agronomic, Microbial Interventions, and Seed Priming

In addition to breeding strategies based on visual selection and genomic approaches, agronomic and management practices play a crucial role in reducing waterlogging stress. Raised-bed planting systems combined with balanced nutrient management have shown significant improvements in plant growth, grain yield, and straw yield compared to traditional flat planting. For example, Chhabra et al. (2025) reported grain yields of 6.97 t/ha under raised-bed planting with recommended NPK application. Conservation agriculture practices, such as applying organic manure and using furrow-irrigated raised-bed systems, have also proven effective, improving soil structure, increasing water infiltration, and reducing crop lodging under saturated conditions (Tiwari et al. 2017). Microbial interventions, especially the use of plant growth-promoting rhizobacteria (PGPR), offer sustainable options to enhance wheat resilience in waterlogged environments. PGPR, including *Azospirillum*, *Pseudomonas*, and *Bacillus*, improve root architecture, reduce ethylene-induced damage through ACC deaminase activity, and boost antioxidant enzyme production. These bacteria also help nutrient uptake and root development under hypoxic conditions. Research has shown that synergistic effects amongst multiple microbial groups can increase stress tolerance, and field trials involving microbial seed treatments or root dips have produced positive results (Maciag and Krzyżanowska 2025). Hormonal and physiological approaches provide additional strategies for managing waterlogging. Treatments with ethylene inhibitors, gibberellins, and cytokinins can reduce stress effects by delaying senescence, encouraging shoot growth, and increasing tolerance to oxidative stress. These methods are especially effective in boosting early stage resilience and can be combined with other management practices. Field trials with microbial seed priming or root-dipping treatments have improved survival rates and yields under waterlogged conditions. This integrated approach offers a sustainable and cost-effective way to support genetic and agronomic solutions.

Seed priming and hormonal treatments are effective strategies for enhancing early stage tolerance to waterlogging. Priming seeds with water, polyethylene glycol (PEG), or osmoprotectants improves germination, root emergence, and seedling growth. Controlled waterlogging during the vegetative stage increases antioxidant capacity and photosynthetic efficiency, thereby reducing yield loss at the reproductive stage (Li et al. 2011). Stress priming promotes adaptive root modifications. For instance, primed wheat seedlings demonstrate higher antioxidant enzyme activity, elevated soluble sugar levels, and increased aerenchyma formation, resulting in reduced yield loss under subsequent waterlogging stress (Jin et al. 2025). Notably, stress memory can be transmitted across generations; progeny of primed plants exhibit enhanced photosynthetic efficiency, greater root aerenchyma formation, and improved carbohydrate allocation to roots, supporting growth and yield under hypoxic conditions (Feng et al. 2022). Seed priming with water or osmoprotectants, such as PEG, improves germination, root development, and seedling vigour under low-oxygen conditions. Additionally, surface seeding under excessively moist conditions at sowing is highly effective for managing early stage stress.

### 11.4.3 QTL Mapping and GWAS for Waterlogging Tolerance in Wheat

Quantitative trait locus (QTL) mapping is an effective approach for understanding the genetic basis of complex traits, such as wheat resistance to waterlogging. This strategy helps identify genomic regions associated with phenotypic variation, supports marker-assisted breeding, and provides insights into the genetic makeup. Over the past two decades, many studies have reported QTL linked to waterlogging tolerance in wheat (Table 11.2). For example, a population of recombinant inbred lines (RILs) derived from a cross between common wheat Forno and spelt wheat Oberkulmer revealed five QTL influencing survival rates under waterlogging stress. A significant QTL, explaining 20.4% of phenotypic variation was located on chromosome 5A, with additional loci on chromosomes 2B, 3B, 5A, and 7S. These findings suggest that chromosome 5A plays a key role in waterlogging tolerance (St Burgos et al. 2001). Similarly, another RIL population developed from wheat Opata 85 and synthetic wheat identified 32 QTL for the waterlogging tolerance index across 14 chromosomes. Chromosome 7A harboured the most prominent QTL for the germination index, explaining 23.9% of the observed phenotypic variation. This study emphasized the polygenic nature of waterlogging tolerance, with both major and minor QTL involved in the trait (Yu et al. 2013). Despite identifying these important loci, none of the QTL has been fine-mapped or cloned yet. The main challenges include the varied genetic pathways underlying waterlogging tolerance and the complex structure of the hexaploid wheat genome, which have limited the success of MAS for this trait. Recent technological advances are beginning to overcome these challenges. The development of high-resolution genome-sequencing methods and the availability of high-quality wheat reference genomes have transformed the analysis of stress-tolerance traits. For example, transcriptome analysis (RNA-seq) of two wheat cultivars, XM 55 (waterlogging-tolerant) and YM 158 (sensitive), conducted by Wei et al. (2021), identified 942 differentially expressed genes, many of which were involved in hormone signalling, brassinosteroid pathways, and steroid metabolism and biosynthesis. These findings have deepened the understanding of the molecular mechanisms underlying waterlogging tolerance and identified candidate genes for further study. Combining QTL mapping with functional genomics and high-throughput sequencing has huge potential to accelerate the development of waterlogging-tolerant wheat varieties.

**Table 11.2** Summary of the genetic, genomic, and molecular basis of waterlogging tolerance in wheat

| Category | Gene/QTL/genotype/population | Chromosome/origin | Trait(s) studied/function | Key findings/mechanism | Reference |
|---|---|---|---|---|---|
| QTL mapping | Oberkulmer (spelt) × Forno (common wheat) (RIL) | 5A | Survival rate under waterlogging | 5 QTLs; major on 5A | St Burgos et al. (2001) |
| QTL mapping | ITMI & SC populations | Multiple | TDWI and its components under waterlogging | 36 QTLs (ITMI) and 10 QTLs (SC) were identified | Yu and Chen (2013) |
| QTL mapping | W7984 × Opata85 (RIL) | 2B | Root/shoot dry weight | Major QTL identified | Yu et al. (2013) |
| | | 7A (GRI-7A) | Germination rate index | Major QTL detected | |
| | | 7A (germination index) | Waterlogging tolerance index and germination index | 32 QTLs; major on 7A | |
| QTL mapping (RIL) | USG3209 × Jaypee | 1BL | Root fresh biomass | A major effect QTL was detected | Ballesteros et al. 2015 |
| | | 1BL | Shoot fresh biomass (QSfbio.ua-1B-WGH) | | |
| | | 1BL | Chlorophyll content (QSpadpost.ua-1B-WF) | | |
| | | 1D (chlorophyll) | Leaf chlorophyll content and vegetative traits | 48 QTLs clustered into 10 regions; major on 1D | |
| Transcriptome (RNA-seq) | XM55 (tolerant) × YM158 (susceptible) | – | Gene expression under waterlogging | 942 DEGs; genes in steroid metabolism, brassinosteroid biosynthesis, and hormone signalling | Wei et al. (2021) |

| | | | | | |
|---|---|---|---|---|---|
| Genome-wide identification of the alcohol dehydrogenase (ADH) gene family | Whole genome data of wheat | 22 ADH genes | Aldehyde dehydrogenase gene | ADH genes were significantly responsive to waterlogging stress at the seed germination stage | Shen et al. (2021) |
| GWAS | GWAS conducted on 329 wheat genotypes | Linked to significant QTLs on chromosome 4B | Adventitious root and root cortical aerenchyma growth | Associated with waterlogging tolerance in wheat | Xu et al. (2022) |
| QTL mapping | qWlg5A, qWlg7B, qWlg2D | 5A, 7B, 2D | Waterlogging tolerance | Three major QTLs; validated by KASP assays in SN135 × SY936 & HM18 × CD1840 | Pang et al. (2022) |
| QTL mapping | QWTC.caas-4AL, QWTC.caas-5AS, QWTC.caas-7DL | 4AL, 5AS, 7DL | Waterlogging tolerance and reduced plant height | Breeder-friendly KASP markers (K_AX_111523809, K_AX_108971224, K_AX_110553316) developed | Ding et al. (2023) |

## 11.5 Evolution in the Contemporary Hi-Tech Era

Researchers and farmers now tackle waterlogging stress in wheat by integrating advanced technologies. Each stage, from trait discovery and genetic prediction to field-scale monitoring and decision support, plays a unique role in this effort. The identification of quantitative trait loci (QTLs) linked to adaptive traits such as adventitious root production and aerenchyma formation has accelerated with the availability of the wheat reference genome and advances in genome-wide association studies. For example, Xu et al. (2022) identified QTLs on chromosome 4B associated with root traits during waterlogging, as shown in Table 11.3. These discoveries have bolstered the use of genomic selection and marker-assisted selection in breeding programs. Beyond genomic data, transcriptome profiling under hypoxic conditions has uncovered thousands of genes differentially expressed, involved in anaerobic metabolism (e.g., ADH), reactive oxygen species detoxification (e.g., SOD and CAT), and ethylene signalling. Xu et al. (2022) highlight that these databases support the identification of candidate genes for targeted modification and functional validation. Moreover, the CRISPR–Cas9 genome-editing system has offered unprecedented precision in breeding. Although direct applications for waterlogging tolerance are still in early stages, proof-of-concept studies show great potential for editing genes related to ethylene biosynthesis, aquaporins, and root plasticity, which could shorten breeding cycles (Zhang et al. 2025).

Early detection of waterlogging stress at the field level is crucial. Integrating remote sensing (RS) technologies, such as satellites and unmanned aerial vehicles (UAVs), with geographic information systems (GIS) allows for spatial and temporal mapping of stress. For instance, Zhang et al. (2023) identified waterlogging hotspots in Anhui, China, by combining multisource RS data with crop–hydrological models. Liu et al. (2018) showed that high-resolution satellite imagery can detect waterlogging during the jointing–booting stage, which is strongly linked to yield losses. These technologies provide real-time advice to farmers and officials. Moreover, decision support is improved by big data platforms and artificial intelligence (AI). Mobile apps that integrate crop models, weather forecasts, and soil moisture data offer actionable suggestions. Yang et al. (2025) developed the Wheat Breeding Large Language Model (WBLM), which uses multimodal datasets to accurately predict yields. These systems turn complex research findings into practical digital tools for farmers.

Phenotyping challenges, particularly those related to concealed root traits, have historically hindered breeding for waterlogging tolerance in wheat. Recent advances in high-throughput phenotyping (HTP) now address these limitations. Large-scale simulators enable rapid screening of breeding populations, whilst 3D imaging systems such as Wheat3DGS capture volumetric growth and root–shoot responses under controlled waterlogging conditions (Zhang et al. 2025). The integration of HTP with genomic prediction models accelerates trait introgression. Additionally, hyperspectral canopy indices provide noninvasive diagnostics for early detection of stress. Yang et al. (2021) showed that indices such as red-band entropy and SRPI

**Table 11.3** Advance technological interventions to mitigate waterlogging stress in wheat

| Technology | Example | Key findings | Reference |
|---|---|---|---|
| High-resolution satellite imagery | Field-scale mapping of waterlogging in winter wheat | Showed severe yield losses when stress occurred at jointing–booting stages | Liu et al. (2018) |
| Hyperspectral canopy indices | SRPI and red-band entropy for stress detection | Accurately classified the severity of waterlogging stress in wheat | Yang et al. (2021) |
| Genomic tools (GWAS/QTL mapping) | QTLs for adventitious root (AR) and root cortical aerenchyma (RCA) on chromosome 4B | Identified genomic regions linked to waterlogging tolerance | Xu et al. (2022) |
| Transcriptomics | Differentially expressed genes (ERFs, ADH, SOD, and CAT) under hypoxia | Insights into signalling pathways and ROS detoxification | Xu et al. (2022) |
| Remote sensing + crop models | HYDRUS-1D + WOFOST + multisource RS assimilation | Increased soil moisture accuracy (NSE 0.75) and mapped waterlogging hotspots in Anhui, China | Zhang et al. (2023) |
| UAV + ML phenotyping | UAV multispectral imaging under water stress | Achieved yield prediction accuracy up to $R^2 = 0.80$ | Sharma et al. (2024) |
| High-throughput phenotyping (HTP) | Wheat3DGS using 3D gaussian splatting | Nondestructive volumetric phenotyping of wheat heads at scale | Zhang et al. (2025) |

effectively distinguish stress intensity, supporting both field management and germplasm evaluation.

The integration of big data analytics is also essential. Combining multi-omics, phenotypic, and environmental data enables the development of predictive models for genotype × environment × management (G × E × M) interactions. Machine learning techniques enhance selection accuracy, allowing breeders to efficiently manage complex datasets. Despite these advances, ideotype design remains dependent on physiological insights. Key mechanisms in breeding include adventitious root formation for nutrient uptake under hypoxic conditions, aerenchyma development for improved oxygen transport, and the activity of ROS-scavenging enzymes such as SOD and CAT for oxidative protection (Setter and Waters 2003). Integrating genetics and physiology provides a comprehensive approach to developing robust wheat variants. The convergence of genomics, transcriptomics, genome editing, phenotyping, remote sensing–geographic information systems (RS–GIS), artificial intelligence (AI), big data analytics, and physiology marks a transformative shift in addressing waterlogging stress in wheat.

Future progress will depend on establishing effective channels to translate laboratory innovations into field applications, integrating multi-omics at the systems level for deeper understanding, and developing digital platforms that deliver location-specific, real-time guidance to farmers. This holistic approach will facilitate the development of climate-resilient wheat genotypes.

## 11.6 Statistical Insights for Mitigating Waterlogging Stress in Wheat

Statistical analyses and data-driven insights are crucial for understanding the extent of the waterlogging problem and assessing the effectiveness of mitigation strategies, as they provide an empirical basis for informed decision-making, research prioritization, and resource allocation. In India, about 11.6 million hectares are affected by waterlogging. This issue is exacerbated by the buildup of soluble salts in the root zone, leading to waterlogging and salinity problems. The eastern part of India has the largest affected area, mainly due to higher rainfall and saucer-shaped topography (Sethi et al. 2022). Projections indicate that the frequency and severity of waterlogging are likely to rise soon, driven by climate change, particularly in regions with heavy and unpredictable rainfall, such as parts of China, the United States, India, Pakistan, Argentina, and Europe (Feyen et al. 2012; Ploschuk et al. 2018). Soils with higher clay content are generally more susceptible to waterlogging because of poor drainage. Additionally, soil compaction from repeated use of agricultural machinery worsens waterlogging issues (Tian et al. 2021). The impacts of soil waterlogging on wheat crop yields have been documented in several studies. Yield losses from waterlogging stress at seedling, jointing, and tillering stages range from 7.75% to 16.3% (Ding et al. 2020). Negative effects at tillering, jointing, booting, and anthesis stages have also been observed, with tillering being the most vulnerable stage. A 12% reduction in grain yield was mainly due to fewer spikes (Wu et al. 2017). To promote sustainable crop production under current conditions, adopting abiotic stress-tolerant varieties is one of the most effective strategies; however, their dissemination remains limited. Addressing this requires farmer-friendly policies to improve access to inputs, effective dissemination of information about these varieties, and ensuring seed availability (Akitaya et al. 2025).

Raised-bed planting improves soil water drainage and decreases soil water content. Research shows that this practice helps wheat seedling establishment and root growth, speeds up stem and tiller development, and delays late-season root and leaf ageing by reducing waterlogging stress, leading to an 11.3%–14.1% increase in yield (Du et al. 2021). Raised beds also reduce lodging and enhance plant stability, with a reported 11.2% yield increase compared to flat planting (Ahmad and Mahmood 2005). Furthermore, cultivating wheat in raised beds has been linked to more tillers per plant, longer spike length, increased grains per spike, and a 26.5% yield advantage over traditional methods (Chhabra et al. 2025). Therefore, the raised bed system is preferable to traditional planting in waterlogged areas. Statistical analyses are crucial for assessing breeding program results, as they help identify promising genetic material and traits for future breeding. As a result, a data-driven feedback loop is essential for the continued progress in research on waterlogging stress.

## 11.7 Future Projections and Global Challenges

Future efforts to mitigate waterlogging stress will be shaped by factors such as climate change, the development of climate-resilient cultivars, and socio-economic barriers to the adoption of new technologies. A major challenge is the expected rise in rainfall variability and extreme precipitation events under climate change scenarios (Mukherjee et al. 2018). The intensification of the hydrological cycle caused by global warming is likely to increase the frequency of extreme precipitation and flooding. As water availability increases from dry to wet regions, the risk of such events will also grow. This unpredictability highlights the need for flexible, climate-smart, and resilient farming systems rather than relying on traditional, static agricultural schedules (Tabari 2020). Modern wheat cultivars tend to have reduced tolerance to abiotic stress because early breeding focused on yield rather than on stress resilience, leading to the loss of beneficial traits from wild relatives. A shift in plant breeding priorities towards climate resilience has renewed interest in utilizing the genetic diversity of wild wheat relatives. This diversity can be harnessed to breed stress-tolerant varieties, and progress may be sped up through advanced techniques such as speed breeding, gene editing, and automated phenotyping (Toulotte et al. 2022). Going forward, developing cultivars that tolerate only waterlogging stress may be inadequate given the changing climate. There is an increasing need for multi-stress-tolerant wheat varieties capable of enduring additional stresses such as salinity and heat (Elhabashy et al. 2025).

Alongside environmental issues, major socio-economic challenges hinder the adoption of better management methods and technologies. For instance, drainage systems are often too costly for smallholder farmers, and the high initial expenses and labour involved in practices like raised-bed farming create additional barriers. Limited access to credit, information, and quality seeds also delays the uptake of newly developed tolerant varieties. These issues highlight the importance of supportive policies and financial mechanisms to bridge the gap between research and practical use. Collaborative projects can speed up the development of tolerant cultivars and management strategies. Improving agricultural extension systems is crucial to effectively transfer scientific knowledge and new technologies from research to the field.

## 11.8 Conclusion

In recent years, the incidence of waterlogging stress has increased significantly due to severe and unpredictable rainfall associated with climate change. This stress is particularly prevalent in irrigated and high-rainfall regions. Waterlogging adversely affects several morphological and physiological traits, resulting in substantial reductions in wheat yield. To address this challenge, researchers have identified several tolerant varieties and genotypes that are both climate-resilient and beneficial to farmers. Recent biotechnological advancements have enabled the development of waterlogging-stress-tolerant varieties through sophisticated methods such as MAS and genome editing. Additionally, agronomic measures, including laser land levelling, improved drainage, raised-bed planting, and nutrient management, have proven effective in mitigating waterlogging stress. Combining diverse approaches—breeding, genetic, genomic, physiological, agronomic, and engineering interventions—will improve the delivery of technologies to farmers and promote the sustainable growth of wheat crops under changing climatic conditions.

## References

Aggarwal P, Kalra N, Chander S et al (2006) InfoCrop: a dynamic simulation model for the assessment of crop yields, losses due to pests, and environmental impact of agro-ecosystems in tropical environments. I. Model description. Agric Syst 89(1):1–25. https://doi.org/10.1016/j.agsy.2005.08.001

Ahmad R, Mahmood N (2005) Impact of raised bed technology on water productivity and lodging of wheat. Pak J Water Resour 9:7–15

Akitaya R, Saito Y, Pandit R (2025) Assessment of factors influencing farmers' adoption of flood-tolerant rice varieties in Nepal. Int J Agric Sustain 23(1):2551998. https://doi.org/10.1080/14735903.2025.2551998

Ali N, Dayal A, Thomas N et al (2018) Effect of different sowing time on seed vigour parameters of wheat (*Triticum aestivum* L.) varieties. Int J Pure Appl Biosci 6(2):1532–1538. https://doi.org/10.18782/2320-7051.6627

Alizadeh-Vaskasi F, Pirdashti H, Cherati Araei A et al (2018) Waterlogging effects on some antioxidant enzymes activities and yield of three wheat promising lines. Acta Agric Slov 111(3):621–631. https://doi.org/10.14720/aas.2018.111.3.10

Arguello MN, Mason RE, Roberts TL et al (2016) Performance of soft red winter wheat subjected to field soil waterlogging: grain yield and yield components. Field Crop Res 194:57–64. https://doi.org/10.1016/j.fcr.2016.04.040

Ballesteros DC, Mason RE, Addison CK et al (2015) Tolerance of wheat to vegetative stage soil waterlogging is conditioned by both constitutive and adaptive QTL. Euphytica 201:329–343. https://doi.org/10.1007/s10681-014-1184-3

Boru G, van Ginkel M, Kronstad W et al (2001) Expression and inheritance of tolerance to waterlogging stress in wheat. Euphytica 117:91–98. https://doi.org/10.1023/A:1003929803920

Chen Y, Zhang H, Chen W et al (2024) The role of ethylene in the regulation of plant response mechanisms to waterlogging stress. Plant Cell Rep 43(12):278. https://doi.org/10.1007/s00299-024-03367-9

Cheng B, Feng X, Buitrago S et al (2024) The wheat *TaSAMS10* improves survival rate under waterlogging stress by inducing ethylene synthesis and early flowering. Plant Growth Regul 102:379–393. https://doi.org/10.1007/s10725-023-01067-0

Chhabra V, Sreethu S, Kaur G et al (2025) Response of wheat crop to water-logged conditions under different land configurations and nutrient management. Sci Rep 15:7168. https://doi.org/10.1038/s41598-024-83752-2

Collaku A, Harrison SA (2002) Losses in wheat due to waterlogging. Crop Sci 42(2):444–450. https://doi.org/10.2135/cropsci2002.4440

Colmer TD (2002) Long-distance transport of gases in plants: a perspective on internal aeration and radial oxygen loss from roots. Plant Cell Environ 26(1):17–36. https://doi.org/10.1046/j.1365-3040.2003.00846.x

Deng C, Xiong Q, Harrison MT et al (2025) Declining frequency yet increasing intensity of crop waterlogging events under future climates. J Hydrol 663:134178. https://doi.org/10.1016/j.jhydrol.2025.134178

Ding J, Liang P, Wu P et al (2020) Effects of waterlogging on grain yield and associated traits of historic wheat cultivars in the middle and lower reaches of the Yangtze River, China. Field Crops Res 246:107695. https://doi.org/10.1016/j.fcr.2019.107695

Ding F, Tong J, Xu R et al (2023) Identification of stable quantitative trait loci underlying waterlogging tolerance post-anthesis in common wheat (*Triticum aestivum*). Crop J 11(4):1163–1170. https://doi.org/10.1016/j.cj.2023.06.004

Du HY, Liu DX, Liu GT et al (2018) Relationship between polyamines and anaerobic respiration of wheat seedling root under water-logging stress. Russ J Plant Physiol 65:874–881. https://doi.org/10.1134/S1021443718060055

Du X, He W, Wang Z et al (2021) Raised bed planting reduces waterlogging and increases yield in wheat following rice. Field Crop Res 265:108119. https://doi.org/10.1016/j.fcr.2021.108119

Elhabashy S, Zhang S, Qiu CW et al (2025) Identification of wheat genotypes with high tolerance to combined salt and waterlogging stresses using biochemical and morpho-physiological insights at the seedling stage. Plants (Basel) 14(9):1268. https://doi.org/10.3390/plants14091268

Feng K, Wang X, Zhou Q et al (2022) Waterlogging priming enhances hypoxia stress tolerance of wheat offspring plants by regulating root phenotypic and physiological adaption. Plants (Basel) 11(15):1969. https://doi.org/10.3390/plants11151969

Feyen L, Dankers R, Bódis K et al (2012) Fluvial flood risk in Europe in present and future climates. Clim Chang 112:47–62. https://doi.org/10.1007/s10584-011-0339-7

Ghobadi ME, Ghobadi M, Zebarjadi A (2017) Effect of waterlogging at different growth stages on some morphological traits of wheat varieties. Int J Biometeorol 61:635–645. https://doi.org/10.1007/s00484-016-1240-x

Govaerts B, Sayre KD, Lichter K et al (2007) Influence of permanent raised bed planting and residue management on physical and chemical soil quality in rain-fed maize/wheat systems. Plant Soil 291:39–54. https://doi.org/10.1007/s11104-006-9172-6

Gupta V, Taliyaan P, Singh G et al (2022) Multi-environment characterization of bread wheat genotypes for waterlogging tolerance. Genetika 54:161–172. https://doi.org/10.2298/GENSR2201161G

Hayashi T, Yoshida T, Fujii K et al (2013) Maintained root length density contributes to the waterlogging tolerance in common wheat (*Triticum aestivum* L.). Field Crop Res 152:27–35. https://doi.org/10.1016/j.fcr.2013.03.020

Herzog M, Striker GG, Colmer TD et al (2016) Mechanisms of waterlogging tolerance in wheat—a review of root and shoot physiology. Plant Cell Environ 39(5):1068–1086. https://doi.org/10.1111/pce.12676

Hossain A, Uddin SN (2011) Mechanisms of waterlogging tolerance in wheat: morphological and metabolic adaptations under hypoxia or anoxia. Aust J Crop Sci 5:1094–1101

Huang B, Johnson JW, Nesmith S et al (1994) Growth, physiological and anatomical responses of two wheat genotypes to waterlogging and nutrient supply. J Exp Bot 45(2):193–202. https://doi.org/10.1093/jxb/45.2.193

Jackson MB (2007) Ethylene-promoted elongation: an adaptation to submergence stress. Ann Bot 101(2):229–248. https://doi.org/10.1093/aob/mcm237

Jiang X, Mao D, Zhu M et al (2022) Evaluating the waterlogging tolerance of wheat cultivars during the early growth stage using the comprehensive evaluation value and digital image analysis. Agriculture 12(3):384. https://doi.org/10.3390/agriculture12030384

Jin N, Cai Z, Ye L et al (2025) Improvement of waterlogging tolerance in wheat by stress priming through inducing aerenchyma formation. Plant Growth Regul 105:245–255. https://doi.org/10.1007/s10725-024-01274-3

Li C, Jiang D, Wollenweber B et al (2011) Waterlogging pretreatment during vegetative growth improves tolerance to waterlogging after anthesis in wheat. Plant Sci 180(5):672–678. https://doi.org/10.1016/j.plantsci.2011.01.009

Liu W, Huang J, Wei C et al (2018) Mapping water-logging damage on winter wheat at parcel level using high spatial resolution satellite data. ISPRS J Photogramm Remote Sens 142:243–256. https://doi.org/10.1016/j.isprsjprs.2018.05.024

Maciag T, Krzyżanowska DM (2025) Microbial enhancement of plant tolerance to waterlogging: mechanisms and interplay with biological control of pathogens. Int J Mol Sci 26(16):8034. https://doi.org/10.3390/ijms26168034

Malik AI, Islam MR, Colmer TD (2011) Transfer of the barrier to radial oxygen loss in roots of Hordeum marinum to wheat (*Triticum aestivum*): evaluation of four *H. marinum*–wheat amphiploids. New Phytol 190(2):499–508. https://doi.org/10.1111/j.1469-8137.2010.03519.x

Mukherjee S, Aadhar S, Stone D et al (2018) Increase in extreme precipitation events under anthropogenic warming in India. Weather Clim Extremes 20:45–53. https://doi.org/10.1016/j.wace.2018.03.005

Pais IP, Moreira R, Coelho AR et al (2024) Unveiling the impact of growth traits on the yield of bread wheat germplasm subjected to waterlogging. Agriculture 14(2):241. https://doi.org/10.3390/agriculture14020241

Pang Y, Wang X, Zhao M et al (2022) Identification and validation of the genomic regions for waterlogging tolerance at germination stage in wheat. Agronomy 12(8):1848. https://doi.org/10.3390/agronomy12081848

Ploschuk RA, Miralles DJ, Colmer TD et al (2018) Waterlogging of winter crops at early and late stages: impacts on leaf physiology, growth and yield. Front Plant Sci 9:1863. https://doi.org/10.3389/fpls.2018.01863

Rautaray SK (2021) Field design for enhancing water productivity in waterlogged areas with efficient water harvesting and farming system. Agric Res 10:255–261. https://doi.org/10.1007/s40003-020-00493-0

Sairam RK, Kumutha D, Ezhilmathi K et al (2008) Physiology and biochemistry of waterlogging tolerance in plants. Biol Plant 52:401–412. https://doi.org/10.1007/s10535-008-0084-6

Sasidharan R, Bailey-Serres J, Ashikari M et al (2017) Community recommendations on terminology and procedures used in flooding and low oxygen stress research. New Phytol 214(4):1403–1407. https://doi.org/10.1111/nph.14519

Sethi RR, Jena SK, Renuka Rani B et al (2022) Advances in water management technology for enhancing agricultural productivity. ICAR-Indian Institute of Water Management & National Institute of Agricultural Extension Management, Bhubaneswar/Hyderabad

Setter T, Waters I (2003) Review of prospects for germplasm improvement for waterlogging tolerance in wheat, barley and oats. Plant Soil 253:1–34. https://doi.org/10.1023/A:1024573305997

Setter TL, Waters I, Sharma SK et al (2009) Review of wheat improvement for waterlogging tolerance in Australia and India: the importance of anaerobiosis and element toxicities associated with different soils. Ann Bot 103(2):221–235. https://doi.org/10.1093/aob/mcn137

Sharma V, Honkavaara E, Hayden M et al (2024) UAV remote sensing phenotyping of wheat collection for response to water stress and yield prediction using machine learning. Plant Stress 12:100464. https://doi.org/10.1016/j.stress.2024.100464

Shen C, Yuan J, Ou X et al (2021) Genome-wide identification of alcohol dehydrogenase (ADH) gene family under waterlogging stress in wheat (*Triticum aestivum*). PeerJ 9:e11861. https://doi.org/10.7717/peerj.11861

Shiferaw B, Smale M, Braun HJ et al (2013) Crops that feed the world 10: past successes and future challenges to the role played by wheat in global food security. Food Sec 5:291–317. https://doi.org/10.1007/s12571-013-0263-y

Singh G, Kulshreshtha N, Singh BN et al (2014) Germplasm characterization, association and clustering for salinity and waterlogging tolerance in bread wheat (*Triticum aestivum*). Indian J Agric Sci 84(9):1102–1110

Singh G, Singh MK, Tyagi BS et al (2017) Germplasm characterization and selection indices in bread wheat (*Triticum aestivum*) for waterlogged soils in India. Indian J Agric Sci 87(9):1139–1148

Singh G, Setter TL, Singh MK et al (2018a) Number of tillers in wheat is an easily measurable index of genotype tolerance to saline waterlogged soils: evidence from 10 large-scale field trials in India. Crop Pasture Sci 69(6):561–573. https://doi.org/10.1071/CP18053

Singh G, Kumar P, Gupta V et al (2018b) Multivariate approach to identify and characterize bread wheat (*Triticum aestivum*) germplasm for waterlogging tolerance in India. Field Crop Res 221:81–89. https://doi.org/10.1016/j.fcr.2018.02.019

St Burgos M, Messmer M, Stamp P et al (2001) Flooding tolerance of spelt (*Triticum spelta* L.) compared to wheat (*Triticum aestivum* L.): a physiological and genetic approach. Euphytica 122:287–295. https://doi.org/10.1023/A:1012945902299

Sundgren TK, Uhlen AK, Waalen W et al (2018) Field screening of waterlogging tolerance in spring wheat and spring barley. Agronomy 8(4):38. https://doi.org/10.3390/agronomy8040038

Tabari H (2020) Climate change impact on flood and extreme precipitation increases with water availability. Sci Rep 10:13768. https://doi.org/10.1038/s41598-020-70816-2

Tian LX, Zhang YC, Chen PL et al (2021) How does the waterlogging regime affect crop yield? A global meta-analysis. Front Plant Sci 12:634898. https://doi.org/10.3389/fpls.2021.634898

Tiwari V, Mamrutha HA, Sareen S et al (2017) Managing abiotic stresses in wheat. In: Abiotic stress management for resilient agriculture. Springer, Singapore, pp 313–337

Toulotte JM, Pantazopoulou CK, Sanclemente MA et al (2022) Water stress resilient cereal crops: lessons from wild relatives. J Integr Plant Biol 64(2):412–430. https://doi.org/10.1111/jipb.13222

Wei M, Li X, Yang R et al (2021) Novel insights into genetic responses for waterlogging stress in two local wheat cultivars in Yangtze River Basin. Front Genet 12:681680. https://doi.org/10.3389/fgene.2021.681680

Wu X, Tang Y, Li C et al (2017) Individual and combined effects of soil waterlogging and compaction on physiological characteristics of wheat in southwestern China. Field Crop Res 215:163–172. https://doi.org/10.1016/j.fcr.2017.10.016

Xu L, Zhao C, Pang J et al (2022) Genome-wide association study reveals quantitative trait loci for waterlogging-triggered adventitious roots and aerenchyma formation in common wheat. Front Plant Sci 13:1066752. https://doi.org/10.3389/fpls.2022.1066752

Yamauchi T, Watanabe K, Fukazawa A et al (2014) Ethylene and reactive oxygen species are involved in root aerenchyma formation and adaptation of wheat seedlings to oxygen-deficient conditions. J Exp Bot 65(1):261–273. https://doi.org/10.1093/jxb/ert371

Yang F, Liu S, Wang Q et al (2021) Assessing waterlogging stress level of winter wheat from hyperspectral imagery based on harmonic analysis. Remote Sens 14(1):122. https://doi.org/10.3390/rs14010122

Yang R, Li M, Harrison MT et al (2022) iTRAQ proteomic analysis of wheat (*Triticum aestivum* L.) genotypes differing in waterlogging tolerance. Front Plant Sci 13:890083. https://doi.org/10.3389/fpls.2022.890083

Yang G, Li Y, He Y et al (2025) Multimodal large language model for wheat breeding: a new exploration of smart breeding. ISPRS J Photogramm Remote Sens 225:492–513. https://doi.org/10.1016/j.isprsjprs.2025.03.027

Yu M, Chen GY (2013) Conditional QTL mapping for waterlogging tolerance in two RIL populations of wheat. SpringerPlus 2(1):245. https://doi.org/10.1186/2193-1801-2-245

Yu M, Mao S, Chen G et al (2013) QTLs for waterlogging tolerance at germination and seedling stages in population of recombinant inbred lines derived from a cross between synthetic and cultivated wheat genotypes. J Integr Agric 13(1):31–39. https://doi.org/10.1016/S2095-3119(13)60354-8

Zhang J, Pan B, Shi W et al (2023) Monitoring waterlogging damage of winter wheat based on HYDRUS-1D and WOFOST coupled model and assimilated soil moisture data of remote sensing. Remote Sens 15(17):4133. https://doi.org/10.3390/rs15174133

Zhang D, Gajardo J, Medic T et al (2025) Wheat3DGS: in-field 3D reconstruction, instance segmentation and phenotyping of wheat heads with Gaussian splatting. In: Proceedings of the IEEE/CVF conference on computer vision and pattern recognition, pp 5360–5370

# Advances in Wheat Breeding for Metal and Metalloid Stress Tolerance

12

Abhishek Dadsena, Anita Kumari, Alka Bharati,
Jyotsana Tilgam, Santosh Kumar, Nitish Ranjan Prakash,
Malemnganbi Keisham, and Sandeep Jaiswal

**Abstract**

Heavy metal and metalloid contamination in soils is detrimental to productivity and food safety, primarily due to the increased bioavailability of toxic elements such as cadmium (Cd), lead (Pb), chromium (Cr), arsenic (As), and aluminium (Al). In *Triticum aestivum*, excessive exposure to these metals impairs root development, nutrient uptake, photosynthetic efficiency, and redox balance, leading to oxidative damage, enzyme inhibition, altered carbon metabolism, and yield reductions that can exceed 50% under severe stress, whilst also increasing dietary health risks through toxic element accumulation in grains. This chapter reviews recent advances in understanding the physiological, biochemical, and molecular mechanisms underlying metal tolerance in wheat, highlighting the roles of antioxidant defence systems, the ascorbate–glutathione cycle, phytochelatin- and metallothionein-mediated chelation, and vacuolar sequestration as primary detoxification strategies. The functions of key transporter families,

A. Dadsena
ICAR – National Institute for Plant Biotechnology, New Delhi, India

A. Kumari
Department of Plant and Soil Sciences, Texas Tech University, Lubbock, TX, USA

A. Bharati · M. Keisham · S. Jaiswal (✉)
ICAR Research Complex for NEH Region, Umiam, Meghalaya, India

J. Tilgam
ICAR – National Bureau of Agriculturally Important Microorganisms,
Mau, Uttar Pradesh, India

S. Kumar
ICAR – Indian Agricultural Research Institute, Gauria Karma, Jharkhand, India

N. R. Prakash
ICAR – Central Soil Salinity Research Institute, Karnal, Haryana, India



R. Tiwari et al. (eds.), *Wheat Stress Management*,
https://doi.org/10.1007/978-981-92-0902-6_12

including NRAMP, HMA, ZIP, ABC, YSL, VIT, and MATE, as well as regulatory transcription factors such as NAC and WRKY, are examined in relation to metal uptake, distribution, and exclusion. Advances in quantitative trait locus (QTL) mapping and genome-wide association studies (GWAS), particularly the identification of major loci such as *Cdu-B1* (*TdHMA3-B1*) that control grain Cd accumulation, have facilitated marker-assisted and genomics-assisted breeding for low-metal cultivars. The integration of conventional screening, high-throughput phenotyping, genomic selection, multi-omics approaches, and CRISPR/Cas-mediated genome editing is accelerating the development of wheat genotypes with metal tolerance, high yield, and grain safety, whilst the strategic use of landraces, wild relatives, synthetic hexaploids, and alien introgressions further broadens the genetic base for durable tolerance and the breeding of cultivars suited to contaminated agroecosystems.

**Keywords**

Heavy metal stress · Cadmium accumulation · Metal transporters · Antioxidant defense system · Quantitative trait loci (QTL) · Genomics-assisted wheat breeding

## 12.1 Introduction

Wheat (*Triticum aestivum* L.) is a principal global staple crop cultivated across a wide range of agroclimatic zones. It provides a substantial proportion of calories and proteins to nearly 60% of the world's population (Zhou and Li 2022). Wheat is particularly susceptible to environmental stressors that compromise grain quality and yield. Amongst these stressors, contamination with heavy metals and metalloids has emerged as a critical concern, primarily driven by rapid industrialization, intensive agricultural practices, urbanization, and increased natural resource extraction. These anthropogenic activities have contributed to the accumulation of toxic elements in agricultural soils, irrigation water, and the broader agroecosystem (Wang et al. 2017; Aprile et al. 2019; Zhang et al. 2022). Heavy metals encompass both essential micronutrients, such as copper (Cu), zinc (Zn), manganese (Mn), nickel (Ni), iron (Fe), cobalt (Co), and molybdenum (Mo), as well as nonessential toxic elements including cadmium (Cd), lead (Pb), mercury (Hg), and chromium (Cr). Essential heavy metals are required in trace amounts for enzymatic activity, redox homeostasis, and photosynthesis (Ali et al. 2013; Page and Feller 2015), but their permissible concentration range is narrow. Elevated levels of these metals induce oxidative stress, protein misfolding, and cellular dysfunction. Conversely, nonessential heavy metals are toxic even at minimal concentrations, disrupting membrane integrity, nutrient transport, signal transduction, and both genetic and metabolic processes (Khan et al. 2009; Emamverdian et al. 2015). In *T. aestivum*, exposure to heavy metals impairs root development, inhibits cell division and elongation, alters water relations, and diminishes the uptake and translocation of essential nutrients, including calcium (Ca), magnesium (Mg), phosphorus (P), iron (Fe), and zinc (Zn). Heavy metals also damage chloroplast ultrastructure, reduce chlorophyll content, impair photosynthetic components, and elevate reactive oxygen species (ROS) production, resulting in oxidative stress, enzyme inhibition, DNA damage, lipid peroxidation, and metabolic failure (Gadd 2007; Jyothsna and Murthy 2016). Together, these effects lead to substantial yield losses, with severe contamination by Cd, Pb, or Cr reducing wheat grain yield by more than 50%, and moderate contamination causing reductions of 10–20% (Choudhary et al. 2018; Rizvi et al. 2020; Haider et al. 2021; Anas et al. 2024). Beyond yield reduction, the accumulation of heavy metals poses significant risks to food safety and public health. Toxic elements present in soil and irrigation water are readily absorbed and translocated to wheat grains, resulting in direct dietary exposure (Meng et al. 2022). Concentrations of Cd and Pb frequently exceed the maximum permissible limit of 0.2 mg $kg^{-1}$ in wheat grains established by the Codex Alimentarius (Aprile et al. 2019). Chronic consumption of contaminated grains is associated with renal dysfunction, skeletal disorders, neurodevelopmental impairment, immunosuppression, metabolic disturbances, and increased cancer risk (Iyengar and Nair 2000; Türkdoğan et al. 2003; El-Kady and Abdel-Wahhab 2018). Maternal exposure to metals such as aluminium (Al), cadmium (Cd), manganese (Mn), and lead (Pb) further elevates the risk of intrauterine growth restriction and developmental abnormalities (Khan et al. 2010). Climate change may intensify these challenges by modifying soil pH, organic matter decomposition, and redox dynamics, thereby enhancing metal solubility and plant uptake. Recent studies also indicate that wheat leaves can directly absorb atmospheric Cd via stomata, with the flag leaf accounting for up to one-third of grain Cd during early grain filling. These observations highlight the need for integrated management strategies that address both soil and atmospheric contamination. Therefore, developing wheat varieties with enhanced tolerance to heavy metals and metalloids, whilst reducing metal accumulation in edible tissues, has become a critical objective.

## 12.2 Sources, Mobility, and Agricultural Impacts of Heavy Metals in Wheat-Growing Soils

Heavy metals enter soils and plants through both natural and anthropogenic sources, with human activities now the primary source in wheat-growing regions (Fig. 12.1). Natural sources include rock weathering, soil formation, volcanic eruptions, mineral decomposition, and wind-blown dust. Under these conditions, metals generally occur at low concentrations, such as cadmium (Cd) at 0.01–0.7 mg $kg^{-1}$, lead (Pb) at approximately 50 mg $kg^{-1}$, and nickel (Ni) at 15–30 mg $kg^{-1}$. These metals are mainly present in forms with limited bioavailability to plants. The mobility of these metals is largely determined by soil pH, organic matter content, and soil aeration or moisture levels. For example, acidic soils promote greater dissolution of aluminium (Al) and Cd. Volcanic ash and dust can also temporarily increase concentrations of Pb, arsenic (As), and mercury (Hg) in local wheat soils (Szolnoki et al. 2013).

**Fig. 12.1** The geogenic and anthropogenic sources of heavy metals in wheat-growing soils

In contrast, anthropogenic activities are the primary contributors to heavy metal accumulation and bioavailability in cultivated soils. Major sources include mining, smelting, industrial effluents, and coal combustion, all of which release Cd, Pb, chromium (Cr), copper (Cu), Ni, and Hg. Wastewater irrigation, particularly prevalent in water-scarce regions, substantially increases metal loading and plant uptake. Agricultural practices further intensify contamination through the use of Cd- and Pb-containing phosphate fertilizers, Cu-based fungicides, and residual metal-containing pesticides. The application of contaminated biosolids, manures, and composts introduces metals in forms that are readily available to plants. Additional sources include atmospheric deposition of industrial and vehicular particulates, emerging contaminants from e-waste recycling and microplastic-associated metals, and postharvest leaching from metallic storage containers. Metals from anthropogenic sources are typically more mobile, persistent, and toxic than those from geogenic origins. These characteristics result in cumulative negative effects on soil biota, nutrient cycling, wheat productivity, and food safety. The risks are further exacerbated by rapid urbanization, increased wastewater reuse, and cultivation on marginal lands. Mitigating heavy metal contamination in wheat systems requires integrated soil and water management strategies, as well as the development of *Triticum aestivum* L. cultivars with reduced metal uptake and enhanced detoxification capacity.

## 12.3 Impacts of Heavy Metals on Wheat Growth and Productivity and the Importance of Breeding Metal-Tolerant Varieties

Heavy metals significantly impair wheat growth and productivity by inhibiting root development, disrupting nutrient uptake, altering root and shoot anatomy, and inducing oxidative stress. These combined effects reduce biomass, yield, and grain quality. Toxic metals are primarily absorbed through the root system, impairing respiration and photosynthesis and causing visible symptoms such as chlorosis, necrosis, wilting, and stunted growth (Gomes et al. 2011; Austruy et al. 2014; Dey and Mondal 2016). At the cellular level, heavy metals compromise antioxidant defences and promote excessive accumulation of reactive oxygen species (ROS). They also cause electrolyte leakage and induce protein and enzyme dysfunction by binding to sulfhydryl groups. The displacement of $Ca^{2+}$ in calmodulin further disrupts signalling pathways and intensifies stress responses (Asati et al. 2016; Alsafran et al. 2022). Metals such as cadmium (Cd), lead (Pb), nickel (Ni), chromium (Cr), and $Al^{3+}$ induce interconnected physiological and metabolic disorders. For example, Cd interferes with nitrogen metabolism, chlorophyll content, photosystem efficiency, antioxidant enzyme activity, and root elongation (Benavides et al. 2005; Gratão et al. 2005; Lux et al. 2011; Ahanger et al. 2020). Pb disrupts nutrient and water transport, whilst Ni suppresses cell division, photosynthesis, and respiration. Cr induces oxidative stress and damages mesophyll, xylem, and phloem tissues (Gomes et al. 2017). $Al^{3+}$ toxicity in acidic soils rapidly inhibits root elongation, reduces nutrient and water uptake, diminishes chlorophyll content and photosystem II (PSII) efficiency, and lowers Ca, Mg, and P accumulation (Jaiswal et al. 2023, 2024). Overall, heavy metal stress reduces photosynthetic pigment levels, seed germination rates, nutrient acquisition, and DNA integrity. Roots exhibit stunting, lignification, discolouration, microtubule disruption, and reduced mitotic activity. Wheat employs adaptive mechanisms, including increased antioxidant activity, regulation of ion homeostasis, vacuolar sequestration, and induction of stress-responsive genes. However, these responses generally limit tissue injury but do not prevent metal accumulation in grains (Rizwan et al. 2016; Murtaza et al. 2019; Li et al. 2022a). As a result, consumption of contaminated wheat poses significant risks to human health, including renal, neurological, developmental, and immune disorders (Debnath et al. 2019; Hu et al. 2021; Ma et al. 2021; Xiong et al. 2021; Hernández-Cruz et al. 2022). Although mitigation strategies such as soil amendments, chelation, phytoremediation, and microbial remediation demonstrate potential, they are often slow, costly, or impractical for large-scale application (Dhaliwal et al. 2020; Yan et al. 2020; Wang et al. 2021). Consequently, breeding wheat varieties with inherently reduced heavy metal uptake and enhanced tolerance is considered the most effective and sustainable solution. This approach can be realized through root-based exclusion, preferential sequestration in nonedible tissues, and improved detoxification pathways. Marker-assisted selection, quantitative trait locus (QTL) mapping, and genomic and transcriptomic approaches facilitate this strategy (Zaid et al. 2018; Mondal et al. 2021).

## 12.4 Early Evidence and Evolution of Breeding Efforts for Heavy Metal Tolerance and Exclusion in Wheat

Heavy metal contamination of agricultural soils has been recognized for several decades, with extensive field studies documenting significant reductions in crop growth, productivity, and grain quality under contaminated conditions (Rizwan et al. 2016). Early research showed that plant species and genotypes vary substantially in their sensitivity to heavy metal stress, revealing exploitable physiological and genetic diversity. In wheat, initial screenings identified differential responses, such as the lower zinc sensitivity of variety W-141 compared to W-142 (Kirmani et al. 2018), and the greater susceptibility of WH-711 to arsenic (As) stress relative to the tolerant genotype C-306. Comparable patterns have been observed in other species. For example, a classic study on *Agrostis tenuis* demonstrated that populations from copper-mining sites exhibited higher tolerance to copper, nickel, lead, and zinc than those from uncontaminated pastures, providing early evidence of local adaptation to metal-rich environments (Gregory and Bradshaw 1965). Furthermore, vegetable crops cultivated in urban areas were found to accumulate higher levels of lead (Pb) and Zinc (Zn) than their rural counterparts, reflecting differences in exposure and adaptive responses (Yang et al. 2011; Fang et al. 2014).

Systematic screening of wheat landraces, cultivars, and wild relatives has confirmed the existence of valuable variation for both heavy metal tolerance and low metal accumulation. For example, the cultivar Pirsabak 2004 exhibited superior germination and seedling vigour under Pb stress, whilst the Canadian durum wheat cultivar Strongfield combined high yield and protein content with naturally low grain cadmium (Cd) accumulation, representing a significant breeding advancement (Clarke et al. 2005). Early breeding and screening efforts primarily relied on visual and growth-based traits, such as germination rate, seedling vigour, root and shoot length, biomass accumulation, and foliar chlorosis or necrosis, due to their simplicity and sensitivity to heavy metal stress (Mathur et al. 2016; Rizwan et al. 2016; Singha et al. 2025). However, variability observed across growth stages and metal concentrations highlighted the need for standardized protocols and more robust physiological and biochemical selection criteria. Despite increased awareness of heavy metal constraints, breeding for heavy metal tolerance received limited attention during the Green Revolution, when priorities focused on yield potential, fertilizer responsiveness, plant stature, and lodging resistance (Koźmińska et al. 2018). This lack of emphasis coincided with rising heavy metal contamination resulting from industrialization, increased use of agrochemicals, and expanded wastewater irrigation, thereby widening the gap between breeding objectives and environmental challenges. Nevertheless, isolated successes such as Strongfield wheat and the As-tolerant rice varieties BRRI dhan29 and BR-26 demonstrate that breeding for heavy metal tolerance is feasible when prioritized (Clarke et al. 2005; Chen et al. 2017). Although significant genetic variation for heavy metal tolerance and exclusion exists in wheat and its wild relatives, this diversity remained underutilized for decades due to the absence of structured breeding frameworks and clear selection indices (Oladzad-Abbasabadi et al. 2018; Li et al. 2021). Recently, growing recognition of heavy metal contamination as both an agronomic and public health concern has shifted research priorities towards developing wheat cultivars that sustain yield whilst minimizing metal accumulation in edible grains.

## 12.5 Scientific Advancements and Principles Developed over the Past Centuries

### 12.5.1 Mechanisms Underlying Heavy Metal Toxicity in Wheat

A comprehensive understanding of the molecular, biochemical, and physiological mechanisms underlying heavy metal toxicity is crucial for developing wheat varieties with enhanced tolerance or reduced metal accumulation. Heavy metals disrupt multiple interconnected metabolic pathways, particularly photosynthesis, redox homeostasis, and enzyme-mediated biochemical reactions. These disruptions impair energy production, reduce biomass accumulation, and diminish both yield and grain quality. Photosynthesis is amongst the earliest and most sensitive processes affected by HM stress. Trace metals such as iron (Fe), manganese (Mn), copper (Cu), and zinc (Zn) serve as essential cofactors in chlorophyll biosynthesis and the photosynthetic electron transport chain; however, excessive metal concentrations destabilize chloroplast structure and function (Paunov et al. 2018). HM exposure rapidly alters stomatal regulation, restricting $CO_2$ uptake and gas exchange (Rizvi et al. 2020). Reductions in chlorophyll and carotenoid contents are attributed to increased chlorophyllase activity, reactive oxygen species (ROS)-induced pigment degradation, and the substitution of $Mg^{2+}$ in the chlorophyll porphyrin ring by competing metal ions (Hegedüs et al. 2001; Gill et al. 2012; Latif et al. 2020). At the ultrastructural level, HM stress induces thylakoid membrane swelling, lamellar disorganization, and lipid peroxidation, impairing both photosystem I and photosystem II (PSII), with PSII being particularly susceptible (Ekmekçi et al. 2009; Paunov et al. 2018). In wheat, exposure to cadmium (Cd) and Zn can reduce $CO_2$ assimilation by approximately 30%. HM stress also impairs electron transport and increases the dissipation of absorbed light energy as heat.

Enzymatic inhibition represents a major mechanism of HM toxicity. Many HMs bind strongly to sulfhydryl (–SH) groups in cysteine-rich proteins, promoting protein misfolding, inactivation, and elevated oxidative stress. For instance, mercury (Hg) binds to cysteine and glutathione (Ajsuvakova et al. 2020). Toxic metals further disrupt metabolism by displacing essential cofactors from metalloproteins. Cd, Cu, Fe, and Mn can substitute for Mg in RuBisCO, thereby reducing carbon fixation. Cd may also replace Mn in PSII, impairing water splitting. Lead (Pb) and aluminium (Al) can displace Fe from ferritin, thereby affecting iron storage, whereas Pb, Cd, and Zn may replace Ca in calmodulin, thereby disrupting calcium-dependent signalling. Substitution of cobalt (Co) or nickel (Ni) for Zn in enzymes results in a loss of catalytic activity (Viehweger 2014). Together, these effects—including damage to photosynthetic machinery, enzyme inhibition, nutrient imbalances, and excess ROS—reduce metabolic efficiency. These disruptions constrain root and shoot growth, ultimately decreasing wheat yield and grain quality (Gill et al. 2012; Mittler 2017).

### 12.5.2 Physiological and Biochemical Responses to Heavy Metal Stress in Wheat

Plants, including wheat, possess integrated physiological and biochemical defence systems that mitigate heavy metal toxicity by maintaining cellular homeostasis, limiting oxidative damage, and detoxifying or compartmentalizing toxic ions. A major consequence of heavy metal exposure is the excessive production of reactive oxygen species (ROS), which disrupts the redox balance and damages cellular components, including lipids, proteins, nucleic acids, and membranes, when not properly regulated (Mittler 2017; Goyal et al. 2020). In response, plants activate a coordinated antioxidant network distributed across chloroplasts, mitochondria, peroxisomes, and the cytosol. Enzymatic antioxidants, including superoxide dismutase (SOD), catalase (CAT), ascorbate peroxidase (APX), peroxidases (POD), and glutathione reductase, convert ROS into less reactive molecules. Nonenzymatic antioxidants, such as ascorbate, glutathione (GSH), carotenoids, phenolics, tocopherols, and alkaloids, further stabilize cellular redox status (Sharma et al. 2012; Emamverdian et al. 2015; Mittler et al. 2022; Mohamed et al. 2025). The ascorbate–glutathione (AsA–GSH) cycle is the key to this defence, with GSH functioning as both a redox buffer and a precursor for metal-chelating phytochelatins (PCs), supported by biosynthetic and regulatory genes such as *VTC2* and various *APX* isoforms. Furthermore, secondary-metabolism enzymes (PAL and TAL) and emerging regulators, such as melatonin, enhance antioxidant capacity, metal chelation, vacuolar sequestration, and stress-responsive gene expression under heavy metal stress (Jańczak-Pieniążek et al. 2022; Altaf et al. 2023).

The accumulation of ROS also functions as a signal that activates calcium-dependent protein kinases (CDPKs) and mitogen-activated protein kinase (MAPK) cascades, including MEKK1–MKK4/5–MPK3/6 and MEKK1–MKK2–MPK4/6. These signalling pathways link stress perception to transcriptional reprogramming and metabolic adjustment (Chmielowska-Bąk et al. 2018). Simultaneously, plants employ biochemical detoxification mechanisms based on chelation and sequestration. Low-molecular-weight thiols, such as GSH and cysteine, serve as substrates for PC synthesis via phytochelatin synthase (PCS). This process produces cysteine-rich peptides that bind metals into stable complexes, which are then transported into vacuoles by ATP-binding cassette (ABC)-type transporters, including HMA2, HMA3, and HMA4 (Vatamaniuk et al. 2000; Clemens 2006; Mirza et al. 2014; Chaudhary et al. 2016). PCs are essential for cadmium (Cd) detoxification and also chelate $Hg^{2+}$, $Pb^{2+}$, $Cu^{2+}$, $Ni^{2+}$, $Zn^{2+}$, and arsenic species. Their biosynthesis is regulated by metal availability, transcription factors such as MYB40, and genes such as *TaPCS1* in wheat (Clemens et al. 1999).

Metallothioneins (MTs) represent an additional and complementary detoxification mechanism. These small, cysteine-rich proteins bind metal ions through thiol groups, thereby reducing cytosolic metal concentrations and contributing to ROS scavenging and redox regulation (Cobbett and Goldsbrough 2002). Functional studies in cereals demonstrate their diverse roles in metal homeostasis and tolerance, including binding Cd, Cu, and Zn, regulating metal transport to reproductive tissues, and modulating oxidative stress. Notable examples include *OsMT1e*, *OsMT2b*, and wheat *TaMT1f-1B*, which confer increased tolerance to Cd and other stresses (Cao et al. 2025). Collectively, the integration of antioxidant defences, stress signalling, metal chelation, and vacuolar sequestration governs plant tolerance to heavy metal stress and highlights key molecular targets for breeding and biotechnological strategies to develop cultivars with enhanced resilience and reduced metal accumulation.

### 12.5.3 Genes and Transcription Factors Affecting Heavy Metal Toxicity Tolerance

A thorough understanding of the genes and transcription factors regulating heavy metal transport, sequestration, detoxification, and homeostasis is critical for breeding wheat varieties with improved tolerance and reduced grain heavy metal accumulation. Plants rely on a complex molecular network that includes membrane transporters responsible for uptake, efflux, and compartmentalization, as well as chelators that immobilize toxic ions. This network further incorporates transcriptional regulators that orchestrate stress responses and signalling pathways, integrating these mechanisms with broader physiological processes. These coordinated systems maintain micronutrient balance and minimize toxicity (Zhou and Zheng 2022). Key transporter families—natural resistance-associated macrophage proteins (NRAMPs), heavy metal ATPases (HMAs), zinc–iron-regulated transporter-like proteins (ZIPs), ATP-binding cassette transporters (ABCs), yellow stripe-like transporters (YSLs), vacuolar iron transporters (VITs), ferroportins (FPNs), cation–chloride cotransporters (CCCs), multidrug and toxic compound extrusion transporters (MATEs), and copper transport proteins (COPTs)—each contribute to distinct stages of metal influx, redistribution, sequestration, and detoxification across plant tissues and organelles. Figure 12.2 depicts the various molecular responses of wheat under heavy metal stress conditions.

NRAMPs mediate the uptake and translocation of divalent cations, transporting both essential metals ($Fe^{2+}$, $Mn^{2+}$, and $Zn^{2+}$) and toxic ions ($Cd^{2+}$, $Pb^{2+}$, and $Hg^{2+}$). In *Arabidopsis thaliana*, AtNRAMP1 is responsible for Cd uptake, whilst AtNRAMP3 and AtNRAMP4 mobilize Mn from vacuoles (Lanquar et al. 2010). In cereals, OsNRAMP5 regulates Cd and Pb uptake in rice, and OsNRAMP5 mutants show reduced metal accumulation in shoots and grains (Chang et al. 2022; Tang et al. 2022). In wheat, loss-of-function mutations in TaNRAMP5 reduce grain Cd content by over 30% (Cheng et al. 2025). TaNRAMP1 and TaHMA2 are upregulated under antimony stress (Zhou and Zheng 2022), and NRAMP homologs in *Aegilops tauschii* exhibit functional diversification in Mn and Cu transport. HMAs (P1B-type ATPases) actively transport metals using ATP, regulating efflux, xylem loading, and vacuolar sequestration. AtHMA2 and AtHMA4 mediate Zn and Cd efflux and root-to-shoot translocation, whilst AtHMA3 sequesters metals into vacuoles (Baekgaard et al. 2010; Haque et al. 2022). In rice, OsHMA2 is involved in xylem loading, and OsHMA3 enhances vacuolar Cd sequestration (Maghrebi et al. 2021). In wheat,

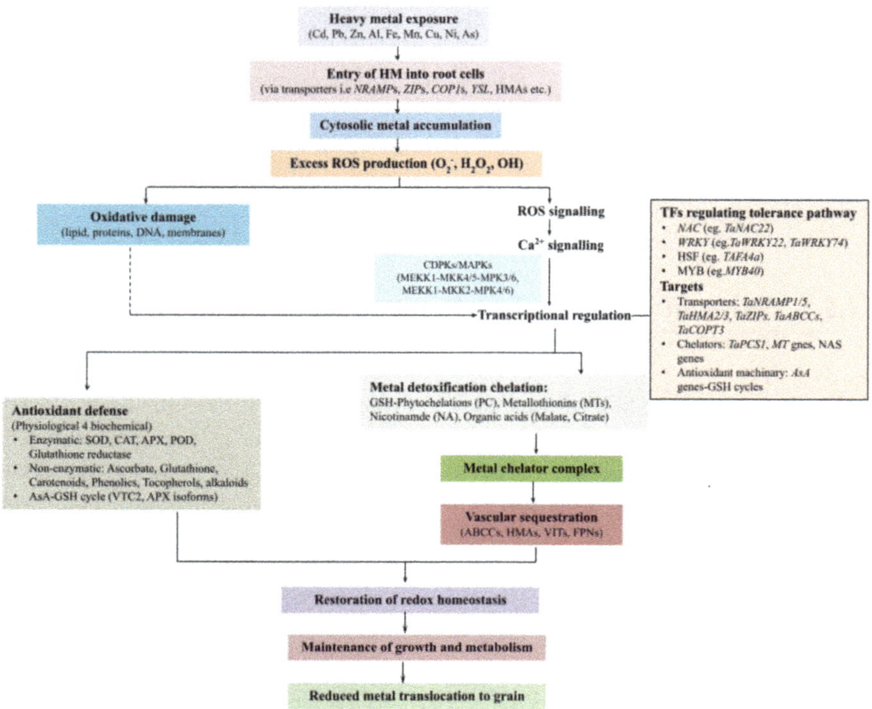


**Fig. 12.2** Molecular response of wheat under heavy metal stress conditions

TaHMA2 promotes Cd export, and TaHMA3 determines root-to-shoot Cd partitioning. DNA methylation regulates the expression of TaHMA2 and TaABCC2/3/4 under Cd stress (Zhu et al. 2020).

ZIP transporters play a central role in micronutrient uptake but can also transport toxic metals due to their broad substrate specificity. AtIRT1 mediates Fe uptake and also transports $Zn^{2+}$, $Mn^{2+}$, and $Cd^{2+}$ (Nishida et al. 2011). In rice, OsZIP5 and OsZIP9 are involved in uptake, whilst OsZIP3, OsZIP7, and OsZIP10 mediate internal translocation. In wheat, TaZIP1 and ZnT1 are induced by Cd, TaZIP8 is induced by Pb, and several ZIPs respond to Zn deficiency, indicating both conserved and diversified functions in cereals (Tan et al. 2020; Ma et al. 2023; Zhang et al. 2021; Li et al. 2020; Zeng et al. 2024). ABC transporters contribute to metal detoxification by sequestering metal–phytochelatin complexes in vacuoles. AtABCC1, AtABCC2, and AtABCC3 are involved in Cd and Hg detoxification (Park et al. 2012; Brunetti et al. 2015), whilst wheat TaABCC genes (including *TaABCC3*, *TaABCC4*, *TaABCC11*, *TaABCC14*, *TaABCC6*, *TaABCC8*, *TaABCC12*, *TaABCC13*, and *TaABCC17*) are induced by Cd and oxidative stress (Wang et al. 2017; Bhati et al. 2015). Additional transporters further refine metal homeostasis: YSLs redistribute metal–nicotianamine complexes (Kumar et al. 2019), VITs store Fe, Zn, and Mn in vacuoles (Sabella et al. 2022), FPNs sequester Ni and Co (Kan et al. 2022),

COPT3 modulates Cu homeostasis and Cd toxicity (Zheng et al. 2023), and CCC and MATE transporters contribute to Fe and Mn balance and Al detoxification (Liu et al. 2025; Duan et al. 2022).

Chelation and detoxification processes are supported by phytochelatins (PCs), metallothioneins (MTs), glutathione (GSH), nicotianamine (NA), and organic acids. PCs, synthesized from GSH, bind Cd and As and are transported into vacuoles by ABC transporters (Gupta et al. 2013; Faizan et al. 2024). MTs bind Zn, Cu, and Cd, thereby contributing to detoxification, redox balance, and reactive oxygen species (ROS) scavenging (Cobbett and Goldsbrough 2002; Pourjalali et al. 2023). GSH functions as both an antioxidant and a precursor for PC synthesis. NA facilitates long-distance metal transport and tolerance; overexpression of NAS enhances Cd tolerance and improves Fe and Zn nutrition, with multiple NAS transcripts strongly induced by Cd in wheat (Sabella et al. 2022; Seregin and Kozhevnikova 2023). Organic acids such as malate and citrate also chelate metals, including Cr, thereby reducing their toxicity.

Transcription factors (TFs) regulate these pathways by controlling the expression of transporter and chelator genes. The NAC (NAM, ATAF1/2, and CUC2), WRKY, and heat shock factor (HSF) families serve as key regulators. For instance, TaNAC22 enhances Cd tolerance by increasing antioxidant and osmolyte levels whilst reducing ROS (Yu and Zhang 2023). TaWRKY22 activates TaCOPT3D, linking Cu and Cd responses (Liu et al. 2022). TaWRKY74 enhances Cd tolerance by activating the ascorbate–glutathione cycle and repressing Cd transporters, including TaNRAMP1, TaNRAMP5, TaHMA2, TaHMA3, and TaIRT1 (Li et al. 2022b). TaHsfA4a induces MTs under Cd stress (Shim et al. 2009). Collectively, NRAMPs, HMAs, ZIPs, ABCs, and auxiliary transporters, together with TF-regulated chelators (PCs, MTs, and NA), constitute the core genetic framework for heavy metal tolerance in wheat.

## 12.6 Breeding Strategies and Genetic Resources for Developing Heavy Metal—Tolerant Wheat Cultivars

Heavy metal stress significantly constrains wheat productivity and food safety. As a result, developing heavy metal-tolerant cultivars of *Triticum aestivum* L. has become a primary focus of breeding programs worldwide (Rizvi et al. 2020). Contemporary breeding strategies prioritize selecting varieties that tolerate high heavy-metal concentrations or limit heavy-metal uptake and accumulation. Figure 12.3 shows a breeding pipeline for developing heavy metal-tolerant wheat cultivars. Whilst no Indian wheat cultivar has yet been officially designated as heavy metal-tolerant, improvement programs are ongoing. These efforts use conventional selection, marker-assisted breeding (MAB), QTL mapping, genome-wide association studies (GWAS), and genome-editing techniques (Zulfiqar et al. 2022). The success of these strategies depends on a thorough characterization of genetic diversity within wheat germplasm, landraces, and wild relatives of *T. aestivum.*

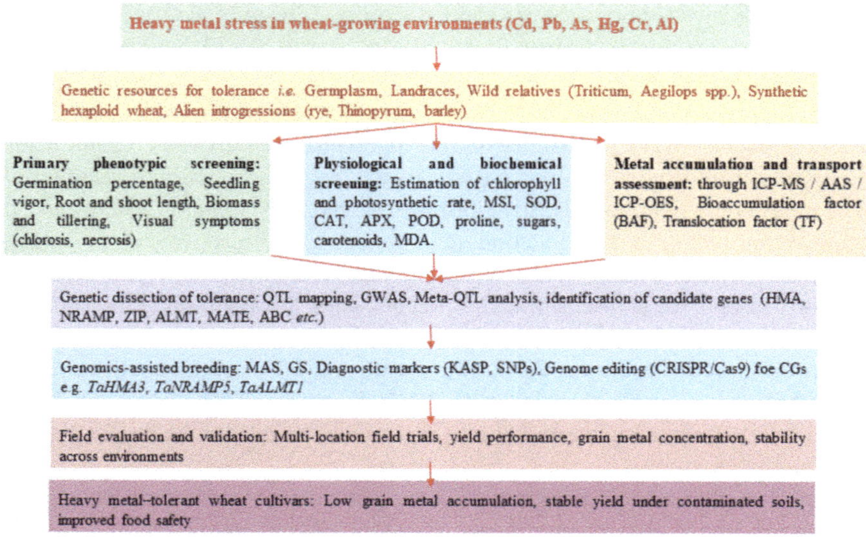


**Fig. 12.3** Breeding pipeline for developing heavy metal-tolerant wheat cultivars

### 12.6.1 Screening for Heavy Metal Toxicity

Before the advent of molecular tools, the identification of heavy metal-tolerant wheat genotypes primarily relied on visual and growth-based assays conducted in contaminated soils or under controlled stress conditions. These cost-effective approaches remain important for preliminary evaluation, as they measure integrative traits such as germination percentage, seedling vigour, root and shoot length, biomass, tillering capacity, and overall plant vigour. Collectively, these traits indicate physiological disruption under heavy metal stress (Rizwan et al. 2016; Singha et al. 2025). Seed germination and early seedling growth are particularly sensitive indicators and are widely used for early stage screening. Diagnostic symptoms such as chlorosis, necrosis, and pigment loss indicate impaired photosynthesis and oxidative balance, whilst growth suppression reflects cumulative metabolic stress. Different metals elicit specific responses: Cadmium (Cd) strongly inhibits germination and root elongation; chromium (Cr) and arsenic (As) cause chlorosis and wilting; mercury (Hg) and lead (Pb) affect chlorophyll and carotenoid contents depending on dose and exposure. These responses form the basis of quantitative indices such as vigour index, tolerance index, and germination percentage, which facilitate genotype classification (Israr et al. 2006). Because symptom expression varies with soil properties, developmental stage, and metal bioavailability, the use of standardized protocols and complementary physiological markers is essential (Israr et al. 2006).

Phenotypic screening is conducted in hydroponic, pot-culture, and field environments at each selection stage (Alybayeva et al. 2016; Jaiswal et al. 2023; Zhai et al. 2023). Tolerant lines typically exhibit elevated levels of proline, soluble proteins, sugars, carotenoids, and membrane stability index (MSI%), along with increased activity of antioxidant enzymes—glutathione reductase (GR), ascorbate peroxidase (APX), peroxidase (POD), catalase (CAT), and superoxide dismutase (SOD)—reflecting enhanced reactive oxygen species (ROS) detoxification capacity (Ali et al. 2018; Murtaza et al. 2019). Exposure to heavy metals generally reduces germination, root and shoot growth, biomass, and yield. For germination toxicity, the order of severity is Zn > Cd > Mn > Cr (Shaikh et al. 2013). Additional physiological parameters, including chlorophyll content, photosynthetic rate, membrane integrity, and malondialdehyde (MDA) levels, improve discrimination amongst genotypes (Jiang et al. 2019). Simultaneous quantification of tissue metal concentrations using atomic absorption spectrometry (AAS), graphite furnace atomic absorption spectrometry (GF-AAS), inductively coupled plasma mass spectrometry (ICP-MS), inductively coupled plasma optical emission spectrometry (ICP-OES), or hydride generation atomic fluorescence spectrometry (HG-AFS) is essential. Indices such as bioaccumulation factor (BAF) and translocation factor (TF) are valuable for identifying genotypes with low uptake or restricted internal transport (Aladesanmi et al. 2019). Genomics-enabled screening approaches, such as RNA sequencing, metabolite profiling, DNA methylation analysis, and expression analysis of metal-chelating genes, further advance mechanistic understanding (Zaid et al. 2023).

Extensive screening has identified heavy metal-tolerant wheat genotypes across diverse regions. Notable examples include Kazakhstan spring wheat lines tolerant to Cd and Zn, winter wheat 'Mironovskaya-808' with minimal heavy metal accumulation and high yield, and C-306, which demonstrates the highest Cr tolerance and the lowest Cr content in grain. The Iranian genotype 'Arg' exhibits the greatest boron (B) tolerance, whereas 'Chamran' is the most sensitive. Under Pb stress, 'Akbar' shows the highest tolerance. 'Gimmiza_11' and 'Benisweif_7' are distinguished by their tin (Sn) and Pb tolerance, respectively. 'Lasani-2008' and 'Iqbal-2000' have the lowest shoot Cd concentrations.

Integrated screening pipelines that incorporate morphological, physiological, biochemical, analytical, and molecular profiling provide a comprehensive strategy for identifying heavy metal-tolerant wheat genotypes. These strategies support the development of cultivars adapted to contaminated agroecosystems, promote stable yields, reduce grain metal content, and improve food safety.

### 12.6.2 Identification of Genetic Variation and QTL Mapping for Heavy Metal Tolerance in Wheat

Genetic variation forms the foundation of crop improvement efforts. Classification of wheat genotypes based on heavy metal deposition, biochemical responses, and health risk assessments has identified tolerant, moderately sensitive, and highly sensitive groups to heavy metal stress (Ali et al. 2018). Significant natural variation in heavy metal tolerance exists in modern bread wheat germplasm, landraces, wild relatives, and synthetic hexaploids. These groups represent a critical genetic resource for breeding programs aimed at improving resilience.

GWAS and QTL mapping have identified multiple loci regulating concentrations of cadmium (Cd), zinc (Zn), copper (Cu), and iron (Fe) in wheat grains and tissues. Genes from the heavy metal ATPase (HMA) and natural resistance-associated macrophage protein (NRAMP) families often co-localize with these QTLs and serve as reliable markers for selection (Ban et al. 2020). Furthermore, wild relatives, including *Triticum* spp. and *Aegilops* spp., contain novel alleles absent in elite cultivars, providing additional sources of tolerance (Pour-Aboughadareh et al. 2021).

### 12.6.3 Role of Landraces and Wild Relatives, Synthetic Wheats, and Alien Introgressions

Landraces constitute a critical yet underexploited source of genetic diversity for wheat improvement. Although typically associated with lower yields, landraces have adapted to a wide range of agroecological environments and exhibit extensive allelic variation that confers stress tolerance, yield stability, and adaptive traits (Lopes et al. 2015). Both bread wheat (*Triticum aestivum* L.) and durum wheat (*Triticum durum* Desf.) landraces exhibit substantially greater genetic diversity than modern cultivars, as demonstrated by broad morphological, physiological, biochemical, and molecular variation in germplasm from regions such as India, Ethiopia, Morocco, Spain, China, and Uzbekistan (Zencirci et al. 1994; Barbosa Neto and de Carvalho 2002; Adhikari et al. 2022). However, processes of domestication and intensive breeding have resulted in considerable genetic erosion, reducing diversity by approximately 84% in durum wheat and 69% in bread wheat (Dvorak et al. 1998). The Green Revolution further reduced the genetic base by approximately 75% through the widespread replacement of landraces with semi-dwarf cultivars (Rufo et al. 2019). As a result, wheat currently exhibits greater genetic vulnerability than most other cereals, posing significant long-term threats to food and nutritional security.

Strategic use of landraces is crucial for wheat improvement. Landraces can be efficiently crossed with elite cultivars and provide valuable sources of novel alleles for pre-breeding programs (Lopes et al. 2015). Advances in high-throughput genotyping and phenotyping have accelerated the identification of beneficial alleles. Technologies such as single-nucleotide polymorphism (SNP) arrays, genotyping-by-sequencing (GBS), QTL and meta-QTL mapping, GWAS, resequencing, and exome capture facilitate the targeted utilization of rare, trait-associated alleles (Adhikari et al. 2022; Jaiswal et al. 2024). Notable examples include the discovery of glutenin subunit genes in Tibetan landraces (Lan et al. 2013) and grain-quality alleles in Mexican, Mediterranean, and Indian germplasm. Furthermore, direct cultivar development has been accomplished using landrace-derived alleles, as demonstrated by drought-tolerant 'Aragon 03' (Spain), cold-tolerant 'Turkey' (USA), and the semi-dwarfing allele *Rht8c* from the Japanese landrace 'Aka Komugi' (Ellis et al. 2007). Beyond landraces, wild relatives, synthetic hexaploid wheats (SHWs), and alien introgressions have been instrumental in broadening the wheat gene pool for abiotic stress tolerance. SHWs (AABBDD), generated by crossing *T. durum* with *Aegilops tauschii* Coss., restore critical D-genome diversity and have proven especially valuable for improving heavy metal tolerance, including aluminium (Al) resistance (Mujeeb-Kazi et al. 1996). The development of diagnostic markers for the major Al-tolerance gene *TaALMT1* on chromosome 4DL and related loci such as *TaMATE1B* has advanced marker-assisted selection. Genome-wide association studies in SHWs have identified numerous additional loci and candidate genes, such as ATP-binding cassette (ABC) transporters, glutathione synthetase, and C2H2-type zinc-finger transcription factors (Ma et al. 2005; Raman et al. 2008; Sasaki et al. 2014, 2016; Emebiri et al. 2020). Evidence from cultivars including BH1146, Atlas 66, and Chinese Spring highlights the importance of both established and novel regions on 4DL (Luo and Dvořák 1996; Riede and Anderson 1996; Raman et al. 2005). Additional improvements have been achieved through alien introgressions, such as rye-derived loci for Al tolerance and NAC transcription factors from *Aegilops markgrafii* (Greuter) Hammer associated with reduced cadmium accumulation, with further potential from *Thinopyrum* spp. and barley (*Hordeum vulgare* L.). The integration of landraces, wild relatives, SHWs, and alien germplasm, combined with advanced molecular tools, provides a comprehensive strategy for developing high-yielding, metal-tolerant wheat cultivars that enhance resilience, food safety, and long-term food security.

### 12.6.4 QTL Mapping and Genomics-Assisted Breeding for Heavy Metal Tolerance

The use of high-density molecular markers has markedly enhanced the resolution of QTL mapping for heavy-metal tolerance in wheat, enabling precise identification of relevant loci and candidate genes. The most thoroughly characterized locus is *Cdu-B1* (*Cdu1*) on chromosome 5BL, which encodes the vacuolar cadmium (Cd) transporter *TdHMA3-B1*. This transporter sequesters Cd within root vacuoles, restricting its translocation to shoots and grain. The nonfunctional allele *TdHMA3-B1b* results in elevated Cd accumulation, whilst homologous regions on chromosome 5A harbour single-nucleotide polymorphisms (SNPs) linked to low-Cd phenotypes. Additional QTLs affecting Cd concentration, such as *QGr_Cd_Conc-4B*, have been identified, with *TpCCX2-4B* suggested as a candidate gene for Cd homeostasis. Major-effect QTLs, including *QCdu.ndsu-5B* and markers like IWA1775, explain a substantial proportion of grain Cd variation and have been incorporated into breeder-friendly assays. The *Cdu1* region remains a central focus for global low-Cd wheat breeding programs.

Beyond chromosome 5B, multiple loci contribute to tolerance against Cd, arsenic (As), and other heavy metals. In bread wheat, *QCdc.4B-kita*, *QCdc.6B-kita1*, and *QCdc.6B-kita2* together account for 9–25% of Cd variation (Ban et al. 2020). Two stable arsenic-associated loci, *QGrAsc.sau-AM-1A* and *QGrAsc.sau-AM-4A*, have been identified in populations derived from wild and cultivated crosses. Multielement mapping has revealed additional loci influencing the accumulation of Cd, zinc (Zn), and iron (Fe), including *SCd-1A*, *qSCd-1D*, *qZn-2B1*, *qZn-2B2*, and *qFe-6D*, underscoring the polygenic basis of heavy metal tolerance. For aluminium toxicity, the major loci are *Qalt.pser-4DL*, *Qalt.pser-3BL*, and *Qalt.pser-2A*, which collectively explain approximately 82% of phenotypic variation in Chinese landraces, with *TaALMT1* on 4DL mediating malate efflux and aluminium detoxification (Ma et al. 2005; Cai et al. 2009a, b). Boron tolerance is conferred by the *Bo1* locus on 7BL, which restricts boron accumulation. Genome-wide analyses have also identified 28 *TaHMA* genes distributed across Cu/Ag and Zn/Co/Cd/Pb subfamilies, illustrating the complexity of heavy metal transport mechanisms (Shao et al. 2025).

Recent population-based mapping studies have expanded the understanding of the genetic architecture underlying heavy metal tolerance. For example, QTLs associated with root biomass under Cd stress were identified in a 300-line recombinant inbred line (RIL) population derived from H461 and Chinese Spring (Zhou et al. 2025). A mixed-model GWAS in a 200-line spring wheat panel (WAMI) identified SNP-trait associations for manganese (Mn), Fe, copper (Cu), Zn, nickel (Ni), and Cd concentrations (El-Soda and Aljabri 2022). These findings reveal considerable allelic diversity amongst landraces, wild relatives, and elite germplasm, providing a robust foundation for genomics-assisted breeding. This strategy supports the integration of mapped loci and diagnostic markers into wheat cultivars with reduced heavy metal accumulation and enhanced tolerance in contaminated environments.

### 12.6.5 Genomics-Assisted Breeding for Heavy Metal Tolerance and Reduced Grain Metal Accumulation

Recent advances in molecular genetics, functional genomics, and precision breeding have substantially improved the development of wheat cultivars with enhanced heavy-metal tolerance and reduced grain contamination. Whilst traditional breeding remains relevant, its effectiveness is constrained by the polygenic nature of heavy-metal tolerance, significant genotype-by-environment interactions, and lengthy selection cycles. The integration of molecular markers, GWAS, genomic selection (GS), and genome editing within genomics-assisted breeding (GAB) has accelerated genetic improvement (Jaiswal et al. 2024). When combined with advanced phenotyping, transcriptomics, and metabolomics, GAB now addresses multiple objectives, including yield, quality, food safety, and environmental sustainability. A principal advantage of GAB is the utilization of loci and markers identified through QTL mapping. Marker-assisted selection (MAS) for *HMA3*-based alleles at the *Cdu-B1* locus on chromosome 5B accounts for up to 20% of the variation in grain cadmium (Cd) content (Liu et al. 2019). Functional and nonfunctional *TdHMA3-B1* alleles are distinguished by a 17-base pair insertion, and Kompetitive Allele Specific PCR (KASP) markers such as Xusw59 achieve over 96% classification accuracy in durum wheat (*Triticum durum* Desf.) (AbuHammad et al. 2016; Knox et al. 2009). These assays are now standard in North American breeding programs. In bread wheat, *TaHMA3* is less influential; instead, major QTLs on chromosomes 4B and 6B (*QCdc.4B-kita*, *QCdc.6B-kita1*, and *QCdc.6B-kita2*) explain up to 25% of the variation (Ban et al. 2020).

In addition to Cd, aluminium tolerance has advanced through genomics-guided strategies. The *TaALMT1* gene on chromosome 4D, first identified in the cultivar Atlas 66, encodes a malate efflux transporter that binds aluminium ions in the rhizosphere. Markers flanking this region enable rapid introgression into sensitive cultivars, reducing breeding time by half. Rye (*Secale cereale* L.) possesses complementary aluminium-tolerance loci that have been incorporated into wheat–rye hybrids, thereby expanding the genetic base. The introgression of *TaALMT1* and *TaMATE1B* from bread wheat into durum wheat has improved tolerance under acidic soil conditions, with *TaMATE1B* facilitating citrate efflux (Han et al. 2016). Similarly, the functional *HMA3* allele from wild emmer wheat (*Triticum dicoccoides* Körn., Zavitan accession) introduced into durum wheat has significantly reduced grain Cd levels (Maccaferri et al. 2019).

GWAS and comparative genomics continue to identify loci and pathways involved in detoxification, chelation, and metal transport (Ma et al. 2005). Cross-species analyses have revealed conserved tolerance genes, facilitating candidate gene validation and translational applications. Genome editing technologies, particularly CRISPR/Cas9, now enable precise modification of these genes. For instance, targeted editing of *TaNRAMP5* has reduced Cd uptake without compromising yield (Cheng et al. 2025). These precision interventions are transforming metal-tolerance breeding.

Modern breeding pipelines combine GAB with high-throughput phenotyping and multi-omics datasets to more effectively capture genotype–environment interactions. This strategy has generated multi-trait indices that balance low grain-metal content with yield, stress tolerance, and disease resistance (Zhou and Zheng 2022). There is also an emphasis on incorporating phytoexclusion and phytostabilization traits, which restrict soil-metal uptake or immobilize metals within roots. Genes such as *TaALMT1* and *TaMATE1B* promote phytoexclusion by chelating aluminium ions in the rhizosphere, whilst *TaHMA3* and its rice (*Oryza sativa* L.) ortholog *OsHMA3* mediate vacuolar sequestration of Cd for phytostabilization (Maccaferri et al. 2019; Zhang et al. 2020). Enhanced expression of phytochelatin synthases and metallothioneins further increases metal immobilization. Low-Cd durum wheat lines carrying functional *HMA3* alleles are now selected directly using molecular markers rather than chemical assays. In bread wheat, new loci such as *QCd.sicau-1D* have been converted into KASP assays for breeding programs (Zhou et al. 2025). Similarly, single-nucleotide polymorphisms (SNPs) on chromosomes 4B and 6B identified by Ban et al. (2020) serve as predictive markers. MAS and GS are now standard molecular approaches for managing Cd stress in wheat. For instance, Pakistani breeding programs use natural low-Cd cultivars such as 'Lasani-2008' and 'Iqbal-2000', which are likely to carry favourable alleles. Manipulation of transcription factors offers additional control; for example, overexpression of *TaWRKY74* reduces Cd translocation by downregulating transporter genes and activating the ascorbate–glutathione detoxification pathway. Field studies have demonstrated genotypic differences: under Cd stress, the cultivar 'Mustang' retains more Cd in roots than 'Lancer', resulting in lower grain Cd whilst maintaining nutrient balance (Halim et al. 2021). This physiological variation is now directly utilized in breeding programs.

In addition to genetic approaches, integrated strategies increasingly employ biological solutions. Metal-tolerant plant-growth-promoting rhizobacteria (PGPR) immobilize metals through exopolysaccharide secretion and redox reactions, complementing plant-based exclusion mechanisms (Rizvi et al. 2020). The combination of PGPR inoculation with tolerant genotypes enhances phytostabilization efficiency and supports sustainable agriculture.

### 12.6.6 Field Evaluation and Cultivar Development for Heavy Metal Tolerance

Recent progress in breeding for heavy metal tolerance in wheat is being translated into field evaluations and the development of candidate cultivars; however, the number of officially released varieties remains limited. Global screening initiatives have identified several low-metal genotypes, particularly those tolerant to cadmium (Cd) and aluminium (Al). Extensive trials in China, Europe, and Americas have evaluated numerous wheat genotypes for low grain Cd accumulation under field conditions. In durum wheat, advanced breeding lines containing the functional *HMA3* allele from the wild emmer accession Zavitan are undergoing multi-location testing. In South America, particularly Brazil, aluminium-tolerant cultivars such as 'Toropi' and 'Carazinho' serve as foundational parents in breeding programs. Both cultivars possess the 4D *TaALMT1* allele, which mediates malate efflux and confers enhanced Al tolerance under acidic soil conditions. Similarly, in North America and Australia, durum cultivars derived from low-Cd parental lines carrying the *Cdu1* locus (*TdHMA3-B1*) are recommended for cultivation in Cd-contaminated regions (Li et al. 2023). These examples demonstrate the successful translation of molecular discoveries into practical breeding outcomes. Despite these advancements, breeding for heavy metal tolerance remains challenging. Phenotyping for metal accumulation is resource-intensive and requires specialized analytical facilities, and natural genetic variation for many metal-tolerance traits is limited. Consequently, marker-assisted selection (MAS) methods are increasingly used to accelerate breeding and the release of varieties. For Cd tolerance, the MAS Wheat program at UC Davis provides validated markers and selection protocols, such as the KASP marker Xusw59, to facilitate rapid identification of low-Cd durum lines (Maccaferri et al. 2019). Although low-arsenic (As) rice cultivars have been released in South and Southeast Asia, comparable wheat varieties have not been developed because wheat grain is less commonly contaminated with As. Ongoing screening of Indian and Bangladeshi germplasm, however, indicates considerable variation in As uptake potential. Currently, most released wheat cultivars are region-specific and display partial rather than complete tolerance to heavy metals. Future breeding programs aim to integrate low-metal accumulation with yield stability, disease resistance, and abiotic stress tolerance through multi-trait selection strategies. A common method involves hybridizing low-Cd parental lines with high-yielding elite cultivars, followed by marker-assisted backcrossing using *Cdu1*-linked markers (Maccaferri et al. 2019). The combined application of genomic selection and MAS is now standard in advanced breeding pipelines, ensuring the incorporation of metal-tolerance traits alongside superior agronomic performance. The development of heavy metal-tolerant cultivars is increasingly facilitated by advances in genomics and genome editing technologies. CRISPR/Cas9-mediated knockout of metal transporter genes, such as *TaNRAMP5*, offers a promising strategy for generating low-Cd wheat lines. Furthermore, GWAS and speed breeding techniques are expediting the pyramiding of tolerance QTLs across multiple metals. These approaches are shortening breeding timelines and enabling the deployment of improved, low-metal cultivars adapted to diverse agroecological environments. In India, heavy-metal stress in wheat has become a priority concern due to soil acidification in parts of the Northeast and the Indo-Gangetic Plain, as well as contamination from industrial effluents and arsenic-laden irrigation water. Surveys in regions such as Bihar and Punjab have demonstrated that wheat plants can accumulate Cd and As from polluted soils and groundwater (Halim et al. 2021). As a result, Indian wheat breeding programs are now incorporating metal tolerance traits into mainstream improvement pipelines. Institutions including the Indian Institute of Wheat and Barley Research (IIWBR), ICAR–IARI, and CSIR laboratories are actively screening Indian cultivars and landraces under both hydroponic and field conditions. Marker-assisted tools developed internationally, such as those linked to *TaHMA3/Cdu1* (Maccaferri et al. 2019) and *TaALMT1* (Han et al. 2016), are being applied to Indian breeding populations. Current efforts focus on identifying local cultivars with inherently low Cd or As accumulation and utilizing exotic donors, including synthetic wheats with *TaALMT1*-mediated Al tolerance, in cross-breeding programs. Although no officially released 'heavy-metal-tolerant' wheat variety has been released in India to date, several breeding programs have adopted these traits as selection criteria. The primary objective is to combine phytoexclusion traits, such as root sequestration and limited translocation, with traditional stress resistance and yield stability. Given India's agroecological diversity, future cultivars are expected to carry multiple tolerance alleles, including *HMA3*, *ALMT1*, and *MATE1B*, stacked with native resilience genes for drought and heat adaptation.

## 12.7 Evolution in the Contemporary Hi-Tech Era

The accumulation of toxic metals and metalloids, such as cadmium (Cd), lead (Pb), and arsenic (As), in wheat grains poses a major challenge to global agriculture by endangering human health and disrupting the balance of essential micronutrients, including iron (Fe), zinc (Zn), manganese (Mn), and copper (Cu). Advanced methodologies are now employed to elucidate the genetic mechanisms underlying tolerance, modify uptake and translocation pathways, and accelerate the breeding of wheat cultivars suited to contaminated environments. The use of next-generation sequencing (NGS) platforms and GWAS has provided substantial understanding of the genetic loci controlling metal accumulation in wheat. The Wheat Association Mapping Initiative (WAMI), for example, has profiled seed concentrations of Fe, Zn, Mn, Cu, Ni, and Cd, identifying over 100 significant marker–trait associations, including pleiotropic loci that affect multiple metals simultaneously (El-Soda and Aljabri 2022). These findings highlight the polygenic nature of metal homeostasis and reveal candidate genomic regions for breeding integration. Recent GWAS on kernel traits under Pb and Sn stress have identified loci such as *TraesCS2A02G080700* and *TraesCS3A02G516400*, which encode ABC transporters involved in kernel development and stress response. Additionally, studies emphasize the significance of genotype-by-environment interactions, as marker–trait associations for micronutrient content often vary under heat or drought stress, highlighting the need to breed for resilience across diverse agroecological conditions. Whilst GWAS enables the identification of natural allelic variation, genome editing offers direct strategies to modify key genes involved in metal uptake and transport. In wheat, CRISPR/Cas9-mediated knockout of *TaNRAMP5* has been shown to reduce Cd accumulation in grains without negatively impacting growth (Cheng et al. 2025), representing a significant advancement in functional editing for food safety. Similar strategies can be applied to members of the Heavy Metal ATPase (HMA) family, such as *TaHMA2* and *TaHMA3*, which mediate xylem loading and vacuolar sequestration of Cd and Zn (Zhu et al. 2020; Shao et al. 2025). The polyploid nature of the wheat genome requires careful management of homoeologous gene copies across the A, B, and D genomes. However, multiplex editing techniques are increasingly facilitating the reengineering of complex pathways (Afroz et al. 2023). Multi-omics approaches, including genomics, transcriptomics, proteomics, metabolomics, miRNAomics, phenomics, and epigenomics, have emerged as transformative tools for deciphering the molecular processes underlying plant defence against heavy metals. Integrating these datasets enables the identification of key genes, proteins, metabolites, and regulatory RNAs that confer tolerance, supporting the development of resilient and sustainable crop varieties for cultivation in contaminated environments. In *Triticum aestivum*, heavy metal-responsive miRNAs regulate transcription factors and detoxification networks under Cd and Pb stress. Metabolomics studies further reveal that treatments such as oxide nanoparticles can reprogram antioxidant metabolism, thereby enhancing Cd resistance in wheat seedlings (Charagh et al. 2024). Systems-level data facilitate network analyses that identify candidate regulators and biomarkers, bridging the gap between gene discovery and functional validation. For example, multi-omics analyses of wheat under Cu stress have identified candidate HMAs and associated redox pathways as central components of tolerance mechanisms (Batool et al. 2023). Efforts to screen wheat germplasm for metal tolerance have also been accelerated by high-throughput phenotyping (HTP). Whilst destructive chemical assays such as ICP–MS remain the gold standard for quantifying grain metal concentrations, hyperspectral imaging and machine-learning-based chemometrics are emerging as nondestructive alternatives for early detection of Pb stress signatures (Zhang et al. 2022). Large panels, including advanced spring wheat lines, have been phenotyped for kernel size, shape, and weight under Pb and Sn stress, revealing stress-specific alleles that can guide pre-breeding. Such phenotyping platforms, when coupled with GWAS and genomic prediction, provide a scalable route to identifying superior genotypes from large germplasm sets. Artificial intelligence (AI) and machine learning (ML) models are increasingly applied to enhance predictive breeding under metal stress conditions. For instance, a boosted regression tree (BRT) model developed to predict heavy metal accumulation in wheat demonstrated that Pb and As inputs from atmospheric dustfall significantly contribute to their accumulation in grains, whilst available Cd in soil is the primary determinant of Cd accumulation. Bioavailable fractions of heavy metals in soil exert a much greater influence on grain content than total concentrations (Meng et al. 2024). Multi-trait ML and deep learning models further support the analysis of genotype–environment interactions, enabling breeders to predict the performance of candidate lines across diverse contaminated environments (Sandhu et al. 2021). Integrating genomics, environmental characterization, and AI accelerates selection cycles and advances the goal of sustainable cultivation in marginal or polluted soils. Despite these advances, significant challenges remain in wheat breeding for metal tolerance. The hexaploid genome complicates editing and fine-mapping, whereas the moderate heritability of metal-accumulation traits necessitates replication across multiple environments. Furthermore, breeding must carefully balance reducing toxic metal uptake with enhancing essential micronutrient content to avoid compromising nutritional quality. Nevertheless, the combined power of NGS-enabled discovery, CRISPR/Cas editing, multi-omics integration, HTP, and AI-driven predictive models provides a robust toolkit for modern wheat improvement.

## 12.8 Statistical Insight into Wheat Breeding for Metal and Metalloid Stress Tolerance

A comprehensive understanding of metal and metalloid stress in wheat necessitates integrating field-level contamination data, yield and food-safety assessments, and adoption statistics to guide breeding priorities. This section offers a statistical analysis of contamination trends, agronomic impacts, varietal responses, and policy-driven mitigation strategies that influence contemporary wheat improvement programs.

### 12.8.1 Regional and Global Soil Contamination and Toxicity Patterns

Regional soil and crop analyses indicate that heavy-metal contamination is primarily concentrated in northwestern and industrially affected regions of India. Soils and wheat grains from Haryana, western Uttar Pradesh, and peri-urban Delhi NCR often exceed permissible limits for cadmium (Cd), lead (Pb), and nickel (Ni). Wastewater irrigation and industrial effluents are identified as the main sources of contamination (Verma et al. 2015; Elbagory et al. 2025). Studies in Saharanpur (Uttar Pradesh) identified Cd and chromium (Cr) as the principal pollutants associated with paper-mill effluent irrigation. Bioaccumulation and translocation factors demonstrate the uptake and movement of these metals into grain (Elbagory et al. 2025). Similar surveys in Punjab and Haryana reported exceedances of WHO/FAO limits. Transfer factors in wheat were comparable to those observed in leafy vegetables (Batool et al. 2023; Kaur et al. 2025). Seedling-based assessments reveal significant inhibition of root growth and cellular damage under combined Cd and Cr stress (Nazir et al. 2021). In contrast, bioaugmentation with the metal-tolerant *Pseudomonas aeruginosa* CPSB1 reduces tissue concentrations of Cd and Cr, highlighting the potential of rhizosphere-based mitigation strategies (Rizvi et al. 2020). Arsenic (As) toxicity, which is particularly prevalent in Indo-Gangetic rice–wheat rotations, remains a significant concern. Differential activation of the ascorbate–glutathione (AsA–GSH) pathway explains the variable tolerance observed amongst Indian and Bangladeshi cultivars (Saeed et al. 2024). Globally, large-scale soil datasets comprising over 790,000 samples indicate that 14–17% of croplands exceed safe thresholds for one or more toxic metals (Hou et al. 2025). Major contamination clusters coincide with wheat belts in Europe, North America, and East Asia, regions affected by legacy mining and smelting. In Japan, approximately 6700 hectares of farmland remain contaminated with Cd despite long-term remediation efforts (Ban et al. 2020). Industrial zones with extreme contamination report Pb and Cd levels reaching 35,000 and 47 mg/kg, respectively, emphasizing the persistence of anthropogenic contamination (Ondrasek et al. 2025).

### 12.8.2 Yield, Food Safety, and Economic Impacts

Heavy-metal contamination leads to substantial agronomic and economic losses. Cadmium (Cd) exposure reduces photosynthetic efficiency, chlorophyll fluorescence (ΦPSII, Fv/Fm), and pigment content, thereby decreasing biomass and yield (Ci et al. 2010). These physiological impairments are associated with oxidative stress, as indicated by increased reactive oxygen species (ROS) accumulation, lipid peroxidation, and membrane damage (Rahman and Upadhyaya 2021). Food safety remains a concern in contaminated regions. The Codex Alimentarius specifies a maximum Cd concentration of 0.2 mg $kg^{-1}$ in wheat grain, and the Joint FAO/WHO Expert Committee on Food Additives (JECFA) recommends a tolerable monthly intake of 25 μg Cd $kg^{-1}$ body weight (Chunhabundit 2016). Continued monitoring in Germany and Japan reveals persistent Cd contamination despite emission controls (Hüsken et al. 2024; Ban et al. 2020). In India, wastewater-irrigated sites frequently report soil and grain Cd and chromium (Cr) concentrations exceeding FAO limits, posing chronic exposure risks (Elbagory et al. 2025). Contamination-induced yield reductions and export restrictions substantially reduce profitability, especially for durum and bread wheat grown near industrial zones. Reducing metal accumulation through targeted genotype selection is critical to protecting farmers' incomes and maintaining compliance with food safety regulations.

### 12.8.3 Trends in Grain Metal Accumulation and Varietal Response

Soil acidity and irrigation source are critical determinants of metal bioavailability. In South Asia, effluent-irrigated soils result in cadmium (Cd) accumulation in wheat in the order root > stem > leaf > grain, whereas borewell irrigation is associated with lower Cd concentrations (Elbagory et al. 2025). Global meta-analyses report that 10–12% of wheat samples surpass Cd and lead (Pb) safety thresholds, a pattern commonly associated with soil pH values below 6.0. Genotype significantly influences metal accumulation in wheat. Low-Cd bread wheat cultivars, such as JM22 and JM229, consistently exhibit 30–40% lower grain Cd concentrations than high-Cd cultivars such as SX828 (Bai et al. 2023). Tolerant genotypes typically limit root uptake and restrict translocation from the xylem to the phloem, thereby helping maintain yield and mineral balance under moderate stress. In aluminium-toxic soils, resistant cultivars such as 'Yecora E' sustain higher photosynthetic performance by promoting ion homeostasis (Moustaka et al. 2016). Remote sensing and spectral reflectance indices are increasingly utilized to statistically predict pre-harvest grain Cd levels, offering a valuable tool for field-based monitoring. This nondestructive phenotyping method integrates molecular selection with practical risk assessment.

### 12.8.4 Adoption of Tolerant Varieties

Field-based adoption statistics indicate significant progress in multiple regions. In Canada, the median export cadmium (Cd) concentration in durum wheat declined from 0.16 to 0.07 mg $kg^{-1}$ between 2003 and 2020, largely attributed to the widespread adoption of *HMA3*-based low-Cd cultivars (Tittlemier and Richter 2022). In China, wheat varieties such as LX99, JM22, and JM229 consistently exhibit low Cd accumulation across various agroecological zones (Bai et al. 2023). In India, low-metal breeding is in its early stages but is gaining momentum. Current germplasm screening under the IIWBR and IARI programs is identifying indigenous cultivars with naturally reduced Cd or arsenic (As) uptake. Although no 'metal-tolerant' variety has been officially released, marker-assisted selection and genomic selection pipelines are integrating low-accumulation traits into region-specific adaptation programs. These efforts align with global strategies that emphasize multi-trait selection, including yield, stress resistance, and food safety.

### 12.8.5 Policy Frameworks and Breeder Initiatives

At the international level, the Codex Committee on Contaminants in Foods initiated the Code of Practice for Cadmium Control in 2024, with completion anticipated by 2027. This initiative seeks to harmonize standards and offers guidance on source monitoring, cultivar selection, and soil management.

In India, the policy framework is evolving substantially. The Central Pollution Control Board (CPCB) and the Central Water Commission (CWC) have revised water-quality standards to restrict the use of untreated effluent for irrigation and to establish trace-metal limits for agricultural reuse. The Draft National Policy on Safe Reuse of Treated Wastewater (2024) introduces end-use-specific benchmarks for irrigation quality and mandates periodic monitoring. These policies align with recommendations from Setia et al. (2021), who advocate for integrated soil–crop–water safety frameworks.

Breeder-led initiatives complement these policy measures. The All India Coordinated Research Project on Wheat and Barley (AICRP–W&B) now incorporates low-accumulation and biofortified traits in regional trials. Programs at the Indian Institute of Wheat and Barley Research (IIWBR) and the Indian Agricultural Research Institute (ICAR–IARI) integrate high-throughput phenotyping and molecular screening, and implement soil amendment strategies to mitigate metal toxicity in affected regions. Collectively, these efforts foster an enabling ecosystem that links molecular breeding, sustainable agronomy, and policy implementation.

Statistical evidence indicates that heavy metal contamination in wheat-growing regions poses significant agronomic and food safety challenges. Soil acidity, irrigation quality, and genotype are primary determinants of metal accumulation. Both global and national breeding programs have achieved measurable reductions in grain cadmium (Cd) through targeted variety adoption, although progress remains inconsistent. Integrating advanced phenotyping, regional contamination data, and coordinated policy frameworks is essential for sustainable mitigation. The integration of genomics, agronomy, and governance is now shaping the pathway towards safe and resilient wheat production in metal-affected agroecosystems.

## 12.9 Future Projections and Global Challenges

Managing metal and metalloid stress in wheat is expected to become increasingly complex. Global soil monitoring programs and meta-analyses demonstrate that toxic-metal contamination already affects a substantial portion of agricultural land. In the absence of coordinated mitigation, contamination is projected to increase. Analysis of nearly 800,000 soil samples indicates that approximately 14–17% of global cropland exceeds safety thresholds for at least one toxic metal, particularly arsenic, cadmium (Cd), and lead (Pb). The most affected regions are South and East Asia, industrial areas of Europe and North America, and landscapes with ongoing mining activities. Many of these regions overlap with major wheat-producing zones (Hou et al. 2025). These contamination patterns collectively threaten the food security and health of more than 1 billion people. Ongoing industrial emissions, the reuse of untreated wastewater, and land degradation are expected to intensify these risks.

The implications for wheat cultivation are multifaceted. Elevated environmental metal concentrations increase the risk of toxic element accumulation in grains, leading to chronic health outcomes such as renal impairment, developmental neurotoxicity, and cancer. Climate-driven soil changes, including acidification and altered hydrological regimes, further enhance metal solubility and bioavailability, which accelerates plant uptake. Therefore, breeding objectives should address resilience to multiple stresses rather than focusing solely on individual stressors. Developing multistress resilience is critical, as metal contamination often co-occurs with drought, salinity, heat stress, and pathogen pressure in degraded agroecosystems (Hou et al. 2025; Vullaganti et al. 2025).

Recent technological advancements have created new opportunities for monitoring and selection. Portable X-ray fluorescence (pXRF) allows for rapid, nondestructive field screening of soils and grains, with calibration against certified reference materials ensuring accuracy at low concentrations (Alqattan et al. 2024; Jenkins et al. 2025). Remote sensing techniques, including hyperspectral and multispectral imaging, utilize machine learning models to map contamination gradients and detect plant stress across extensive areas (Su et al. 2025; Wang et al. 2025). Gene-editing strategies, such as *TaNRAMP5* knockouts, have reduced Cd accumulation in wheat without significant yield penalties. Combining these approaches with genomic selection and conventional breeding offers a promising pathway to develop low-accumulating germplasm (Cheng et al. 2025).

Despite these advances, several structural barriers hinder widespread adoption. Phenotyping capacity is a persistent limitation. Accurate quantification of metals in plant tissues using inductively coupled plasma mass spectrometry (ICP–MS) or inductively coupled plasma optical emission spectrometry (ICP–OES) remains expensive and is not universally accessible. Although pXRF and spectral proxies provide scalable alternatives, their reliability varies across laboratories and environmental conditions, highlighting the need for standardized calibration and inter-institutional validation (Jenkins et al. 2025). Additionally, germplasm screening is incomplete. Global collections of landraces, wild relatives, and synthetic wheats harbour valuable alleles for metal exclusion, vacuolar sequestration, and antioxidant defence, yet only a small proportion have been evaluated under metal stress (Yan et al. 2020).

Governance and implementation efforts are inconsistent. Initiatives such as the Codex Alimentarius Code of Practice for Cadmium Reduction and national wastewater reuse policies represent significant progress. However, effective contamination control requires coordinated enforcement, comprehensive monitoring networks, and context-specific remediation strategies (Hou et al. 2025). Remediation and breeding programs are often disconnected. Interventions, including soil amendments, liming, organic matter application, and phytoremediation, can reduce bioavailable metal pools, but these improvements occur gradually. Integrating these interventions with the breeding and deployment of low-accumulating cultivars would accelerate risk reduction, especially when supported by regional soil–crop surveillance systems.

Advancement in this field requires systems-level integration. Standardizing phenotyping protocols to connect laboratory and field screening is essential. Sharing genotype–phenotype databases will facilitate genome-wide association studies and genomic prediction. Rapid field-based mapping should be implemented to inform the placement of breeding trials. Breeding priorities must be aligned with regulatory frameworks and market incentives. The deployment of gene-edited or low-accumulating cultivars should incorporate transparent risk assessment, comprehensive farmer support, and equitable access to seed and testing services to ensure that smallholders in contaminated regions are not marginalized. The scientific capacity to reduce grain contamination is now available; however, its effectiveness depends on integrating breeding, governance, remediation, monitoring, and farmer outreach at both local and global scales (Yan et al. 2020; Hou et al. 2025).

## 12.10 Conclusion

Recent advances in molecular biology, physiology, and genomics have significantly advanced understanding and breeding for wheat tolerance to metal and metalloid stress. These developments have elucidated the mechanisms governing the uptake, translocation, detoxification, and compartmentalization of toxic elements, including cadmium (Cd), lead (Pb), and aluminium (Al). The identification of key transporter genes (*HMA3*, *NRAMP5*, *ALMT1*, and *MATE1B*) and major-effect QTLs across the A, B, and D genomes has provided robust genetic targets for selection and introgression. The integration of genomics, transcriptomics, metabolomics, and ionomics with high-throughput phenotyping has enabled the development of diagnostic markers and predictive breeding tools. Consequently, approaches such as marker-assisted selection (MAS), GWAS, genomic selection, and CRISPR/Cas-based genome editing have shifted wheat improvement from empirical phenotypic selection to a mechanism-driven, data-centric discipline. The combination of precision breeding with rapid field diagnostics, including portable X-ray fluorescence and hyperspectral imaging, as well as genomic prediction, speed breeding, and multi-environment testing, has accelerated the transition from controlled discovery to field application. This progress is demonstrated by the release of low-Cd durum wheat in North America and Al-tolerant bread wheat in South America. Nevertheless, genetic improvement alone is insufficient to fully mitigate contamination risks. Advancements in soil and water management, phytoremediation, microbial bioaugmentation, organic carbon enhancement, pH regulation, and enforcement of wastewater and effluent standards are also necessary, supported by policy frameworks such as the Codex Code of Practice for Cadmium Reduction and India's National Policy on the Safe Reuse of Treated Wastewater (2024). Sustained progress will depend on coordinated efforts amongst breeders, soil scientists, agronomists, policymakers, and farmers to integrate physiological understanding, scalable remediation, optimized agronomy, and regulatory compliance. Looking ahead, the development of multi-stress-tolerant wheat cultivars that maintain yield, grain safety, and nutritional quality under combined metal toxicity, drought, heat, and salinity—supported by genomics-guided breeding, soil restoration, digital monitoring, and farmer outreach—offers a promising pathway toward resilient and sustainable wheat production in contaminated agroecosystems.

## References

AbuHammad WA, Mamidi S, Kumar A et al (2016) Identification and validation of a major cadmium accumulation locus and closely associated SNP markers in North Dakota durum wheat cultivars. Mol Breed 36:112. https://doi.org/10.1007/s11032-016-0536-1

Adhikari S, Kumari J, Jacob SR et al (2022) Landraces—potential treasure for sustainable wheat improvement. Genet Resour Crop Evol 69:499–523. https://doi.org/10.1007/s10722-021-01310-5

Afroz N, Ansary MWR, Islam T (2023) CRISPR-Cas genome editing for the development of abiotic stress-tolerant wheat. In: Khan MK, Pandey A, Hamurcu M et al (eds) Abiotic stresses in wheat. Academic Press, pp 195–207. https://doi.org/10.1016/B978-0-323-95368-9.00014-X

Ahanger MA, Aziz U, Sahli AA et al (2020) Combined kinetin and spermidine treatments ameliorate growth and photosynthetic inhibition in *Vigna angularis* by up-regulating antioxidant and nitrogen metabolism under cadmium stress. Biomolecules 10:147. https://doi.org/10.3390/biom10010147

Ajsuvakova OP, Tinkov AA, Aschner M et al (2020) Sulfhydryl groups as targets of mercury toxicity. Coord Chem Rev 417:213343. https://doi.org/10.1016/j.ccr.2020.213343

Aladesanmi OT, Oroboade JG, Osisiogu CP et al (2019) Bioaccumulation factor of selected heavy metals in *Zea mays*. J Health Pollut 9:191207. https://doi.org/10.5696/2156-9614-9.24.191207

Ali M, Abbasi BH, Ul Haq I (2013) Production of commercially important secondary metabolites and antioxidant activity in cell suspension cultures of *Artemisia absinthium* L. Ind Crop Prod 49:400–406. https://doi.org/10.1016/j.indcrop.2013.05.033

Ali Z, Mujeeb-Kazi A, Quraishi UM et al (2018) Deciphering adverse effects of heavy metals on diverse wheat germplasm on irrigation with urban wastewater of mixed municipal-industrial origin. Environ Sci Pollut Res Int 25:18462–18475. https://doi.org/10.1007/s11356-018-1996-0

Alqattan ZA, Artiola JF, Walls D et al (2024) Evaluating the portable X-ray fluorescence reliability for metal(loids) detection and soil contamination status. Environ Monit Assess 196:765. https://doi.org/10.21203/rs.3.rs-3414584/v1

Alsafran M, Saleem MH, Rizwan M et al (2022) An overview of heavy metals toxicity in plants, tolerance mechanism, and alleviation through lysine-chelation with micronutrients—a novel approach. Plant Growth Regul 100:337–354. https://doi.org/10.1007/s10725-022-00940-8

Altaf MA, Hao Y, Shu H et al (2023) Melatonin enhanced the heavy metal-stress tolerance of pepper by mitigating oxidative damage and reducing heavy metal accumulation. J Hazard Mater 454:131468. https://doi.org/10.1016/j.jhazmat.2023.131468

Alybayeva R, Kruzhaeva V, Alenova A et al (2016) The genetic potential of wheat resistance to heavy metals. Bioeng Biosci 4:34–41. https://doi.org/10.13189/bb.2016.040302

Anas M, Bashir MS, Saleem K et al (2024) Adaptive mechanisms of wheat cultivars to lead toxicity through enhanced oxidative defense, ionomic redistribution, and anatomical modifications. J Plant Physiol 303:154370. https://doi.org/10.1016/j.jplph.2024.154370

Aprile A, Sabella E, Francia E et al (2019) Combined effect of cadmium and lead on durum wheat. Int J Mol Sci 20:5891. https://doi.org/10.3390/ijms20235891

Asati A, Pichhode M, Nikhil K (2016) Effect of heavy metals on plants: an overview. Int J Appl Innov Eng Manag 5:56–66

Austruy A, Shahid M, Xiong T et al (2014) Mechanisms of metal-phosphates formation in the rhizosphere soils of pea and tomato: environmental and sanitary consequences. J Soils Sediments 14:666–678. https://doi.org/10.1007/s11368-014-0862-z

Baekgaard L, Mikkelsen MD, Sørensen DM et al (2010) A combined zinc/cadmium sensor and zinc/cadmium export regulator in a heavy metal pump. J Biol Chem 285:31243–31252. https://doi.org/10.1074/jbc.M110.111260

Bai L, Ding S, Huang X et al (2023) Prediction of the cadmium content in grains of low-accumulating wheat cultivars and soil cadmium threshold for safe production. J Clean Prod 417:138081. https://doi.org/10.1016/j.jclepro.2023.138081

Ban Y, Ishikawa G, Ueda H et al (2020) Novel quantitative trait loci for low grain cadmium concentration in common wheat (*Triticum aestivum* L.). Breed Sci 70:331–341. https://doi.org/10.1270/jsbbs.19150

Barbosa Neto JF, de Carvalho FIF (2002) Genetic variability in common wheat germplasm based on COP. Genet Mol Biol 25:211–215. https://doi.org/10.1590/S1415-47572002000200015

Batool TS, Aslam R, Gul A et al (2023) Genome-wide analysis of heavy metal ATPases (HMAs) in Poaceae species and their potential role against copper stress in *Triticum aestivum*. Sci Rep 13:7551. https://doi.org/10.1038/s41598-023-32023-7

Benavides MP, Gallego SM, Tomaro ML (2005) Cadmium toxicity in plants. Braz J Plant Physiol 17:21–34. https://doi.org/10.1590/S1677-04202005000100003

Bhati KK, Sharma S, Aggarwal S et al (2015) Genome-wide identification and expression characterization of ABCC-MRP transporters in hexaploid wheat. Front Plant Sci 6:488. https://doi.org/10.3389/fpls.2015.00488

Brunetti P, Zanella L, De Paolis A et al (2015) Cadmium-inducible expression of the ABC-type transporter AtABCC3 increases phytochelatin-mediated cadmium tolerance in *Arabidopsis*. J Exp Bot 66:3815–3829. https://doi.org/10.1093/jxb/erv185

Cai Q, Long ML, Zhu M et al (2009a) Food chain transfer of cadmium and lead to cattle in a lead–zinc smelter in Guizhou, China. Environ Pollut 157:3078–3082. https://doi.org/10.1016/j.envpol.2009.05.048

Cai L, Xu Z, Ren M et al (2009b) Source identification of eight hazardous heavy metals in agricultural soils of Huizhou, Guangdong Province, China. Ecotoxicol Environ Saf 78:2–8. https://doi.org/10.1016/j.ecoenv.2011.07.004

Cao HW, Gao XY, Wang K et al (2025) Metallothionein TaMT1f-1B has dual functions in enhancing cadmium and salt tolerance of wheat (*Triticum aestivum* L.). J Hazard Mater 495:138936. https://doi.org/10.1016/j.jhazmat.2025.138936

Chang JD, Gao W, Wang P et al (2022) OsNRAMP5 is a major transporter for lead uptake in rice. Environ Sci Technol 56:17481–17490. https://doi.org/10.1021/acs.est.2c06384

Charagh S, Wang H, Wang J et al (2024) Leveraging multi-omics tools to comprehend responses and tolerance mechanisms of heavy metals in crop plants. Funct Integr Genomics 24:194. https://doi.org/10.1007/s10142-024-01481-1

Chaudhary K, Jan S, Khan S (2016) Heavy metal ATPase (HMA2, HMA3, and HMA4) genes in hyperaccumulation mechanism of heavy metals. In: Ahmad P (ed) Plant metal interaction. Elsevier, pp 545–556

Chen Y, Han Y-H, Cao Y et al (2017) Arsenic transport in rice and biological solutions to reduce arsenic risk from rice. Front Plant Sci 8:268. https://doi.org/10.3389/fpls.2017.00268

Cheng Z, Wei J, Zhu B et al (2025) Mutation of TaNRAMP5 impacts cadmium transport in wheat. Plant Physiol Biochem 223:109879. https://doi.org/10.1016/j.plaphy.2025.109879

Chmielowska-Bąk J, Izbiańska K, Ekner-Grzyb A et al (2018) Cadmium stress leads to rapid increase in RNA oxidative modifications in soybean seedlings. Front Plant Sci 8:2219. https://doi.org/10.3389/fpls.2017.02219

Choudhary KM, Jat HS, Nandal DP et al (2018) Evaluating alternatives to rice–wheat system in western Indo-Gangetic Plains: crop yields, water productivity and economic profitability. Field Crop Res 218:1–10. https://doi.org/10.1016/j.fcr.2017.12.023

Chunhabundit R (2016) Cadmium exposure and potential health risk from foods in contaminated area, Thailand. Toxicol Res 32:65–72. https://doi.org/10.5487/TR.2016.32.1.065

Ci D, Jiang D, Wollenweber B et al (2010) Cadmium stress in wheat seedlings: growth, cadmium accumulation and photosynthesis. Acta Physiol Plant 32:365–373. https://doi.org/10.1007/s11738-009-0414-0

Clarke JM, Clarke FR, Pozniak CJ et al (2005) Strongfield durum wheat: combining high yield, protein quality, and low cadmium accumulation. Can J Plant Sci 85:651–654. https://doi.org/10.4141/P04-119

Clemens S (2006) Toxic metal accumulation, responses to exposure and mechanisms of tolerance in plants. Biochimie 88:1707–1719. https://doi.org/10.1016/j.biochi.2006.07.003

Clemens S, Kim EJ, Neumann D et al (1999) Tolerance to toxic metals by a gene family of phytochelatin synthases from plants and yeast. EMBO J 18:3325–3333. https://doi.org/10.1093/emboj/18.12.3325

Cobbett C, Goldsbrough P (2002) Phytochelatins and metallothioneins: roles in heavy metal detoxification and homeostasis. Annu Rev Plant Biol 53:159–182. https://doi.org/10.1146/annurev.arplant.53.100301

Debnath B, Singh WS, Manna K (2019) Sources and toxicological effects of lead on human health. Indian J Med Spec 10:66. https://doi.org/10.4103/INJMS.INJMS_30_18

Dey U, Mondal NK (2016) Ultrastructural deformation of plant cell under heavy metal stress in gram seedlings. Cogent Environ Sci 2:1196472. https://doi.org/10.1080/23311843.2016.1196472

Dhaliwal SS, Singh J, Taneja PK et al (2020) Remediation techniques for removal of heavy metals from the soil contaminated through different sources: a review. Environ Sci Pollut Res 27:1319–1333. https://doi.org/10.1007/s11356-019-06967-1

Duan W, Lu F, Cui Y et al (2022) Genome-wide identification and characterization of wheat MATE genes reveal their roles in aluminium tolerance. Int J Mol Sci 23:4418. https://doi.org/10.3390/ijms23084418

Dvorak J, Luo MC, Yang ZL (1998) Genetic evidence on the origin of *Triticum aestivum* L. In: Damania AB et al (eds) The origins of agriculture and crop domestication. ICARDA, pp 235–251

Ekmekçi Y, Tanyolaç D, Ayhan B (2009) A crop tolerating oxidative stress induced by excess lead: maize. Acta Physiol Plant 31:319–330. https://doi.org/10.1007/s11738-008-0238-3

Elbagory M, Zayed A, El-Khateeb N et al (2025) Risk assessment of potentially toxic heavy metals in wheat (*Triticum aestivum* L.) grown in soils irrigated with paper mill effluent. Toxics 13:497. https://doi.org/10.3390/toxics13060497

El-Kady AA, Abdel-Wahhab MA (2018) Occurrence of trace metals in foodstuffs and their health impact. Trends Food Sci Technol 75:36–45. https://doi.org/10.1016/j.tifs.2018.03.001

Ellis MH, Bonnett DG, Rebetzke GJ (2007) A 192 bp allele at the Xgwm261 locus is not always associated with the *Rht8* dwarfing gene in wheat (*Triticum aestivum* L.). Euphytica 157:209–214. https://doi.org/10.1007/s10681-007-9413-7

El-Soda M, Aljabri M (2022) Genome-wide association mapping of grain metal accumulation in wheat. Genes 13:1052. https://doi.org/10.3390/genes13061052

Emamverdian A, Ding Y, Mokhberdoran F et al (2015) Heavy metal stress and some mechanisms of plant defense response. Sci World J 2015:756120. https://doi.org/10.1155/2015/756120

Emebiri LC, Raman H, Ogbonnaya FC (2020) Synthetic hexaploid wheat as a source of novel genetic loci for aluminium tolerance. Euphytica 216:135. https://doi.org/10.1007/s10681-020-02669-9

Faizan M, Alam P, Hussain A et al (2024) Phytochelatins: key regulator against heavy metal toxicity in plants. Plant Stress 11:100355. https://doi.org/10.1016/j.stress.2024.100355

Fang Y, Nie Z, Liu F et al (2014) Concentration and health risk evaluation of heavy metals in market-sold vegetables and fishes based on questionnaires in Beijing, China. Environ Sci Pollut Res 21:11401–11408. https://doi.org/10.1007/s11356-014-3127-x

Gadd GM (2007) Geomycology: biogeochemical transformations of rocks, minerals, metals and radionuclides by fungi, bioweathering and bioremediation. Mycol Res 111:3–49. https://doi.org/10.1016/j.mycres.2006.12.001

Gill SS, Anjum NA, Ahmad I et al (2012) Metal hyperaccumulation and tolerance in *Alyssum*, *Arabidopsis* and *Thlaspi*: an overview. In: Anjum N, Ahmad I, Pereira M et al (eds) The plant family Brassicaceae. Environmental pollution, vol 21. Springer, Dordrecht. https://doi.org/10.1007/978-94-007-3913-0_4

Gomes MP, Nogueira MDOG, Castro EM et al (2011) Ecophysiological and anatomical changes due to uptake and accumulation of heavy metal in *Brachiaria decumbens*. Sci Agric 68:566–573. https://doi.org/10.1590/S0103-90162011000500009

Gomes MA, Hauser-Davis RA, Suzuki MS et al (2017) Plant chromium uptake and transport, physiological effects and recent advances in molecular investigations. Ecotoxicol Environ Saf 140:55–64. https://doi.org/10.1016/j.ecoenv.2017.01.042

Goyal D, Yadav A, Prasad M et al (2020) Effect of heavy metals on plant growth: an overview. In: Naeem M, Ansari A, Gill S (eds) Contaminants in agriculture. Springer, Cham. https://doi.org/10.1007/978-3-030-41552-5_4

Gratão PL, Polle A, Lea PJ et al (2005) Making the life of heavy metal-stressed plants a little easier. Funct Plant Biol 32:481–494. https://doi.org/10.1071/FP05016

Gregory RPG, Bradshaw AD (1965) Heavy metal tolerance in populations of *Agrostis tenuis* Sibth. and other grasses. New Phytol 64:131–143. https://doi.org/10.1111/j.1469-8137.1965.tb05381.x

Gupta DK, Vandenhove H, Inouhe M (2013) Role of phytochelatins in heavy metal stress and detoxification mechanisms in plants. In: Gupta D, Corpas F, Palma J (eds) Heavy metal stress in plants. Springer, Berlin. https://doi.org/10.1007/978-3-642-38469-1_4

Haider I, Raza MAS, Iqbal R et al (2021) Alleviating the drought stress in wheat (*Triticum aestivum* L.) by foliar application of amino acid and yeast. Pak J Agric Res 34:239–246. https://doi.org/10.17582/journal.pjar/2021/34.1.239.246

Halim MA, Rahman MM, Mondal D et al (2021) Bioaccumulation and tolerance indices of cadmium in wheat plants grown in cadmium-spiked soil: health risk assessment. Front Environ Sci 9:779588. https://doi.org/10.3389/fenvs.2021.779588

Han C, Zhang P, Ryan PR et al (2016) Introgression of genes from bread wheat enhances the aluminium tolerance of durum wheat. Theor Appl Genet 129:729–739. https://doi.org/10.1007/s00122-015-2661-3

Haque AM, Gohari G, El-Shehawi AM et al (2022) Genome-wide identification, characterization and expression profiles of heavy metal ATPase 3 (HMA3) in plants. J King Saud Univ - Sci 34:101730. https://doi.org/10.1016/j.jksus.2021.101730

Hegedüs A, Erdei S, Horváth G (2001) Comparative studies of $H_2O_2$ detoxifying enzymes in green and greening barley seedlings under cadmium stress. Plant Sci 160:1085–1093. https://doi.org/10.1016/S0168-9452(01)00330-2

Hernández-Cruz EY, Arancibia-Hernández YL, Loyola-Mondragón DY et al (2022) Oxidative stress and its role in Cd-induced epigenetic modifications: use of antioxidants as a possible preventive strategy. Oxygen 2:177–210. https://doi.org/10.3390/oxygen2020015

Hou D, Jia X, Wang L et al (2025) Global soil pollution by toxic metals threatens agriculture and human health. Science 388:316–321. https://doi.org/10.1126/science.adr5214

Hu Y, Wang C, Song Z et al (2021) Heavy metal in rice and vegetables and human exposure near a large Pb/Zn smelter in Central China. Int J Environ Res Public Health 18(23):12631. https://doi.org/10.3390/ijerph182312631

Hüsken A, Arent L, Lohmayer R (2024) Cadmium maximum levels and residue situation in the German wheat and rye harvest from 1975 to 2021. J Agric Food Chem 72(29):16496–16505. https://doi.org/10.1021/acs.jafc.4c03560

Israr M, Sahi S, Datta R et al (2006) Bioaccumulation and physiological effects of mercury in *Sesbania drummondii*. Chemosphere 65(4):591–598. https://doi.org/10.1016/j.chemosphere.2006.02.016

Iyengar GV, Nair PP (2000) Global outlook on nutrition and the environment: meeting the challenges of the next millennium. Sci Total Environ 249(1–3):331–346. https://doi.org/10.1016/S0048-9697(99)00529-X

Jaiswal S, Kumar A, Choudhury BU et al (2023) Development of ammonium nitrate-free nutrient media for aluminium toxicity tolerance screening of rice genotypes from North-Eastern India. J Plant Nutr 46(8):1766–1776. https://doi.org/10.1080/01904167.2022.2099892

Jaiswal S, Kumari A, Kumar K et al (2024) Deciphering genetic mechanisms of Al toxicity tolerance through meta-QTL analysis in rice (*Oryza sativa* L.). Environ Exp Bot 228:106030. https://doi.org/10.1016/j.envexpbot.2024.106030

Jańczak-Pieniążek M, Cichoński J, Michalik P et al (2022) Effect of heavy metal stress on phenolic compounds accumulation in winter wheat plants. Molecules 28(1):241. https://doi.org/10.3390/molecules28010241

Jenkins EM, Galbraith D, Paltseva A (2025) Portable X-ray fluorescence as a tool for urban soil assessment: review of accuracy, precision and best practice. Soil 11:565–588. https://doi.org/10.5194/soil-11-565-2025

Jiang K, Wu B, Wang C et al (2019) Ecotoxicological effects of metals with different concentrations and types on the morphological and physiological performance of wheat. Ecotoxicol Environ Saf 167:345–353. https://doi.org/10.1016/j.ecoenv.2018.10.048

Jyothsna P, Murthy SDS (2016) A review on bioremediation of arsenic from contaminated groundwater. Octa J Environ Res 4:155–166

Kan M, Fujiwara T, Kamiya T (2022) Golgi-localized *OsFPN1* is involved in Co and Ni transport and their detoxification in rice. Rice 15:36. https://doi.org/10.1186/s12284-022-00583-3

Kaur N, Singh J, Sharma NR et al (2025) Heavy metal contamination in wastewater-irrigated vegetables: assessing food safety challenges in developing Asian countries. Environ Sci: Processes Impacts 27:1747. https://doi.org/10.1039/d4em00565a

Khan MS, Zaidi A, Wani PA et al (2009) Role of plant growth promoting rhizobacteria in the remediation of metal contaminated soils. Environ Chem Lett 7:1–19. https://doi.org/10.1007/s10311-008-0155-0

Khan S, Rehman S, Khan AZ et al (2010) Soil and vegetables enrichment with heavy metals from geological sources in Gilgit, northern Pakistan. Ecotoxicol Environ Saf 73:1820–1827. https://doi.org/10.1016/j.ecoenv.2010.08.016

Kirmani HF, Hussain M, Ahmad F et al (2018) Impact of zinc uptake on morphology, physiology and yield attributes of wheat in Pakistan. Cercet Agron Mold 51:29–36. https://doi.org/10.2478/cerce-2018-0002

Knox RE, Pozniak CJ, Clarke FR et al (2009) Chromosomal location of the cadmium uptake gene (*Cdu1*) in durum wheat. Genome 52:741–747. https://doi.org/10.1139/g09-042

Koźmińska A, Wiszniewska A, Hanus-Fajerska E et al (2018) Recent strategies of increasing metal tolerance and phytoremediation potential using genetic transformation of plants. Plant Biotechnol Rep 12:1–14. https://doi.org/10.1007/s11816-017-0467-2

Kumar A, Kaur G, Goel P et al (2019) Genome-wide analysis of oligopeptide transporters and detailed characterization of yellow stripe transporter genes in hexaploid wheat. Funct Integr Genomics 19:75–90. https://doi.org/10.1007/s10142-018-0629-5

Lan Q, Feng B, Xu Z et al (2013) Molecular cloning and characterization of five novel low molecular weight glutenin subunit genes from Tibetan wheat landraces (*Triticum aestivum* L.). Genet Resour Crop Evol 60:799–806. https://doi.org/10.1007/s10722-012-9877-8

Lanquar V, Ramos MS, Lelièvre F et al (2010) Export of vacuolar manganese by AtNRAMP3 and AtNRAMP4 is required for optimal photosynthesis and growth under manganese deficiency. Plant Physiol 152:1986–1999. https://doi.org/10.1104/pp.109.150946

Latif U, Farid M, Rizwan M et al (2020) Physiological and biochemical response of *Alternanthera bettzickiana* under acetic acid-assisted phytoextraction of lead. Plants 9:1084. https://doi.org/10.3390/plants9091084

Li XB, Suo HC, Liu JT et al (2020) Genome-wide identification and expression analysis of the potato ZIP gene family under Zn deficiency. Biol Plant 64:845–855. https://doi.org/10.32615/bp.2020.125

Li L, Peng Z, Mao X et al (2021) Genetic insights into natural variation underlying salt tolerance in wheat. J Exp Bot 72:1135–1150. https://doi.org/10.1093/jxb/eraa500

Li D, Zheng X, Lin L et al (2022a) Remediation of soils co-contaminated with cadmium and DDTs by king grass associated with *Piriformospora indica*. J Hazard Mater 422:126936. https://doi.org/10.1016/j.jhazmat.2021.126936

Li GZ, Zheng YX, Liu HT et al (2022b) WRKY74 regulates cadmium tolerance through glutathione-dependent pathway in wheat. Environ Sci Pollut Res 29:68191–68201. https://doi.org/10.1007/s11356-022-20672-6

Li S, Li X, Li S et al (2023) Variation in the tonoplast cadmium transporter heavy metal ATPase 3 (HMA3) homolog gene in *Aegilops tauschii*. PLoS One 18:e0279707. https://doi.org/10.1371/journal.pone.0279707

Liu H, Wang H, Zhang Y et al (2019) Comparison of heavy metal accumulation and cadmium phytoextraction rates among ten leading tobacco cultivars in China. Int J Phytoremediation 21:699–706. https://doi.org/10.1080/15226514.2018.1556589

Liu X, Wang H, He F et al (2022) The TaWRKY22–TaCOPT3D pathway governs cadmium uptake in wheat. Int J Mol Sci 23:10379. https://doi.org/10.3390/ijms231810379

Liu H, Ma MZ, Huang H et al (2025) Research progress on the mechanism of magnesium nutrition alleviating aluminum toxicity in plants. J Plant Nutr Fertil 31:810–818. https://doi.org/10.11674/zwyf.2024218

Lopes MS, El-Basyoni I, Baenziger PS et al (2015) Exploiting genetic diversity from landraces in wheat breeding for adaptation to climate change. J Exp Bot 66:3477–3486. https://doi.org/10.1093/jxb/erv122

Luo MC, Dvořák J (1996) Molecular mapping of an aluminum tolerance locus on chromosome 4D of Chinese Spring wheat. Euphytica 91:31–35. https://doi.org/10.1007/BF00035273

Lux A, Martinka M, Vaculík M et al (2011) Root responses to cadmium in the rhizosphere: a review. J Exp Bot 62:21–37. https://doi.org/10.1093/jxb/erq281

Ma HX, Bai GH, Carver BF et al (2005) Molecular mapping of a quantitative trait locus for aluminum tolerance in wheat cultivar Atlas 66. Theor Appl Genet 112:51–57. https://doi.org/10.1007/s00122-005-0101-5

Ma JJ, Yan Y, Chen XJ et al (2021) Incorporating bioaccessibility and source apportionment into human health risk assessment of heavy metals in urban dust of Xiamen, China. Ecotoxicol Environ Saf 228:112985. https://doi.org/10.1016/j.ecoenv.2021.112985

Ma Y, Wen Y, Wang C et al (2023) ZIP genes are involved in the retransfer of zinc ions during the senescence of zinc-deficient rice leaves. Int J Mol Sci 24:13989. https://doi.org/10.3390/ijms241813989

Maccaferri M, Harris NS, Twardziok SO et al (2019) Durum wheat genome highlights past domestication signatures and future improvement targets. Nat Genet 51:885–895. https://doi.org/10.1038/s41588-019-0381-3

Maghrebi M, Baldoni E, Lucchini G et al (2021) Analysis of cadmium root retention for two contrasting rice accessions suggests an important role for OsHMA2. Plants 10:806. https://doi.org/10.3390/plants10040806

Mathur R, Balaram V, Satyanarayanan M et al (2016) Assessment of heavy metal contamination of road dusts from industrial areas of Hyderabad, India. Environ Monit Assess 188:514. https://doi.org/10.1007/s10661-016-5496-8

Meng C, Wang P, Hao Z et al (2022) Ecological and health risk assessment of heavy metals in soil and Chinese herbal medicines. Environ Geochem Health 44:817–828. https://doi.org/10.1007/s10653-021-00978-z

Meng L, Sheng A, Cao L et al (2024) Contribution assessment and accumulation prediction of heavy metals in wheat grain in a smelting-affected area using machine learning methods. Sci Total Environ 951:175461. https://doi.org/10.1016/j.scitotenv.2024.175461

Mirza N, Mahmood Q, Maroof Shah M et al (2014) Plants as useful vectors to reduce environmental toxic arsenic content. Sci World J 2014:921581. https://doi.org/10.1155/2014/921581

Mittler R (2017) ROS are good. Trends Plant Sci 22:11–19. https://doi.org/10.1016/j.tplants.2016.08.002

Mittler R, Zandalinas SI, Fichman Y et al (2022) Reactive oxygen species signalling in plant stress responses. Nat Rev Mol Cell Biol 23:663–679. https://doi.org/10.1038/s41580-022-00499-2

Mohamed HI, Ullah I, Toor MD et al (2025) Heavy metals toxicity in plants: understanding mechanisms and developing coping strategies for remediation: a review. Bioresour Bioprocess 12:95. https://doi.org/10.1186/s40643-025-00930-4

Mondal S, Sallam A, Sehgal D et al (2021) Advances in breeding for abiotic stress tolerance in wheat. In: Kole C (ed) Genomic designing for abiotic stress resistant cereal crops. Springer, Cham. https://doi.org/10.1007/978-3-030-75875-2_2

Moustaka J, Ouzounidou G, Bayçu G et al (2016) Aluminum resistance in wheat involves maintenance of leaf $Ca^{2+}$ and $Mg^{2+}$ content, decreased lipid peroxidation and Al accumulation, and low photosystem II excitation pressure. Biometals 29:611–623. https://doi.org/10.1007/s10534-016-9938-0

Mujeeb-Kazi A, Rosas V, Roldan S (1996) Conservation of the genetic variation of *Triticum tauschii* in synthetic hexaploid wheats (*T. turgidum* × *T. tauschii*) and its potential utilization for wheat improvement. Genet Resour Crop Evol 43:129–134. https://doi.org/10.1007/BF00126756

Murtaza B, Naeem F, Shahid M et al (2019) A multivariate analysis of physiological and antioxidant responses and health hazards of wheat under cadmium and lead stress. Environ Sci Pollut Res 26:362–370. https://doi.org/10.1007/s11356-018-3605-7

Nazir A, Rafique F, Ahmed K et al (2021) Evaluation of heavy metals effects on morpho-anatomical alterations of wheat (*Triticum aestivum* L.) seedlings. Microsc Res Tech 84:2517–2529. https://doi.org/10.1002/jemt.23801

Nishida S, Tsuzuki C, Kato A et al (2011) AtIRT1, the primary iron uptake transporter in the root, mediates excess nickel accumulation in *Arabidopsis thaliana*. Plant Cell Physiol 52:1433–1442. https://doi.org/10.1093/pcp/pcr089

Oladzad-Abbasabadi A, Kumar A, Pirseyedi S et al (2018) Identification and validation of a new source of low grain cadmium accumulation in durum wheat. G3 8:923–932. https://doi.org/10.1534/g3.117.300370

Ondrasek G, Shepherd J, Rathod S et al (2025) Metal contamination—a global environmental issue: sources, implications & advances in mitigation. RSC Adv 15:3904–3927. https://doi.org/10.1039/D4RA04639K

Page V, Feller U (2015) Heavy metals in crop plants: transport and redistribution processes on the whole plant level. Agronomy 5:447–463. https://doi.org/10.3390/agronomy5030447

Park J, Song WY, Ko D et al (2012) The phytochelatin transporters AtABCC1 and AtABCC2 mediate tolerance to cadmium and mercury. Plant J 69:278–288. https://doi.org/10.1111/j.1365-313X.2011.04789.x

Paunov M, Koleva L, Vassilev A et al (2018) Effects of different metals on photosynthesis: cadmium and zinc affect chlorophyll fluorescence in durum wheat. Int J Mol Sci 19:787. https://doi.org/10.3390/ijms19030787

Pour-Aboughadareh A, Kianersi F, Poczai P et al (2021) Potential of wild relatives of wheat: ideal genetic resources for future breeding programs. Agronomy 11:1656. https://doi.org/10.3390/agronomy11081656

Pourjalali Z, Shahpiri A, Golkar P (2023) Barley metallothionein isoforms, MT2b2 and MT4, differentially respond to photohormones in barley aleurone layer and their recombinant forms show different affinity for binding to zinc and cadmium. Biometals 36:3–18. https://doi.org/10.1007/s10534-022-00452-y

Rahman R, Upadhyaya H (2021) Aluminium toxicity and its tolerance in plant: a review. J Plant Biol 64:101–121. https://doi.org/10.1007/s12374-020-09280-4

Raman H, Zhang K, Cakir M et al (2005) Molecular characterization and mapping of ALMT1, the aluminium-tolerance gene of bread wheat (*Triticum aestivum* L.). Genome 48:781–791. https://doi.org/10.1139/g05-054

Raman H, Ryan PR, Raman R et al (2008) Analysis of TaALMT1 traces the transmission of aluminum resistance in cultivated common wheat (*Triticum aestivum* L.). Theor Appl Genet 116:343–354. https://doi.org/10.1007/s00122-007-0672-4

Riede CR, Anderson JA (1996) Linkage of RFLP markers to an aluminum tolerance gene in wheat. Crop Sci 36:905–909. https://doi.org/10.2135/cropsci1996.0011183X003600040015x

Rizvi A, Zaidi A, Ameen F et al (2020) Heavy metal induced stress on wheat: phytotoxicity and microbiological management. RSC Adv 10:38379–38403. https://doi.org/10.1039/d0ra05610c

Rizwan M, Ali S, Adrees M et al (2016) Cadmium stress in rice: toxic effects, tolerance mechanisms, and management: a critical review. Environ Sci Pollut Res 23:17859–17879. https://doi.org/10.1007/s11356-016-6436-4

Rufo R, Alvaro F, Royo C et al (2019) From landraces to improved cultivars: assessment of genetic diversity and population structure of Mediterranean wheat using SNP markers. PLoS One 14:e0219867. https://doi.org/10.1371/journal.pone.0219867

Sabella E, Aprile A, Tenuzzo BA et al (2022) Effects of cadmium on root morpho-physiology of durum wheat. Front Plant Sci 13:936020. https://doi.org/10.3389/fpls.2022.936020
Saeed W, Mubeen S, Pan J et al (2024) Integrated physiological and metabolomic responses reveal mechanisms of Cd tolerance and detoxification in kenaf (*Hibiscus cannabinus* L.) under Cd stress. Front Plant Sci 15:1332426. https://doi.org/10.3389/fpls.2024.1332426
Sandhu K, Patil SS, Pumphrey M et al (2021) Multitrait machine- and deep-learning models for genomic selection using spectral information in a wheat breeding program. Plant Genome 14:e20119. https://doi.org/10.1002/tpg2.20119
Sasaki T, Tsuchiya Y, Ariyoshi M et al (2014) A domain-based approach for analyzing the function of aluminum-activated malate transporters from wheat (*Triticum aestivum*) and *Arabidopsis thaliana* in xenopus oocytes. Plant Cell Physiol 55:2126–2138. https://doi.org/10.1093/pcp/pcu143
Sasaki T, Tsuchiya Y, Ariyoshi M et al (2016) A chimeric protein of aluminum-activated malate transporter generated from wheat and Arabidopsis shows enhanced response to trivalent cations. Biochim Biophys Acta 1858:1427–1435. https://doi.org/10.1016/j.bbamem.2016.03.026
Seregin IV, Kozhevnikova AD (2023) Phytochelatins: sulfur-containing metal(loid)-chelating ligands in plants. Int J Mol Sci 24:2430. https://doi.org/10.3390/ijms24032430
Setia R, Dhaliwal SS, Singh R et al (2021) Phytoavailability and human risk assessment of heavy metals in soils and food crops around Sutlej River, India. Chemosphere 263:128321. https://doi.org/10.1016/j.chemosphere.2020.128321
Shaikh IR, Shaikh PR, Shaikh RA et al (2013) Phytotoxic effects of heavy metals (Cr, Cd, Mn and Zn) on wheat (*Triticum aestivum* L.) seed germination and seedlings growth in black cotton soil of Nanded. Res J Chem Sci 3:14–23
Shao L, Guo X, Wang Y et al (2025) Genome-wide survey of the HMA gene family in wheat (*Triticum aestivum* L.) and its potential role in cadmium stress. BMC Genomics 26:619. https://doi.org/10.1186/s12864-025-11746-z
Sharma P, Jha AB, Dubey RS et al (2012) Reactive oxygen species, oxidative damage, and antioxidative defense mechanism in plants under stressful conditions. J Bot 2012:217037. https://doi.org/10.1155/2012/217037
Shim D, Hwang JU, Lee J et al (2009) Orthologs of the class A4 heat shock transcription factor HsfA4a confer cadmium tolerance in wheat and rice. Plant Cell 21:4031–4043. https://doi.org/10.1105/tpc.109.066902
Singha L, Singh RS, Priyadharshini B et al (2025) Heavy metal stress in plants: mechanisms, impacts, and mitigation strategies—a comprehensive review. Int J Environ Sci 11:2968–2977. https://doi.org/10.64252/yjv3t467
Su Y, Li B, Li J et al (2025) Hyperspectral remote sensing for soil heavy metal inversion: insights and applications. Int J Digit Earth 18:1. https://doi.org/10.1080/17538947.2025.2520474
Szolnoki Z, Farsang A, Puskás I (2013) Cumulative impacts of human activities on urban garden soils: origin and accumulation of metals. Environ Pollut 177:106–115. https://doi.org/10.1016/j.envpol.2013.02.007
Tan L, Qu M, Zhu Y et al (2020) ZINC TRANSPORTER5 and ZINC TRANSPORTER9 function synergistically in zinc/cadmium uptake. Plant Physiol 183:1235–1249. https://doi.org/10.1104/pp.19.01569
Tang L, Dong J, Qu M et al (2022) Knockout of OsNRAMP5 enhances rice tolerance to cadmium toxicity in response to varying external cadmium concentrations via distinct mechanisms. Sci Total Environ 832:155006. https://doi.org/10.1016/j.scitotenv.2022.155006
Tittlemier SA, Richter A (2022) Cadmium concentrations in Canadian durum exports decreased with the adoption of low accumulating cultivars. Food Addit Contam A 39:1953–1962. https://doi.org/10.1080/19440049.2022.2130441
Türkdoğan MK, Kilicel F, Kara K et al (2003) Heavy metals in soil, vegetables and fruits in the endemic upper gastrointestinal cancer region of Turkey. Environ Toxicol Pharmacol 13:175–179. https://doi.org/10.1016/S1382-6689(02)00156-4
Vatamaniuk OK, Mari S, Lu YP et al (2000) Mechanism of heavy metal ion activation of phytochelatin (PC) synthase: blocked thiols are sufficient for PC synthase-catalyzed transpepti-

dation of glutathione and related thiol peptides. J Biol Chem 275:31451–31459. https://doi.org/10.1074/jbc.M002997200
Verma P, Agrawal M, Sagar R (2015) Assessment of potential health risks due to heavy metals through vegetable consumption in a tropical area irrigated by treated wastewater. Environ Syst Decis 35:375–388. https://doi.org/10.1007/s10669-015-9558-1
Viehweger K (2014) How plants cope with heavy metals. Bot Stud 55:35. https://doi.org/10.1186/1999-3110-55-35
Vullaganti N, Ram BG, Sun X (2025) Precision agriculture technologies for soil site-specific nutrient management: a comprehensive review. Artif Intell Agric 15:147–161. https://doi.org/10.1016/j.aiia.2025.02.001
Wang X, Wang C, Sheng H et al (2017) Transcriptome-wide identification and expression analyses of ABC transporters in dwarf Polish wheat under metal stresses. Biol Plant 61:293–304. https://doi.org/10.1007/s10535-016-0697-0
Wang L, Rinklebe J, Tack FM et al (2021) A review of green remediation strategies for heavy metal contaminated soil. Soil Use Manag 37:936–963. https://doi.org/10.1111/sum.12717
Wang K, Wang Z, Li Z et al (2025) Oriented object detection in optical remote sensing images using deep learning: a survey. Artif Intell Rev 58:350. https://doi.org/10.1007/s10462-025-11256-0
Xiong B, Li R, Johnson D et al (2021) Spatial distribution, risk assessment, and source identification of heavy metals in water from the Xiangxi River, three gorges reservoir region, China. Environ Geochem Health 43:915–930. https://doi.org/10.1007/s10653-020-00614-2
Yan A, Wang Y, Tan SN et al (2020) Phytoremediation: a promising approach for revegetation of heavy metal-polluted land. Front Plant Sci 11:359. https://doi.org/10.3389/fpls.2020.00359
Yang QW, Xu Y, Liu SJ et al (2011) Concentration and potential health risk of heavy metals in market vegetables in Chongqing, China. Ecotoxicol Environ Saf 74:1664–1669. https://doi.org/10.1016/j.ecoenv.2011.05.006
Yu Y, Zhang L (2023) The wheat NAC transcription factor TaNAC22 enhances cadmium stress tolerance in wheat. Cereal Res Commun 51:867–877. https://doi.org/10.1007/s42976-023-00354-w
Zaid IU, Zheng X, Li X (2018) Breeding low-cadmium wheat: progress and perspectives. Agronomy 8:249. https://doi.org/10.3390/agronomy8110249
Zaid IU, Faheem M, Zia MA et al (2023) Temporal comparative transcriptome analysis on wheat response to acute Cd toxicity at the seedling stage. Plants 12:642. https://doi.org/10.3390/plants12030642
Zencirci N, Aktan B, Atli A (1994) Genetic relationships of Turkish durum wheat cultivars. Turk J Agric For 18:187–192
Zeng X, Yang S, Li F et al (2024) Genome-wide identification of OsZIPs in rice and gene expression analysis under manganese and selenium stress. Genes 15:696. https://doi.org/10.3390/genes15060696
Zhai Y, Zhou L, Qi H et al (2023) Application of visible/near-infrared spectroscopy and hyperspectral imaging with machine learning for high-throughput plant heavy metal stress phenotyping: a review. Plant Phenomics 5:0124. https://doi.org/10.34133/plantphenomics.0124
Zhang L, Gao C, Chen C et al (2020) Overexpression of rice OsHMA3 in wheat greatly decreases cadmium accumulation in wheat grains. Environ Sci Technol 54:10100–10108. https://doi.org/10.1021/acs.est.0c02877
Zhang T, Xiao J, Zhao Y et al (2021) Comparative physiological and transcriptomic analyses reveal ascorbate and glutathione coregulation of cadmium toxicity resistance in wheat genotypes. BMC Plant Biol 21:459. https://doi.org/10.1186/s12870-021-03225-w
Zhang XY, Geng LP, Gao PP et al (2022) Bioimaging of Pb by LA-ICP-MS and Pb isotopic compositions reveal distributions and origins of Pb in wheat grain. Sci Total Environ 802:149729. https://doi.org/10.1016/j.scitotenv.2021.149729
Zheng S, Qi J, Fu T et al (2023) Novel mechanisms of cadmium tolerance and Cd-induced fungal stress in wheat: transcriptomic and metagenomic insights. Ecotoxicol Environ Saf 256:114842. https://doi.org/10.1016/j.ecoenv.2023.114842

Zhou M, Li Z (2022) Recent advances in minimizing cadmium accumulation in wheat. Toxics 10:187. https://doi.org/10.3390/toxics10040187

Zhou M, Zheng S (2022) Multi-omics uncover the mechanism of wheat under heavy metal stress. Int J Mol Sci 23:15968. https://doi.org/10.3390/ijms232415968

Zhou X, Wang Z, Chen X et al (2025) Identification, validation, and candidate gene analysis of major QTL for cadmium tolerance in common wheat. Euphytica 221:106. https://doi.org/10.1007/s10681-025-03559-8

Zhu J, Zhao P, Nie Z et al (2020) Selenium supply alters the subcellular distribution and chemical forms of cadmium and the expression of transporter genes involved in cadmium uptake and translocation in winter wheat (*Triticum aestivum*). BMC Plant Biol 20:550. https://doi.org/10.1186/s12870-020-02763-z

Zulfiqar U, Jiang W, Xiukang W et al (2022) Cadmium phytotoxicity, tolerance, and advanced remediation approaches in agricultural soils; a comprehensive review. Front Plant Sci 13:773815. https://doi.org/10.3389/fpls.2022.773815

# Advances and Challenges in Using Plant Biomarkers for Early Stress Detection in Wheat

13

Saumya Agrawal, Yukti Singh, Sandhya Mehrotra, and Rajesh Mehrotra

**Abstract**

Early stress detection in wheat (*Triticum aestivum*) is critical for minimizing yield losses under increasing biotic and abiotic pressures. This chapter provides a comprehensive review of advances in biomarker-based diagnostics, encompassing molecular, biochemical, physiological, and imaging-derived indicators. It integrates evidence from multi-omics strategies, high-throughput phenotyping, and sensor technologies, including transcriptomics, proteomics, metabolomics, chlorophyll fluorescence, hyperspectral and thermal imaging, Raman and infrared (IR) spectroscopy, biosensors, lab-on-chip platforms, and CRISPR-based diagnostics for pre-symptomatic detection of drought, heat, salinity, and pathogen-induced stress. The focus is on integrating molecular signatures such as stress-responsive transcription factors, specific miRNAs, variations in phytohormones, and reactive oxygen species (ROS) dynamics with proximal and distal sensing platforms, further enhanced by machine learning (ML) for improved specificity and scalability. Pathway-guided breeding and phenotyping approaches, UAV-based monitoring, wearable sensors, and artificial intelligence (AI-facilitated data fusion are also discussed. Critical issues related to genotype-by-environment interactions, cross-location validation, cost constraints, and the laboratory-to-field translation gap are highlighted. Suggestions are made for developing standardized protocols, interoperable data systems, and farmer-centric sensor designs. The chapter concludes with future perspectives, emphasizing that stable biosensors, integrated multi-omics approaches, and AI-powered monitoring can accelerate the development of climate-resilient wheat and safeguard global food security.

S. Agrawal · Y. Singh · S. Mehrotra · R. Mehrotra (✉)
Department of Biological Sciences, BITS- Pilani, K.K. Birla Goa Campus, Goa, India
e-mail: rajeshm@goa.bits-pilani.ac.in



R. Tiwari et al. (eds.), *Wheat Stress Management*,
https://doi.org/10.1007/978-981-92-0902-6_13

**Keywords**

Wheat biomarker · Early stress detection · Multi-omics integration · High-throughput phenotyping · Remote sensing · Artificial Intelligence

## 13.1 Introduction

Wheat (*Triticum aestivum*) is a crucial global crop that contributes significantly to food security for a growing world population. However, its productivity is increasingly threatened by a complex array of biotic (pests, diseases, and weeds) and abiotic (drought, salinity, and heat) stresses, many of which are being intensified by climate change and sustainable intensification. These stress factors, often acting concurrently, can cause severe yield losses ranging from 30% to more than 70%, particularly in developing countries, and in extreme cases may even result in complete crop failure (Munaweera et al. 2022). Conventional stress detection methods rely on visible symptoms and are typically cumbersome, subjective, and reactive. By the time symptoms become apparent, crops may already have suffered irreversible biochemical or physiological damage. Therefore, there is a pressing need to develop early, accurate, and nondestructive diagnostic tools for timely intervention and sustainable crop management (Mahlein 2016).

Indicators such as remote and proximal sensing, Raman and infrared (IR) spectroscopy, and high-throughput phenotyping (HTP) are relatively easy to implement. These methods can detect early stress-induced physiological responses before visible symptoms appear. For instance, RS demonstrates high precision in distinguishing between drought stress, nitrogen deficiency, and aphid infestation. Such technologies support precision farming by enabling targeted interventions that optimize input use and protect crop yield. Among these innovations, plant biomarkers have emerged as effective early warning signals of stress. Biomarkers—quantifiable biochemical, molecular, or physiological indicators—reflect the plant's internal response to stress well before external symptoms become apparent. They include phytohormones, transcription factors, stress-inducible genes, proteins, small RNAs, metabolites, and changes in photosynthetic activity (Mahlein 2016; Payne and Kurouski 2021).

Plant biomarkers are broadly classified into four categories based on their function (Dong et al. 2024):

1. Molecular biomarkers consist of genes and regulatory factors, including transcription factors (e.g., WRKY, NAC, and DREB) and small RNAs that regulate defense mechanisms. For example, *TaNAC47* in wheat is associated with drought tolerance. These biomarkers are typically analyzed using RNA-seq, qPCR, or Northern blotting.
2. Biochemical biomarkers include stress-responsive proteins, enzymes, and metabolites such as aquaporins (e.g., *PIP1;5*), dehydrins (e.g., *DHN7*), and anti-

oxidants (e.g., SOD, CAT). They are commonly detected through ELISA, Western blotting, or mass spectrometry.
3. Physiological biomarkers are represented by functional changes such as stomatal conductance, photosynthetic rate, and chlorophyll content, which can be measured using gas-exchange systems and fluorometers.
4. Imaging-based biomarkers capture spatial and spectral signatures through non-invasive techniques such as chlorophyll fluorescence (Fv/Fm), hyperspectral imaging, thermal imaging, or MALDI-mass spectrometry imaging.

Integrating biomarkers with nanotechnology, genetic engineering, and climate-smart agriculture can accelerate the development of stress-resilient crops. Advances in omics platforms—including genomics, transcriptomics, proteomics, metabolomics, and phenomics—enable data-driven approaches to elucidate and exploit biomarker responses. Research has demonstrated the feasibility of combining single-nucleotide polymorphism (SNP) genotyping, metabolite profiling, and remote sensing for early stress detection and wheat breeding (Raza et al. 2024).

In precision agriculture, physiological and biochemical biomarkers such as abscisic acid content, aquaporins, and chlorophyll fluorescence provide real-time indicators of plant health. Combined with remote imaging technologies, they enable field-scale monitoring and site-specific management practices that enhance sustainability and resource-use efficiency. In crop breeding, molecular biomarkers such as stress-responsive transcription factors and signaling genes (e.g., *MPKs* and *OsbZIP23*) facilitate high-throughput screening and marker-assisted selection. Their early detection shortens breeding cycles and supports the development of climate-resilient cultivars by addressing complex genotype–environment interactions. Table 13.1 summarizes universal, specific, and emerging biomarkers, along with their examples, associated stress types, and functions. Overall, biomarkers act as lead indicators, enabling early, accurate, and cost-effective identification of plant stress responses. They support timely decisions regarding irrigation, fertilization, and pest control, thereby stabilizing yields and reducing the environmental footprint. By promoting proactive rather than reactive strategies, biomarkers represent a breakthrough in crop stress management (Kumar et al. 2024; Khan et al. 2025).

With rising environmental stresses and growing global concern over food security, the ability to detect stress in crops such as wheat before permanent damage occurs is crucial. This chapter provides a historical overview of biomarker discovery and characterization, highlighting key developments that have advanced the field. It also discusses how progress in molecular biology, omics technologies, and imaging approaches has enhanced our capacity to identify stress responses at early stages. Furthermore, the chapter examines the applications of plant biomarkers in precision agriculture and breeding, while addressing the technical and biological challenges that limit their widespread adoption.

**Table 13.1** Detailed classification of plant stress biomarkers into universal, specific, and emerging categories, along with their types, examples, associated stress factors, and functions

| Category | Biomarker type | Example(s) | Stress type | Function | Reference |
|---|---|---|---|---|---|
| Universal | Heat shock proteins (HSPs) | HSP70, HSP90, and sHSPs | Heat, drought, and oxidative | Protein folding, protection, and repair of denatured proteins | Ali et al. (2023) |
| | Transcription factors (TFs) | DREB, bZIP, WRKY, and MYB, NAC | Drought, cold, salt, and pathogen | Regulate stress-responsive gene expression | Rolly et al. (2021) |
| | Osmoregulators/compatible solutes | Proline, glycine betaine, and trehalose | Drought, salinity, and cold | Maintain cell turgor, protect enzymes and membranes | Ali et al. (2023) |
| | Reactive oxygen species (ROS) scavengers | SOD, CAT, APX, and GPX | Oxidative, drought, and salinity | Detoxification of ROS, signaling in stress adaptation | Jan et al. (2024) |
| | Phytohormones | ABA, SA, JA, ethylene, and cytokinin | Abiotic (ABA) and biotic (SA and JA) | Signal transduction, stomatal closure, and defense activation | Das et al. (2025) |
| | Glutathione system | GSH, and glutathione reductase (GR) | Oxidative and heavy metal | Redox buffering and detoxification | Myers et al. (2024) |
| | Thioredoxins and peroxiredoxins | TRX and PRX | Oxidative stress | Regulate protein redox state and antioxidant defense | Şen et al. (2024) |
| | MAPK cascade components | MAPK3, MAPK6, and MPK1 | Biotic, drought, and oxidative | Transmit stress signals to activate TFs | Aina et al. (2024) |
| | Calcium signaling components | CDPKs (calcium-dependent protein kinases), CBLs, and CIPKs | Cold, drought, and pathogen | Calcium-dependent signal transduction | Aina et al. (2024) |
| | Ubiquitin-proteasome components | E3 ligases and 26S proteasome | Broad stress types | Protein degradation and turnover | Steinfath et al. (2010) |
| Specific | Dehydrins/LEA proteins | DHN1 and DHN5 | Cold and drought | Membrane stabilization and water retention | Nakashima et al. (2014) |
| | Aquaporins | PIP1 and PIP2 | Drought and salinity | Regulate water movement across membranes | Dong et al. (2024) |
| | Pathogenesis-related proteins (PR proteins) | PR-1, PR-2 (β-1,3-glucanase) and PR-5 (thaumatin) | Biotic stress | Antifungal/antibacterial roles and systemic acquired resistance | Nakashima et al. (2014) |
| | Secondary metabolites | Phenolics, flavonoids and lignin | Biotic, UV, and wounding | Antioxidant activity and structural reinforcement | Pérez-Llorca et al. (2023) |
| | Cold-responsive genes | CBFs (C-repeat binding factors) | Cold | Regulate cold acclimation pathways | Nakashima et al. (2014) |

| | | | | | |
|---|---|---|---|---|---|
| | Protease inhibitors | Serine protease inhibitors | Insect and fungal attack | Inhibit pest and pathogen proteases | Steinfath et al. (2010) |
| | Lipid-derived signals | Oxylipins and Jasmonates | Wounding and biotic | Signal transduction and defense gene activation | Alonso et al. (2024) |
| | R genes (resistance genes) | *Lr34*, *Sr2*, and *Yr36* | Pathogen-specific (rusts) | Recognize specific pathogens and activate defense pathways | Yousfi et al. (2016) |
| | Metabolite markers | GABA and polyamines | Drought, heat, and oxidative | Stress buffering, signaling, and membrane stability | Raza et al. (2024) |
| | Chaperones (non-HSP) | DnaJ, ClpB, and BiP | Heat, oxidative, and ER stress | Assist in protein folding and prevent aggregation | Ali et al. (2023) |
| | Metallothionein | MT1 and MT2 | Heavy metal and oxidative | Metal ion sequestration and ROS scavenging | Raza et al. (2024) |
| | ABC transporters | ABCG and ABCB | Salinity, xenobiotic, and pathogen | Transport of secondary metabolites, hormones, and toxins | Kajla et al. (2023) |
| | Ion transporters | HKT1, NHX1, and SOS1 | Salinity | $Na^+/K^+$ homeostasis and ion compartmentalization | Bashir et al. (2019) |
| | Cell wall-associated kinases (WAKs) | WAK1 and WAK2 | Pathogen and mechanical stress | Perception of cell wall integrity | Juárez and Kurouski (2024) |
| | Lipid transfer proteins (LTPs) | LTP1 and LTP3 | Pathogen and drought | Lipid trafficking and defense signaling | Alonso et al. (2024) |
| | Volatile organic compounds (VOCs) | Methyl jasmonate and green leaf volatiles | Biotic and mechanical stress | Long-distance stress signaling and priming | Alonso et al. (2024) |
| | Alternative oxidase (AOX) | AOX1a and AOX1b | Mitochondrial/oxidative stress | Maintains respiration under stress and ROS balance | Wang et al. (2024) |
| | Polyamine biosynthesis enzymes | SAMDC, ADC, and SPDS | Drought, heat, and salt | Stress signal molecules and membrane protection | Tahmasebi et al. (2019) |
| | Stress-associated proteins (SAPs) | SAP1 and SAP5 | Drought, salt, and oxidative | Zinc-finger proteins involved in gene regulation | Ali et al. (2023) |
| Emerging | miRNAs | miR398, miR156, and miR319 | Drought, cold, and pathogens | Post-transcriptional regulation of stress genes | Tahmasebi et al. (2019) |
| | Epigenetic regulators | Histone modifiers and DNA methyltransferases | Drought, heat, and memory responses | Stress memory and transcriptional regulation | Abdulraheem et al. (2024) |
| | Small RNAs/lncRNAs | lncRNA145, miR164, and miR171 | Abiotic and biotic stresses | Post-transcriptional and epigenetic regulation | Roychowdhury et al. (2023) |
| | Ribosomal stress markers | Ribosomal protein S6 kinase (RPS6K) | Heat and nutrient starvation | Protein translation regulation under stress | Dong et al. (2024) |

## 13.2 History

For most of its history, wheat stress was diagnosed based on visible plant symptoms. Stressed wheat typically exhibits leaf chlorosis (yellowing), wilting, and eventual tissue necrosis. Chlorosis has long been recognized as "the first visible warning of diminishing plant health" (*NASA Technical Reports Server, NTRS* 2025). For instance, under drought or nutrient stress, older wheat leaves often show yellowing and abscission, while severe moisture stress leads to wilting and browning or necrosis of leaf tips and margins. However, these visual symptoms provide only late and indirect indications of stress and are neither specific nor quantitative.

In the mid-twentieth century, researchers began investigating internal biochemical indicators of plant stress. A key finding was that certain metabolites tend to accumulate under stress conditions. In wheat, for instance, free proline—an osmoprotectant amino acid—was shown to increase dramatically during drought stress. Bates et al. (1973) developed a rapid colorimetric ninhydrin assay for proline, which became a standard marker for drought. Similarly, oxidative stress in plants was monitored using antioxidant and peroxidation assays. Heath and Packer (1968) pioneered the thiobarbituric acid (TBA) technique for determining malondialdehyde (MDA), a lipid peroxidation byproduct that yields a pink chromogen indicative of oxidative damage. By the late twentieth century, it had also become standard practice to quantify reductions in ascorbate and enzymatic antioxidants (superoxide dismutase, catalase, and peroxidase) in stressed wheat tissues (Şen et al. 2024). These early colorimetric biomarker assays, though widely applied, had notable limitations. While the MDA and proline tests produced readily quantifiable colored products, they required destructive tissue grinding and laboratory-based chemical processing, making real-time or in-field deployment impractical. In particular, the TBA–MDA assay is now recognised to suffer from low specificity and methodological artefacts, since other plant aldehydes also react with TBA, leading to overestimation of lipid peroxidation (Morales and Munné-Bosch 2019).

More broadly, numerous stress metabolites (e.g., sugars, organic acids, and proline) may accumulate under various stresses or even during normal senescence, making single markers insufficiently specific for stress diagnosis. At the same time, electrophoretic protein profiling began to reveal stress-specific patterns. With the introduction of one-dimensional SDS–PAGE and later two-dimensional gel electrophoresis, researchers were able to compare whole-leaf protein profiles from stressed and control plants, leading to the identification of classical stress-responsive proteins. For example, heat-shock proteins (HSPs) were identified as novel or intensified bands under heat and other stress conditions. It is now well established that HSP levels increase dramatically when cells are exposed to elevated temperatures, where they function as molecular chaperones to protect proteins during stress (Ul Haq et al. 2019). Early approaches to wheat stress detection evolved from visual symptom assessment to biochemical and proteomic markers such as dehydrins. However, these methods were constrained by delayed detection, lack of specificity, and labor-intensive procedures. These limitations highlighted the need for faster, real-time sensor-based and molecular technologies for accurate stress monitoring (Morales and Munné-Bosch 2019; Ul Haq et al. 2019).

## 13.3 Scientific Advancements and Principles Developed Over the Past Centuries

The scientific understanding of stress detection in wheat has advanced remarkably over the past three centuries, evolving from empirical observation to sophisticated digital and molecular technologies. In the twentieth century, physiological parameters such as stomatal conductance, transpiration, and leaf water potential became standard indicators of drought and salinity stress. These were complemented by biochemical markers, including the accumulation of proline and glycine betaine, as well as elevated activity of antioxidant enzymes such as SOD, CAT, and POD (Jan et al. 2024). The molecular era introduced gene-level biomarkers such as LEA proteins, dehydrins, and heat-shock proteins, while cloning and EST libraries enabled the identification of cold- and drought-responsive genes. Since the 2000s, high-throughput genomics approaches (GWAS, RNA-seq, and QTL mapping) have facilitated the discovery of adaptive alleles. Proteomic and metabolomic studies have further revealed stress-induced cellular changes. More recently, systems biology platforms, including WheatNet and TWAS, have mapped complex gene networks underlying stress tolerance. Real-time, noninvasive sensing technologies such as hyperspectral imaging, nano-biosensors, and CRISPR-based diagnostics now enable rapid in-field stress detection, supporting both monitoring and the breeding of climate-resilient wheat (Jan et al. 2024).

### 13.3.1 Plant Stress Response Pathways

In detecting external stress, plants possess a remarkable internal defense system that senses, communicates, and responds to environmental challenges long before visible damage occurs. At the core of this system lies a highly coordinated interaction among reactive oxygen species (ROS), phytohormones, and gene-expression cascades. This dynamic three-way interaction enables plants to withstand unfavorable conditions and provides a valuable molecular window for early stress detection (Pérez-Clemente et al. 2013). Figure 13.1 illustrates this interplay between ROS, phytohormones, and gene-expression cascades in regulating stress-induced responses within plant cells.

#### 13.3.1.1 Reactive Oxygen Species (ROS) Signaling Pathway

When exposed to stress, one of the plant's earliest responses is the rapid generation of reactive oxygen species (ROS), including superoxide ($O_2\cdot^-$), hydrogen peroxide ($H_2O_2$), singlet oxygen ($^1O_2$), and hydroxyl radicals ($\cdot OH$). Once considered merely toxic byproducts, ROS are now recognized as central signaling molecules. They are synthesized in various cellular compartments—such as chloroplasts, mitochondria, peroxisomes, and the apoplast—often via the enzymatic activity of respiratory burst oxidase homologs. At low concentrations, particularly $H_2O_2$, ROS regulate protein activities through reversible oxidation, activate kinase cascades, and interact with other signaling molecules such as $Ca^{2+}$ and $K^+$ ions. These interactions control key processes including stomatal closure, ion transport, and gene expression. A remarkable feature of ROS signaling is the ROS wave—a systemic, cell-to-cell propagation of ROS signals that underlies systemic acquired acclimation (SAA). This wave is mediated by aquaporins, calcium signaling, and MAP kinase pathways, enabling remote tissues to acclimate to stress (Devireddy et al. 2021; Wang et al. 2024; Yu et al. 2024).

To prevent oxidative damage, plants activate antioxidant defenses. Enzymatic systems such as superoxide dismutase (SOD), catalase (CAT), and ascorbate peroxidase (APX), along with small molecules such as ascorbate and glutathione, maintain ROS homeostasis. ROS abundance is tightly regulated and depends on stress type, intensity, tissue, and developmental stage, making them powerful early biomarkers of stress (Myers et al. 2024).

#### 13.3.1.2 Phytohormonal Regulation

Phytohormones are chemical messengers that regulate plant growth, development, and stress responses. Under stress conditions, plants undergo hormonal reprogramming involving major hormones such as abscisic acid (ABA), ethylene (ET), jasmonic acid (JA), salicylic acid (SA), auxins, cytokinins (CKs), gibberellins (GAs), brassinosteroids (BRs), strigolactones (SLs), and melatonin (Das et al. 2025). Among these, ABA is the primary regulator of abiotic stress responses, particularly under drought and salinity. Elevated ABA levels induce stomatal closure, osmotic adjustment, and ABA-responsive gene expression. Genes such as *NCED* (*9-cis-epoxycarotenoid dioxygenase*), which are involved in ABA biosynthesis, are also widely used as molecular markers (Singh and Roychoudhury 2023). Other hormones, such as JA and SA, play critical roles during pathogen attack and oxidative stress, often acting synergistically or antagonistically with ABA. Ethylene interacts with JA to enhance defense and also mediates ROS signaling and senescence. Growth-regulating hormones such as auxins, CKs, and GAs are typically downregulated to conserve energy, while also reprogramming plant architecture to facilitate acclimation (Pérez-Llorca et al. 2023; Alonso et al. 2024). Crosstalk among hormones fine-tunes stress responses. For example, ABA antagonizes GA to repress growth during drought, while JA and ET synergize to strengthen antioxidant defenses. These hormonal changes serve as measurable biomarkers, detectable by RT-qPCR, ELISA, and LC–MS, and provide indication of the stress type and severity even before visible phenotypic symptoms appear (Panozzo et al. 2025).

#### 13.3.1.3 Gene Expression Cascade

The gene expression cascade is the third major component of the stress signaling pathway and constitutes a tightly regulated transcriptional network that reprograms plant metabolism in response to stress perception (Aina et al. 2024). Stress cues are initially sensed by receptor-like kinases (RLKs) or intracellular sensors that detect changes in ROS, phytohormones, or mechanical stress. This sensing activates intracellular cascades involving secondary messengers such as $Ca^{2+}$ and $H_2O_2$, which in turn induce protein kinases, particularly calcium-dependent protein kinases (CDPKs) and mitogen-activated protein kinases (MAPKs) (Aina et al. 2024).

These kinases phosphorylate transcription factors (TFs), thereby activating them to bind stress-responsive cis-elements (e.g., ABRE, DRE) within gene promoters. Key TF families include:

(a) DREB/CBF: ABA-independent cold and dehydration responses
(b) AREB/ABF: ABA-dependent drought and salinity responses
(c) WRKY and ERF: Pathogen defense and oxidative stress responses

These TFs regulate genes encoding LEA proteins, dehydrins, antioxidant enzymes (SOD, CAT, and APX), pathogenesis-related proteins, heat shock proteins (HSPs), and ion transporters. The cascade culminates in metabolic reprogramming, including the production of osmolytes (e.g., proline), secondary metabolites, and other protective molecules (Nakashima et al. 2014). Importantly, transcriptional regulation is shaped by feedback loops, hormonal crosstalk, and epigenetic modifications (DNA methylation and histone modifications), which contribute to stress memory—an adaptive trait whereby prior stress exposure enhances future resilience (Abdulraheem et al. 2024). Gene expression profiling using RNA-seq, RT-qPCR, and epigenomic assays enables early detection of major TFs (e.g., DREB2, AREB1, and AtbZIP62) and their downstream targets, supporting diagnostics of stress and the selection of tolerant genotypes (Rolly et al. 2021). Collectively, ROS signaling, phytohormonal dynamics, and the gene expression cascade form a multilayered, interconnected network that regulates plant stress responses. These systems do not function in isolation but interact to fine-tune responses according to stress duration, type, and intensity. Their interplay ensures rapid yet coordinated activation of defense mechanisms (Fig. 13.1). This integrated model provides multiple entry points for early stress detection. Monitoring ROS levels, hormonal signatures, and transcriptional activity enables pre-symptomatic diagnosis, making these pathways particularly valuable for precision agriculture, crop management, and breeding for resilience (Aina et al. 2024).

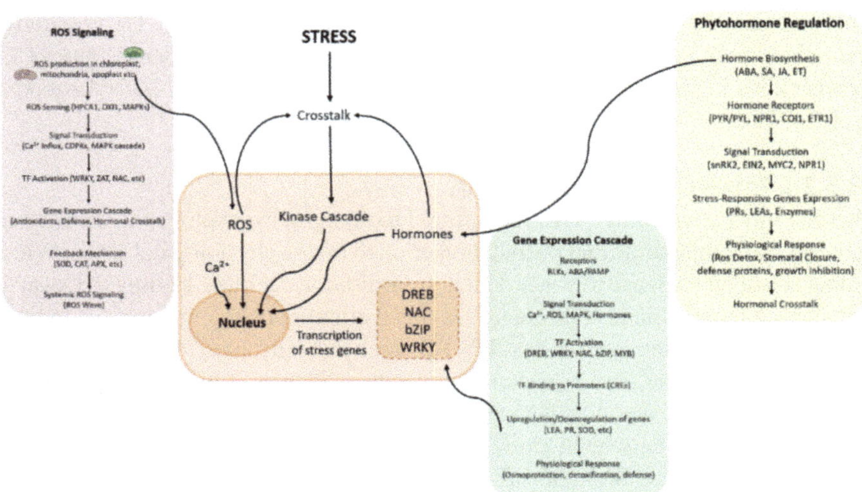


**Fig. 13.1** Illustration of the interconnected signaling pathways of plant stress responses, depicting the crosstalk among ROS signaling, phytohormone regulation, and gene-expression cascades

### 13.3.2 Noninvasive Physiological Techniques

Early and accurate stress sensing is crucial for maintaining crop resilience in the face of a changing climate. Conventional techniques often rely on destructive sampling or visual assessments, which limit real-time measurement and scalability. In contrast, noninvasive methods provide real-time, high-resolution information on plant physiology without damaging tissues. These approaches enable the early detection of stress-induced changes well before visible symptoms appear and are widely applied in phenotyping, precision agriculture, and biomarker discovery. Common noninvasive methods include chlorophyll fluorescence imaging, thermal infrared imaging, spectral sensing, vibrational spectroscopy, and fluorescent biosensors (Chaerle et al. 2009).

#### 13.3.2.1 Chlorophyll Fluorescence Imaging

Chlorophyll fluorescence imaging (CFI) is a well-established, nondestructive method for assessing photosynthetic performance, particularly Photosystem II (PSII) efficiency. During photosynthesis, most light energy is utilized in photochemistry or dissipated as heat, while a small fraction is re-emitted as fluorescence. CFI measures this chlorophyll-induced fluorescence, primarily in the far-red range, to estimate parameters such as Fv/Fm (maximum PSII quantum efficiency), ΦPSII (effective quantum yield), and NPQ (non-photochemical quenching). These indices are highly sensitive to abiotic stress-induced perturbations in photosynthesis, serving as early indicators of both abiotic and biotic stress (Pérez-Bueno et al. 2019; Park et al. 2024). CFI allows quantitative, spatially resolved imaging of stress responses at the whole-plant or tissue level. It is well-suited for detecting the effects of drought, salinity, heat, pathogen infection, and herbicide exposure, often before morphological symptoms appear. Unlike conventional gas exchange techniques, CFI is scalable and adaptable for high-throughput phenotyping. Recent advances have produced pulse amplitude modulation-based fluorescence devices that are portable and suitable for field use, enhancing real-time crop monitoring. The sensitivity and reliability of the Fv/Fm ratio (typically ~0.78–0.84 in healthy plants) make CFI a cornerstone in physiological biomarker studies (McAusland et al. 2019; Legendre et al. 2021).

#### 13.3.2.2 Infrared Thermal Imaging

Thermal infrared imaging (TIR), or infrared thermography, exploits the correlation between leaf temperature and stomatal function to assess physiological stress. When stomata close in response to water deficit, transpiration declines, leading to elevated leaf surface temperatures. TIR quantifies this thermal radiation and derives indices such as the crop water stress index (CWSI), a reliable surrogate for plant water status, transpiration rate, and stomatal conductance (Pineda et al. 2021; Yang et al. 2025). TIR is highly versatile, with applications ranging from leaf-level diagnosis in the laboratory to canopy-level monitoring via drones and satellites. It has been successfully used to track drought, salinity, heat, and pathogen infection. Its utility is further enhanced when combined with other imaging methods, such as hyperspectral sensing, for comprehensive stress diagnosis. Although TIR lacks specificity for individual stressors, integration with AI and machine learning enables robust automated interpretation, supporting high-throughput phenotyping and precision irrigation management (Zhou et al. 2021; He et al. 2025).

#### 13.3.2.3 Hyperspectral and Multispectral Imaging

Multispectral and hyperspectral imaging are powerful, nondestructive methods that measure light reflectance across specific wavelengths to assess plant health and detect stress. Multispectral platforms capture data from a limited number of broad bands, whereas hyperspectral platforms scan hundreds of contiguous narrow spectral bands, providing high-resolution biochemical and structural information (Lowe et al. 2017; Terentev et al. 2022). Key vegetation indices such as NDVI, PRI, PSRI, and SIPI correlate well with chlorophyll content, pigment composition, water status, and leaf senescence, serving as effective indicators of nutrient deficiency, drought, salinity, and pathogen infection. Hyperspectral imaging enables early stress detection by revealing subtle spectral changes in pigments and water absorption features, often days before visual symptoms appear (Feng et al. 2020). Advanced data modeling approaches, including support vector machines (SVMs), quadratic discriminant analysis (QDA), and convolutional neural networks (CNNs), enhance the quantification and classification of stress responses. These imaging platforms are highly scalable, ranging from drone- and satellite-based systems to handheld sensors, facilitating high-throughput phenotyping, crop monitoring, and precision input management (Nansen et al. 2024).

#### 13.3.2.4 Raman and Infrared Spectroscopy

Raman and infrared (IR) spectroscopy, including Fourier-transform infrared (FTIR) and near-infrared (NIR) modes, provide molecular-level characterization of plant biochemistry by detecting chemical bond vibrational modes. These label-free, nondestructive methods allow analysis of live tissues—such as leaves, stems, roots, and seeds—without the need for sectioning or staining (Kolašinac et al. 2025). FTIR and NIR spectroscopy are particularly sensitive to polar bonds (e.g., O–H, C=O, N–H), enabling detection of variations in water content, proteins, carbohydrates, and cell wall components. Raman spectroscopy (RS), on the other hand, is more effective for analyzing nonpolar molecules such as lignin, carotenoids, lipids, and aromatic compounds. RS is less affected by water interference, allowing more accurate analysis of aqueous plant tissues (Payne and Kurouski 2021; Saletnik et al. 2024). These techniques can detect biochemical changes induced by oxidative stress, nutrient imbalances, pathogen infection, and salinity. For instance, reduced carotenoid peaks in Raman spectra may indicate ROS-mediated damage, while FTIR can reveal protein degradation or cell wall reorganization. When combined with chemometric and machine learning approaches (e.g., PCA, SVM, and CNN), Raman and FTIR/NIR spectroscopy achieve high accuracy in stress type identification, varietal discrimination, and trait prediction. Technological advancements such as surface-enhanced Raman spectroscopy (SERS) and spectroscopic imaging further enhance detection sensitivity and spatial resolution for plant stress diagnostics (Beć et al. 2020; Juárez and Kurouski 2024).

#### 13.3.2.5 Fluorescence-Based Biosensors

Fluorescence-based biosensors are at the forefront of noninvasive molecular imaging in plant biology. Genetically encoded biosensors, for example, use fluorescent proteins (FPs) such as GFP, YFP, or FRET-based pairs to report real-time changes in hormone levels, ion fluxes, redox states, and reactive oxygen species (ROS) within living cells (Balcerowicz et al. 2021). Biosensors can be broadly categorized into two groups: single-FP sensors, which alter fluorescence intensity or wavelength upon ligand binding, and FRET-based sensors, which detect changes in energy transfer between two fluorophores in response to molecular interactions. Examples include ABACUS and ABAleon for ABA, DII-Venus for auxin, and HyPer for hydrogen peroxide. These tools enable high-resolution visualization of key signaling molecules involved in stress responses, facilitating the study of hormone dynamics, redox regulation, and cell-type-specific signaling events in real time (Okumoto 2012; Rowe et al. 2023). Compared to conventional biochemical assays, biosensors offer spatiotemporal resolution, reversibility, and subcellular specificity, allowing longitudinal monitoring of stress biomarkers without harming the plant. Technological advances have improved brightness, spectral range, and photobleaching resistance, enabling multicolor imaging, ratiometric quantification, and use in intact, complex tissues (Okumoto 2012; Jones et al. 2014). Sensors such as ABACUS2, with enhanced affinity and higher signal-to-noise ratios, have revealed previously unseen phenomena, such as root-to-shoot ABA signaling under drought, highlighting their indispensable role in digital phenotyping and precision breeding. Their integration with imaging platforms and deployment in transgenic lines highlights their potential for early stress detection, physiological mapping, and crop improvement programs (Balcerowicz et al. 2021; Rowe et al. 2023). Together, these nondestructive sensing and imaging technologies provide a comprehensive toolkit for monitoring physiological and biochemical responses in stressed plants. By capturing early, often pre-symptomatic, changes in photosynthesis, temperature, spectral reflectance, molecular composition, and intracellular signaling, they offer a robust platform for identifying and validating physiological biomarkers (Okumoto 2012).

### 13.3.3 Pathway-Guided Breeding and Phenotyping

The rising intensity and frequency of droughts and other abiotic stresses due to climate change demand innovative strategies for wheat improvement. As one of the most widely cultivated cereals globally, wheat can benefit from the integration of stress-responsive biochemical pathways into breeding and phenotyping approaches, offering a transformative route to developing climate-resilient cultivars. This strategy exploits the plant's inherent biochemical stress responses—such as osmolyte accumulation, antioxidant synthesis, and stress protein expression—as early biomarkers to guide genotype selection and trait evaluation under adverse conditions (Rose et al. 2022). During drought stress, wheat plants activate a suite of biochemical processes that protect cellular integrity and maintain physiological stability. A prominent response is the accumulation of osmoprotectants such as proline, glycine betaine, and trehalose, which safeguard proteins and membranes by maintaining osmotic balance. Concurrently, antioxidant systems are upregulated to detoxify reactive oxygen species (ROS), including enzymatic components such as superoxide dismutase and ascorbate peroxidase, as well as nonenzymatic molecules like flavonoids and ascorbate. These stress-induced biochemical changes often precede visible phenotypic symptoms, making them excellent early detection biomarkers. Quantifying these metabolites allows breeders and researchers to screen genotypes based on their intrinsic biochemical tolerance, accelerating the selection of drought-resistant lines (Raza et al. 2023).

#### 13.3.3.1 Wheat Breeding Pipelines

Biochemical pathway markers are most effective when integrated with modern genomic and transcriptomic technologies. Marker-assisted selection (MAS) can exploit genes encoding key enzymes, such as *Δ$^1$-pyrroline-5-carboxylate synthetase* (*P5CS*), which is involved in proline biosynthesis, to select lines exhibiting enhanced metabolic responses to drought. Similarly, expression quantitative trait loci (eQTLs) and metabolite quantitative trait loci (mQTLs) can link metabolite profiles to specific genomic regions, enabling the development of molecular markers associated with stress-specific biochemical traits. Integration of multi-omics data—including metabolomics, proteomics, and transcriptomics—can reveal complex regulatory networks governing stress responses. This system-level knowledge supports ideotype breeding, in which wheat lines are designed with an optimized combination of traits such as effective osmotic adjustment, ROS detoxification capacity, and carbon partitioning efficiency, enhancing performance under water-limited conditions (Ali et al. 2023).

#### 13.3.3.2 Phenotypic Protocols

Parallel to genetic selection, stress-specific biochemical traits are increasingly integrated into phenotyping pipelines to improve accuracy and throughput. Traditional phenotyping relies on visible traits such as wilting, chlorosis, or biomass reduction, which typically appear after stress damage. In contrast, biochemical phenotyping captures early molecular changes that reflect a plant's intrinsic stress response mechanisms (Ali et al. 2023). Greenhouse and field experiments now routinely quantify biomarkers such as proline content (an osmotic stress marker), malondialdehyde (MDA; a lipid peroxidation indicator), antioxidant enzyme activities (e.g., catalase, peroxidase), and total phenolic and flavonoid contents (reflecting ROS scavenging capacity). These measurements, when combined with high-throughput phenotyping (HTP) platforms such as hyperspectral imaging and thermal cameras—which nondestructively monitor physiological traits like canopy temperature, stomatal conductance, and chlorophyll content—allow correlations between external sensor-derived data and internal metabolic states. This enables rapid screening of large wheat populations (Fiorani and Schurr 2013; Trapp et al. 2016). Applications in wheat and other cereals confirm the utility of biochemical markers in phenotyping and breeding. For example, drought-induced proline accumulation has been correlated with yield stability under water-limited conditions, making it a reliable selection trait. Co-proteomic and metabolomic studies in barley identified root metabolites such as aspartic and glutamic acids that influence grain weight under drought (Trapp et al. 2016). Selective phenotyping provides another powerful strategy, wherein extreme phenotypes (e.g., highly drought-tolerant versus susceptible lines) are biochemically profiled to identify metabolic differences associated with performance. These informations can guide large-scale phenotyping or inform genomic prediction models. By enabling early detection of stress responses and linking them with genetic selection, this approach accelerates the development of wheat varieties capable of thriving under increasingly erratic environmental conditions (Fiorani and Schurr 2013).

## 13.4 Recent Developments in the Use of Plant Biomarkers for Early Stress Detection in Wheat

### 13.4.1 Omics-Integrated Approach for Early Stress Marker Discovery in Wheat

It is important to identify stable molecular biomarkers that can predict and enhance stress tolerance. Comprehensive omics strategies, including transcriptomics, proteomics, and metabolomics, have proven to be valuable tools for dissecting gene expression changes and identifying stress-responsive molecular signatures in plants (Varadharajan et al. 2025). Table 13.2 summarizes plant stress detection tools, ranging from molecular markers and biochemical assays to biosensors, lab-on-chip devices, and in silico approaches.

Transcriptomic profiling has been instrumental in uncovering gene expression signatures associated with stress in wheat. High-throughput platforms such as RNA sequencing (RNA-seq) and microarrays enable the identification of differentially expressed genes (DEGs), long noncoding RNAs (lncRNAs), and microRNAs (miRNAs) that are activated in response to specific stressors. For example, under drought stress, wheat exhibits elevated levels of LEA proteins, dehydrins, and key transcription factors such as DREB, WRKY, MYB, and NAC, which are often involved in abscisic acid (ABA)-mediated signaling and osmoprotective pathways. Salinity stress is accompanied by upregulation of ion transporters (*HKT1*, *NHX1*, and *SOS1*) and antioxidant enzymes (SOD, CAT, and APX). Heat stress induces heat shock proteins (HSPs), heat shock transcription factors (HSFs), chaperones, and detoxifying enzymes, some of which overlap with drought-responsive gene networks. During pathogen stress, particularly from fungi like *Fusarium graminearum* or *Puccinia* spp., transcriptomic analyses reveal rapid activation of pattern-recognition receptors (PRRs), R genes, and defense-related hormones such as salicylic acid (SA), jasmonic acid (JA), and ethylene (ET). These transcriptomic signatures serve as molecular biomarkers and inform strategies for marker-assisted selection, transgenic development, and CRISPR-based genome engineering (Yousfi et al. 2016; Bashir et al. 2019; Li et al. 2022; Mikołajczak et al. 2023). While transcriptomics

**Table 13.2** Summary of diverse categories of tools and approaches used for detecting and analysing plant biomarkers

| Category | Type/tool | Application | Reference |
|---|---|---|---|
| Molecular biomarkers | microRNAs (miRNAs) | Stress-responsive regulation, biomarker discovery (e.g., *miR1119–MYC2* module) | Shamloo-Dashtpagerdi et al. (2023) |
| | Transcriptomics (RNA-seq, EST mining) | Identification of stress-inducible genes/miRNAs | Pandey et al. (2013) |
| | DNA markers (SSR, SCAR, and SNP) | Genotype differentiation, marker-assisted selection (MAS) | Singh et al. (2018) |
| | Epigenetic markers (methylation profiling) | Stress memory and gene expression regulation | Abdulraheem et al. (2024) |
| Biochemical tools | qPCR (SYBR green and TaqMan) | Gene-specific stress marker validation (e.g., *NAC, MYB,* and *HSP* genes) | Ali et al. (2023) |
| | ELISA (Sandwich and competitive) | Quantification of stress-related proteins (e.g., DON, WGA, HSPs, and PR proteins) | Ali et al. (2023) |
| | Western blot/immunodot | Stress-related protein expression analysis | Dong et al. (2024) |
| Biosensors | Ethylene biosensors (ArM and enzyme-based) | Monitoring abiotic/biotic stress signals (e.g., heat, flooding, pathogens) | Vong et al. (2019) |
| | ROS biosensors (fluorescent/redox probes) | Real-time detection of oxidative stress | Kim and Lee (2022) |
| | Electrochemical/ nanobiosensors | Detection of hormones, nutrients, and toxins (e.g., DON) | Kim and Lee (2022) |
| Lab-on-chip (LOC) | Microfluidic biosensing platforms | High-throughput screening for stress biomarkers (low-volume samples) | Chaturvedi et al. (2025) |
| | Paper-based LOC devices | Portable diagnostics in field conditions | Chaturvedi et al. (2025) |
| Remote sensing/ field-based | NDVI, NDWI, PRI, MSI, NDRE, and PSRI. | Real-time monitoring of water, pigment, senescence, and nutrient stress | Feng et al. (2020) |
| | CWSI, CTD (thermal indices) | Drought stress detection and irrigation scheduling | Pineda et al. (2021) |
| | Hyperspectral and multispectral imaging | Disease surveillance and early stress detection | Hernández-Clemente et al. (2019) |
| | Thermal cameras, UAVs, and Sentinel-2 data | Precision agriculture and phenotyping | Zhou et al. (2021) |

(continued)

**Table 13.2** (continued)

| Category | Type/tool | Application | Reference |
|---|---|---|---|
| In silico tools | miRNA target prediction (psRNATarget, miRanda) | Identification of miRNA-regulated stress pathways | Tahmasebi et al. (2019) |
| | Protein-protein interaction (STRING, BioGRID) | Mapping stress-responsive pathways | Jan et al. (2024) |
| | Machine learning (ML/DL models) | Stress classification, phenotype prediction | Juárez and Kurouski (2024) |
| | FFCNN (fast Fourier CNN) | Accurate detection of yellow rust, N-deficiency | Shi et al. (2023) |
| Multi-omics approaches | Integrative omics (transcriptomics + proteomics + metabolomics) | Comprehensive understanding of stress adaptation mechanisms | Jamil et al. (2020) |

provides understanding of gene regulation, proteomics and metabolomics offer functional validation by reflecting real-time molecular responses at the protein and metabolite levels. Metabolomics involves the measurement of small-molecule metabolites that indicate physiological changes in the plant. During the early phases of stress, wheat accumulates osmoprotectants such as proline and glycine betaine, exhibits increased antioxidant levels like ascorbate and glutathione, and undergoes extensive alterations in carbon and nitrogen metabolism. These metabolomic changes occur rapidly and often precede visible symptoms, making them reliable early response markers.

Proteomics, on the other hand, identifies stress-responsive proteins, their post-translational modifications, and interactions. Techniques such as LC–MS/MS, iTRAQ, and label-free quantification have enabled the detection of heat shock proteins (HSPs), ROS-scavenging enzymes, kinases, 14-3-3 proteins, and pathogenesis-related (PR) proteins as early indicators of abiotic and biotic stresses. Additionally, proteomic analyses have revealed detoxification enzymes and cell wall-modifying proteins that contribute to resistance against Fusarium head blight (FHB), a devastating wheat disease (Ali et al. 2019; Zenda et al. 2021; Fanelli et al. 2023).

The integration of transcriptomics, metabolomics, and proteomics provides a system-level view of stress responses. Multi-omics approaches address issues of biomarker reliability and are instrumental in the development of diagnostic tools, precision phenotyping platforms, and climate-smart breeding strategies. Combined, these technologies accelerate the improvement of wheat cultivars for enhanced resilience and performance under environmental stress conditions (Roychowdhury et al. 2023).

### 13.4.2 Molecular and Biochemical Biomarker Detection Tools

Early and precise identification of wheat stress responses is crucial to effective crop management and the development of stress-resistant cultivars. Modern techniques such as biosensors, qPCR, ELISA, and microRNA profiling provide efficient platforms for molecular and biochemical screening under both laboratory and field conditions.

#### 13.4.2.1 Biosensors and Lab-on-Chip Devices

Biosensors and lab-on-chip (LOC) technologies enable noninvasive, real-time monitoring of stress markers such as ethylene and reactive oxygen species (ROS). These devices couple biological recognition components (e.g., enzymes, antibodies, and nucleic acids) with transducers for highly sensitive analyte detection. Ethylene biosensors, including artificial metalloenzyme-based systems, can detect early stress signals associated with drought, heat, or pathogen infection. Such biosensors, including ArM ethylene probes or receptor-engineered systems, transduce ethylene levels into measurable optical or electrochemical signals. ROS biosensors detect oxidizing species such as hydrogen peroxide ($H_2O_2$) and superoxide radicals, which accumulate rapidly during stress. These sensors often employ redox-sensitive fluorescent proteins or nanomaterial-based probes that exhibit observable fluorescence or conductivity changes upon ROS interaction. Lab-on-chip (LOC) platforms miniaturize these assays into microfluidic devices capable of multiplexed analysis with minimal sample volumes. LOC devices offer rapid, high-throughput diagnostics and are portable and easy to use, making them ideal for on-site field screening. Over the past decade, advances in nanomaterials, such as carbon nanotubes and graphene oxide, have improved sensor stability, specificity, and digital integration, establishing these technologies as key tools in precision phenotyping and stress diagnostics in wheat (Li and Suslick 2019; Vong et al. 2019; Kim and Lee 2022; Chaturvedi et al. 2025).

#### 13.4.2.2 qPCR and ELISA-Based Screening Kits

Quantitative PCR (qPCR) is the gold standard for measuring gene expression changes under stress, particularly for stress-inducible transcripts identified through transcriptomics or candidate gene approaches, allowing verification of key transcription factors. For example, in durum and bread wheat, *TaNAC29* and *TaNAC6* expression levels were accurately quantified using reference genes such as *GAPDH*, *EF1α*, and *β-tubulin*, which were experimentally validated for stability under salinity and water-deficit stress conditions. Advances like SYBR Green-based qPCR platforms enable gel-free and cost-effective detection of sequence-characterized amplified region (SCAR) markers. These systems have proven particularly useful in tracking resistance genes such as *Sr24*, *Sr26*, and *Sr31* against stem rust, as well as seed quality loci (*Pina*, *Pinb*, and *Glu-D1d*). qPCR is reported to be approximately 22% less expensive and more than tenfold faster than traditional agarose gel-based protocols, significantly improving efficiency for marker-assisted selection (MAS). Additionally, simple sequence repeat (SSR) markers associated with salt tolerance have been developed using genome-anchored information in tolerant and susceptible wheat lines. Phylogenetic and population structure analyses using these markers facilitated genotype discrimination based on salinity tolerance and supported their use in MAS programs (Vishwakarma et al. 2016; Mehta et al. 2021; Pfrieme et al. 2022). Concurrently, complementary protein-level assays, such as ELISA, provide measurement of stress-related proteins. Direct competitive ELISA kits have been developed for detecting deoxynivalenol (DON), a mycotoxin associated with Fusarium stress, while sandwich ELISAs target proteins such as wheat germ agglutinin (WGA). The toolkit is being further extended to detect heat shock proteins, pathogenesis-related (PR) proteins, and ROS-detoxifying enzymes, providing protein-level information alongside gene-level data. Together, qPCR and ELISA platforms offer a holistic strategy for marker-assisted selection, enabling both genetic and phenotypic validation of stress responses (Güll et al. 2007; Singh et al. 2018; Han et al. 2020).

#### 13.4.2.3 MicroRNAs as Regulatory Biomarkers

MicroRNAs (miRNAs) are a class of small noncoding RNAs that regulate gene expression post-transcriptionally and play central roles in wheat responses to abiotic and biotic stresses. Over 150 wheat miRNAs, including both conserved and novel families, have been identified through high-throughput sequencing and are responsive to drought, salinity, heat, and pathogen challenges. These miRNAs modulate key regulators such as antioxidant enzymes, hormone biosynthesis genes, and transcription factors. Notably, the *miR1119–MYC2* module contributes to drought adaptation by regulating ABA levels, ROS detoxification, and water retention (Kajla et al. 2023; Shamloo-Dashtpagerdi et al. 2023). Expression analyses show that drought-resistant wheat genotypes exhibit lower *miR1119* and higher *MYC2* expression under water-deficit conditions, correlating with enhanced physiological responses, including improved water retention and stress alleviation. Similar regulatory functions have been reported for *tae-miR9674a* (stomatal regulation and ROS homeostasis) and *tae-miR1120c/miR9664* (salt-alkali tolerance). Differential expression of other miRNAs, such as *tae-miR395* (involved in sulphur metabolism and drought response) and *tae-miR156* (linked to hormone signaling and secondary metabolism), further highlights their involvement in diverse stress pathways. Stress-inducible miRNAs, being early, specific, and genotype-associated markers, hold immense potential for precision breeding. Incorporating them into molecular screening can improve the selection of climate-resilient wheat genotypes (Pandey et al. 2013; Wang et al. 2023b; Sharma et al. 2025).

### 13.4.3 Physiological and Field-Based Monitoring Approaches

Early and accurate detection of wheat physiological stress is essential for minimizing yield loss and guiding targeted agricultural interventions. Remote sensing techniques across spectral, thermal, and computational domains are powerful, nondestructive tools for continuous stress assessment under field conditions. These techniques rely on measurements of canopy reflectance, temperature, and spectral changes associated with stress-driven physiological processes (Sun et al. 2019). Spectral indices such as the Normalized Difference Vegetation Index (NDVI), Photochemical Reflectance Index (PRI), and Normalized Difference Water Index (NDWI) enable monitoring of photosynthetic activity, chlorophyll content, and water status. NDVI, for example, measures canopy greenness and biomass, while NDWI and the Moisture Stress Index (MSI) detect water stress by exploiting NIR and SWIR absorption features. The Transformed Chlorophyll Absorption in Reflectance Index (TCARI/OSAVI) and Normalized Difference Red Edge Index (NDRE) are effective for early detection of nitrogen and pigment stress, especially over heterogeneous soil backgrounds. Thermal sensing complements spectral data by providing a direct window into transpiration and canopy temperature. Indices such as the Crop Water Stress Index (CWSI) and canopy temperature depression (CTD) serve as reliable indicators for drought, irrigation scheduling, and stomatal closure. Hyperspectral and multispectral imaging systems, delivered via satellites (e.g., Sentinel-2), unmanned aerial vehicles (UAVs or drones), or proximal spectrometers, enable high-resolution monitoring of biotic and abiotic stress (Hernández-Clemente et al. 2019). These systems can detect subtle reflectance changes caused by infections or nutrient deficiencies well before visible symptoms appear. For instance, the Plant Senescence Reflectance Index (PSRI) and Normalized Difference Angle Index (NDAI) have been applied to detect senescence and disease pressure in wheat. Advances in machine learning and deep learning have further improved the accuracy and interpretability of remote sensing-based diagnostics. Notably, fast Fourier convolutional neural networks (FFCNNs) can discriminate visually similar stresses, such as yellow rust and nitrogen deficiency, by combining spectral features with physiological priors like photochemical vegetation indices. This approach reduces misclassification and computational burden, enabling high-throughput, explainable stress detection. Remote sensing-associated physiological indicators therefore provide scalable, noninvasive techniques for early stress detection in wheat (Shi et al. 2023).

## 13.5 Evolution in the Contemporary Hi-Tech Era

High-throughput phenotyping (HTP) technologies are increasingly transforming contemporary plant science by enabling real-time, scalable, and noninvasive stress detection in crops. Integration with biomarker data is vital for accelerating plant breeding and precision agriculture. This complementarity facilitates the development of stress-tolerant crop varieties by speeding the discovery and deployment of genotypes with desirable traits, as illustrated in Fig. 13.2. HTP platforms, ranging from satellite and UAV remote sensing to proximal sensing systems, provide extensive evaluation of physiological and morphological characteristics under diverse field conditions. Although these platforms share the common goal of phenotypic data assessment, their applications differ in research and crop production. In plant breeding, HTP enables the evaluation of many small experimental plots, allowing breeders to identify superior genotypes and accelerate genetic gain through improved trait selection. In contrast, precision agriculture applications survey entire fields, detecting early symptoms of stress caused by nutrient deficiency, pest infestation, or drought, thereby enabling timely, site-specific management interventions (Fig. 13.2). Each HTP strategy has distinct strengths. Satellite imagery offers broad spatial coverage, making it suitable for large-scale field monitoring and breeding trials, though its utility can be limited by weather conditions, resolution, and revisit frequency. UAV-based phenotyping provides higher temporal and spatial resolution, ideal for small-plot measurements and detailed stress detection. Proximal sensing systems, including hand-held sensors and mobile devices, deliver high-resolution quantification of plant traits and are essential for monitoring extensive breeding nurseries and accurately estimating specific physiological attributes. The next step involves integrating HTP data with molecular biomarker information—genomics, transcriptomics, and metabolomics—as outlined in Fig. 13.2. This integration enhances quantitative trait predictive models, improves genomic selection accuracy, and supports decision-making in precision agriculture. Interoperability of phenotypic and biomarker data provides a robust foundation for sustainable crop improvement through early, accurate, and cost-effective screening of abiotic and biotic stress responses on a large scale (Chawade et al. 2019).

Its integration with artificial intelligence (AI) further enhances its functionality, providing a data-intensive and scalable platform for evaluating plant responses to environmental stresses, particularly high-temperature stress, as illustrated in Fig. 13.2. Leveraging advanced imaging modalities—such as multispectral, hyperspectral, thermal, fluorescence, and 3D imaging—these platforms enable accurate, nondestructive quantification of morphological and physiological traits throughout plant development. Traits associated with thermotolerance, including canopy temperature, transpiration efficiency, flowering duration, and root architecture, serve as reliable indicators of heat-resistant genotypes (Kundu et al. 2024).

Deep learning algorithms and AI play a central role in analyzing large-scale phenotypic data, mapping genotype–phenotype relationships, and modeling predictive traits. When integrated with robotic greenhouse systems, UAVs, environmental sensors, and software platforms such as VOC-SCREEN and CronoRoot, they enable in situ phenotyping and real-time stress monitoring. This integrative approach significantly enhances genetic gain, increases breeding program throughput and accuracy, and facilitates the development of heat-tolerant cultivars suitable for changing climatic conditions, as demonstrated in Fig. 13.2. Both indoor (e.g., LemnaTec and Phenovator) and field-based (e.g., LeasyScan, CropQuant, and UAVs) high-throughput plant phenotyping (HT3P) platforms provide rapid, nondestructive measurements of plant traits. These infrastructures combine imaging technologies—including RGB, hyperspectral, and thermal cameras—to monitor dynamic characteristics such as plant height, canopy temperature, and visible stress symptoms. By correlating phenotypic observations with genomic information and predicting genotype-by-environment interactions, HT3P platforms allow for

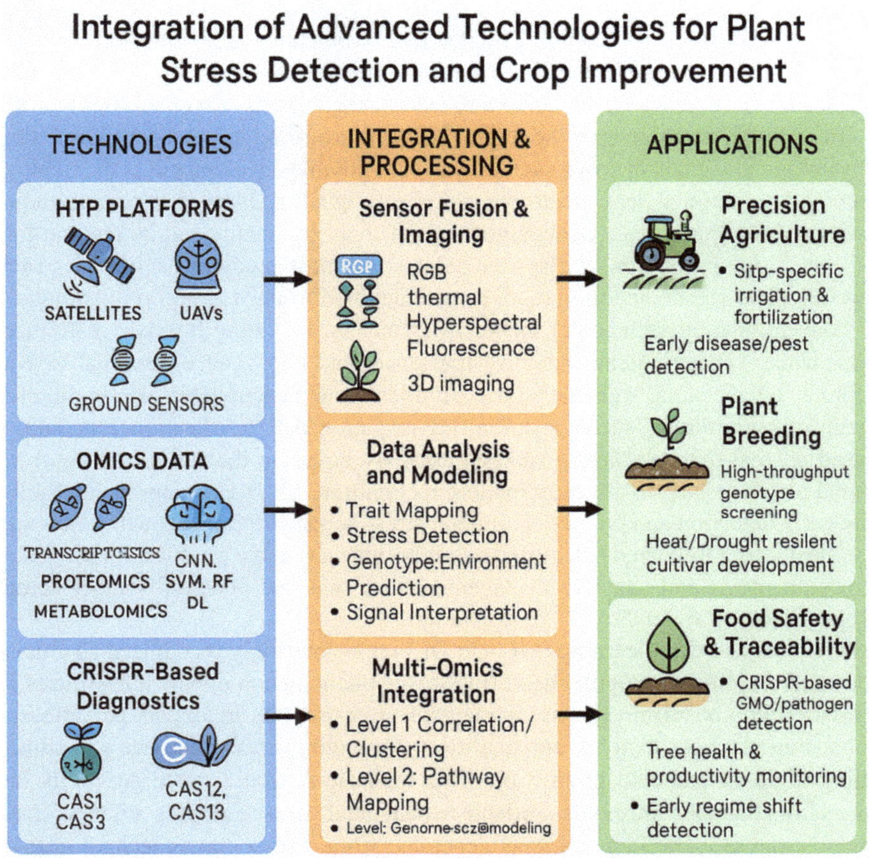


**Fig. 13.2** Integration of advanced high-throughput platforms, multi-omics data, and CRISPR-based diagnostics with sensor fusion, imaging, and data analysis enables precision agriculture, plant breeding, and food safety applications, enhancing stress detection and crop improvement

accurate trait predictions, thereby improving breeding efficiency. Advances in sensor fusion, automation, and simulation-based phenotyping continue to enhance data accuracy while enabling climate-smart crop development at scale (Li et al. 2021).

Beyond phenotyping, machine learning (ML) and deep learning (DL) are increasingly applied to stress detection and trait selection, addressing the urgent challenge of global food security. With the global population projected to reach 10 billion by 2050, early detection of plant stress using AI-driven tools is critical for sustaining agricultural productivity. ML and DL techniques analyze large datasets generated by contemporary high-throughput phenotyping (HTP) platforms, including drones, satellite systems, and ground-based sensors. Algorithms such as convolutional neural networks (CNNs), support vector machines (SVMs), and random forests (RFs) are employed to classify, recognize, and predict plant stress scenarios. These models automatically process and interpret data, shortening breeding cycles and enhancing crop yields through early detection of pests and diseases. Advanced phenotyping infrastructures, such as TERRA-REF, when integrated with ML/DL pipelines, achieve unprecedented accuracy and speed, transforming raw data into actionable intelligence for breeding and field management (Gou et al. 2024).

In addition, machine learning (ML) platforms are being tailored for interpreting physiological signals in crops such as wheat. A notable development is the deployment of multimodal leaf electrophysiological signal systems that can decouple water and nitrogen stress. Using algorithms such as gradient boosting, random forest, and ridge regression, the system selects the most predictive model for each physiological signal, enabling more precise and noninvasive monitoring compared to conventional environmental sensors. Parameters, including intrinsic resistance, impedance, capacitive and inductive reactance, and capacitance, are analyzed to capture specific stress signatures. These models provide actionable information for precision agriculture, such as optimized irrigation and fertilization scheduling, based on real-time physiological responses. By covering the full workflow from signal preprocessing to advanced pattern recognition, ML-based approaches facilitate early detection and prediction of plant stress across different growth stages and leaf positions. Optimized ML systems thus provide a reliable platform for precision stress diagnosis and adaptive management in wheat and potentially other cereal crops (Zhang et al. 2025b).

Machine and deep learning (ML and DL) are increasingly recognized as crucial for understanding and regulating hormesis, the phenomenon in which low doses of stressors elicit positive responses in plants. Hormesis holds significant potential for enhancing plant productivity and quality by promoting stress tolerance and stimulating higher metabolite production. This biphasic response paradigm shifts the focus from damage thresholds to adaptive, beneficial stress responses, often referred to as "eustress." The large datasets generated from diverse omics technologies—phenomics, proteomics, genomics, metabolomics, and transcriptomics—require ML and DL models for accurate interpretation and integration. These computational tools are essential for characterizing the complex dose-response behaviours associated with hormesis. Frameworks such as identification, classification, quantification, and prediction (ICQP), ML models including SVMs, random forests, and CNNs have been applied to image-based plant stress classification and quantitative modeling. Deep learning, in particular, is well-suited for capturing the nonlinear and high-dimensional nature of hermetic responses. Together, these technologies enhance early stress detection, trait performance prediction, and precise, sustainable crop management. The integration of physiological, phenotypic, and molecular data through AI-based approaches represents a paradigm shift in breeding, management, and climate-resilient crop development (Mahood et al. 2020).

CRISPR/Cas diagnostic platforms represent a groundbreaking advancement in plant biotechnology, providing efficient, rapid, and portable systems for the detection of plant diseases and genetic analysis. Leveraging the high specificity of Cas endonucleases and guide RNAs, these platforms enable accurate identification of nucleic acid targets without the complexity of conventional methods such as RT-PCR or sequencing. Cas12 and Cas13 enzymes are of particular interest: Cas12a exhibits collateral cleavage of single-stranded DNA, while Cas13 targets RNA, forming the basis of highly sensitive detection platforms like SHERLOCK and DETECTR. When combined with isothermal amplification techniques such as RPA or LAMP, these platforms deliver rapid and robust diagnostics suitable for point-of-care applications. Notable uses include the early detection of plant pathogens and the differentiation of genetically modified organisms (GMOs), both critical for effective disease management and regulatory compliance. Future developments toward more portable, multiplexed, and robust CRISPR-based systems will further advance crop protection and accelerate breeding programs, ultimately enhancing agricultural productivity and food security (Tanny et al. 2023).

CRISPR-based diagnostic protocols offer a promising solution for point-of-care (POC) detection of plant diseases, addressing the limitations of traditional approaches such as ELISA, PCR, and genome sequencing, which require specialized equipment and skilled personnel. These diagnostics exploit the target nucleic acid specificity and collateral cleavage properties of CRISPR–Cas systems. Integration with lateral flow assays, microfluidics, and smartphone technology enhances field usability by simplifying readouts and enabling real-time data sharing. Their portability and ease of use make CRISPR diagnostics particularly suitable for agriculture, allowing farmers to rapidly and accurately detect plant diseases (Kasfy et al. 2024).

CRISPR/Cas technology has revolutionary potential to enhance food traceability, ensuring food safety and authenticity. Beyond its renowned genome-editing applications, CRISPR/Cas is now being explored for non-editing functions such as molecular traceability. Its precision and specificity make it ideal for identifying genetic markers in food products, detecting adulteration, allergens, misbranding, and verifying crop species. Compared to conventional PCR-based systems, CRISPR-based approaches are faster, less complex, and more sensitive, enabling real-time, on-site analysis. When combined with DNA barcoding and omics technologies, CRISPR/Cas becomes an even more versatile and high-throughput platform for next-generation food traceability systems (Farinati et al. 2024).

Multi-omics integration (MOI) approaches are emerging as a central focus in plant systems biology, providing a unifying framework for integrating and modeling large-scale data from transcriptomics, proteomics, and metabolomics. Table 13.2 lists plant stress detection tools, including computational in silico approaches and integrative multi-omics strategies, for understanding stress responses. MOI strategies can be classified into three levels based on analytical complexity. Level 1 involves correlation, clustering, and multivariate analyses, suitable for preliminary data exploration. Level 2 incorporates pathway mapping and co-expression analyses to highlight molecular interactions. Level 3, the most advanced, involves genome-scale differential analysis, facilitating metabolic flux prediction and dynamic modeling. While each level offers distinct advantages, challenges remain, such as data variability and incomplete annotation. The review emphasizes the importance of intuitive tools and machine learning algorithms in enhancing annotation and MOI utility. These approaches provide a robust platform for elucidating complex biological systems and informing crop improvement programs, particularly for newcomers to MOI pipelines. Multi-omics integration will continue to drive advances in plant biology and precision agriculture (Jamil et al. 2020).

Chloroplasts play a critical role in regulating plant growth, defense, and adaptability, placing them at the forefront of strategies to address global agricultural challenges. Beyond photosynthesis, chloroplasts are central to hormone biosynthesis and stress signaling, regulating development and immunity. Recent research in chloroplast genomics, transcriptomics, translation science, and proteomics has further clarified their regulatory functions and cross-talk with nuclear-encoded proteins. Targeted interventions, such as the expression of elongation factor EF-2 and flavodiol proteins, enhance heat and drought tolerance, respectively. Integrating omics information with nanotechnology and synthetic biology provides effective strategies for engineering stress-tolerant crops. By highlighting the dynamic role of chloroplasts in growth-defense trade-offs, this research highlights their importance in sustaining agriculture through advanced biotechnological approaches (Zhang et al. 2025a).

Multilayer integration of multi-omics data—including genomic, transcriptomic, proteomic, and metabolomic layers—enables a holistic understanding of biological systems. Tools such as iCluster, SNF, MOFA, and PARADIGM facilitate disease subtyping, biomarker discovery, and pathway analysis. Despite challenges like heterogeneity and complex preprocessing requirements, integrative approaches are essential in precision agriculture and systems biology, driving functional investigation, diagnostics, and crop improvement (Subramanian et al. 2020).

Technological advances in forest and crop monitoring are central to climate-smart forestry (CSF) and agriculture, providing real-time, in situ measurements. Wireless sensor networks (WSNs), Internet of Things (IoT) devices, UAVs, and satellites, often complemented by citizen science, enhance spatial and temporal resolution. These technologies support early warning of ecological changes, analysis of plant and tree growth, and climate-adaptive management (Torresan et al. 2021).

Unmanned aerial vehicles (UAVs) equipped with thermal and multispectral sensors provide real-time, high-resolution monitoring of plant water stress, disease, and yield potential. Key applications include the use of the Crop Water Stress Index (CWSI) and integration with RGB and multispectral imagery to improve stress detection and optimize irrigation scheduling (Messina and Modica 2020). Thermal sensor-equipped UAVs detect canopy temperature variations, enabling precise water stress assessment and irrigation management. Combining thermal data with spectral indices such as NDVI and advanced image processing further enhances accuracy. UAV-based systems offer scalable, cost-effective solutions for improving crop yields under varying environmental conditions (Awais et al. 2023; Barbedo 2019).

## 13.6 Statistical Insight in Developing and Using Plant Biomarkers

Meta-analysis provides a robust framework to evaluate biomarker efficacy across genotypes and environments by integrating data from multiple studies, thereby identifying markers with stable predictive power for precision breeding, diagnostics, and stress monitoring. A key advance is metaGE, a tool for analyzing genotype-by-environment (G × E) interactions in multi-environment GWAS that adjusts for QTL variability across environments and identifies allelic effects linked to environmental factors, as shown in Arabidopsis (flowering QTLs under competition) and maize (yield QTLs under drought) (De Walsche et al. 2025). In rice, a computational network-based approach integrating co-expression, protein interactions, and pathways revealed drought tolerance mechanisms, with ABA signaling, respiration, and cell wall reorganization upregulated and photosynthesis and growth suppressed. A cross-species meta-analysis, CSA: Drought, integrating 17 datasets from Arabidopsis, rice, wheat, and barley, identified 225 conserved drought-adaptive genes, validated in Brachypodium and highlighting osmotic adjustment as a conserved response (Shaar-Moshe et al. 2015). Similarly, meta-analysis in Salicaceae showed that chemical defenses and herbivore performance were primarily genetically controlled, while growth and nitrogen composition were environmentally regulated, with limited plasticity in defenses (Barker et al. 2019). A tissue-specific meta-analysis of 12 medicinal plants identified 880 DEGs and 31 miRNAs regulating secondary metabolite pathways, providing information on hormone signaling, epigenetic regulation, and compound enhancement strategies (Tahmasebi et al. 2019).

Advances in statistical and computational tools have strengthened biomarker discovery and plant disease research. Multivariate, Bayesian, and nonparametric methods, alongside decision theory and epidemic modeling, have advanced plant pathology, though adoption in Asian rice research remains limited (Nayak et al. 2018). Untargeted metabolomics combined with machine learning identified metabolites predicting potato tuber traits such as bruising susceptibility and chip quality, validated across environments (Steinfath et al. 2010). Bayesian Gaussian and Binomial models have improved diagnostic evaluation for *Sclerotinia* in oilseed rape, using ROC-based thresholds to estimate disease probability and account for sampling effects (Makowski et al. 2008). In sunflower, a gene expression-based biomarker enabled accurate, genotype-independent monitoring of plant water status across field trials (Marchand et al. 2013). Complementing these, wearable sensors now provide in situ, real-time monitoring of growth, stress, and environmental conditions, with advances in lightweight, flexible, and highly sensitive materials expanding their role in precision agriculture (Li et al. 2024).

Such innovations are especially relevant for wheat, a crop facing multiple stresses in South Asia. The Indian Wheat Genomics Initiative exploits the National Gene Bank to identify genes and QTLs associated with resistance, stress tolerance, and nutritional quality, while integrating genomic selection, rapid breeding pipelines, and germplasm evaluation to develop climate-resilient wheat (Kumar et al. 2022). Septoria nodorum blotch (SNB), caused by *Parastagonospora nodorum*, remains a major challenge, with integrated disease management, precise fungicide timing, and forecasting models essential for yield protection, although resistance breeding is constrained by complex genetic control (Ficke et al. 2018). Rising demand and climate pressures underscore the importance of pre-breeding strategies, including trait-based crossing, genomic prediction, and physiological screening for heat and drought tolerance. Global programs such as *bred to adapt* and CIMMYT's stress adapted trait yield nurseries (SATYNs), together with international gene bank resources, support the development of durable germplasm for local breeding programs and help ensure regional food security (Sukumaran et al. 2021; Joshi et al. 2007).

## 13.7 Future Projections and Global Challenges

The application of biomarkers for stress detection in wheat holds great promise, but several global challenges must be addressed to realize their full potential. A major limitation is the lack of universal protocols for biomarker identification, validation, and deployment across diverse agro-climatic regions, as biomarkers identified in one genotype or environment may not perform consistently elsewhere due to genotype-by-environment interactions. Developing standardized procedures for sample collection, data analysis, and cross-environment validation is essential to improve reproducibility and reliability (Ali et al. 2023). Environmental heterogeneity further complicates their application, as variations in soil type, temperature, rainfall, and disease pressure can influence gene or protein expression, potentially limiting the predictive power of biomarkers beyond their initial trial conditions. Consequently, stress biomarkers must be validated across multiple wheat cultivars and field environments, ideally through multilocation trials and integrated databases (Grzesiak et al. 2018). Another critical challenge is translating laboratory-based assays into scalable, affordable, and accessible tools, particularly in low-resource agricultural settings. Portable, user-friendly biosensor platforms such as paper-based assays, lateral-flow devices, and smartphone-connected readers offer potential solutions but must be designed to function reliably under field conditions with minimal technical expertise (Bergua et al. 2022). Adoption also depends on fostering interdisciplinary collaboration, as biomarker development spans molecular biology, plant physiology, agronomy, breeding, and data science. Coordination among these disciplines—from discovery to field validation and digital integration—is vital, and training programs and collaborative platforms can help bridge gaps and accelerate innovation (Bergua et al. 2022). Emerging technologies promise to transform biomarker systems: AI-based digital phenotyping, plant-wearable sensors, and nano-biosensors enable dynamic, real-time monitoring of plant volatiles, sap chemistry, temperature fluctuations, and stress signals, complementing conventional molecular markers (Wang et al. 2023a, b). Successful deployment requires farmer-centric design prioritizing affordability, simplicity, and compatibility with local infrastructure, alongside safeguards for data privacy, connectivity, and user training. By overcoming practical, environmental, and technological challenges, such integrated approaches can advance the development of climate-resilient wheat systems.

## 13.8 Conclusion

Plant biomarkers are quantitative indicators of wheat's physiological status and serve as early warning signals of impending stress before visible damage occurs, directly reflecting gene, protein, or metabolite changes induced by stress and enabling timely interventions such as triggered irrigation or foliar applications. Monitoring stress-responsive molecules allows early diagnosis and precision management, with significant advances in identifying specific markers for various stresses: biochemically, osmolytes and oxidative products like proline and malondialdehyde (MDA) consistently accumulate under drought and other stresses, while physiological traits such as chlorophyll fluorescence and stomatal conductance change predictably under heat or water stress. At the molecular level, numerous stress-responsive proteins and genes have been characterized; for example, overexpression of ***HVA1*** preserves membrane stability and reduces electrolyte leakage under drought or salinity, and metabolomics and proteomics have identified sets of metabolites and enzymes whose patterns are diagnostic of specific stresses. Emerging technologies are transforming biomarker applications, as multimodal analytics integrating hyperspectral imaging, fluorescence sensing, and AI can rapidly scan large fields and detect stress signals more sensitively than manual approaches, while field-portable devices like hand-held Raman leaf-clip spectrometers enable in situ analysis of metabolites for real-time stress diagnosis prior to symptom manifestation. Parallel advances in high-throughput genomics, transcriptomics, and metabolomics are accelerating the discovery of new stress markers across the wheat genome and metabolome, dramatically enhancing the speed, sensitivity, and scale of stress phenotyping from lab-scale datasets to farm-scale monitoring. Nevertheless, translating laboratory findings to field practice remains challenging, as most biomarkers require validation across diverse environments due to heterogeneity in soil, climate, and genotype, and stress phenotyping is still "costly and time-consuming," limiting its application in breeding and management. Bridging this lab-to-field gap will necessitate rugged, low-cost sensors, automated platforms capable of operating under real growing conditions, and integration of biomarker data into breeding programs and decision-support tool. By addressing these challenges through interdisciplinary collaboration and extensive field testing, biomarker-based approaches can be mainstreamed to improve wheat stress resilience, productivity, and climate-adaptive agricultural management.

## References

Abdulraheem MI, Xiong Y, Moshood AY et al (2024) Mechanisms of plant epigenetic regulation in response to plant stress: recent discoveries and implications. Plants (Basel) 13:163. https://doi.org/10.3390/plants13020163

Aina O, Bakare OO, Fadaka AO et al (2024) Plant biomarkers as early detection tools in stress management in food crops: a review. Planta 259:60. https://doi.org/10.1007/s00425-024-04333-1

Ali A, Cao J, Jiang H et al (2019) Unraveling molecular and genetic studies of wheat (*Triticum aestivum* L.) resistance against factors causing pre-harvest sprouting. Agronomy 9:117. https://doi.org/10.3390/agronomy9030117
Ali I, Anwar S, Ali A et al (2023) Biochemical and phenological characterization of diverse wheats and their association with drought tolerance genes. BMC Plant Biol 23:326. https://doi.org/10.1186/s12870-023-04278-9
Alonso S, Gautam K, Iglesias-Moya J et al (2024) Crosstalk between ethylene, jasmonate and ABA in response to salt stress during germination and early plant growth in *Cucurbita pepo*. Int J Mol Sci 25:8728. https://doi.org/10.3390/ijms25168728
Awais M, Li W, Cheema MJM et al (2023) UAV-based remote sensing in plant stress imaging using high-resolution thermal sensor for digital agriculture practices: a meta-review. Int J Environ Sci Technol 20:1135–1152. https://doi.org/10.1007/s13762-021-03801-5
Balcerowicz M, Shetty KN, Jones AM (2021) Fluorescent biosensors illuminating plant hormone research. Plant Physiol 187:590–602. https://doi.org/10.1093/plphys/kiab278
Barbedo JGA (2019) A review on the use of unmanned aerial vehicles and imaging sensors for monitoring and assessing plant stresses. Drones 3:40. https://doi.org/10.3390/drones3020040
Barker HL, Holeski LM, Lindroth RL (2019) Independent and interactive effects of plant genotype and environment on plant traits and insect herbivore performance: a meta-analysis with Salicaceae. Funct Ecol 33:422–435. https://doi.org/10.1111/1365-2435.13249
Bashir K, Matsui A, Rasheed S et al (2019) Recent advances in the characterization of plant transcriptomes in response to drought, salinity, heat, and cold stress. F1000Res 8:F1000 Faculty Rev-658. https://doi.org/10.12688/f1000research
Bates LS, Waldren RP, Teare ID (1973) Rapid determination of free proline for water-stress studies. Plant Soil 39:205–207. https://doi.org/10.1007/BF00018060
Beć KB, Grabska J, Bonn GK et al (2020) Principles and applications of vibrational spectroscopic imaging in plant science: a review. Front Plant Sci 11:1226. https://doi.org/10.3389/fpls.2020.01226
Bergua JF, Álvarez-Diduk R, Idili A et al (2022) Low-cost, user-friendly, all-integrated smartphone-based microplate reader for optical-based biological and chemical analyses. Anal Chem 94:1271–1285. https://doi.org/10.1021/acs.analchem.1c04491
Chaerle L, Lenk S, Leinonen I et al C (2009) Multi-sensor plant imaging: towards the development of a stress-catalogue. Biotechnol J 4:1152–1167. https://doi.org/10.1002/biot.200800242
Chaturvedi A, Tripathi D, Ranjan R (2025) Nano-enabled biosensors in early detection of plant diseases. Front Nanotechnol 7:1545792. https://doi.org/10.3389/fnano.2025.1545792
Chawade A, van Ham J, Blomquist H et al (2019) High-throughput field-phenotyping tools for plant breeding and precision agriculture. Agronomy 9:258. https://doi.org/10.3390/agronomy9050258
Das S, Shil S, Rime J et al (2025) Phytohormonal signaling in plant resilience: advances and strategies for enhancing abiotic stress tolerance. Plant Growth Regul 105:329–360. https://doi.org/10.1007/s10725-025-01279-6
De Walsche A, Vergne A, Rincent R et al (2025) metaGE: investigating genotype × environment interactions through GWAS meta-analysis. PLoS Genet 21:e1011553. https://doi.org/10.1371/journal.pgen.1011553
Devireddy AR, Zandalinas SI, Fichman Y et al (2021) Integration of reactive oxygen species and hormone signaling during abiotic stress. Plant J 105:459–476. https://doi.org/10.1111/tpj.15010
Dong B, Liu Y, Huang G et al (2024) Plant NAC transcription factors in the battle against pathogens. BMC Plant Biol 24:958. https://doi.org/10.1186/s12870-024-05636-x
Fanelli G, Kuzmanović L, Giovenali G et al (2023) Untargeted metabolomics reveals a multifaceted resistance response to Fusarium head blight mediated by the *Thinopyrum elongatum Fhb7E* locus transferred via chromosome engineering into wheat. Cells 12:1113. https://doi.org/10.3390/cells12081113
Farinati S, Devillars A, Gabelli G et al (2024) How helpful may be a CRISPR/Cas-based system for food traceability? Foods 13:3397. https://doi.org/10.3390/foods13213397

Feng X, Zhan Y, Wang Q et al (2020) Hyperspectral imaging combined with machine learning as a tool to obtain high-throughput plant salt-stress phenotyping. Plant J 101:1448–1461. https://doi.org/10.1111/tpj.14597

Ficke A, Cowger C, Bergstrom G et al (2018) Understanding yield loss and pathogen biology to improve disease management: *Septoria nodorum* blotch—a case study in wheat. Plant Dis 102:696–707. https://doi.org/10.1094/PDIS-09-17-1375-FE

Fiorani F, Schurr U (2013) Future scenarios for plant phenotyping. Annu Rev Plant Biol 64:267–291. https://doi.org/10.1146/annurev-arplant-050312-120137

Gou C, Zafar S, Fatima et al (2024) Machine and deep learning: artificial intelligence application in biotic and abiotic stress management in plants. Front Biosci (Landmark Ed) 29:20. https://doi.org/10.31083/j.fbl2901020

Grzesiak S, Hordyńska N, Szczyrek P et al (2018) Variation among wheat (*Triticum aestivum* L.) genotypes in response to the drought stress: I—selection approaches. J Plant Interact 14:30–44. https://doi.org/10.1080/17429145.2018.1550817

Güll I, Wirth M, Gabor F (2007) Development of a sensitive and reliable ELISA for quantification of wheat germ agglutinin. J Immunol Methods 318:20–29. https://doi.org/10.1016/j.jim.2006.07.025

Han L, Li YT, Jiang JQ et al (2020) Development of a direct competitive ELISA kit for detecting deoxynivalenol contamination in wheat. Molecules 25:50. https://doi.org/10.3390/molecules25010050

He J, Ning K, Naznin A et al (2025) Technological advances in imaging and modelling of leaf structural traits: a review of heat stress in wheat. J Exp Bot:eraf070. https://doi.org/10.1093/jxb/eraf070

Heath RL, Packer L (1968) Photoperoxidation in isolated chloroplast I. Kinetics and stoichiometry of fatty acid peroxidation. Arch Biochem Biophys 125:189–198. https://doi.org/10.1016/0003-9861(68)90654-1

Hernández-Clemente R, Hornero A, Mottus M et al (2019) Early diagnosis of vegetation health from high-resolution hyperspectral and thermal imagery: lessons learned from empirical relationships and radiative transfer modelling. Curr Forestry Rep 5:169–183. https://doi.org/10.1007/s40725-019-00096-1

Jamil IN, Remali J, Azizan KA et al (2020) Systematic multi-omics integration (MOI) approach in plant systems biology. Front Plant Sci 11:944. https://doi.org/10.3389/fpls.2020.00944

Jan S, Rustgi S, Barmukh R et al (2024) Advances and opportunities in unraveling cold-tolerance mechanisms in the world's primary staple food crops. Plant Genome 17:e20402. https://doi.org/10.1002/tpg2.20402

Jones AM, Danielson JA, Manojkumar SN et al (2014) Abscisic acid dynamics in roots detected with genetically encoded FRET sensors. Elife 3:e01741. https://doi.org/10.7554/eLife.01741

Joshi AK, Mishra B, Chatrath R et al (2007) Wheat improvement in India: present status, emerging challenges and future prospects. Euphytica 157:431–446. https://doi.org/10.1007/s10681-007-9385-7

Juárez ID, Kurouski D (2024) Contemporary applications of vibrational spectroscopy in plant stresses and phenotyping. Front Plant Sci 15:1411859. https://doi.org/10.3389/fpls.2024.1411859

Kajla M, Roy A, Singh IK et al (2023) Regulation of the regulators: transcription factors controlling biosynthesis of plant secondary metabolites during biotic stresses and their regulation by miRNAs. Front Plant Sci 14:1126567. https://doi.org/10.3389/fpls.2023.1126567

Kasfy SH, Hia FT, Islam T (2024) Do CRISPR-based disease diagnosis methods qualify as point-of-care diagnostics for plant diseases? Nucleus 67:11–24. https://doi.org/10.1007/s13237-024-00489-z

Khan MA, Anwar S, Abbas M et al (2025) Impacts of climate change on cotton production and advancements in genomic approaches for stress resilience enhancement. J Cotton Res 8:17. https://doi.org/10.1186/s42397-025-00223-3

Kim MY, Lee KH (2022) Electrochemical sensors for sustainable precision agriculture—a review. Front Chem 10:848320. https://doi.org/10.3389/fchem.2022.848320

Kolašinac SM, Mladenović M, Pećinar I et al (2025) Raman and FT-IR spectroscopy coupled with machine learning for the discrimination of different vegetable crop seed varieties. Plants 14:1304. https://doi.org/10.3390/plants14091304
Kumar S, Jacob SR, Mir RR et al (2022) Indian wheat genomics initiative for harnessing the potential of wheat germplasm resources for breeding disease-resistant, nutrient-dense, and climate-resilient cultivars. Front Genet 13:834366. https://doi.org/10.3389/fgene.2022.834366
Kumar R, Das SP, Choudhury BU et al (2024) Advances in genomic tools for plant breeding: harnessing DNA molecular markers, genomic selection, and genome editing. Biol Res 57:80. https://doi.org/10.1186/s40659-024-00562-6
Kundu S, Saini DK, Meena RK et al (2024) High-throughput phenotyping and AI technologies for deciphering crop resilience to heat stress. Plant Physiol Rep 29:699–715. https://doi.org/10.1007/s40502-024-00821-4
Legendre R, Basinger NT, van Iersel MW (2021) Low-cost chlorophyll fluorescence imaging for stress detection. Sensors 21:2055. https://doi.org/10.3390/s21062055
Li Z, Suslick KS (2019) Colorimetric sensor array for monitoring CO and ethylene. Anal Chem 91:797–802. https://doi.org/10.1021/acs.analchem.8b04321
Li D, Quan C, Song Z et al (2021) High-throughput plant phenotyping platform (HT3P) as a novel tool for estimating agronomic traits from the lab to the field. Front Bioeng Biotechnol 8:623705. https://doi.org/10.3389/fbioe.2020.623705
Li N, Liu T, Guo F et al (2022) Identification of long non-coding RNA–microRNA–mRNA regulatory modules and their potential roles in drought stress response in wheat (*Triticum aestivum* L.). Front. Plant Sci 13:1011064. https://doi.org/10.3389/fpls.2022.1011064
Li XH, Li MZ, Li JY et al (2024) Wearable sensor supports in-situ and continuous monitoring of plant health in precision agriculture era. Plant Biotechnol J 22:1516–1535. https://doi.org/10.1111/pbi.14283
Lowe A, Harrison N, French AP (2017) Hyperspectral image analysis techniques for the detection and classification of the early onset of plant disease and stress. Plant Methods 13:80. https://doi.org/10.1186/s13007-017-0233-z
Mahlein AK (2016) Plant disease detection by imaging sensors—parallels and specific demands for precision agriculture and plant phenotyping. Plant Dis 100:241–251. https://doi.org/10.1094/PDIS-03-15-0340-FE
Mahood EH, Kruse LH, Moghe GD (2020) Machine learning: a powerful tool for gene function prediction in plants. Appl Plant Sci 8:e11376. https://doi.org/10.1002/aps3.11376
Makowski D, Denis J-B, Ruck L et al (2008) A Bayesian approach to assess the accuracy of a diagnostic test based on plant disease measurement. Crop Prot 27:1187–1193. https://doi.org/10.1016/j.cropro.2008.02.006
Marchand G, Mayjonade B, Varès D et al (2013) A biomarker based on gene expression indicates plant water status in controlled and natural environments. Plant Cell Environ 36:2175–2189. https://doi.org/10.1111/pce.12127
McAusland L, Atkinson JA, Lawson T et al (2019) High throughput procedure utilising chlorophyll fluorescence imaging to phenotype dynamic photosynthesis and photoprotection in leaves under controlled gaseous conditions. Plant Methods 15:109. https://doi.org/10.1186/s13007-019-0485-x
Mehta G, Muthusamy SK, Singh GP et al (2021) Identification and development of novel salt-responsive candidate gene based SSRs (cg-SSRs) and MIR gene based SSRs (mir-SSRs) in bread wheat (*Triticum aestivum*). Sci Rep 11:2210. https://doi.org/10.1038/s41598-021-81698-3
Messina G, Modica G (2020) Applications of UAV thermal imagery in precision agriculture: state of the art and future research outlook. Remote Sens 12:1491. https://doi.org/10.3390/rs12091491
Mikołajczak K, Kuczyńska A, Krajewski P et al (2023) Transcriptome profiling disclosed the effect of single and combined drought and heat stress on reprogramming of gene expression in barley flag leaf. Front Plant Sci 13:1096685. https://doi.org/10.3389/fpls.2022.1096685
Morales M, Munné-Bosch S (2019) Malondialdehyde: facts and artifacts. Plant Physiol 180:1246–1250. https://doi.org/10.1104/pp.19.00405

Munaweera TIK, Jayawardana NU, Rajaratnam R et al (2022) Modern plant biotechnology as a strategy in addressing climate change and attaining food security. Agric Food Secur 11:26. https://doi.org/10.1186/s40066-022-00369-2

Myers RJ, Peláez-Vico MÁ, Fichman Y (2024) Functional analysis of reactive oxygen species-driven stress systemic signalling, interplay and acclimation. Plant Cell Environ 47:2842–2851. https://doi.org/10.1111/pce.14894

Nakashima K, Yamaguchi-Shinozaki K, Shinozaki K (2014) The transcriptional regulatory network in the drought response and its crosstalk in abiotic stress responses including drought, cold, and heat. Front Plant Sci 5:170. https://doi.org/10.3389/fpls.2014.00170

Nansen C, Savi PJ, Mantri A (2024) Methods to optimize optical sensing of biotic plant stress—combined effects of hyperspectral imaging at night and spatial binning. Plant Methods 20:163. https://doi.org/10.1186/s13007-024-01292-2

NASA Technical Reports Server (NTRS) Available at: https://ntrs.nasa.gov/. Accessed 1 Aug 2025

Nayak P, Mukherjee AK, Pandit E et al (2018) Application of statistical tools for data analysis and interpretation in rice plant pathology. Rice Sci 25:1–18. https://doi.org/10.1016/j.rsci.2017.07.001

Okumoto S (2012) Quantitative imaging using genetically encoded sensors for small molecules in plants. Plant J 70:108–117. https://doi.org/10.1111/j.1365-313X.2012.04910.x

Pandey B, Gupta OP, Pandey DM et al (2013) Identification of new stress-induced microRNA and their targets in wheat using computational approach. Plant Signal Behav 8:e23932. https://doi.org/10.4161/psb.23932

Panozzo A, Bolla PK, Barion G et al (2025) Phytohormonal regulation of abiotic stress tolerance, leaf senescence and yield response in field crops: a comprehensive review. Biotech 14:14. https://doi.org/10.3390/biotech14010014

Park B, Wi S, Chung H et al (2024) Chlorophyll fluorescence imaging for environmental stress diagnosis in crops. Sensors 24:1442. https://doi.org/10.3390/s24051442

Payne WZ, Kurouski D (2021) Raman spectroscopy enables phenotyping and assessment of nutrition values of plants: a review. Plant Methods 17:78. https://doi.org/10.1186/s13007-021-00781-y

Pérez-Bueno ML, Pineda M, Barón M (2019) Phenotyping plant responses to biotic stress by chlorophyll fluorescence imaging. Front Plant Sci 10:1135. https://doi.org/10.3389/fpls.2019.01135

Pérez-Clemente RM, Vives V, Zandalinas SI et al (2013) Biotechnological approaches to study plant responses to stress. Biomed Res Int 2013:654120. https://doi.org/10.1155/2013/654120

Pérez-Llorca M, Pollmann S, Müller M (2023) Ethylene and jasmonates signaling network mediating secondary metabolites under abiotic stress. Int J Mol Sci 24:5990. https://doi.org/10.3390/ijms24065990

Pfrieme AK, Ruckwied B, Habekuß A et al (2022) Identification and validation of quantitative trait loci for wheat dwarf virus resistance in wheat (*Triticum* spp.). Front Plant Sci 13:828639. https://doi.org/10.3389/fpls.2022.828639

Pineda M, Barón M, Pérez-Bueno ML (2021) Thermal imaging for plant stress detection and phenotyping. Remote Sens 13:68. https://doi.org/10.3390/rs13010068

Raza A, Mubarik MS, Sharif R et al (2023) Developing drought-smart, ready-to-grow future crops. Plant Genome 16:e20279. https://doi.org/10.1002/tpg2.20279

Raza A, Salehi H, Bashir S et al (2024) Transcriptomics, proteomics, and metabolomics interventions prompt crop improvement against metal(loid) toxicity. Plant Cell Rep 43:80. https://doi.org/10.1007/s00299-024-03153-7

Rolly NK, Mun BG, Yun BW (2021) Insights into the transcriptional regulation of branching hormonal signaling pathways genes under drought stress in Arabidopsis. Genes 12:298. https://doi.org/10.3390/genes12020298

Rose T, Wilkinson M, Lowe C et al (2022) Novel molecules and target genes for vegetative heat tolerance in wheat. Plant Environ Interact 3:264–289. https://doi.org/10.1002/pei3.10096

Rowe J, Grangé-Guermente M, Exposito-Rodriguez M et al (2023) Next-generation ABACUS biosensors reveal cellular ABA dynamics driving root growth at low aerial humidity. Nat Plants 9:1103–1115. https://doi.org/10.1038/s41477-023-01447-4

Roychowdhury R, Das SP, Gupta A et al (2023) Multi-omics pipeline and omics-integration approach to decipher plant's abiotic stress tolerance responses. Genes 14:1281. https://doi.org/10.3390/genes14061281

Saletnik A, Saletnik B, Zaguła G et al (2024) Raman spectroscopy for plant disease detection in next-generation agriculture. Sustainability 16:5474. https://doi.org/10.3390/su16135474

Şen A, Gümüş T, Temel A et al (2024) Biochemical and proteomic analyses in drought-tolerant wheat mutants obtained by gamma irradiation. Plants 13:2702. https://doi.org/10.3390/plants13192702

Shaar-Moshe L, Hübner S, Peleg Z (2015) Identification of conserved drought-adaptive genes using a cross-species meta-analysis approach. BMC Plant Biol 15:111. https://doi.org/10.1186/s12870-015-0493-6

Shamloo-Dashtpagerdi R, Shahriari AG, Tahmasebi A et al (2023) Potential role of the regulatory *miR1119-MYC2* module in wheat (*Triticum aestivum* L.) drought tolerance. Front Plant Sci 14:1161245. https://doi.org/10.3389/fpls.2023.1161245

Sharma P, Mishra S, Kaur A et al (2025) Novel and conserved drought-responsive microRNAs expression analysis in root tissues of wheat (*Triticum aestivum* L.) at reproductive stage. Front Plant Sci 16:1581542. https://doi.org/10.3389/fpls.2025.1581542

Shi Y, Han L, González-Moreno P et al (2023) A fast Fourier convolutional deep neural network for accurate and explainable discrimination of wheat yellow rust and nitrogen deficiency from Sentinel-2 time series data. Front Plant Sci 14:1250844. https://doi.org/10.3389/fpls.2023.1250844

Singh A, Roychoudhury A (2023) Abscisic acid in plants under abiotic stress: crosstalk with major phytohormones. Plant Cell Rep 42:961–974. https://doi.org/10.1007/s00299-023-03013-w

Singh AK, Chaurasia S, Kumar S et al (2018) Identification, analysis and development of salt responsive candidate gene based SSR markers in wheat. BMC Plant Biol 18:249. https://doi.org/10.1186/s12870-018-1476-1

Steinfath M, Strehmel N, Peters R et al (2010) Discovering plant metabolic biomarkers for phenotype prediction using an untargeted approach. Plant Biotechnol J 8:900–911. https://doi.org/10.1111/j.1467-7652.2010.00516.x

Subramanian I, Verma S, Kumar S et al (2020) Multi-omics data integration, interpretation, and its application. Bioinform Biol Insights 14:1177932219899051. https://doi.org/10.1177/1177932219899051

Sukumaran S, Krishna H, Singh K et al (2021) Progress and prospects of developing climate resilient wheat in South Asia using modern pre-breeding methods. Curr Genomics 22:440–449. https://doi.org/10.2174/1389202922666210705125006

Sun H, Feng M, Xiao L et al (2019) Assessment of plant water status in winter wheat (*Triticum aestivum* L.) based on canopy spectral indices. PLoS One 14:e0216890. https://doi.org/10.1371/journal.pone.0216890

Tahmasebi A, Ebrahimie E, Pakniyat H et al (2019) Tissue-specific transcriptional biomarkers in medicinal plants: application of large-scale meta-analysis and computational systems biology. Gene 691:114–124. https://doi.org/10.1016/j.gene.2018.12.056

Tanny T, Sallam M, Soda N et al (2023) CRISPR/Cas-based diagnostics in agricultural applications. J Agric Food Chem 71:11765–11788. https://doi.org/10.1021/acs.jafc.3c00913

Terentev A, Dolzhenko V, Fedotov A et al (2022) Current state of hyperspectral remote sensing for early plant disease detection: a review. Sensors 22:757. https://doi.org/10.3390/s22030757

Torresan C, Garzón MB, O'Grady M et al (2021) A new generation of sensors and monitoring tools to support climate-smart forestry practices. Can J For Res 51:1751–1765. https://doi.org/10.1139/cjfr-2020-0295

Trapp JJ, Urrea CA, Zhou J et al (2016) Selective phenotyping traits related to multiple stress and drought response in dry bean. Crop Sci 56:1460–1472. https://doi.org/10.2135/cropsci2015.05.0281

Ul Haq S, Khan A, Ali M et al (2019) Heat shock proteins: dynamic biomolecules to counter plant biotic and abiotic stresses. Int J Mol Sci 20:5321. https://doi.org/10.3390/ijms20215321

Varadharajan V, Rajendran R, Muthuramalingam P et al (2025) Multi-omics approaches against abiotic and biotic stress—a review. Plants 14:865. https://doi.org/10.3390/plants14060865

Vishwakarma G, Saini A, Das BK et al (2016) Rapid and convenient gel-free screening of SCAR markers in wheat using SYBR green-based melt-profiling. Plant Breed 135:643–653. https://doi.org/10.1111/pbr.12415

Vong K, Eda S, Kadota Y et al (2019) An artificial metalloenzyme biosensor can detect ethylene gas in fruits and Arabidopsis leaves. Nat Commun 10:5746. https://doi.org/10.1038/s41467-019-13758-2

Wang C, He T, Zhou H et al (2023a) Artificial intelligence enhanced sensors: enabling technologies to next-generation healthcare and biomedical platform. Bioelectron Med 9:17. https://doi.org/10.1186/s42234-023-00118-1

Wang L, Bai X, Qiao Y et al (2023b) *tae-miR9674a*, a microRNA member of wheat, confers plant drought and salt tolerance through modulating the stomata movement and ROS homeostasis. Plant Biotechnol Rep 17:471–488. https://doi.org/10.1007/s11816-022-00787-5

Wang P, Liu WC, Han C et al (2024) Reactive oxygen species: multidimensional regulators of plant adaptation to abiotic stress and development. J Integr Plant Biol 66:330–367. https://doi.org/10.1111/jipb.13601

Yang CY, Zhang YC, Hou YL (2025) Assessing water status in rice plants in water-deficient environments using thermal imaging. Bot Stud 66:6. https://doi.org/10.1186/s40529-025-00452-4

Yousfi S, Márquez AJ, Betti M et al (2016) Gene expression and physiological responses to salinity and water stress of contrasting durum wheat genotypes. J Integr Plant Biol 58:48–66. https://doi.org/10.1111/jipb.12359

Yu H, Xiao H, Abou-Elwafa SF et al (2024) Molecular evolution and interaction of ROS with ion transport for plant abiotic stresses. New Plant Prot 1:22. https://doi.org/10.1002/npp2.22

Zenda T, Liu S, Dong A et al (2021) Omics-facilitated crop improvement for climate resilience and superior nutritive value. Front Plant Sci 12:774994. https://doi.org/10.3389/fpls.2021.774994

Zhang C, Li W, Wu Y et al (2025a) Chloroplast functionality at the interface of growth, defense, and genetic innovation: a multi-omics and technological perspective. Plants 14:978. https://doi.org/10.3390/plants14060978

Zhang S, Du X, Zhang B et al (2025b) Impedance-driven decoupling water–nitrogen stress in wheat: a parallel machine learning framework leveraging leaf electrophysiology. Agronomy 15:1612. https://doi.org/10.3390/agronomy15071612

Zhou Z, Majeed Y, Naranjo GD et al (2021) Assessment for crop water stress with infrared thermal imagery in precision agriculture: a review and future prospects for deep learning applications. Comput Electron Agric 182:106019. https://doi.org/10.1016/j.compag.2021.106019

# Integrated Strategies for Enhancing Drought Stress Tolerance in Wheat (*Triticum aestivum* L.)

# 14

Ian Moot and Vijay Tiwari

**Abstract**

Bread wheat (*Triticum aestivum*; $2n = 6x = 42$; AABBDD) is a key component of global food security. Nevertheless, it faces significant challenges from abiotic stresses, particularly heat and drought. Climate change is projected to increase the frequency of drought on agricultural land, further threatening wheat yields and intensifying food insecurity. Drought stress, caused by limited water availability, elevated soil salinity, and extreme temperatures, disrupts key developmental stages in wheat, such as tillering, flowering, and grain filling, resulting in substantial yield reductions. Given that drought tolerance is a complex, polygenic trait, a comprehensive approach that integrates advanced breeding with omics technologies is required. Approaches such as genome-wide association studies (GWAS), transcriptomics, and genome engineering enable the identification of genetic determinants underlying drought resilience in *Triticum* species. Investigating the broad genetic diversity present in wild wheat relatives facilitates the discovery of novel drought-resistance genes. Introgression of these genes into modern bread wheat can improve productivity and stability under increasingly arid conditions. The combination of conventional breeding with omics-based strategies is critical for developing drought-resilient wheat varieties, thereby enhancing global food security and meeting the challenges of contemporary agriculture.

## 14.1 Introduction

Global food security is increasingly challenged by the projected rise in world population to 9.7 billion by 2050 (UN DESA 2022; FAO 2022). The concurrent intensification of climate change and urbanization threatens staple crops vital to global

I. Moot (✉) · V. Tiwari
Department of Plant Science and Landscape Architecture, University of Maryland, College Park, MD, USA
e-mail: imoot@umd.edu



R. Tiwari et al. (eds.), *Wheat Stress Management*,
https://doi.org/10.1007/978-981-92-0902-6_14

diets. Wheat (*Triticum aestivum*), which provides a primary food source for over 35% of the global population, has experienced a decline in yield growth from 5.7% annually in the mid-twentieth century to an estimated 1.7%, with this rate projected to persist through 2050 (FAO 2022). Worsening drought conditions, especially in arid and semi-arid wheat-producing regions, further heighten the urgency of achieving stable yield improvements.

Drought stress represents one of the most significant constraints on global crop production, a challenge intensified by climate change-induced reductions in water availability. Recent studies indicate a 29% increase in global drought frequency over the past two decades, with extreme drought events causing substantial yield losses in wheat-producing regions. In the United States alone, drought-related economic losses average $8 billion annually, with repercussions extending throughout global food supply chains. The impact is particularly severe in regions such as East Africa, where recurrent drought cycles result in chronic food insecurity for millions (FAO 2022). Advanced satellite monitoring technologies, including high-resolution vegetation health (VH) indices, have become essential for early drought detection and proactive management (Liu et al. 2021).

Projections suggest a substantial increase in severe water scarcity (SWS) events across major wheat-producing regions. Without targeted climate mitigation under the RCP8.5 scenario, up to 60% of global wheat-growing areas may experience concurrent SWS events by 2100, a sharp rise from the current 15%, with the frequency expected to double between 2041 and 2070 (Trnka et al. 2019; Williams et al. 2020). Drought stress impairs photosynthetic efficiency, chlorophyll synthesis, and cellular expansion, resulting in reduced biomass, smaller leaves, and diminished root growth. Wheat is particularly susceptible to drought during critical developmental stages, including germination, tillering, and grain filling, highlighting the need to develop drought-resilient varieties to address these persistent challenges (Wang et al. 2021a).

Addressing these challenges requires adopting innovative strategies to enhance drought tolerance in wheat (Fig. 14.1). Current approaches include breeding for adaptive root architecture, applying genomics-assisted selection, and employing advanced phenotyping tools to identify drought-resilient traits. Through these advanced methodologies, researchers seek to develop wheat varieties with improved drought resilience (Singh et al. 2025). This chapter examines recent progress in drought tolerance strategies for wheat, highlighting their potential to strengthen global food security amid escalating environmental pressures.

## 14.2 Physiological and Biochemical Responses to Drought Stress

### 14.2.1 Physiological Basis of Drought Tolerance in Wheat

Drought tolerance in wheat results from coordinated physiological and biochemical processes that influence plant perception, responses, and survival under

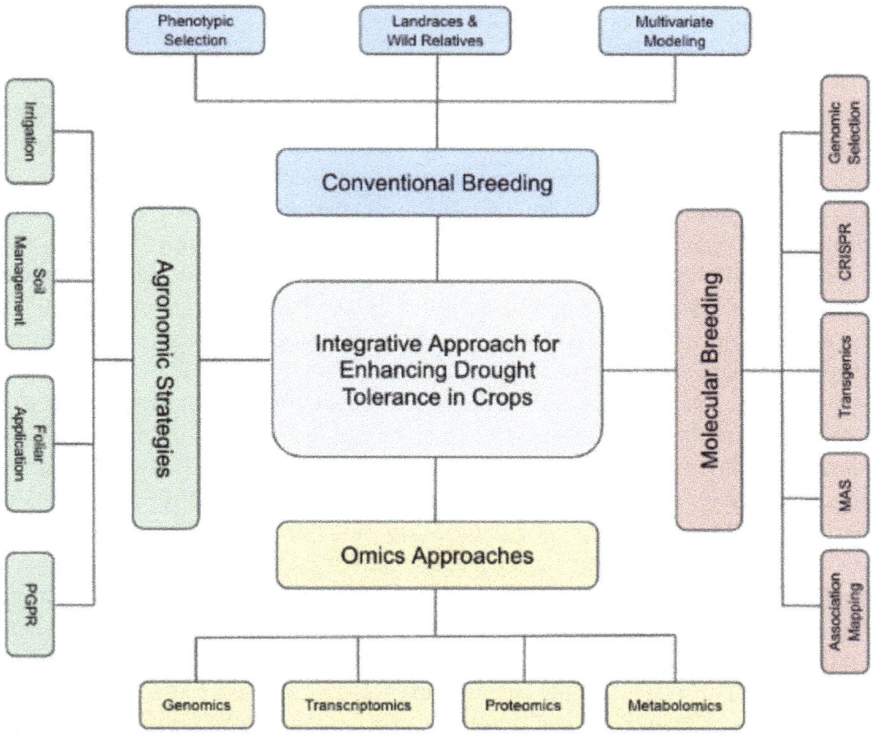


**Fig. 14.1** A comprehensive strategy for improving crop drought tolerance through the integration of conventional breeding, molecular techniques, omics technologies, and agronomic strategies

water-deficit conditions (Abid et al. 2018; Akter et al. 2023). These mechanisms regulate water balance, cellular homeostasis, and metabolic stability, collectively determining the capacity to sustain growth and productivity during drought (Guo et al. 2018; Itam et al. 2020). Characterizing these fundamental responses offers critical insight into the biological basis of drought resilience and forms the basis for agronomic management and genetic improvement strategies.

## 14.2.2 Mechanisms of Drought Tolerance

### 14.2.2.1 Stomatal Regulation and Water Use Efficiency

Stomatal regulation in higher plants constitutes a primary physiological response to drought stress and is essential for balancing water conservation with carbon assimilation. During periods of water deficit, plants typically reduce transpiration by closing stomata, which restricts $CO_2$ entry into the leaf and limits Calvin cycle activity. This process results in decreased photosynthetic rates and growth. However, reduced stomatal conductance can enhance water use efficiency (WUE) under drought by minimizing water loss relative to carbon gain (Ullah et al. 2023a, b). Therefore, stomatal behavior serves as a fundamental mechanism by which plants adapt to reduced water availability.

Water-use efficiency is a crucial factor in drought adaptation, especially for crops grown in rainfed and semi-arid environments. Enhanced WUE is linked to physiological and morphological traits, including reduced transpiration due to smaller or wax-coated leaves, deeply embedded stomata, and increased stomatal responsiveness to environmental signals. Beyond leaf-level adaptations, effective water acquisition through extensive root systems, characterized by deep, highly branched, and hairy roots, promotes higher WUE by improving access to soil moisture. An extensive root network not only optimizes water uptake but also strengthens drought perception and signalling, thereby facilitating integrated whole-plant stress responses (Wang et al. 2024).

Stomatal regulation in bread wheat under drought conditions involves both isohydric and anisohydric strategies, with stomatal conductance serving as a key determinant of drought tolerance. Drought-tolerant wheat lines frequently demonstrate an anisohydric response during moderate drought by maintaining elevated stomatal conductance through the repression of abscisic acid (ABA) synthesis in spikes and flag leaves. This approach supports sustained photosynthesis and growth during the initial stages of stress. As drought severity increases, these genotypes shift to an isohydric response, closing stomata at higher drought thresholds to conserve water. In contrast, drought-sensitive wheat lines generally exhibit an isohydric response from the onset of stress, closing stomata early to minimize water loss, which results in reduced gas exchange and photosynthetic capacity. Delayed stomatal closure during early drought stages has been associated with enhanced drought tolerance and increased grain yield in wheat (Onyemaobi et al. 2021).

At the molecular level, stomatal behavior in wheat is regulated by complex hormonal and signaling pathways. Water deficit typically induces ABA accumulation, which reduces stomatal conductance and photosynthetic activity. Wheat plants can partially offset these drought-induced limitations through coordinated osmotic adjustment and signaling mechanisms. Recent research has identified E3 ubiquitin ligases, including seven in Absentia (SINA) proteins, as important modulators of drought responses by influencing stomatal movement and promoting lateral root development, thereby enhancing drought tolerance (Ma et al. 2023c). Additionally, OB buffer-based stomatal closure assays provide a controlled method for quantifying genotype-specific differences in stomatal sensitivity. When combined with ABA or salt treatments, this approach effectively distinguishes wheat genotypes based on their stomatal responsiveness to drought and salinity stress (Guo et al. 2024).

#### 14.2.2.2 Osmotic Adjustment and Accumulation of Osmolytes

Osmotic adjustment is a fundamental physiological process across plant species that enables cells to maintain water status and sustain metabolic activity during drought stress. This process involves the regulated accumulation of solutes, which lowers cellular osmotic potential, preserves turgor pressure, and prevents dehydration-induced damage (Lv et al. 2022a, b). As a result, effective osmotic adjustment allows plant tissues to maintain physiological functions despite declining external water availability.

In wheat, osmotic potential is a key indicator of drought tolerance, reflecting the plant's capacity to maintain cellular turgor through regulated solute accumulation under water-limited conditions. This relationship was demonstrated by Verbeke et al. (2022), who examined drought-sensitive wheat subjected to controlled pre- and post-anthesis drought stress. By quantifying relative water content (RWC), xylem water potential, osmotic potential, turgor pressure, and relative osmolyte content (ROC) in stem and flag leaf tissues, the authors captured dynamic shifts in water potential gradients, illustrating how osmotic adjustment buffers wheat tissues against desiccation during drought.

At the molecular level, osmotic adjustment relies on the synthesis and accumulation of compatible solutes, which are small, nontoxic molecules that increase cellular osmolarity without disrupting metabolic processes. These osmolytes include proline, glycine betaine, soluble sugars, polyols, and selected inorganic ions. Collectively, these compounds lower osmotic potential, stabilize membranes and proteins, and maintain enzymatic activity under stress conditions (Yu et al. 2022). Beyond their role in osmotic balance, many osmolytes also contribute to redox homeostasis by scavenging reactive oxygen species (ROS), thereby mitigating oxidative damage associated with drought-induced cellular dehydration. This coordinated response is regulated by drought-responsive gene networks that control both osmolyte biosynthesis and antioxidant defense pathways (Ozturk et al. 2021a; b).

Several osmo-protectants, particularly proline, exhibit multifunctional roles in plant drought responses. In addition to acting as osmolytes and antioxidants, these compounds serve as molecular chaperones and signaling molecules that influence gene expression and stress-response pathways (Kolupaev et al. 2022). The integration of osmotic regulation, redox buffering, and signaling functions through osmolyte accumulation enhances whole-plant drought resilience by conserving water, protecting photosynthetic machinery, and stabilizing biomass production during extended periods of water deficit.

#### 14.2.2.3 Reactive Oxygen Species (ROS) Scavenging and Antioxidant Defense

Drought stress induces the production of reactive oxygen species (ROS) in plants, resulting in oxidative damage and the initiation of programmed cell death. The primary ROS generated during drought conditions include singlet oxygen ($1O_2$), superoxide radicals ($O_2^-$), hydrogen peroxide ($H_2O_2$), and hydroxyl radicals ($OH^-$), all of which disrupt cellular homeostasis by oxidizing lipids, proteins, and nucleic acids. The accumulation of ROS is strongly associated with drought-induced stomatal closure, which restricts $CO_2$ uptake for photosynthesis. This reduction in $CO_2$ fixation limits Calvin cycle activity and disrupts the photosynthetic electron transport chain due to a shortage of terminal electron acceptors, thereby increasing ROS production in chloroplasts (Kopecká et al. 2023).

To counteract ROS-induced damage, plants initiate a coordinated antioxidant defense system in response to disruptions in redox homeostasis. Moderate elevations in ROS function as signaling molecules that induce both enzymatic and nonenzymatic antioxidant pathways, facilitating the detoxification of excess ROS and the restoration of cellular redox equilibrium (Gupta et al. 2024a). This regulatory mechanism enables plants to withstand temporary oxidative stress and prevents irreversible cellular injury.

In wheat, drought tolerance is primarily determined by the efficiency and responsiveness of antioxidant defense mechanisms, rather than by morphological traits such as trichome development. Wheat employs a range of biochemical and physiological strategies, including the maintenance of chlorophyll content, osmotic adjustment via accumulation of compatible solutes such as proline, optimized root-to-shoot allocation, and robust activation of antioxidant enzymes (Abid et al. 2018; Qayyum et al. 2021). The principal enzymatic components of the wheat antioxidant system, superoxide dismutase (SOD), catalase (CAT), peroxidase (POD), ascorbate peroxidase (APX), monodehydroascorbate reductase (MDHAR), dehydroascorbate reductase (DHAR), glutathione reductase (GR), and glutathione peroxidase (GPX) work collectively to detoxify ROS and safeguard cellular structures under water-deficit conditions (Mahmood et al. 2020; Sallam et al. 2019; Nyaupane et al. 2024a, b).

The capacity for reactive oxygen species (ROS) scavenging in wheat has been extensively characterized through integrated biochemical assays and molecular analyses. Superoxide accumulation is visualized and quantified using nitro blue tetrazolium (NBT) staining, whereas malondialdehyde (MDA) content serves as a widely accepted biochemical marker for lipid peroxidation and oxidative damage. Additionally, transcriptional regulation of antioxidant pathways is evaluated by quantitative real-time PCR (qRT-PCR) targeting genes encoding key enzymes, including ascorbate peroxidase (APX), iron superoxide dismutase (FeSOD), catalase (CAT), and aldo-keto reductase (AKR). Collectively, these methodologies provide comprehensive insights into the regulation and efficacy of ROS detoxification mechanisms in wheat under drought stress (Vijayaraghavareddy et al. 2022). The regulation of ROS production and activation of efficient antioxidant defenses constitute essential components of drought tolerance in wheat. Elucidating the genetic and physiological regulation of these systems is critical to enhancing oxidative stress resilience.

## 14.2.3 Biochemical Markers of Drought Tolerance

### 14.2.3.1 Osmo-Protectants

Proline functions as an osmo-protectant and is essential for the pentose phosphate pathway, which regulates cellular redox balance and sustains antioxidants in their active forms. Consequently, proline is critical for preventing oxidative damage, including scavenging $O_2^-$ radicals within thylakoid membranes (Raeisi et al. 2022). Its biosynthesis is regulated by enzymes such as $\Delta^1$-pyrroline-5-carboxylate synthetase (P5CS), which catalyzes the conversion of glutamate to pyrroline-5-carboxylate (P5C), and $\Delta^1$-pyrroline-5-carboxylate reductase (P5CR), which reduces P5C to facilitate proline accumulation. Glycine betaine, another major osmo-protectant, is synthesized via betaine aldehyde dehydrogenase (BADH) and fulfils a similar function in osmotic adjustment (Li et al. 2023). The coordinated biosynthesis of proline and glycine betaine alleviates drought-induced cellular dehydration by preserving osmotic balance and safeguarding cellular structures (Wang et al. 2021b; Du et al. 2022).

A combination of physiological, biochemical, and molecular methods is used to evaluate the contribution of osmo-protectants to drought tolerance in wheat. Proline accumulation is quantified through spectrophotometric colorimetric assays, offering a direct measure of osmolyte buildup under water deficit conditions (Nardino et al. 2022; Ghaffar et al. 2023). Glycine betaine content is similarly assessed using spectrophotometric techniques, which are associated with high osmotic adjustment in drought-tolerant wheat genotypes (Ud-din et al. 2024). Ionic balance, essential for osmotic regulation, is determined by measuring sodium and potassium concentrations using flame photometry or atomic absorption spectroscopy. Antioxidant activities related to osmo-protectant function are characterized using total phenolic content (TPC) assays, ABTS radical scavenging activity, and ferric reducing antioxidant power (FRAP) methods. These assays collectively evaluate the plant's ability to mitigate oxidative damage and maintain redox homeostasis during drought stress (Nardino et al. 2022). Together, these integrated approaches establish a comprehensive pathway for understanding the regulation of osmo-protectants such as proline and glycine betaine and their role in enhancing drought resilience in wheat.

#### 14.2.3.2 Antioxidant Enzymes

The elimination of reactive oxygen species (ROS) from plant cells is essential for maintaining membrane stability and cellular function during drought stress. This defense relies on the coordinated activation of enzymatic antioxidant systems that detoxify ROS and sustain redox homeostasis. Figure 14.2 demonstrates that drought perception triggers signal transduction cascades involving calcium signaling, ROS, and phytohormones, which together activate downstream transcriptional regulators responsible for antioxidant enzyme expression.

Principal antioxidant enzymes, such as superoxide dismutase (SOD), catalase (CAT), and peroxidase (POD), are integral to the elimination of ROS and serve as physiological markers for evaluating stress tolerance in crops (Wang et al. 2022a, 2022b, 2022c). Under drought conditions, drought-tolerant winter wheat cultivars consistently exhibit increased SOD and CAT activities, as well as higher sugar and flavonoid contents, indicating enhanced antioxidant capacity and improved stress adaptation (Kolupaev et al. 2022). These physiological changes correspond to the transcriptional regulation shown in Fig. 14.2, in which stress-responsive transcription factor families, including bZIP, NAC, WRKY, and MYB, regulate the expression of antioxidant and protective genes.

At the regulatory level, multiple transcription factors directly modulate antioxidant defenses in wheat. For instance, bZIP transcription factors, including TaFDL2-1A, enhance reactive oxygen species (ROS) scavenging by increasing superoxide dismutase (SOD) and glutathione peroxidase (GPX) activity under

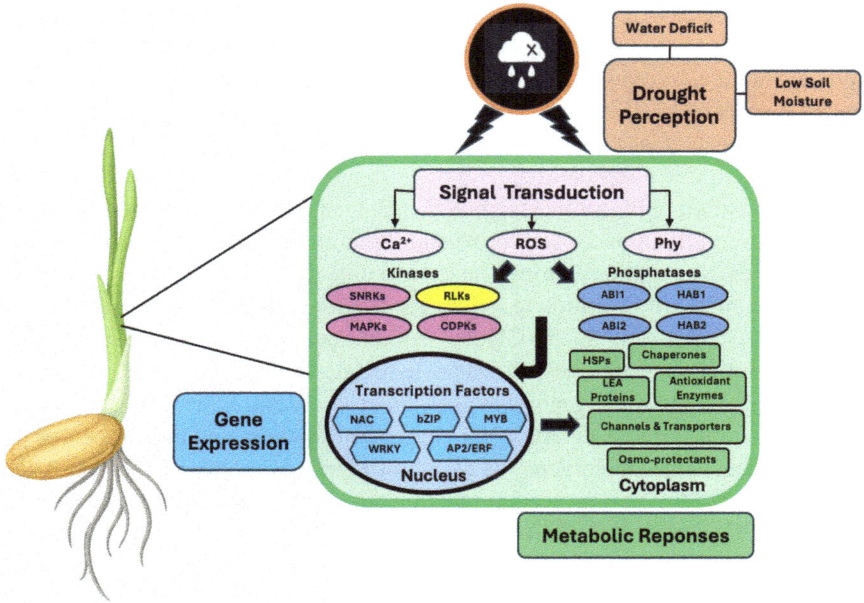


**Fig. 14.2** Overview of drought stress signaling and transcriptional regulation in plants, highlighting ROS-mediated signaling pathways and antioxidant gene regulation

drought stress (Wang et al. 2022a, b, c). Additionally, catalase-associated proteins contribute to the precise regulation of ROS detoxification. Genes such as *TaWD40-4B.1* and *NRX1* interact with canonical catalases, promoting enzyme oligomerization and activity, which reduces hydrogen peroxide ($H_2O_2$) accumulation during drought stress (Zhang et al. 2021; Tian et al. 2023). These molecular interactions illustrate the integration of antioxidant defense pathways within broader drought-responsive signaling networks, as summarized in Fig. 14.2.

Figure 14.2 summarizes that drought stress, initiated by environmental factors such as water deficit and low soil moisture, is detected by plant cells through various sensors, including receptor-like kinases (RLKs). RLKs function as either positive or negative regulators depending on the signaling context. This detection triggers a complex signal transduction cascade involving calcium ions ($Ca^{2+}$), reactive oxygen species (ROS), and changes in phytohormone levels. These signals subsequently activate key kinases, such as mitogen-activated protein kinases (MAPKs), calcium-dependent protein kinases (CDPKs), and SNF1-related kinases (SnRKs), which serve as positive regulators of stress signaling.

In contrast, protein phosphatases such as *ABI1*, *ABI2*, *HAB1*, and *HAB2* function as negative regulators, primarily by inhibiting *SnRK2* activity in the absence of abscisic acid (ABA). The activated kinases regulate the activity of several transcription factor families, including NAC, WRKY, and bZIP (notably AREB/ABF), AP2/ERF, and MYB, through direct or indirect mechanisms. These transcription factors induce the expression of drought-responsive genes that facilitate metabolic and cellular adaptation, including heat shock proteins (HSPs), late embryogenesis abundant (LEA) proteins, osmoprotectants, chaperones, antioxidant enzymes, and membrane transporters. Collectively, these molecular components enhance stress tolerance and promote adaptation to drought conditions.

## 14.3 Agronomic Strategies for Mitigating Drought Stress in Wheat

Agronomic strategies are essential management tools for mitigating drought stress in wheat, providing short-term improvements in yield stability and water-use efficiency (WUE) under water-limited conditions. These methods primarily function by modifying growing conditions or temporarily supporting plant responses, making them suitable for short-term management in drought-prone systems. Nevertheless, the effectiveness of agronomic strategies is often limited by environmental variability, substantial input requirements, and labor intensity, which restrict their scalability and consistency across diverse agroecological regions. Therefore, while agronomic practices help mitigate drought, they cannot replace the need for durable genetic improvement.

Among agronomic interventions, optimized irrigation management is a prominent agronomic intervention aimed at balancing yield stability with water conservation. Research comparing various irrigation regimes, from no irrigation to limited and ample irrigation, has shown that targeted water application during sensitive developmental stages such as jointing and grain filling can maintain grain yield and enhance WUE under drought stress (Zhou et al. 2022). Selection in restricted irrigation and rainfed environments has demonstrated strong correlations among grain yield, biomass accumulation, normalized difference vegetation index (NDVI), and drought tolerance, highlighting the importance of synchronizing irrigation timing with critical reproductive stages to maximize productivity while reducing water inputs (Bagwan et al. 2023). Furthermore, integrating drought tolerance indices with genotype–environment interaction analyses has proven effective for identifying wheat varieties with stable performance across variable moisture regimes (Roostaei et al. 2022). However, irrigation-based strategies remain highly resource-dependent and are frequently impractical or economically unsustainable in water-scarce regions.

Soil and nutrient management practices help mitigate drought by increasing water retention and nutrient availability in the root zone. Techniques such as mulching and surface residue retention reduce evaporative losses and promote soil moisture conservation (Merrium et al. 2022). The architecture of root systems is critical for drought resilience, as deeper and more extensive roots facilitate access to water and nutrients in lower soil layers. Agronomic interventions that encourage deep rooting, such as optimized nitrogen management, modified irrigation scheduling, and deep tillage, have demonstrated improvements in both drought tolerance and nitrogen uptake efficiency (Odone et al. 2024). The use of organic amendments and biofertilizers can further stimulate rhizosphere enzyme activity and microbial functions, thereby enhancing nutrient availability and water-use efficiency (WUE) in drought-tolerant wheat cultivars (Hosseini et al. 2024). However, these approaches require ongoing management and site-specific adjustments, which can restrict their widespread and consistent implementation.

Biological interventions, such as plant growth-promoting rhizobacteria (PGPR), offer environmentally sustainable strategies to enhance drought resilience by supporting root development, nutrient uptake, and water retention. Exopolysaccharide (EPS)-producing strains, including *Klebsiella* sp. (LEW16), have been shown to improve seedling growth and root morphology under drought conditions. Other strains, such as *Chryseobacterium* sp. (LEW3) and *Acinetobacter* sp. (LEW9), facilitate phosphate and zinc solubilization in water-limited soils (Latif et al. 2022). Further research indicates that PGPR strains like *Bacillus rugosus* (MMH101) enhance shoot and root growth, chlorophyll content, and proline accumulation, thereby contributing to drought adaptation (Salem et al. 2024). Despite these benefits, the effectiveness of PGPR varies across environments due to differences in soil properties, microbial competition, and host–microbe interactions, which can limit their large-scale reliability (Ahmed et al. 2023).

Foliar application of bio-stimulants constitutes a management-based strategy to mitigate drought-induced physiological stress in wheat. Compounds including selenium nanoparticles, salicylic acid, zinc oxide nanoparticles, calcium, and sulfur dioxide have been shown to enhance photosynthetic efficiency, facilitate osmotic adjustment, and increase antioxidant capacity under drought conditions. For instance, selenium nanoparticles promote growth and induce stress-responsive gene expression, thereby conferring tolerance to both drought and heat stress (Omar et al. 2023). Salicylic acid and strigolactones facilitate proline accumulation and regulate stomatal function, whereas zinc oxide nanoparticles increase antioxidant enzyme activity and upregulate drought-responsive genes involved in osmolyte biosynthesis (Sedaghat et al. 2020; Raeisi et al. 2022). Combined foliar treatments further improve chlorophyll content and antioxidant defences, thereby reducing oxidative damage during water deficit (Akin et al. 2024). Despite these benefits, foliar biostimulant applications often require repeated treatments and precise timing, and they incur additional costs, which can limit their long-term economic viability in large-scale wheat production systems.

In summary, agronomic strategies can mitigate drought stress and enhance short-term wheat performance, especially when integrated and adapted to specific local conditions. However, their dependence on ongoing inputs, labor, and site-specific management highlights the need for additional approaches to improve wheat drought tolerance. By comparison, conventional breeding, molecular breeding, and omics-enabled gene discovery provide scalable and durable solutions by identifying and utilizing genetic variation that imparts stable drought resilience across diverse environments. Therefore, agronomic practices should be regarded as supportive measures that complement, rather than substitute for, genetic and genomic strategies to enhance drought resistance in wheat.

## 14.4 Conventional Breeding Approaches

### 14.4.1 Selection of Drought-Tolerant Varieties

Traditional breeding for drought tolerance in wheat relies on the systematic selection of plants exhibiting key traits, such as an extensive root system, optimized leaf area, and a waxy cuticle. These characteristics contribute to yield stability and enhanced water-use efficiency (Li et al. 2022a). Focusing on these criteria has enabled the development of wheat varieties that are resilient and can thrive under severe drought conditions. Notably, drought-tolerant genotypes often display a minimized root system relative to a large aboveground biomass. Addressing this trait, the development and evaluation of genotypes based on morpho-physiological criteria allow for more precise selection of drought-tolerant cultivars (Ghanem et al. 2024).

A range of physiological indicators can be identified and correlated with morphological differences under drought stress. The application of multivariate models enables the determination of whether traits are regulated independently or in a dependent manner and highlights covariance among physiological indicators to assess drought tolerance under both drought and normal conditions (Wang et al. 2022a, b, c). Principal component analysis (PCA) captures total variance within datasets across diverse conditions, while hierarchical cluster analysis groups wheat genotypes based on performance (Ayyub et al. 2024). Integrating PCA with stepwise multiple linear regression (SMLR) provides a robust method for identifying drought-tolerant wheat genotypes by analyzing complex trait interactions, including chlorophyll content, polyphenol oxidase activity, glycine betaine levels, relative turgidity, stomatal conductance, flag leaf area, plant height, grain-filling duration, and grain yield. This comprehensive multivariate approach has proven effective in identifying drought-tolerant bread wheat genotypes and offers a valuable framework for enhancing crop resilience under variable environmental conditions (Sallam et al. 2024a, 2024b). A range of morphological traits has been widely utilized as indirect indicators of drought resilience and yield stability in wheat to standardize phenotypic selection under drought stress. Table 14.1 summarizes the most commonly applied morphological indices and their corresponding calculations.

These traits underlie trait-based selection strategies and are often incorporated into multivariate models to identify drought-tolerant genotypes. Alongside morphological traits, yield-based stress tolerance indices are widely employed to quantify genotype performance under drought compared to non-stress conditions. These indices provide a comparative framework for identifying stable, high-performing genotypes across diverse environments (Table 14.2).

**Table 14.1** Morphological trait indices for assessing drought resilience in wheat

| Morphological indices | Statistical formula | References |
|---|---|---|
| Spike length (cm) | = length of single spike (cm)/n | – |
| Spike number | = # spikes counted/area ($m^2$) | – |
| Grain number/spike | = # grains in each spike/n | – |
| 1000 kernel weight (g) | = weight of 100 seeds x 10 | – |
| Grain yield (g/$m^2$) | = total weight of harvested grains/harvested area ($m^2$) | Donald and Hamblin (1976) |
| Straw yield (g/$m^2$) | = total weight of straw/harvested area ($m^2$) | – |
| Harvest index | = grain yield/total biomass | Donald and Hamblin (1976) |
| Specific leaf area | = leaf area/dry mass | Beadle (1993) |
| Degree of succulence | = water amount/leaf area | Delf (1912) |
| Shoot distribution | = fresh mass/length | Arduini et al. (1994) |
| Shoot density | = dry mass/length | Arduini et al. (1994) |
| Relative water content | = (fresh weight-dry weight)/dry weight | Weatherley (1950) |

**Table 14.2** Stress tolerance indices used to evaluate drought response in wheat

| Stress tolerance indices | Statistical formula | References |
|---|---|---|
| Drought susceptibility index | $= [1-(Y_d/Y_p)]/D$ | Fischer and Maurer (1978) |
| Stress tolerance index | = (Ypi*Ysi)/Yp2 | Fernandez (1992) |
| Yield stability index | = Ys/Yp | Bouslama and Schapaugh (1984) |
| Mean productivity | = (Yp + Ys)/2 | Bouslama and Schapaugh (1984) |
| Tolerance index | = Yp–Ys | Rosielle and Hamblin (1981) |

### 14.4.2 Leveraging Genetic Diversity from Landraces and Wild Relatives

Ancient landraces and wild relatives play a vital role in maintaining the genetic diversity needed to improve drought-stress tolerance in wheat breeding germplasm worldwide. In contrast to elite cultivars, which have experienced repeated domestication and improvement bottlenecks, landraces and wild species retain adaptive traits shaped by prolonged exposure to diverse and challenging environments. Consequently, these genetic resources have been extensively studied to elucidate the physiological, biochemical, and molecular mechanisms underlying drought tolerance and to identify alleles suitable for introgression into modern wheat varieties.

Multiple studies have shown that wheat landraces possess considerable drought-adaptive variation, which can be identified through comprehensive morpho-physiological screening. For instance, Chaouachi et al. (2023) evaluated Tunisian durum wheat landraces under drought stress and observed significant genotype-dependent differences in relative water content, chlorophyll stability, canopy temperature, and yield components, thereby identifying accessions with advantageous drought-response profiles. Similarly, Gaikwad et al. (2025) examined ancient Indian dwarf wheat (*Triticum sphaerococcum*) and reported enhanced drought and heat resilience based on several stress tolerance indices, highlighting the importance of underutilized landraces for consistent performance under combined abiotic stresses.

Wild relatives of wheat broaden the range of drought-adaptive mechanisms beyond those typically found in cultivated germplasm. Pour-Aboughadareh et al. (2017) demonstrated that several wild wheat relatives display stronger physiological responses to drought than modern wheat, such as improved osmotic adjustment, increased membrane stability, and heightened antioxidant activity. Additional comparative studies have confirmed these observations by identifying species-specific drought tolerance strategies. For example, Pantha et al. (2025) reported that *Triticum araraticum* exhibits distinct drought response mechanisms compared to modern bread wheat, including differences in stomatal regulation, osmolyte accumulation, and redox homeostasis. Similarly, Jadidi et al. (2025) documented contrasting physiological and molecular responses between bread wheat and its wild relatives, highlighting differential transcriptional regulation of stress-responsive pathways under water deficit.

Among diploid wheat species, einkorn wheat (*Triticum monococcum*) occupies a unique evolutionary position as the earliest domesticated wheat and the only diploid species existing in both wild and domesticated forms ($2n = 2x = 14$; $A^mA^m$ genome) (Saripalli et al. 2023; Ahmed et al. 2023). This reduced genomic complexity makes einkorn a powerful model for dissecting drought tolerance mechanisms while remaining directly relevant to wheat improvement. In addition, einkorn kernels contain higher levels of antioxidants, such as carotenoids and tocols, than those of other wheat species, suggesting the presence of alleles that may contribute to enhanced oxidative stress tolerance under drought conditions (Brandolini and Hidalgo 2011). Wild emmer wheat (*Triticum dicoccoides*), the direct progenitor of durum and bread wheat, represents another critical genetic resource for drought tolerance. Certain wild emmer genotypes maintain full fertility under arid desert conditions, and transcriptomic analyses have identified drought-adaptive genes involved in osmoregulation, cuticular wax biosynthesis, lignification, and redox and hormonal homeostasis. These pathways collectively contribute to optimized cellular homeostasis, bioenergetics, and carbon metabolism under water-limited conditions (Abdelrahman et al. 2024). In durum wheat (*Triticum turgidum* subsp. *durum*), which is predominantly cultivated in drought-prone Mediterranean environments, breeding efforts increasingly rely on landrace and wild germplasm to enhance drought tolerance and yield stability, given the economic and nutritional importance of the crop (Mohammadi and Geravandi 2024). The identification and introgression of drought-adaptive traits from ancient landraces and wild relatives can substantially improve drought tolerance in modern bread and durum wheat varieties.

## 14.5 Molecular Breeding and Biotechnology

### 14.5.1 Marker-Assisted Selection

Marker-assisted selection (MAS) is extensively utilized in wheat breeding programs to enhance drought tolerance by targeting genomic regions that control key morphological and physiological traits related to stress adaptation. MAS employs DNA markers that are closely linked to quantitative trait loci (QTL) or major-effect marker–trait associations (MTAs) to expedite the selection and introgression of advantageous alleles into elite genetic backgrounds (Maulana et al. 2020; Paudel et al. 2024). The effectiveness of MAS for complex traits such as drought tolerance is highly dependent on the identification of stable, large-effect QTL that are consistently observed across diverse environments and genetic backgrounds.

Multiple drought-related QTL and MTAs have demonstrated sufficient robustness to facilitate MAS-based introgression. QTL associated with canopy temperature (CT) and grain yield under drought conditions have been consistently identified in multi-environment trials and across diverse germplasm panels. For instance, stable QTL located on chromosomes 1B, 3B, 4A, and 7D are linked to canopy temperature depression and yield stability under water-limited conditions, establishing them as reliable targets for selection (Suzuky et al. 2010; Pinto and Reynolds 2015; Lepekhov 2022; Pankaj et al. 2022a, b). Rathan et al. (2022) effectively utilized KASP markers associated with these QTL in elite Indian wheat lines, resulting in measurable improvements in grain yield and drought performance across various environments. Additionally, MTAs governing normalized difference vegetation index (NDVI) and stay-green phenotypes, which are strongly correlated with drought tolerance, have proven reproducible and are increasingly incorporated into MAS pipelines.

Root-related traits are promising targets for MAS when governed by consistent QTL. QTL associated with root depth, root length density, and root angle, particularly on chromosomes 2A, 2D, and 5B, have been shown to enhance water acquisition and yield stability under terminal drought conditions (Iquebal et al. 2019; Soriano and Álvaro 2019; Salarpour et al. 2020; Li et al. 2021). These loci have been validated across diverse mapping populations and environments, which increases their suitability for marker-assisted introgression. Similarly, QTL linked to osmotic adjustment and relative water content demonstrate moderate to high stability and are considered viable candidates for MAS when integrated with physiological screening.

Recent advances in genotyping platforms have enhanced the implementation of MAS. Kompetitive allele-specific PCR (KASP) genotyping facilitates cost-effective, high-throughput screening for single-nucleotide polymorphisms (SNPs) associated with drought-relevant QTLs and MTAs, thereby improving selection accuracy and reducing false positives (Negisho et al. 2022; Eltaher et al. 2023; Liu et al. 2023). High-density SNP-based MAS approaches are particularly effective when applied to well-validated loci, as opposed to genome-wide associations with small or environment-specific effects. Additionally, presence–absence variation (PAV) has emerged as a valuable marker for identifying structural variation underlying drought-adaptive traits, further expanding the MAS toolkit (Zhao et al. 2023).

Despite these advancements, MAS remains limited in its effectiveness for drought tolerance due to the polygenic nature of the trait and strong genotype-by-environment interactions. Many MTAs identified through genome-wide association studies have small effects or lack reproducibility, which diminishes their value for introgression. Gudi et al. (2024) emphasize that stringent filtering criteria, including multi-environment validation, effect-size thresholds, and biological relevance, are necessary to eliminate spurious associations while retaining major, actionable QTL. Therefore, MAS is the most effective when focused on a small number of stable, high-confidence loci and is optimally used in conjunction with complementary strategies such as genomic selection and omics-enabled gene discovery.

### 14.5.2 Genomic Selection

Accelerating genetic gain through genomic selection and modern breeding technologies is essential for maximizing crop improvement rates and ensuring future food security (Abdelrahman et al. 2024). Over recent decades, advances in high-throughput, cost-effective genotyping technologies have established genomic selection (GS) as a practical and effective tool in both animal and plant breeding (Budhlakoti et al. 2022). Genomic selection utilizes all available markers within a population to estimate genome-estimated breeding values (GEBVs), enabling the identification of elite lines without the need for phenotyping (Seleiman et al. 2021). Using multivariate statistical models, GEBVs can be estimated independently of phenotypic data (Paudel et al. 2024). For example, a GS model trained on genomic and high-throughput phenomic data has demonstrated significantly improved predictive accuracy for drought-associated traits, such as grain yield in wheat under field conditions (Montesinos-López et al. 2023). Based on the principles of GS, predictions for untested individuals are generated using genotype–phenotype models developed from training populations. Nevertheless, it remains standard practice to first summarize phenotypic data collected across multiple environments before analyzing their association with genotypic data (Li et al. 2021). GS is employed to capture all minor-effect QTL, identify individuals with the highest GEBVs for target traits, and reduce the number of generations required to select superior phenotypes (Bapela et al. 2022).

### 14.5.3 Genome-Wide Association Study

In recent years, genome-wide association studies (GWAS) have emerged as essential tools for identifying QTL associated with abiotic and biotic stress tolerance in various crop species, including wheat (Maulana et al. 2020). GWAS connects genotype and phenotype by utilizing linkage disequilibrium (LD) that accumulates over multiple generations within a population (Rabieyan et al. 2023). By leveraging genome-wide markers and historical recombination events, GWAS enables precise identification of marker-trait associations (MTAs) linked to specific traits. This approach has demonstrated effectiveness in detecting significant MTAs for traits such as grain yield, chlorophyll fluorescence, and stress tolerance in wheat (Gill et al. 2022; Gudi et al. 2023). For example, a recent GWAS conducted on a diverse panel of Iranian wheat accessions identified novel loci on chromosomes 2A and 5D associated with root length and shoot biomass under drought conditions, thereby expanding the genetic resources available for root-based drought tolerance (Zarei et al. 2022). To ensure robust detection of these associations, GWAS utilizes various statistical models, including the general linear model (GLM), mixed linear model (MLM), and FarmCPU, which are evaluated for their effectiveness and statistical power (Sallam et al. 2022). In wheat breeding programs targeting drought tolerance, phenotypic selection alone may be insufficient to predict genotype performance, highlighting the importance of genomic approaches such as GWAS for identifying tolerant alleles and selecting promising parental lines (Sallam et al. 2024a, b). Furthermore, integrating GWAS with gene expression-based GWAS (eGWAS) improves the dissection of the genetic basis of complex traits by not only identifying genetic variants but also linking them to gene expression levels, which facilitates the prioritization of candidate genes for drought tolerance (Yang et al. 2024).

### 14.5.4 QTL Mapping

Improving drought tolerance through conventional breeding remains difficult because it is governed by multiple QTLs (Maulana et al. 2020). QTL mapping enables the identification of specific genomic regions in wheat associated with drought tolerance. By analyzing genetic variation and drought response across diverse wheat lines, researchers can determine critical loci for breeding and MAS (Paudel et al. 2024). The combined use of GWAS and QTL mapping substantiates the lack of correlation between trait groups by identifying distinct genes, QTL, and markers linked to each trait category (Sallam et al. 2022). For example, a recent QTL mapping study utilizing a recombinant inbred line (RIL) population derived from "Drysdale" × "Waagan" detected stable QTL for chlorophyll content and grain yield under drought conditions across multiple environments (Gaurav et al. 2020). Structural variations (SVs), particularly presence/absence variations (PAVs) of tandem repeats, significantly influence wheat drought tolerance by affecting meiotic recombination and segregation patterns (Zhao et al. 2023). QTL mapping and gene expression analyses are critical for identifying and utilizing key genes in wheat that confer drought tolerance by encoding proteins and enzymes such as LEA, RAB, rubisco, helicase, proline, GST, and various carbohydrates, which collectively mitigate drought stress (Gupta et al. 2024b). The integration of near-isogenic lines (NILs) with RNA-seq analysis has proven effective in identifying genes responsible for drought tolerance and in revealing underlying QTL for important quantitative traits.

### 14.5.5 Genetic Engineering and Overexpression Analysis

Contemporary genetic strategies for assessing plant adaptability to changing climatic conditions prioritize genetic engineering and transgenic techniques. These methods focus on identifying and introducing advantageous genes into target plants to generate new varieties with improved adaptability. By screening diverse genotypes and identifying novel trait combinations, researchers facilitate the development of resilient plant varieties (Nyaupane et al. 2024a, b). For example, CRISPR-Cas9-mediated editing of the *TaDREB2* gene in bread wheat has led to increased expression of drought-inducible genes and higher survival rates under water-deficit conditions (Zhang et al. 2021). Genes encoding abiotic stress-responsive transcription factors are key tools in genetic engineering to enhance plant tolerance to adverse environmental conditions. These transcription factors activate the expression of downstream genes involved in responses to various abiotic stresses, making them essential targets in efforts to improve plant resilience (Trono and Pecchioni 2022).

Overexpression analysis provides valuable insights into how increased gene activity affects phenotypes, especially when genetic redundancy obscures the impact of individual genes. This method determines whether the activity of a single gene is sufficient to confer specific traits, thereby elucidating gene function. Noteworthy examples include the overexpression of the ethylene-responsive factor TaERF87, which enhances drought tolerance by activating the proline biosynthesis pathway (Du et al. 2022); TaPRX-2A, which augments antioxidant enzyme activity (Su et al. 2023); TaWRKY46, which improves osmotic stress tolerance, resulting in higher germination rates and increased root length and TaEXPA2, which elevates antioxidant capacity by increasing enzyme activity and the expression of ROS scavenging genes (Yang et al. 2020).

## 14.6 Omics Approaches

### 14.6.1 Genomics

Enhancing drought stress tolerance in plants requires targeted breeding strategies combined with molecular and genomic approaches. Omics technologies, including metabolomics, proteomics, genomics, transcriptomics, glycomics, and phenomics, are central to these advancements (Seleiman et al. 2021). Genomics provides high-resolution mapping resources by leveraging the wheat reference genome and skim-sequencing of large genetic populations. This methodology enables precise identification of crossover breakpoints and fine-scale mapping of key traits, thereby facilitating the discovery of small physical intervals that contain a manageable number of candidate genes (Saripalli et al. 2023). The integration of high-throughput phenotyping tools with advanced genomic techniques is essential for elucidating the genetic and regulatory architectures of wheat under stress conditions. Such genomic approaches allow for more accurate identification of critical genetic factors, thereby supporting the development of targeted and efficient breeding strategies to enhance wheat resilience (Abdelrahman et al. 2024). Recent fine-mapping of a major QTL for drought tolerance on chromosome 7B, using skim sequencing and bulked segregant analysis, has identified novel candidate genes associated with root elongation and water retention (Liu et al. 2023). Employing less genetically complex wheat progenitors, such as diploid A-genome wheat (einkorn), represents a valuable strategy for improving bread wheat's stress resistance. Reference genome assemblies for both wild and domesticated einkorn accessions have been developed to facilitate this potential. The Einkorn genome database offers a comprehensive platform for the cereals research community, supporting comparative genomics, applied genetics, and breeding research. It includes searchable annotated genes, a genome browser, a BLAST interface for sequence alignment, and tools for genomic synteny visualization. This resource is expected to significantly advance wheat research and breeding. Database URL https://wheat.pw.usda.gov/GG3/pangenome (Sharma et al. 2023).

### 14.6.2 Transcriptomics

Transcriptomics has transformed the study of drought tolerance in plants by elucidating gene expression changes in response to stress. Research in Arabidopsis and other species has identified key drought-responsive genes and pathways, including those associated with phytohormones and Golgi-targeted proteins (Kopecká et al. 2023). In wheat, transcriptome analyses indicate that NAC, WRKY, and MYB transcription factors are frequently upregulated under drought conditions (Liu et al. 2021; Xi et al. 2023; Barratt et al. 2023; Li et al. 2022b). Integration of transcriptomics with metabolomics provides detailed insights into co-regulated pathways, such as amino acid biosynthesis and secondary metabolite production, which are essential for adaptive responses during drought and subsequent rehydration (Lv et al. 2022b). Comparative transcriptomics leverages natural genetic variation among genotypes with differing drought responses to identify differentially regulated genes that represent promising candidates for breeding drought-tolerant crops (Baldoni et al. 2021). For instance, analysis of wheat genotypes "DEGUO 2" and "Truman" has identified critical genes involved in flavonoid biosynthesis, abscisic acid (ABA) signaling, and antioxidant pathways, with melatonin shown to enhance drought tolerance. Additionally, the involvement of noncoding RNAs, including long noncoding RNAs (lncRNAs) and microRNAs (miRNAs), in regulating drought stress responses highlights the complexity of the underlying genetic networks, as demonstrated by the identification of thousands of differentially expressed genes associated with drought resistance (Li et al. 2022c). In summary, transcriptomics provides a robust framework for dissecting the molecular mechanisms underlying drought tolerance, thereby facilitating targeted breeding and crop improvement.

### 14.6.3 Proteomics

Integrating transcriptomic analyses of gene expression with comprehensive proteomic approaches provides a more complete understanding of drought responses in wheat (Langridge and Reynolds 2021). Proteomic studies demonstrate that drought stress significantly disrupts the leaf proteome in drought-susceptible wheat cultivars such as Bahar, particularly affecting photosynthetic proteins and enzymes involved in sugar and nitrogen metabolism. In contrast, drought-tolerant cultivars like Kavir display only minor changes at the protein level. Proteomic profiling and molecular analyses of flag leaf tissues in drought-tolerant wheat genotypes have revealed increased abundance of dehydrins and LEA proteins, supporting their role in stress adaptation (Deng et al. 2018; Duvnjak et al. 2024). When combined with metabolomic data, these proteomic findings identify critical biochemical pathways involved in the drought response and suggest valuable targets for improving crop resilience (Michaletti et al. 2018). Additionally, a proteomic study of wheat roots identified key molecular responses to combined drought and low nitrogen (LN) stress in the drought-tolerant cultivar Chang6878, highlighting 163 differentially abundant proteins (DAPs) uniquely upregulated under these conditions. These DAPs are enriched in pathways such as MAPK signaling, phenylpropanoid biosynthesis, and ethylene metabolism. The results emphasize the importance of altered ethylene synthesis and signal transduction in enhancing drought and LN tolerance, offering valuable insights for breeding stress-resistant wheat varieties (Yan et al. 2023).

### 14.6.4 Metabolomics

Metabolomics has become an essential approach for elucidating the biochemical mechanisms underlying drought stress tolerance in wheat. Analysis of metabolite profiles in wheat lines with varying drought resilience has revealed key pathways that contribute to stress adaptation. For example, a drought-tolerant mutant demonstrated elevated levels of metabolites associated with osmotic regulation, reactive oxygen species (ROS) scavenging, and abscisic acid (ABA) signalling compared to common and drought-susceptible wheat varieties (Lv et al. 2022a, b). Metabolomic studies have also emphasized the significance of polyamines, such as spermidine, in promoting drought tolerance during wheat germination, with substantial metabolic changes observed in varieties such as Rakhshan and Mihan under stress conditions (Gholizadeh et al. 2022). Additionally, metabolomic analyses have shown that drought-susceptible Bahar and drought-tolerant Kavir exhibit distinct metabolic responses to water deficiency, highlighting the involvement of amino acids, organic acids, and sugars in drought adaptation (Michaletti et al. 2018). The identification of metabolites such as raffinose and galactinol has further clarified the role of metabolic pathways, including galactinol biosynthesis, in enhancing drought tolerance (Seleiman et al. 2021). In drought-stressed seedlings, targeted metabolomic profiling has identified the accumulation of raffinose, trehalose, and galactinol as key metabolic markers of drought adaptation (Zhou et al. 2022). Collectively, these findings demonstrate that metabolomics facilitates the discovery of novel genes and pathways involved in drought resilience, providing critical insights for breeding wheat varieties with improved drought tolerance. Recent progress in genomics, transcriptomics, proteomics, and metabolomics has enabled the identification and functional validation of numerous drought-responsive genes in wheat. The principal candidate genes identified through these integrative approaches are summarized in Table 14.3.

Together, these discoveries illustrate how multi-omics strategies can accelerate trait and gene discovery, providing a foundation for integrated breeding and management approaches.

## 14.7 Integrated Approaches for Enhancing Drought Tolerance

Escalating challenges from climate change, including altered rainfall patterns, rising atmospheric $CO_2$, and rising temperatures, require integrated strategies to improve drought tolerance in wheat. Nyaupane et al. (2024a, b) report that drought stress induces oxidative damage, reducing yield and impairing physiological functions. Wheat responds to these stresses through adaptive mechanisms, including increased trichome density and higher root-to-shoot ratios. Advanced genomic tools, as described by Paudel et al. (2024), enable the identification of candidate genes for drought resistance using methods such as QTL mapping and genome editing. Combining molecular techniques with agronomic practices, such as applying melatonin to regulate stress, offers a promising approach to breeding resilient wheat varieties. The integration of omics technologies (Seleiman et al. 2021) further advances the understanding of drought responses and supports the development of molecular markers for marker-assisted selection. A holistic strategy that incorporates genetic engineering and traditional breeding, as emphasized by Langridge and Reynolds (2021), can optimize root and shoot traits critical for water efficiency. The application of drought resistance indices, as demonstrated by Guizani et al. (2024), refines genotype selection. Ultimately, a comprehensive approach that unites genetic, physiological, and agronomic perspectives is essential for developing high-yielding, drought-tolerant wheat varieties suitable for arid environments (Banoth et al. 2023).

## 14.8 Challenges and Future Directions

### 14.8.1 Current Challenges in Enhancing Drought Tolerance

Enhancing drought tolerance in wheat remains a significant challenge for plant biologists and geneticists, primarily due to the complex genetic architecture of this trait. Numerous QTL and diverse environmental factors influence drought tolerance,

**Table 14.3** Candidate genes associated with drought tolerance in wheat identified through an integrative approach

| Gene | Chromosome location | Discovery method | Reference |
|---|---|---|---|
| *TaWRKY31* | 3A | Virus-induced gene silencing (VIGS) and overexpression | Ge et al. (2024) |
| *TaSnRK2.9-5A* | 5A | Genetic mapping, KASP markers, and MAS | Ur Rehman et al. (2019) |
| *TaSNAC4-3A* | 3A | Overexpression and transcriptomic analysis | Mei et al. (2021) |
| *TabHLH27-A1* | 1A, 1B, 1D | GWAS, transcriptome profiling, and knockout studies | Wang et al. (2024) |
| *TaWRKY24-1D* | 1D | Y2H, BiFC, BSMV-VIGS, qPCR, and LUC assays | Li et al. (2024a) |
| *TaTIP41* | 5A, 5B, 5D | Bioinformatics analysis and GFP subcellular localization | Ma et al. (2023a) |
| *TaHsfC3-4* | – | Omics analysis, overexpression in Arabidopsis, and protein interaction assays | Ma et al. (2024) |
| *TaSINA2B, Ta SINA1D* | 1D, 2B | Yeast two-hybrid, BiFC assays | Ma et al. (2023b) |
| *TaNAC071-A* | 2A | GWAS and gene expression analysis | Mao et al. (2022) |
| *TaWD40-4B.1* | 4B | GWAS and functional analysis | Tian et al. (2023) |
| *DIW1/ TaPP2C158* | 5A | Mutant screening and CRISPR/Cas9 | Wang et al. (2023a, b) |
| *TaNHX2, TaGAD1* | 2D, 3B | Protein interaction assays and transgenic validation | Li et al. (2024b) |
| *MYC2, ASMT* | 3B, 4A | RNA-Seq and promoter analysis | – |
| *TaERF87, TaASK1* | 6A, 2D | RNA-Seq and biochemical assays | Du et al. (2023) |
| *TaCER1-6A* | 6A | CRISPR/Cas9 and expression analysis | He et al. (2022) |
| *TaSNAC8-6A* | 6A | Candidate gene association analysis | Mao et al. (2020) |
| *TaPRX-2A* | 2A | Comparative transcriptomic analysis | Su et al. (2023) |
| *TaBADH-A1* | 3A | Gene cloning, allele mining, and marker development | Yu et al. (2022) |
| *TaWRKY46* | 5D | Functional characterization in transgenic Arabidopsis | – |
| *TaP5CS* | 3A, 3B, 3D | DNA methylation analysis | Li et al. (2023) |
| *TaBADH* | 4A, 4B, 4D | DNA methylation analysis | Li et al. (2023) |
| *TaNRX1* | 2A, 2B, 2D | Overexpression and RNAi studies in wheat | Zhang et al. (2021) |
| *TaFDL2-1A* | 1A | Overexpression and functional analysis | Wang et al. (2022a) |
| *TaSNAC4-3D* | 3D | Expression analysis and functional assays | Ma et al. (2022) |
| *TaEXPA2* | – | Overexpression, RNAi analysis, and transcriptional regulation | Yang et al. (2020) |

which is inherently multifaceted and frequently associated with heat and other abiotic stresses. These stresses induce distinct morphological and physiological changes. Comprehensive research is required to elucidate the genetic interactions underlying drought adaptation and to identify the specific genes involved in tolerance mechanisms (Sallam et al. 2019).

A major obstacle in developing drought-tolerant wheat varieties is identifying and selecting genotypes that perform consistently across all growth stages. Effective phenotyping is crucial for breeding success and offers direct benefits to producers. However, existing phenotyping methods are frequently labor-intensive and inefficient, highlighting the need for high-throughput, nondestructive approaches to assess root traits. In wheat-dependent regions such as China, recurrent drought and low nitrogen availability significantly constrain crop growth and yield (Bapela et al. 2022; Yan et al. 2023).

Recent research has demonstrated that drought-adaptive traits in wheat are polygenic and strongly influenced by genotype–environment interactions, which reduce selection efficiency and contribute to inconsistencies in identifying robust drought-tolerant varieties. The large genome size of wheat further complicates the identification of genetic loci associated with drought tolerance, highlighting the need to establish stable QTL and develop new mapping populations suited to drought conditions.

Genetic engineering offers promising strategies to develop wheat germplasm with enhanced tolerance to drought and heat stress. Current research focuses on identifying candidate genes and evaluating transgenic lines under controlled conditions. Despite progress in gene editing technologies, significant barriers remain to the adoption of transgenic wheat in breeding programs, including high regulatory costs and limited field testing (Langridge and Reynolds 2021). Integration of transgenic drought-tolerant wheat into breeding pipelines requires substantial yield improvements, ideally exceeding 10% under stress conditions. The recent approval of HB4 wheat in Argentina, which reportedly achieves a 20% yield increase under drought without compromising performance under normal conditions (Rafiei et al. 2024), represents a major advancement. Collaborative efforts are necessary to bridge the gap between research and practical application, enabling the agricultural sector to address the ongoing challenges posed by drought (Sallam et al. 2019; Langridge and Reynolds 2021; Bapela et al. 2022; Yan et al. 2023).

### 14.8.2 Future Research Prospects

The development of robust, drought-tolerant wheat varieties requires implementing several innovative strategies. Advanced genomic tools, such as GWAS and QTL mapping, are essential for identifying novel alleles and candidate genes associated with drought tolerance. These genomic findings can be integrated into breeding programs through MAS and genomic selection, allowing breeders to target specific loci that improve water-use efficiency, root architecture, photosynthetic capacity, antioxidant defense, hormonal regulation, early flowering, and leaf surface traits.

In addition to conventional breeding, targeting induced local lesions in genomes (TILLING) offers a robust reverse genetics approach for discovering and validating drought-tolerant alleles. Screening extensive mutagenized populations enables the identification of mutations in candidate genes that confer drought resilience. As a non-transgenic technique, TILLING supports the discovery of novel gene variants and aligns with regulatory and consumer preferences in regions where genetically modified crops face resistance.

Whereas TILLING enables modification of endogenous genes, transgenic approaches allow the introduction of novel traits from other species. For example, overexpression of drought-responsive genes, including transcription factors and osmo-protectant-related enzymes, has shown potential to enhance drought tolerance. Future research should expand the range of such genes and systematically evaluate their effects on drought tolerance under diverse field conditions.

CRISPR–Cas9 gene editing provides a precise method for genetic modification without introducing foreign DNA, making it particularly suitable for improving drought tolerance. This technology enables targeted knockouts or insertions in genes regulating water-use efficiency, stomatal conductance, and root system architecture. For example, CRISPR can be used to disrupt genes that negatively influence drought tolerance or to enhance genes involved in osmotic adjustment and antioxidant pathways. Future studies should explore multiplex CRISPR strategies to edit multiple drought-responsive genes simultaneously, thereby optimizing genetic architecture for improved performance in arid environments.

Overexpression analysis can also be employed to elucidate the functions of key drought-related genes. Overexpressing specific genes in wheat or model organisms enables assessment of their effects on physiological traits such as root length, leaf water retention, and photosynthetic efficiency under water-deficit conditions. These findings will facilitate the identification of the most promising genes for inclusion in breeding programs or gene-editing strategies.

Research should prioritize root system architecture, as deeper or more extensive root systems allow wheat plants to access water from deeper soil layers. Recent studies demonstrate that breeding for deeper roots can significantly enhance drought tolerance. Further investigation into the genetic regulation of root traits and their interactions with soil moisture remains necessary. Integrating phenotypic and genotypic data using high-throughput phenotyping tools will improve the selection of optimal root traits.

Another critical research area involves the physiological response to drought, with particular emphasis on mechanisms of stomatal regulation and osmotic adjustment. Stomatal conductance plays a key role in balancing water loss and carbon assimilation during drought conditions. Examining the genetic regulation of stomatal closure and reopening may identify new targets for improving water-use efficiency. Additionally, genes involved in osmotic adjustment, such as those responsible for the biosynthesis of compatible solutes like proline and glycine betaine, may be overexpressed or edited using CRISPR to enhance plant resilience to drought-induced osmotic stress.

Expanding field trials across diverse environments is essential to validate the stability of drought tolerance traits under real-world conditions. Traits that perform well in controlled environments may not consistently result in field success, particularly in regions with variable environmental stressors. Multisite and multilocation trials are required to confirm that drought-tolerant varieties maintain high yields across different climates, soil types, and stress levels. Incorporating multi-environmental testing ensures that newly developed varieties are resilient and adaptable, and can thrive amid increasing climate variability.

In conclusion, future research should adopt a holistic, multidisciplinary approach that integrates genomic tools, gene-editing technologies such as TILLING and CRISPR, physiological studies, and field trials. Integrating these strategies will accelerate the development of drought-tolerant wheat varieties that can help ensure global food security amid climate change.

## 14.9 Conclusion

In conclusion, improving drought stress tolerance in wheat necessitates an integrative framework that bridges physiological understanding, genetic diversity, and modern breeding innovation. The strategies discussed in this review, including the exploitation of wild wheat relatives, deployment of marker-assisted and genomic selection, and the application of high-resolution mapping approaches such as GWAS and QTL analysis, collectively accelerate the discovery and utilization of adaptive alleles. Advances in functional genomics and gene-editing technologies, particularly CRISPR-based platforms, further enable precise validation and manipulation of candidate genes, strengthening the pipeline from discovery to cultivar development. As climate variability intensifies the frequency and severity of drought events, leveraging these complementary approaches will be essential for stabilizing wheat productivity and safeguarding global food security. Ultimately, the convergence of genetic resources, biotechnological innovation, and translational breeding strategies provides a sustainable pathway toward developing resilient wheat cultivars capable of thriving under increasingly water-limited environments.

## References

Abdelrahman M, Gorafi YSA, Sulieman S et al (2024) Wild grass-derived alleles represent a genetic architecture for the resilience of modern common wheat to stresses. Plant J. https://doi.org/10.1111/tpj.16887

Abid M, Ali S, Qi L et al (2018) Physiological and biochemical changes during drought and recovery periods at tillering and jointing stages in wheat (*Triticum aestivum* L.). Sci Rep 8:21441. https://doi.org/10.1038/s41598-018-21441-7

Ahmed HI, Heuberger M, Schoen A et al (2023) Einkorn genomics sheds light on history of the oldest domesticated wheat. Nature 620:830–838. https://doi.org/10.1038/s41586-023-06389-7

Akin S, Kaya C, Ashraf M (2024) Salicylic acid and sodium hydrosulfide synergy improves wheat drought tolerance by boosting chlorophyll, antioxidants and osmolytes. Food Energy Secur 13:e532. https://doi.org/10.1002/fes3.532

Akter N, Brishty T, Karim M et al (2023) Leaf water status and biochemical adjustments as a mechanism of drought tolerance in two contrasting wheat (*Triticum aestivum* L.) varieties. Acta Physiol Plant 45:1–13. https://doi.org/10.1007/s11738-023-03530-x

Arduini I, Masoni A, Ercoli L et al (1994) Grain yield, dry matter and nitrogen accumulation and remobilization in durum wheat as affected by variety and seeding rate. Eur J Agron 3:113–121. https://doi.org/10.1016/S1161-0301(14)80128-7

Ayyub MU, Khan MA, Iqbal R et al (2024) Characterization of bread wheat genotypes for drought stress adaptation. Pol J Environ Stud 33(4):4525–4537. https://doi.org/10.15244/pjoes/177611

Bagwan NB, Shah K, Bhise V et al (2023) Genetic variability and identification of drought tolerant genotypes in wheat under restricted irrigation. Agric Update 18:35–41

Baldoni E, Frugis G, Martinelli F et al (2021) A comparative transcriptomic meta-analysis revealed conserved key genes and regulatory networks involved in drought tolerance in cereal crops. Int J Mol Sci 22:13062. https://doi.org/10.3390/ijms222313062

Banoth M, Nunavath US, Bhimireddy S et al (2023) Conventional and molecular breeding strategies for improvement of drought tolerance cultivars in rice: recent approaches and outlooks. Environ Conserv J 24:367–381. https://doi.org/10.36953/ECJ.18202550

Bapela T, Shimelis H, Tsilo TJ et al (2022) Genetic improvement of wheat for drought tolerance: progress, challenges and opportunities. Plants 11:1331. https://doi.org/10.3390/plants11101331

Barratt LJ, Reynolds IJ, Franco Ortega S et al (2023) Transcriptomic and co-expression network analyses on diverse wheat landraces identify candidate master regulators of the response to early drought. Front Plant Sci 14:1212559. https://doi.org/10.3389/fpls.2023.1212559

Beadle CL (1993) Growth analysis. In: Hall DO, Scurlock JMO, Bolhàr-Nordenkampf HR et al (eds) Photosynthesis and production in a changing environment. Springer, pp 36–46. https://doi.org/10.1007/978-94-011-1566-7_3

Bouslama M, Schapaugh WT (1984) Stress tolerance in soybeans. Part 1: evaluation of three screening techniques for heat and drought tolerance. Crop Sci 24:933–937. https://doi.org/10.2135/cropsci1984.0011183X002400050026x

Brandolini A, Hidalgo A (2011) Einkorn (*Triticum monococcum*) flour and bread. In: Flour and breads and their fortification in health and disease prevention, pp 79–88. https://doi.org/10.1016/B978-0-12-380886-8.10008-X

Budhlakoti N, Tyagi S, Singh D et al (2022) Genomic selection: a tool for accelerating the efficiency of breeding programs in plants and animals. Front Genet 13:832153. https://doi.org/10.3389/fgene.2022.832153

Chaouachi L, Marín-Sanz M, Kthiri Z et al (2023) The opportunity of using durum wheat landraces to tolerate drought stress: screening morpho-physiological components. AoB Plants 15:plad022. https://doi.org/10.1093/aobpla/plad022

Delf E (1912) Experiments on succulence in plants. New Phytol 11:289–305. https://doi.org/10.1111/j.1469-8137.1912.tb05691.x

Deng X, Liu Y, Xu X et al (2018) Comparative proteome analysis of wheat flag leaves and developing grains under water deficit. Front Plant Sci 9:425. https://doi.org/10.3389/fpls.2018.00425

Donald CM, Hamblin J (1976) The biological yield and harvest index of cereals as agronomic and plant breeding criteria. In: Brady NC (ed) Advances in agronomy, vol 28. Academic Press, New York, pp 361–405. https://doi.org/10.1016/S0065-2113(08)60559-3

Du Y, Zhao Q, Chen L et al (2022) TaERF87 enhances drought tolerance in wheat by activating proline biosynthesis. Plant Physiol 190:2129–2144. https://doi.org/10.1093/plphys/kiac404

Du L, Huang X, Ding L et al (2023) TaERF87 and TaAKS1 synergistically regulate TaP5CS1/TaP5CR1-mediated proline biosynthesis to enhance drought tolerance in wheat. New Phytol 237:232–250. https://doi.org/10.1111/nph.18549

Duvnjak J, Sarcevic H, Vukovic R et al (2024) Effects of drought at anthesis on flag leaf physiology and gene expression in diverse wheat genotypes. Agronomy 14:1522. https://doi.org/10.3390/agronomy14071522

Eltaher S, Hashem M, Ahmed AAM et al (2023) Effectiveness of TaDreb-B1 and 1-FEH w3 KASP markers in spring and winter wheat populations for marker-assisted selection to improve drought tolerance. Int J Mol Sci 24:8986. https://doi.org/10.3390/ijms24108986

Fernandez GCJ (1992) Effective selection criteria for assessing plant stress tolerance. In: Proc Int Symp Adapt Vegetables and Other Food Crops in Temperature and Water Stress. AVRDC, In, pp 257–270

Fischer RA, Maurer R (1978) Drought resistance in spring wheat cultivars. I. Grain yield responses. Aust J Agric Res 29:897–912. https://doi.org/10.1071/AR9780897

Food and Agriculture Organization (FAO) of the United Nations (2022) The state of food security and nutrition in the world 2022: repurposing food and agricultural policies to make healthy diets more affordable. FAO. https://doi.org/10.4060/cc0639en

Gaikwad K, Mazumder A, Kumar M et al (2025) Evaluating heat and drought resilience in ancient Indian dwarf wheat *Triticum sphaerococcum* Percival using stress tolerance indices. Sci Rep 15:2502. https://doi.org/10.1038/s41598-025-02502-0

Gaurav S, Nandakishore T, Sharma D et al (2020) Mapping QTLs for chlorophyll content and grain yield under drought using RILs derived from 'Drysdale × Waagan'. Euphytica 216:95. https://doi.org/10.1007/s10681-020-02635-8

Ge M, Tang Y, Guan Y et al (2024) TaWRKY31, a novel WRKY transcription factor in wheat, participates in regulation of plant drought stress tolerance. BMC Plant Biol 24:27. https://doi.org/10.1186/s12870-023-04709-7

Ghaffar A, Aslam Y, Ali S et al (2023) Photosynthetic activity and metabolic profiling of bread wheat cultivars under drought stress. Sci Rep 13:6525. https://doi.org/10.1038/s41598-023-33593-7

Ghanem A, Abou El-Yazied A, Abbas M (2024) Wheat drought tolerance: morpho-physiological criteria, stress indexes, and yield responses in newly sandy soils. J Plant Growth Regul 43:2234–2250. https://doi.org/10.1007/s00344-024-11259-1

Gholizadeh F, Janda T, Gondor OK et al (2022) Improvement of drought tolerance by exogenous spermidine in germinating wheat (*Triticum aestivum* L.) plants is accompanied with changes in metabolite composition. Int J Mol Sci 23:9047. https://doi.org/10.3390/ijms23169047

Gill HS, Halder J, Zhang J et al (2022) Genome-wide association analysis of drought tolerance traits in wheat. Plant Genome 15:e20164. https://doi.org/10.1002/tpg2.20164

Gudi S, Saini DK, Halladakeri P et al (2023) Genome-wide association study unravels genomic regions associated with chlorophyll fluorescence parameters in wheat (*Triticum aestivum* L.) under different sowing conditions. Plant Cell Rep 42:1453–1472. https://doi.org/10.1007/s00299-023-03041-6

Gudi S, Halladakeri P, Singh G et al (2024) Deciphering the genetic landscape of seedling drought stress tolerance in wheat (*Triticum aestivum* L.) through genome-wide association studies. *Front.* Plant Sci 15:1351075. https://doi.org/10.3389/fpls.2024.1351075

Guizani A, Babay E, Askri H et al (2024) Screening for drought tolerance and genetic diversity of wheat varieties using agronomic and molecular markers. Mol Biol Rep 51:432. https://doi.org/10.1007/s11033-024-09340-9

Guo R, Shi L, Jiao Y et al (2018) Metabolic responses to drought stress in the tissues of drought-tolerant and drought-sensitive wheat genotype seedlings. AoB Plants 10:ply016. https://doi.org/10.1093/aobpla/ply016

Guo C, Memon A, Zhang X et al (2024) An enhanced method for studying wheat stomata physiology. J Plant Growth Regul 43:4886–4893. https://doi.org/10.1007/s00344-024-11441-5

Gupta P, Kaur H, Sharma P et al (2024a) Reactive oxygen species signaling and scavenging in wheat under drought stress. Front Plant Sci 15:1359425. https://doi.org/10.3389/fpls.2024.1359425

Gupta AK, Agrawal M, Yadav H et al (2024b) Drought stress and its tolerance mechanism in wheat. Int J Environ Clim Change 14:529–544. https://doi.org/10.9734/ijecc/2024/v14i13866

He J, Duan Q, Chen F et al (2022) ECERIFERUM1-6A is required for the synthesis of cuticular wax alkanes and improves drought tolerance in wheat. Plant Physiol 190:141–157. https://doi.org/10.1093/plphys/kiac347

Hosseini SS, Razavi BS, Lakzian A (2024) Drought tolerance of wheat genotypes is associated with rhizosphere size and enzyme system. Plant Soil 502:671–685. https://doi.org/10.1007/s11104-024-06576-z

Iquebal M, Sharma P, Jasrotia R et al (2019) RNAseq analysis reveals drought-responsive molecular pathways with candidate genes and putative molecular markers in root tissue of wheat. Sci Rep 9:49915. https://doi.org/10.1038/s41598-019-49915-2

Itam M, Mega R, Tadano S et al (2020) Metabolic and physiological responses to progressive drought stress in bread wheat. Sci Rep 10:74303. https://doi.org/10.1038/s41598-020-74303-6

Jadidi O, Pour-Aboughadareh A, Bocianowski J et al (2025) Physiological and molecular responses of bread wheat and its wild relative species to drought stress. Mol Biol Rep 52:10742. https://doi.org/10.1007/s11033-025-10742-6

Kolupaev YE, Yastreb TO, Salii AM et al (2022) State of antioxidant and osmo-protective systems in etiolated winter wheat seedlings of different cultivars due to their drought tolerance. Zemdirbyste-Agriculture 109:313–322. https://doi.org/10.13080/z-a.2022.109.040

Kopecká J, Pšenčík J, Sobotka R (2023) Drought stress and photosynthetic electron transport: causes of ROS overproduction and damage. Plant Physiol Biochem 196:107558. https://doi.org/10.1016/j.plaphy.2023.107558

Langridge P, Reynolds M (2021) Breeding for drought and heat tolerance in wheat. Theor Appl Genet 134:1753–1769. https://doi.org/10.1007/s00122-021-03795-1

Latif M, Hussain A, Tahir M et al (2022) Exopolysaccharide-producing rhizobacteria mitigate drought in wheat. Sci Rep 12:7462. https://doi.org/10.1038/s41598-022-11446-8

Lepekhov S (2022) Canopy temperature depression for drought- and heat stress tolerance in wheat breeding. Vavilov J Genet Breed 26:196–201. https://doi.org/10.18699/vjgb-22-24

Li L-H, Yi H-L, Liu X-P et al (2021) Sulfur dioxide enhances drought tolerance of wheat seedlings through $H_2S$ signaling. Ecotoxicol Environ Saf 207:111248. https://doi.org/10.1016/j.ecoenv.2020.111248

Li P-F, Ma B-L, Palta JA et al (2022a) Distinct contributions of drought avoidance and drought tolerance to yield improvement in dryland wheat cropping. J Agron Crop Sci 208:265–282. https://doi.org/10.1111/jac.12574

Li M, Li X, Guo X et al (2022b) Transcriptome analysis reveals key drought-stress genes in wheat. Front Genet 13:1060529. https://doi.org/10.3389/fgene.2022.1060529

Li N, Liu T, Guo F et al (2022c) Identification of long non-coding RNA–microRNA–mRNA regulatory modules and their potential roles in drought stress response in wheat (*Triticum aestivum* L.). *Front*. Plant Sci 13:1011064. https://doi.org/10.3389/fpls.2022.1011064

Li Q, Wang X, Sun Z et al (2023) DNA methylation levels of *TaP5CS* and *TaBADH* are associated with enhanced tolerance to PEG-induced drought stress triggered by drought priming in wheat. Plant Physiol Biochem 200:107769. https://doi.org/10.1016/j.plaphy.2023.107769

Li G, Liu J, Wang Y et al (2024a) Wheat WRKY24-1D interacts with TaERFL1a and regulates DHAR-mediated AsA–GSH biosynthesis to enhance drought tolerance in wheat. Res Sq. https://doi.org/10.21203/rs.3.rs-4172975/v1

Li J, Wang T, Ma X et al (2024b) The potassium transporter TaNHX2 interacts with TaGAD1 to modulate ABA sensitivity and stomatal closure in wheat. Nat Commun 15:2877. https://doi.org/10.1038/s41467-024-47941-1

Liu S, Wang X, Wang H et al (2021) Global transcriptome and weighted gene co-expression network analyses reveal key drought-responsive genes in wheat leaves. Front Plant Sci 12:663180. https://doi.org/10.3389/fpls.2021.663180

Liu G, Liu D, Zhang A et al (2023) Identification of KASP markers and candidate genes for drought tolerance in wheat using 90K SNP array genotyping of near-isogenic lines targeting a 4BS quantitative trait locus. Theor Appl Genet 136:190. https://doi.org/10.1007/s00122-023-04438-3

Lv L, Chen X, Li H et al (2022a) Different adaptive patterns of wheat with different drought tolerance under drought stresses and rehydration revealed by integrated metabolomic and transcriptomic analysis. Front Plant Sci 13:1008624. https://doi.org/10.3389/fpls.2022.1008624

Lv X, Zhao Y, Wang C et al (2022b) Integrated metabolomic and transcriptomic analyses reveal mechanisms of osmotic adjustment and antioxidant defense during drought in wheat. BMC Plant Biol 22:377. https://doi.org/10.1186/s12870-022-03709-5

Ma J, Zhang M, Lv W et al (2022) Overexpression of *TaSNAC4-3D* in common wheat (*Triticum aestivum* L.) negatively regulates drought tolerance. Front Plant Sci 13:945272. https://doi.org/10.3389/fpls.2022.945272

Ma J, Geng Y, Liu H et al (2023a) *TaTIP41* and *TaTAP46* positively regulate drought tolerance in wheat by inhibiting PP2A activity. J Integr Plant Biol 65:2056–2070. https://doi.org/10.1111/jipb.13542

Ma J, Wang Y, Tang X et al (2023b) *TaSINA2B*, interacting with *TaSINA1D*, positively regulates drought tolerance and root growth in wheat (*Triticum aestivum* L.). Plant Cell Environ 46:3760–3774. https://doi.org/10.1111/pce.14708

Ma J, Wang W, Yang P et al (2023c) SINA E3 ubiquitin ligases regulate stomatal movement and lateral root development to enhance drought tolerance in wheat. Plant Physiol Biochem 197:107680. https://doi.org/10.1016/j.plaphy.2023.107680

Ma Z, Zhao B, Zhang H et al (2024) Upregulation of wheat heat shock transcription factor *TaHsfC3-4* by ABA contributes to drought tolerance. Int J Mol Sci 25:977. https://doi.org/10.3390/ijms25020977

Mahmood T, Li X, Khan MIR (2020) Incredible role of osmotic adjustment in grain yield sustainability under drought stress in wheat. Plants 9:1208. https://doi.org/10.3390/plants9091208

Mao H, Li S, Wang Z et al (2020) Regulatory changes in *TaSNAC8-6A* are associated with drought tolerance in wheat seedlings. Plant Biotechnol J 18:1078–1092. https://doi.org/10.1111/pbi.13277

Mao H, Li S, Chen B et al (2022) Variation in cis regulation of a NAC transcription factor contributes to drought tolerance in wheat. Mol Plant 15:276–292. https://doi.org/10.1016/j.molp.2021.11.007

Maulana F, Ayalew H, Anderson JD (2020) Genome-wide association mapping of drought tolerance in spring wheat. PLoS One 15:e0229608. https://doi.org/10.1371/journal.pone.0229608

Mei F, Chen B, Li F et al (2021) Overexpression of the wheat NAC transcription factor *TaSNAC4-3A* confers drought tolerance in transgenic *Arabidopsis*. Plant Physiol Biochem 160:37–50. https://doi.org/10.1016/j.plaphy.2021.01.004

Merrium S, Hakeem S, Ali Z (2022) Leaf rolling and leaf angle improve fog capturing and transport in wheat: adaptation for drought stress in an arid climate. Environ Sci Pollut Res 29:40567–40583. https://doi.org/10.1007/s11356-022-18936-2

Michaletti A, Naghavi MR, Toorchi M et al (2018) Metabolomics and proteomics reveal drought-stress responses of leaf tissues from spring wheat. Sci Rep 8:5710. https://doi.org/10.1038/s41598-018-24012-y

Mohammadi R, Geravandi M (2024) Multi-trait selection for agronomic performance and drought tolerance among durum wheat genotypes evaluated under rainfed and irrigated environments. Exp Agric 60:e3. https://doi.org/10.1017/S0014479723000273

Montesinos-López OA, Saint Pierre C, Bentley AR et al (2023) Genomics combined with UAV data enhances prediction of grain yield in winter wheat. Front Genet 14:1130247. https://doi.org/10.3389/fgene.2023.1130247

Nardino M, Alves CR, Ferreira EG et al (2022) Understanding drought response mechanisms in wheat and multi-trait selection under water-deficit conditions. PLoS One 17:e0266368. https://doi.org/10.1371/journal.pone.0266368

Negisho K, Shibru S, Matros A et al (2022) Association mapping of drought tolerance indices in Ethiopian durum wheat (*Triticum turgidum* ssp. *durum*). Front Plant Sci 13:838088. https://doi.org/10.3389/fpls.2022.838088

Nyaupane R, Ali M, Subramanian S (2024a) Physiological and molecular mechanisms of drought tolerance in wheat: a review. Front Plant Sci 15:1372222. https://doi.org/10.3389/fpls.2024.1372222

Nyaupane S, Poudel MR, Panthi B et al (2024b) Drought stress effect, tolerance, and management in wheat: a review. Cogent Food Agric 10:2296094. https://doi.org/10.1080/23311932.2023.2296094

Odone A, Popovic O, Thorup-Kristensen K (2024) Deep roots: implications for nitrogen uptake and drought tolerance among winter wheat cultivars. Plant Soil 500:13–32. https://doi.org/10.1007/s11104-023-06255-5

Omar AA, Heikal YM, Zayed EM et al (2023) Conferring drought and heat stress tolerance in wheat (*Triticum aestivum* L.) genotypes and their response to selenium nanoparticles application. Nano 13:998. https://doi.org/10.3390/nano13060998

Onyemaobi O, Liu H, Siddique KHM et al (2021) Reproductive stage drought tolerance in wheat: importance of stomatal conductance and plant growth regulators. Plants 10:2346. https://doi.org/10.3390/plants10112346

Ozturk M, Turkyilmaz Unal B, Budak H (2021a) Lignin and osmo-protectant metabolism under drought stress in plants. Front Plant Sci 12:645382. https://doi.org/10.3389/fpls.2021.645382

Ozturk M, Turkyilmaz Unal B, García-Caparrós P et al (2021b) Osmoregulation and its actions during drought stress in plants. Physiol Plant 172:1321–1335. https://doi.org/10.1111/ppl.13297

Pankaj Y, Kumar R, Gill K et al (2022a) Unravelling QTLs for non-destructive and yield-related traits under timely, late and very late sown conditions in wheat (*Triticum aestivum* L.). Plant Mol Biol Report 42:369–382. https://doi.org/10.1007/s11105-022-01357-3

Pankaj Y, Kumar R, Pal L et al (2022b) Mapping QTLs for morpho-physiological traits related to grain yield under late sown conditions in wheat (*Triticum aestivum* L.). Cereal Res Commun 50:779–788. https://doi.org/10.1007/s42976-021-00234-1

Pantha S, Kilian B, Özkan H et al (2025) A comparative analysis of drought tolerance mechanisms in *Triticum araraticum* Jakubz. and modern bread wheat (*Triticum aestivum* L.) cultivars. J Agron Crop Sci 211(4):e70097. https://doi.org/10.1111/jac.70097

Paudel P, Pandey MK, Subedi M et al (2024) Genomic approaches for improving drought tolerance in wheat (*Triticum aestivum* L.): a comprehensive review. Plant Arch 24. https://doi.org/10.51470/PLANTARCHIVES.2024.v24.no.1.180

Pinto R, Reynolds M (2015) Common genetic basis for canopy temperature depression under heat and drought stress associated with optimized root distribution in bread wheat. Theor Appl Genet 128:575–585. https://doi.org/10.1007/s00122-015-2453-9

Pour-Aboughadareh A, Ahmadi J, Mehrabi A et al (2017) Physiological responses to drought stress in wild relatives of wheat: implications for wheat improvement. Acta Physiol Plant 39:1–12. https://doi.org/10.1007/s11738-017-2403-z

Qayyum A, Ayoubi S, Sher A et al (2021) Improvement in drought tolerance in bread wheat is related to improved osmolyte production, antioxidant enzyme activities and gaseous exchange. Saudi J Biol Sci 28:5238–5249. https://doi.org/10.1016/j.sjbs.2021.05.040

Rabieyan E, Bihamta MR, Esmaeilzadeh Moghaddam M et al (2023) Analysis of genetic diversity and genome-wide association study for drought tolerance-related traits in Iranian bread wheat. BMC Plant Biol 23:431. https://doi.org/10.1186/s12870-023-04416-3

Raeisi M, Karimi N, Eslami H (2022) Exogenous application of zinc oxide nanoparticles improves drought tolerance through enhanced antioxidant defense and proline metabolism in wheat. Plant Stress 6:100116. https://doi.org/10.1016/j.stress.2022.100116

Rafiei F, Wiersma J, Scofield S et al (2024) Facts, uncertainties, and opportunities in wheat molecular improvement. Heredity 133:371–380. https://doi.org/10.1038/s41437-024-00721-1

Rathan ND, Singh R, Ibba MI et al (2022) Identification and validation of KASP markers linked to canopy temperature and grain yield QTLs for improving drought tolerance in wheat. Theor Appl Genet 135:4213–4230. https://doi.org/10.1007/s00122-022-04167-7

Roostaei M, Rajabi R, Jafarzadeh J et al (2022) Assessment of drought tolerance and grain yield stability of rainfed winter bread wheat genotypes. Crop Breed J. https://doi.org/10.22092/cbj.2022.359076.1075

Rosielle AA, Hamblin J (1981) Theoretical aspects of selection for yield in stress and non-stress environments. Crop Sci 21:943–946. https://doi.org/10.2135/cropsci1981.0011183X002100060033x

Salarpour M, Pakniyat H, Abdolshahi R et al (2020) Mapping QTL for agronomic and root traits in the Kukri/RAC875 wheat population under drought stress conditions. Euphytica 216:1–14. https://doi.org/10.1007/s10681-020-02627-5

Salem MA, Ismail MA, Radwan KH et al (2024) Unlocking the potential of plant growth-promoting rhizobacteria to enhance drought tolerance in Egyptian wheat. Sustainability 16:4605. https://doi.org/10.3390/su16114605

Sallam A, Alqudah AM, Dawood MFA et al (2019) Drought stress and wheat: a review of physiological and molecular responses. Planta 249:815–827. https://doi.org/10.1007/s00425-018-02912-9

Sallam A, Martsch R, Shafei AA et al (2022) Comparative performance of GWAS models (GLM, MLM, FarmCPU) for drought traits in wheat. Euphytica 218:78. https://doi.org/10.1007/s10681-022-03050-8

Sallam A, Dawood MFA, Jarquín D et al (2024a) Genome-wide scanning to identify and validate single nucleotide polymorphism markers associated with drought tolerance in spring wheat seedlings. Plant Genome 17:e20444. https://doi.org/10.1002/tpg2.20444

Sallam M, Ghazy A, Al-Doss A et al (2024b) Combining genetic and phenotypic analyses for detecting bread wheat genotypes of drought tolerance through multivariate analysis techniques. Life 14:183. https://doi.org/10.3390/life14020183

Saripalli G, Adhikari L, Amos C et al (2023) Integration of genetic and genomics resources in einkorn wheat enables precision mapping of important traits. Commun Biol 6:835. https://doi.org/10.1038/s42003-023-05189-z

Sedaghat M, Tahmasebi Sarvestani Z, Emam Y et al (2020) Foliar-applied GR24 and salicylic acid enhanced wheat drought tolerance. Russ J Plant Physiol 67:733–739. https://doi.org/10.1134/S1021443720040159

Seleiman MF, Al-Suhaibani N, Ali N et al (2021) Drought stress impacts on plants and different approaches to alleviate its adverse effects. Plants 10:259. https://doi.org/10.3390/plants10020259

Sharma PK, Ahmed HI, Heuberger M et al (2023) An online database for einkorn wheat to aid in gene discovery and functional genomics studies. Database 2023:baad079. https://doi.org/10.1093/database/baad079

Singh C, Yadav S, Khare V et al (2025) Wheat drought tolerance: unveiling a synergistic future with conventional and molecular breeding strategies. Plants 14:1053. https://doi.org/10.3390/plants14071053

Soriano JM, Álvaro F (2019) Discovering consensus genomic regions in wheat for root-related traits by QTL meta-analysis. Sci Rep 9:47038. https://doi.org/10.1038/s41598-019-47038-2

Su L, Shen X, Liang C et al (2023) TaPRX-2A enhances drought tolerance by improving antioxidant enzyme activity in wheat. BMC Plant Biol 23:160. https://doi.org/10.1186/s12870-023-04120-7

Suzuky R, Reynolds M, Ky B et al (2010) Heat and drought adaptive QTL in a wheat population designed to minimize confounding agronomic effects. Theor Appl Genet 121:1001–1021. https://doi.org/10.1007/s00122-010-1351-4

Tian G, Wang S, Wu J et al (2023) Allelic variation of *TaWD40-4B.1* contributes to drought tolerance by modulating catalase activity in wheat. Nat Commun 14:1200. https://doi.org/10.1038/s41467-023-36901-6

Trnka M, Olesen JE, Eitzinger J et al (2019) Mitigation efforts will not fully alleviate the increase in water scarcity occurrence probability in wheat-producing areas. Proc Natl Acad Sci USA 116:2549–2554. https://doi.org/10.1073/pnas.1817380116

Trono D, Pecchioni N (2022) Candidate genes associated with abiotic stress response in plants as tools to engineer tolerance to drought, salinity and extreme temperatures in wheat: an overview. Plants 11:3358. https://doi.org/10.3390/plants11233358

Ud-din MM, Shah SH, Sial MA et al (2024) Multi-trait index-based selection of drought-tolerant wheat. Plants 14:35. https://doi.org/10.3390/plants14010035

Ullah A, Al-Sadi AM, Farooq M (2023a) Higher intercellular $CO_2$ concentration is associated with improved water use efficiency and drought tolerance in bread wheat. Gesunde Pflanzen 75:1679–1687. https://doi.org/10.1007/s10343-023-00839-y

Ullah A, Shahzad K, Iqbal M et al (2023b) Stomatal regulation, water relations and drought tolerance in wheat: an overview. Physiol Mol Biol Plants 29:1317–1332. https://doi.org/10.1007/s12298-023-01346-3

UN DESA (2022) World population prospects 2022: summary of results. United Nations, Department of Economic and Social Affairs, Population Division. https://www.un.org/development/desa/pd/sites/www.un.org.development.desa.pd/files/wpp2022_summary_of_results.pdf

Ur Rehman S, Wang J, Chang X et al (2019) A wheat protein kinase gene *TaSnRK2.9-5A* associated with yield contributing traits. Theor Appl Genet 132:907–919. https://doi.org/10.1007/s00122-018-3247-7

Verbeke M et al (2022) Osmotic adjustment in wheat (*Triticum aestivum* L.) during pre- and post-anthesis drought. Front Plant Sci 13:775652. https://doi.org/10.3389/fpls.2022.775652

Vijayaraghavareddy P, Seepaul R, Narayanan S (2022) High-throughput phenotyping tools and approaches for drought studies in crops. Front Plant Sci 13:833780. https://doi.org/10.3389/fpls.2022.833780

Wang X, Chen J, Ge J et al (2021a) The different root apex zones contribute to drought priming-induced tolerance to a reoccurring drought stress in wheat. Crop J 9:1088–1097. https://doi.org/10.1016/j.cj.2020.11.008

Wang X, Li Q, Xie J et al (2021b) Abscisic acid and jasmonic acid are involved in drought priming-induced tolerance to drought in wheat. Crop J 9:120–132. https://doi.org/10.1016/j.cj.2020.06.002

Wang B, Li L, Liu M et al (2022a) *TaFDL2-1A* confers drought stress tolerance by promoting ABA biosynthesis, ABA responses, and ROS scavenging in transgenic wheat. Plant J 112:722–737. https://doi.org/10.1111/tpj.15975

Wang J, Zhang X, Han Z et al (2022b) Analysis of physiological indicators associated with drought tolerance in wheat under drought and re-watering conditions. Antioxidants 11:2266. https://doi.org/10.3390/antiox11112266

Wang Y, Zhang J, Liu Y et al (2022c) bZIP transcription factor *TaFDL2-1A* enhances ROS scavenging and drought tolerance in wheat. Environ Exp Bot 201:104943. https://doi.org/10.1016/j.envexpbot.2022.104943

Wang J, Li C, Li L et al (2023a) *DIW1* encoding a clade I PP2C phosphatase negatively regulates drought tolerance by de-phosphorylating *TaSnRK1.1* in wheat. J Integr Plant Biol 65:1918–1936. https://doi.org/10.1111/jipb.13504

Wang N, Chen J, Gao Y et al (2023b) Genomic analysis of isopentenyl transferase genes and functional characterization of *TaIPT8* indicates positive effects of cytokinins on drought tolerance in wheat. Crop J 11:46–56. https://doi.org/10.1016/j.cj.2022.04.010

Wang D, Chen H, Hao X et al (2024) *TabHLH27* orchestrates root growth and drought tolerance to enhance water-use efficiency in wheat. J Integr Plant Biol 66:1634–1651. https://doi.org/10.1111/jipb.14651

Weatherley PE (1950) Studies in the water relations of the cotton plant. I. The field measurement of water deficits in leaves. New Phytol 49:81–97. https://doi.org/10.1111/j.1469-8137.1950.tb05146.x

Williams AP, Cook BI, Smerdon JE et al (2020) Large contribution from anthropogenic warming to an emerging North American megadrought. Science 368:314–318. https://doi.org/10.1126/science.aaz9600

Xi W, Liu Y, Wang W et al (2023) Transcriptome analysis of roots from wheat (*Triticum aestivum*) cultivars under PEG-induced drought stress. Int J Mol Sci 24:7245. https://doi.org/10.3390/ijms24087245

Yan J, Wang D, Kang S et al (2023) Ethylene regulates combined drought and low nitrogen tolerance in wheat: Proteomic analysis. Agronomy 13:1950. https://doi.org/10.3390/agronomy13071950

Yang J, Zhang G, An J et al (2020) Expansin gene *TaEXPA2* positively regulates drought tolerance in transgenic wheat (*Triticum aestivum* L.). Plant Sci 298:110596. https://doi.org/10.1016/j.plantsci.2020.110596

Yang G, Pan Y, Pan W et al (2024) Combined GWAS and eGWAS reveals the genetic basis underlying drought tolerance in emmer wheat (*Triticum turgidum* L.). New Phytol 242:2115–2131. https://doi.org/10.1111/nph.19589

Yu M, Yu Y, Guo S et al (2022) Identification of *TaBADH-A1* allele for improving drought resistance and salt tolerance in wheat (*Triticum aestivum* L.). Front Plant Sci 13:942359. https://doi.org/10.3389/fpls.2022.942359

Zarei L, Ghobadi M, Mahfoozi S et al (2022) Genome-wide association mapping reveals loci on 2A and 5D for root length and shoot biomass under drought in wheat. Mol Biol Rep 49:10455–10467. https://doi.org/10.1007/s11033-022-07347-8

Zhang Y, Li R, Chen H et al (2021) Overexpression and RNAi of *TaNRX1* modulate drought tolerance in wheat by regulating catalase activity and $H_2O_2$ homeostasis. Plant Biotechnol J 19:1010–1026. https://doi.org/10.1111/pbi.13536

Zhao J, Li X, Qiao L et al (2023) Identification of structural variations related to drought tolerance in wheat (*Triticum aestivum* L.). Theor Appl Genet 136:37. https://doi.org/10.1007/s00122-023-04283-4

Zhou Y, Liu J, Guo J et al (2022) *GmTDN1* improves wheat yields by inducing dual tolerance to both drought and low-N stress. Plant Biotechnol J 20:1606–1621. https://doi.org/10.1111/pbi.13836

# Nanotechnology Applications for Enhancing Stress Tolerance in Wheat: Past, Present, and Future Prospects

15

M. D. Oak and J. M. Rajwade

**Abstract**

Wheat (*Triticum aestivum* L.) is a staple crop for more than 35% of the global population. It experiences yield reductions exceeding 50% due to abiotic stresses such as drought, salinity, heat, cold, waterlogging, and heavy metal toxicity. These challenges are further compounded by biotic threats, including fungal pathogens, insects, and viruses, particularly under the pressures of climate change. This chapter critically examines the evolution, mechanisms, and applications of nanotechnology to enhance wheat stress tolerance. It traces developments from Feynman's 1959 vision to recent advances in nanomaterial synthesis (physical, chemical, and green methods), characterization techniques (TEM, DLS, and XRD), and delivery systems (seed nanopriming, foliar sprays, and soil amendments). Metal and metal-oxide nanoparticles (ZnO, $TiO_2$, Ag, and $Fe_3O_4$) have been shown to modulate reactive oxygen species (ROS) homeostasis, upregulate antioxidant enzymes (SOD, CAT, and POD), improve nutrient uptake (such as the $K^+/Na^+$ ratio), stabilize photosynthetic machinery, and reprogram stress-responsive genes through omics-based pathways. Pot and greenhouse studies report enhanced germination, biomass, chlorophyll content, and grain yield under salinity, drought, and freezing conditions. However, silver nanoparticles (AgNPs) exhibit dose-dependent phytotoxicity and genotype-specific effects. While synergies with plant growth-promoting rhizobacteria and meta-analyses confirm the benefits of Fe-, Ti-, and Zn-based nanomaterials, several limitations remain. These include challenges in scaling up from laboratory to field conditions, uncertainties regarding nanomaterial fate (uptake, translocation, and persistence), impacts on soil microbiomes, regulatory gaps (such as EFSA guidelines), and toxicity risks. Addressing these issues requires standardized

M. D. Oak (✉)
Genetics and Plant Breeding Group, Agharkar Research Institute, Pune, Maharashtra, India
e-mail: mdoak@aripune.org

J. M. Rajwade
Nanobioscience Group, Agharkar Research Institute, Pune, Maharashtra, India



R. Tiwari et al. (eds.), *Wheat Stress Management*,
https://doi.org/10.1007/978-981-92-0902-6_15

field trials, comprehensive life-cycle assessments, safe-by-design formulations, and the integration of "nano-breeding" with CRISPR technologies to ensure equitable and sustainable deployment.

**Keywords**

Wheat stress tolerance · Nanotechnology applications · Seed nanopriming · Metal oxide nanoparticles · Abiotic stress mitigation · Nano-enabled agriculture

## 15.1 Introduction

Wheat (*Triticum aestivum* L.) is a critical component of global food security, nutrition, and economic stability. As one of the most widely cultivated and traded crops, wheat is a staple food for over 35% of the global population (Shewry 2025). Wheat cultivation supports the livelihoods of millions, particularly smallholder farmers in rural regions of major producing countries such as China and India. Its high-yielding potential and capacity for long-term storage help stabilize food availability and prices in international markets. Its adaptability to diverse climates, including temperate, subtropical, and semi-arid zones, enhances its reliability and resilience, ensuring stable food supplies despite climate variability (Chen et al. 2023b; Mao et al. 2023). Wheat also serves as a primary source of essential nutrients, including carbohydrates, proteins, vitamins, and minerals, for billions of individuals. In addition to basic food products, wheat is used in a variety of goods, including noodles, pastries, cereals, alcoholic beverages, and nonfood items such as adhesives and biofuels (Shewry and Hey 2015). However, global wheat productivity is significantly limited by both abiotic and biotic stresses, which together can cause yield losses exceeding 50% (Li et al. 201; Aryal et al. 2021). Abiotic stresses, or environmental factors, are the primary contributors to these losses, although biotic factors, including various living organisms, continue to present persistent and evolving challenges (Budak et al. 2015; Yashavanthakumar et al. 2026).

### 15.1.1 Abiotic Stresses

Abiotic stresses adversely affect plant growth, development, and yield. These stresses frequently interact, with their impacts becoming particularly pronounced during critical developmental stages such as flowering and grain filling (Mondal et al. 2016). Among these, drought and salinity represent the most significant constraints, inducing morphological, physiological, and molecular changes that disrupt metabolic processes and can reduce yields by more than 50% (Table 15.1). Temperature extremes, including both high-temperature heat waves during grain filling and low-temperature chilling or freezing, substantially impair wheat membrane integrity and photosynthetic capacity (Asseng et al. 2015; Table 15.1). Additionally, heavy metal toxicity from elements such as cadmium (Cd), chromium

**Table 15.1** Abiotic stresses—their effects on the wheat crop and estimated losses

| Stress type | Primary effects on wheat | Estimated losses (general) | References |
|---|---|---|---|
| Drought (water scarcity) | Inhibits root development, disrupts photosynthesis, causes premature leaf senescence, and reduces grain number and weight | Can decrease yield by up to 50% or more, depending on severity and timing | Dong et al. (2017), Kulkarni et al. (2017), Li et al. (2019), Ateş Sönmezoğlu et al. (2022) and Alam et al. (2025) |
| Heat (high temperature) | Accelerates development, shortens the crop cycle, causes pollen sterility, reduces photosynthesis, and leads to shrivelled grains with lower weight | A 1 °C increase in temperature can cause a 4.1% to 6% yield loss; terminal heat can reduce yield by 20–40% | Kajla et al. (2015), Maulana et al. (2018), Dhakal et al. (2021) and Porras et al. (2023) |
| Salinity (high soil salt) | Creates osmotic stress and ion toxicity ($Na^+$ and $Cl^-$ accumulation), which impairs water and nutrient uptake, damages membranes, and reduces germination | Affects about 20% of irrigated land globally and can reduce yields by up to 50% in highly saline soils | Dhakal et al. (2021), EL Sabagh et al. (2021) and Malipatil et al. (2023) |
| Cold/frost (low temperature) | Causes chilling injury or freezing damage, leading to yellowing (chlorosis), reduced leaf expansion, and pollen sterility, especially during the reproductive stage | Freezing during the anthesis stage can lead to up to 98% sterility in affected spikes | Dhakal et al. (2021) and Yu et al. (2025) |
| Waterlogging | Leads to oxygen deficiency in roots, impairing nutrient and water uptake, which can reduce grain yield by 50–70% if it occurs for 30 days during sensitive periods | | Pais et al. (2023), Pramanick et al. (2023) and Alam et al. (2025) |
| Heavy metal toxicity | Strongly inhibits seed germination and grain filling, disrupts enzyme activities and chlorophyll synthesis, and reduces the biomass | Heavy metal toxicity can reduce wheat production by 15–60% by disrupting overall growth | Balali-Mood et al. (2021), Rashid et al. (2023) and Ilya et al. (2025) |

(Cr), mercury (Hg), and arsenic (As) presents a dual threat by inhibiting wheat growth and increasing the risk of bioaccumulation within the human food chain.

### 15.1.2 Biotic Stresses

Biotic stresses result from living organisms, including pests and pathogens, that compete with or damage wheat plants (Savary et al. 2019; Wang et al. 2023a, b, c; Singh et al. 2025). These stresses encompass fungal pathogens such as *Fusarium*

**Table 15.2** Biotic stresses—their effect on the wheat crop and estimated loss

| Stress | Key pathogens/pests | Primary effects on wheat | References |
|---|---|---|---|
| Fungal diseases | Rusts (stem, leaf, and stripe), powdery mildew, Fusarium head blight (FHB), and Karnal bunt | Fungal diseases can cause significant yield losses, with rust alone capable of affecting millions of hectares and resulting in major economic damage (up to 50–100% loss in severe, untreated cases) | Sharma and Duveiller (2004), Roy et al. (2023), Wang et al. (2023a) and Singh et al. (2025) |
| Insects and pests | Aphids (Russian wheat aphid, bird cherry-oat aphid), Hessian fly, and stem sawflies | Aphid infestations can cause yield losses of up to 40%, with early attacks causing losses as high as 65%. Pests also serve as vectors for viral diseases | Savary et al. (2019) |
| Viruses and other pathogens | Wheat streak mosaic virus (WSMV), *Triticum mosaic virus* (TriMV), nematodes, and weeds | Viruses can cause devastating damage, while weeds compete vigorously with wheat for resources (light, water, and nutrients), resulting in losses of up to 23% | |

*graminearum*, which causes Fusarium head blight, as well as bacterial blight and rusts. These diseases contribute to significant yield losses and grain contamination with mycotoxins (Table 15.2). Additionally, pest infestations by aphids, Hessian fly, stem sawflies, and stored-grain beetles (*Sitophilus oryzae*) threaten both productivity and postharvest storage security (Savary et al. 2019; Table 15.2).

Abiotic and biotic stresses pose significant threats to global food security. Climate change increases the frequency and intensity of these stresses, often leading to their simultaneous occurrence, which has a more severe impact than each stressor alone. Approaches such as developing resistant cultivars through traditional breeding, modern genome-editing technologies, and nanotechnology are essential for achieving sustainable wheat production under changing climatic conditions (Hussain et al. 2021).

Nanotechnology, an interdisciplinary field encompassing physics, chemistry, biology, and materials science, has demonstrated transformative potential in agriculture. By utilizing nanoscale materials (1–100 nm) with unique physicochemical properties, such as a high surface-area-to-volume ratio, tunable surface chemistry, and enhanced reactivity, nanotechnology enables targeted and efficient interventions at the cellular or molecular level. Figure 15.1 summarizes the mechanisms through which nanomaterials enhance plant growth, nutrient uptake, stress tolerance, and defence.

Several of these can be explored in the context of sustainable wheat production amid mounting global challenges.

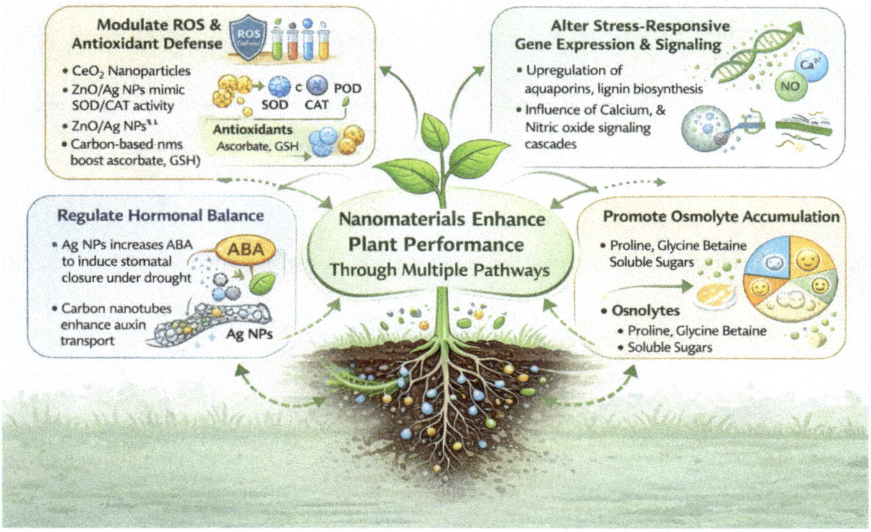


**Fig. 15.1** Role of nanomaterials in enhancing plant growth: Underlying mechanisms

## 15.2 Historical Perspectives on Nanotechnology Applications for Enhancing Stress Tolerance in Wheat

Sir Richard Feynman's 1959 lecture introduced the concept of controlling matter at the atomic scale, widely regarded as the beginning of nanotechnology, although the term itself was not coined until 1974 by N. Taniguchi. The ideas of molecular assembly and nanoscale engineering gained broader recognition through Eric Drexler's 1986 publication, Engines of Creation: The Coming Era of Nanotechnology (https://calteches.library.caltech.edu/47/2/1960Bottom.pdf; https://www.nano.gov; Bayda et al. 2019). Major developments in agri-nanotechnology over the past 25 years are summarized in Fig. 15.2.

## 15.3 Scientific Advancements and Principles Developed Over the Past Centuries

Physical, chemical, and biological or "green" methods are utilized for the synthesis of nanoparticles. The synthesis route determines particle size, shape, crystallinity, surface chemistry, and scalability. For agricultural applications, methods that produce nanoparticles with enhanced colloidal stability and biocompatibility are preferred. The synthesis approach also influences nanoparticle dissolution, reactivity, and interactions with soil or root exudates, which are critical for agronomic outcomes. Figure 15.3 provides a schematic overview of nanotechnology applications in agriculture. Physicochemical characterization is conducted using electron microscopy (TEM/SEM) for size and morphology, dynamic light scattering (DLS), and

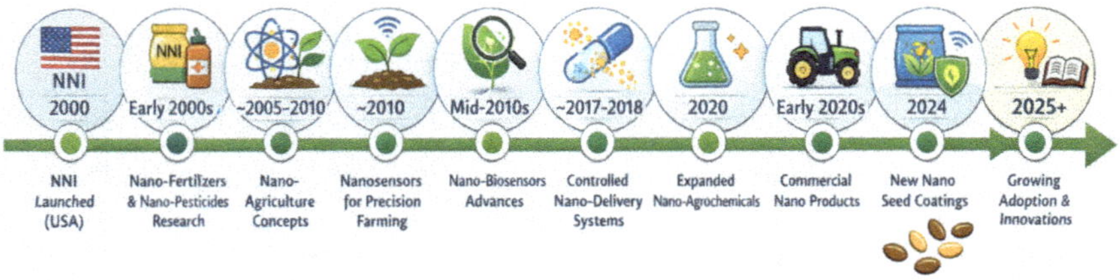


**Fig. 15.2** Timeline indicating the major developments in agri-nanotechnology over the last 25 years

**Fig. 15.3** Summary of studies on applications of Nanotechnology in agriculture

nanoparticle tracking analysis (NTA) for hydrodynamic size and aggregation state, X-ray diffraction (XRD) for crystal phase, X-ray photoelectron spectroscopy (XPS), and energy dispersive X-ray (EDX) for surface chemistry and elemental composition, and spectroscopic techniques (FTIR/UV–Vis) for surface functional groups and optical properties. These properties facilitate the prediction of nanoparticle behavior. Nanoparticle properties frequently change, such as through corona formation, dissolution, or agglomeration, upon contact with environmental matrices. Therefore, characterization should also include matrices such as soil leachate and plant surfaces. Nanoparticle delivery systems range from simple bulk suspensions to engineered carriers. Approaches such as seed nanopriming (coating seeds with nanoparticles or nanoparticle-loaded polymers), foliar nano-sprays (with optimized droplet size and adjuvants), soil amendments (nanoparticles embedded in biochar or nanocomposites), and encapsulated formulations (including polymeric nanoparticles, mesoporous silica, lipid-based carriers, and stimulus-responsive matrices) have been developed to enable controlled release, protect active ingredients, and improve target specificity (Ijaz et al. 2020; Altammar 2023; Sidhu et al. 2025).

Metal and metal-oxide nanoparticles (NPs) have been extensively investigated for their potential to enhance plant performance under abiotic and biotic stress due to their intrinsic reactivity and multifunctional properties. For instance, ZnO NPs serve as both a nano-source of zinc, an essential micronutrient, and modulators of reactive oxygen species (ROS), which can upregulate plant antioxidant enzymes such as SOD, CAT, and POD, stabilize cellular membranes, and improve photosynthetic pigment content under drought and salinity stress. $TiO_2$ NPs are linked to improved light harvesting, increased photochemical efficiency, and activation of antioxidant systems that mitigate photooxidative damage during heat or salinity stress. Silver nanoparticles (AgNPs) have been explored for antimicrobial and seed-treatment applications, where they can reduce pathogen load on seeds and seedlings and modulate host defence signaling; however, dose-dependent phytotoxicity is frequently reported. Magnetite ($Fe_3O_4$) NPs function as iron nano-fertilizers and, in some formulations, enhance redox buffering and ROS scavenging, as well as enable magnetically directed delivery in experimental settings.

Mechanistically, metal and metal-oxide NPs function by (i) serving as nanonutrient sources or altering nutrient bioavailability, (ii) directly interacting with ROS, either generating or scavenging ROS depending on their redox chemistry and surface coating, (iii) inducing transcriptional reprogramming of stress-responsive pathways such as hormone signalling, osmoprotectant biosynthesis, and aquaporin expression, and (iv) protecting photosynthetic machinery by stabilizing chlorophyll content and electron transport to maintain carbon assimilation under stress. Reviews summarizing these effects and their biochemical bases emphasize the strong dependence of beneficial versus adverse outcomes on both material type and dosage (Silva et al. 2022; Wang et al. 2023a, b, c).

A significant advancement in the past decade is the mechanistic understanding of how nanoparticles traverse plant barriers and are distributed within tissues. Uptake pathways depend on the route of exposure. Root uptake involves passage through the rhizosphere and root epidermis via apoplastic movement, entry at root tips and lateral roots, or endocytosis-mediated mechanisms. Foliar uptake relies on stomatal or cuticular penetration, surfactant-assisted wetting, and droplet microphysics. Once internalized, nanoparticles may translocate through the apoplast and symplast or be loaded into the xylem for root-to-shoot movement; phloem mobility is generally limited and typically associated with small ions or nanoparticle dissolution products. Within plant tissues, nanoparticles can transform into new species after interacting with biomolecules, potentially altering their mobility and bioactivity. For cellular localization, techniques such as transmission electron microscopy (TEM) with elemental mapping, synchrotron X-ray fluorescence, laser ablation inductively coupled plasma mass spectrometry (LA–ICP–MS), and isotopic labeling are employed. These methods provide valuable information for regulatory studies (Malejko et al. 2021; Wang et al. 2023b, c; Djanaguiraman et al. 2024).

Recent advancements in nano-enabled fertilizers, nano-pesticides, and stress protectants for targeted action have demonstrated increased fertilizer and nutrient use efficiency, controlled release of active ingredients, and matrices that reduce leaching and volatilization. These technologies also enhance adherence to leaf surfaces. While promising results have been observed in pot and expanded laboratory-scale experiments, further agronomic trials, residue and safety studies, and ecological toxicity assessments are necessary to address public concerns and meet regulatory requirements.

Nanobiotechnology is a multidisciplinary field that facilitates the translation of laboratory discoveries into scientific solutions for food safety and environmental sustainability (Tang et al. 2023). The nanotool box, which includes metal, metal oxide, polymeric, inorganic, and organic composites, can be utilized to probe plant physiology, sense environmental conditions, prime cellular cascades, and deliver various agrochemicals (Goyal et al. 2025). Omics approaches, such as transcriptomics, proteomics, and metabolomics, have been applied to study plant biology following nanomaterial treatment, revealing both conserved and material-specific signatures. These include altered expression of antioxidant genes, osmolyte biosynthesis, hormone signaling components, and photosynthesis-related transcripts, which support mechanistic hypotheses for nanoparticle-mediated tolerance (Malejko et al. 2021; Djanaguiraman et al. 2024). Future implementation will require long-term field validation across diverse soils, climates, and cropping systems, the establishment of standardized exposure and ecotoxicology protocols, comprehensive residue and food safety data, and life-cycle and cost–benefit analyses by interdisciplinary teams to address abiotic and biotic stress in wheat and other crops (El-Saadony et al. 2022; Wei et al. 2024).

## 15.4 Recent Developments in Nanotechnology Applications for Enhancing Stress Tolerance in Wheat

Researchers have investigated the effects of foliar, soil, and seed-applied nanomaterials to mitigate abiotic stress (Fig. 15.4). Among these approaches, seed nanopriming demonstrates the greatest potential; however, additional field-scale studies

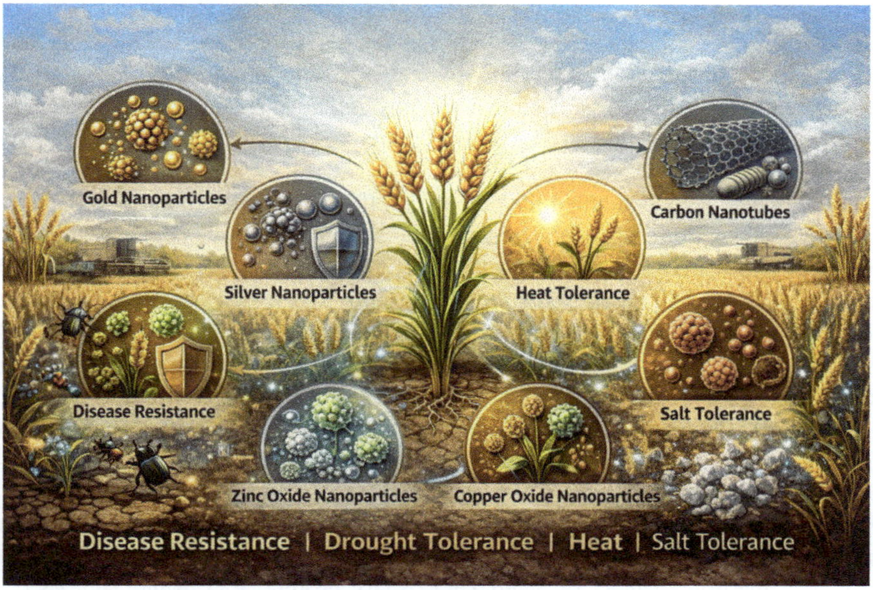


**Fig. 15.4** Applications of nanomaterials in ameliorating stress in wheat

across diverse cultivars and agroclimatic conditions are required. Shelar et al. (2024) reviewed the use of nanomaterials as seed priming agents that confer resistance to drought, salinity, pests, and diseases during seed germination. Their assessment of plant physiological and biochemical responses indicated that seed priming with nanomaterials enhances seedling vigor and stress resilience, ultimately improving crop yields.

Wheat seeds treated with mesoporous silica nanoparticles (MSiNPs) and zinc-loaded mesoporous silica nanoparticles (Zn-MSiNPs) at 1 g/L and 5 g/L, respectively, showed improved germination and seedling growth under salinity stress (200 mM NaCl). Increases in chlorophyll, total protein, and sugar content were accompanied by significant reductions in malondialdehyde (MDA), electrolyte leakage, and ROS accumulation. Nanoparticle treatment also elevated the $K^+/Na^+$ ratio in shoots, thereby enhancing stress tolerance (Ghosh et al. 2024). In a pot experiment, the wheat variety "Gemmieza" exhibited significant increases in chlorophyll A and B contents, plant height at both vegetative and maturity stages, shoot and spike lengths, root fresh and dry weights, and grain yield following soil application of ZnO nanoparticles (0.12 g/pot) under salt stress. These approaches may contribute to improved crop yields under abiotic stress conditions (Adil et al. 2022).

Noble metal nanoparticles have also been evaluated for their effects on wheat. Venzhik et al. (2022) investigated seed priming of winter wheat (*Triticum aestivum* L., var. Moskovskaya 39) with a gold nanoparticle solution. Enhanced freezing tolerance in 7-day-old seedlings was attributed to modifications in the photosynthetic apparatus, oxidative processes, and increased accumulation of soluble sugars. Biogenic silver nanoparticles (B-AgNPs) demonstrated efficacy against ozone phytotoxicity in two wheat cultivars (HD-2967 and DBW-17). Foliar application of B-AgNPs significantly improved grain weight per plant, thousand-grain weight, and harvest index in DBW-17. The enhancement of ozone tolerance by B-AgNPs is associated with increased biochemical and physiological responses and appears to be cultivar-specific, as indicated by the differential response of HD-2967 (Kannaujia et al. 2021).

Researchers have examined the combined effects of nanoparticles and bacteria on stress alleviation in wheat. Treatment of wheat with both copper (II) oxide nanoparticles (CuO-Nps > 100 nm) and plant growth-promoting bacteria (*Bacillus subtilis*, *Lactobacillus casei*, and *Bacillus pumilis*) mitigated genetic impairment under NaCl stress. The study utilized coupled restriction enzyme digestion-iPBS (CRED-iPBS) to assess changes in cytosine methylation and inter-primer binding site (iPBS) primers to evaluate genotoxic effects and genomic stability (Hosseinpour et al. 2022). Co-application of rhizobacteria (*Azospirillum brasilense*) and ZnO nanoparticles may be effective in mitigating drought stress, particularly at the grain-filling stage. This synergistic physiological enhancement from nanoparticle and rhizobacteria interactions could support sustainable wheat crop management in semi-arid regions (Muhammad et al. 2022). In a pot-scale study, Ayman et al. (2024) evaluated the effectiveness of Nano-ZnO (NZn), proline (PA), Nano-$TiO_2$ (NTi), Nano-$SiO_2$ (NSi), and biochar inoculated with *Bacillus subtilis* (OSBS) in enhancing wheat tolerance to salinity stress. The combination of 2% OSBS with NZn and NSi significantly improved wheat growth, leaf area, and nutrient levels, thereby reducing the adverse effects of salinity. Further data from multiple locations are anticipated to validate these promising results.

Environmental pollution, soil and water degradation, and public health concerns are increasingly intensified by globalization, necessitating the adoption of more sustainable and efficient agricultural practices. Nanotechnology represents an innovative strategy to enhance nutrient delivery and minimize environmental impact. Studies indicate that nano fertilizers can increase wheat yields by up to 55%. These nano fertilizers provide an environmentally friendly alternative to conventional methods, improve nutrient delivery, and promote seed growth and overall plant health (Goyal et al. 2025).

Omics studies, including transcriptomics (RNA-seq), proteomics, and metabolomics, on plants treated with nanoparticles have demonstrated modulation of pathways related to antioxidant defence, osmolyte biosynthesis, hormone signaling (abscisic acid, auxin, and ethylene), aquaporin expression, and maintenance of the photosynthetic apparatus. Nanoparticles have been shown to prime stress-responsive genes and directly alter the basal expression of detoxification enzymes and transporters. Additionally, miRNA-mediated regulatory changes following nano-treatments have been reported. These molecular findings provide mechanistic support for observed physiological effects, such as osmotic adjustment, ROS scavenging, and improved photosynthesis, and inform the rational design of safe-by-design nanomaterials with targeted modes of action (García-Sánchez et al. 2021; Akhter Bhat et al. 2025).

The positive effects of nanomaterials in mitigating plant stress have been demonstrated in small-scale experiments, with outcomes influenced by factors such as wheat genotype, nanomaterial type, soil characteristics, and agroclimatic conditions. Therefore, standardized and agronomically relevant dose–response studies, residue analyses, multi-season field trials across diverse soil and climate gradients, comprehensive life-cycle and risk–benefit assessments, and robust regulatory frameworks for nano-agroinputs are essential (Maaz et al. 2025). Integrating molecular phenotyping with field agronomy will facilitate the rational design of nanoparticles and the identification of biomarkers for efficacy and safety, thereby advancing nano-enabled seed technologies and precision delivery systems that can sustainably enhance wheat resilience to climate change (Goyal et al. 2025; Zhang et al. 2025).

## 15.5 Evolution in the Contemporary Hi-Tech Era

In the current era of advanced technology, agricultural nanotechnology is anticipated to address both abiotic and biotic stresses in wheat through multiple converging innovations. These innovations are designed to provide precision delivery platforms for agrochemicals and are expected to support environmental sustainability. Nanoparticles can be administered as (a) foliar sprays, enabling direct uptake through leaves and rapid physiological responses; (b) root-dipping treatments, facilitating seedling establishment and targeted delivery to the root zone; (c) seed coatings or nanopriming, which enhance early vigor and stress resilience; and (d) hydroponic or fertigation systems, allowing nutrient delivery in soil-less cultivation. Consequently, nanoparticles can be engineered for site-specific applications. Ullah et al. (2023) reviewed the penetration and transport of nanoparticles in plants to assess the potential of nanotechnology in agriculture, with particular attention to heavy metal contamination in soil and the role of nanomaterials in mitigating this abiotic stress. Khan et al. (2023) suggest that seed priming with nanomaterials is a rapidly developing, cost-effective, and environmentally friendly research area, as the unique anatomy of the seed coat may reduce biosafety concerns. This emerging technology is projected to enhance crop production under climate change-related stress factors (Khan et al. 2023). Seed treatments with carbon nanotubes (CNTs) and graphene have been shown to promote early seed germination and increase root and shoot growth in vinca (*Catharanthus roseus*), as well as improve overall germination rates in cotton (*Gossypium hirsutum*). In soil, nanomaterial treatments alleviated salt stress, increased the number of flowers and leaves in *Catharanthus, and enhanc*ed fibre biomass in cotton (Pandey et al. 2019). Three types of nanomaterials—Nanoplant Ultra (Co, Mn, Cu, Fe, Zn, Mo, and Se), Nanoplant Ca–Si (Ca, Si, B, and Fe), and Nanoplant Sulfur (S)—used as seed priming agents in peas, were found to improve seed germination, initial plant growth, and crop productivity. Notably, Nanoplant Ca–Si and Nanoplant Sulfur provided protection against drought stress in pea (Tamindžić et al. 2024). Graphene oxide (GO) increased germination percentage and radicle length in melon (*Cucumis melo* L. cvs. "Hasanbey", "Kirkagac-589", and "Kirkagac-637") seeds under salinity stress (Kaymak et al.

2022). Ding et al. (2025) demonstrated that manganese ferrite nanomaterials ($MnFe_2O_4$ NMs) mitigated drought stress in soybean (*Glycine max* L.) seedlings, both through foliar and soil applications, by enhancing photosynthetic efficiency and chlorophyll content, reducing oxidative stress, and optimizing nutrient uptake and translocation.

In maize, silver nanoparticles (AgNPs) used as a seed priming agent enhanced resistance to multiple simultaneous stresses, including drought and cold, drought and salt, and salt and cold. Transcriptomic analyses indicated activation of plant hormone signal transduction and mitogen-activated protein kinase (MAPK) signalling pathways. Comparable outcomes were observed with ROS-generating iron-based nanomaterials ($Fe_2O_3$ and $Fe_3O_4$ nanoparticles). The authors suggest that seed priming with nanomaterials may serve as a straightforward strategy to improve crop climate resilience (Chen et al. 2023a).

Green-synthesized nanoparticles represent eco-friendly and biocompatible alternatives. For example, green-synthesized zinc oxide nanoparticles (ZnO-NPs) promoted seedling growth in quinoa (*Chenopodium quinoa* L. cv Atlas) by providing a bioavailable zinc source. These nanoparticles reduced oxidative damage and enhanced enzyme activities under salinity stress (Türkoğlu et al. 2024). Similarly, seeds of *Zea mays* cv. Balu 787 treated with green-synthesized copper-based nanoparticles (ascorbic acid/starch, ACu) exhibited increased germination rates and improved root-to-shoot ratios, contributing to enhanced tolerance to water deficit (Gomes et al. 2024).

The application of selenium nanoparticles reduced the impact of salinity on tomato growth by altering growth and antioxidant responses (Morales-Espinoza et al. 2019). Graphene oxide (GO) has been shown to be a promising and safe application material for reducing salt stress during germination of Persian clover seeds (Stefanello et al. 2024). Poly(acrylic) acid-coated $Mn_3O_4$ was found to enhance soybean germination by strengthening the antioxidant defense system and reducing malondialdehyde content. The nanomaterial regulated ROS homeostasis under salt stress and reduced ATP consumption (Mao et al. 2024).

The findings of Fan et al. (2023) showed that soil-applied Si QDs regulated radish taproot resistance to the root-chewing insect white grub (*Holotrichia oblita*) and increased radish taproot biomass. The nanoparticles treatment resulted in (i) increased physical resistance and production of compounds such as JA, SA, aucubin, gluconasturtiin, and pipecolic acid; (ii) increased taproot antioxidant enzymes (SOD and POD) and metabolite (flavonoids and glutathione) contents; (iii) no disturbance in rhizosphere bacterial community diversity, and enrichment of several rhizosphere beneficial bacteria (including *Acidobacteria*, *Nitrospirae*, *Verrucomicrobia*, *Hydrogenophaga*, *Sphingomonas*, *Ramlibacter*, and *Nitrospira*).

Several examples of nanomaterial applications for stress mitigation have been discussed. However, research on wheat remains limited. The use of artificial intelligence (AI) and machine learning (ML) for abiotic and biotic stress management is still in its early stages, but these technologies are advancing rapidly and may be integrated with nanosensors to detect crop health and facilitate mitigation. Although specific stimuli-responsive materials have not yet been tested with wheat, they have the potential to release nutrients precisely, thereby minimizing waste (Fig. 15.5).

**Fig. 15.5** AI and nanomaterials for mitigating stress and increasing productivity

The integration of cutting-edge materials science with digital agriculture tools ensures that interventions become smarter, safer, and more context-responsive—paving the way for resilient, high-tech wheat cultivation amid evolving environmental challenges.

## 15.6 Statistical Insight into Nanotechnology Applications in Wheat Stress Management

Recent studies have investigated the application of nanomaterials to mitigate abiotic and biotic stress in wheat, with most experiments conducted at the pot scale. For example, soil application of ZnO nanoparticles (0.12 g per pot) in the wheat variety "Gemmieza" significantly increased chlorophyll A and B contents, plant height at both vegetative and maturity stages, shoot and spike lengths, root fresh and dry weights, and grain yield under salt stress. These approaches may contribute to improved crop yields under abiotic stress conditions (Adil et al. 2022). Venzhik et al. (2022) examined seed priming of winter wheat (*Triticum aestivum* L., var. Moskovskaya 39, Poaceae) with a gold nanoparticle solution, resulting in enhanced freezing tolerance in 7-day-old seedlings. This improvement was attributed to changes in the photosynthetic apparatus, oxidative processes, and increased accumulation of soluble sugars. Additionally, biogenic silver nanoparticles (B-AgNPs) demonstrated efficacy against ozone phytotoxicity in two wheat cultivars (HD-2967 and DBW-17). Foliar application of B-AgNPs significantly increased grain weight per plant, thousand-grain weight, and harvest index in DBW-17, suggesting that B-AgNPs enhance ozone tolerance through biochemical and physiological mechanisms. This effect appears to be cultivar-specific, as indicated by the response of HD-2967 (Kannaujia et al. 2021). Furthermore, treating wheat seeds with mesoporous silica nanoparticles (MSiNPs) and zinc-loaded mesoporous silica nanoparticles (Zn-MSiNPs) at 1 g/L and 5 g/L, respectively, improved seedling germination and growth under salinity stress (200 mM NaCl). Increases in chlorophyll, total protein, and sugar content were accompanied by reductions in malondialdehyde (MDA), electrolyte leakage, and ROS accumulation. Nanoparticle treatments also increased the $K^+/Na^+$ ratio in shoots, thereby enhancing stress tolerance (Ghosh et al. 2024).

Two studies investigated the combined effects of nanoparticles and bacteria on wheat subjected to abiotic stress. Application of copper (II) oxide nanoparticles (CuO-Nps >100 nm) together with plant growth-promoting bacteria (*Bacillus subtilis*, *Lactobacillus casei*, and *Bacillus pumilis*) alleviated genetic impairment in wheat under NaCl-induced stress. The researchers utilized coupled restriction enzyme digestion-iPBS (CRED-iPBS) and inter-primer binding site (iPBS) primers to assess changes in cytosine methylation and genomic stability (Hosseinpour et al. 2022). Additionally, the co-application of rhizobacteria (*Azospirillum brasilense*) and ZnO nanoparticles (NPs) showed potential to mitigate drought stress, particularly by enhancing physiological parameters during the grain-filling stage. This synergistic interaction between NPs and rhizobacteria may support sustainable wheat crop management in semi-arid regions (Muhammad et al. 2022).

Huang et al. (2024) conducted a meta-analysis of interactions between nanomaterials (NMs) and wheat health, synthesizing data from 144 studies and approximately 200 paired observations. The analysis revealed that Fe-, Ti-, and Zn-based nanomaterials significantly increased wheat biomass and photosynthetic activity, whereas Ag NMs had significant negative effects on wheat growth and physiology. A dose-dependent relationship was observed between wheat shoot biomass and nanomaterial concentration, and nanomaterials smaller than 25 nm were found to be toxic to wheat growth.

Studies on seed nano-priming and nanoparticle application in cereals consistently report positive effects on germination, biomass accumulation, and physiological indices under abiotic stress. Nano-priming enhances germination speed, seedling vigor, antioxidant enzyme activities (SOD, CAT, and POD), and chlorophyll content under salinity and drought conditions. While data from pot trials are available, multi-season field experiments across diverse locations remain necessary. Treatments with nanoscale oxides ($TiO_2$, ZnO, and CuO) have generally improved photosynthetic parameters and biomass under controlled stress conditions. However, further standardization and expanded field data are required (Donia and Carbone 2023; Faizan et al. 2024; Kumari et al. 2024). Most studies compare bulk salts with nanomaterial applications, attributing positive effects to nanoparticle size, increased bioavailability, and controlled release. Nevertheless, considerable heterogeneity exists regarding material type, dosage, exposure route, growth medium, and wheat genotype.

Phytotoxic effects of metal-oxide nanoparticles (NPs), such as ZnO and CuO, on cereals typically manifest at concentrations ranging from tens to hundreds of $mg \cdot L^{-1}$ in hydroponic systems. In contrast, effective concentrations for seed priming are often in the low $mg \cdot L^{-1}$ range or lower. When applied at levels above the threshold, nanomaterials lose their beneficial effects and induce phytotoxicity. Antimicrobial nanoparticles, particularly AgNPs, demonstrate lower toxicity thresholds for soil microorganisms, raising concerns regarding potential impacts on the soil microbiome. Thresholds established in simple media cannot be directly extrapolated to soils, as adsorption and transformation processes significantly influence exposure (Das et al. 2025; Zhang et al. 2025). Nanoparticle size, surface coating, plant developmental stage, and partial dissolution (for example, ZnO to $Zn^{2+}$) all contribute to tissue accumulation. High-resolution mapping techniques, such as synchrotron XRF and LA–ICP–MS, reveal that intact particles often accumulate in roots with limited phloem mobility; translocation to grains is generally low but has been documented for certain materials depending on dose and chemical properties (Afzal et al. 2024; Das et al. 2025). The persistence and bioavailability of nanomaterials are influenced by aggregation, dissolution, complexation with organic matter, and redox chemistry. Some metal-oxide NPs may persist as reactive phases for several months, while high dissolution rates decrease particulate persistence but increase ionic exposure. There is a scarcity of long-term fate studies, highlighting the need for systematic experiments and modeling to quantify environmental half-lives and bioaccumulation potential of nanomaterials (Afzal et al. 2024; Das et al. 2025). Studies that correlate nanomaterial effects on plant physiology are essential for informing the future design and safe use of nanomaterials in agriculture, as well as for enhancing public acceptance.

Interest in using nanoparticles to mitigate abiotic stressors in agriculture is increasing, as evidenced by a rapid rise in relevant publications and citations. Researchers from Asian countries are making substantial contributions to this field. Bibliometric studies are valuable for tracking the evolution of research, identifying knowledge gaps, and supporting ongoing scientific activity (Liu et al. 2023). Over the past decade, the number of publications on nano-agriculture has increased, reflecting heightened research interest. A recent review by Kaur et al. (2025) highlights the potential of nanotechnology to enhance wheat resilience. Application-driven research, including nano-priming, nano-fertilizers, nano-pesticides, and nanosensors, is gaining momentum, as evidenced by increased intellectual property generation (Cano et al. 2025; Luo et al. 2025). However, commercialization remains limited and is subject to regulatory oversight.

## 15.7 Future Projections and Global Challenges

Conventional breeding strategies aimed at enhancing agronomic traits and phenotype expression under stress in wheat can be augmented by nano-formulated agro-inputs, as well as by modern breeding techniques utilizing molecular tools such as CRISPR–Cas. This integrated approach, often referred to as "nano-breeding," offers several advantages. Nano-enabled methods in wheat breeding and seed technology can improve seed performance, including faster germination and early vigor under salinity or drought conditions. They also facilitate the controlled, bioavailable delivery of micronutrients and regulators that support stress physiology, such as ZnO and

Si nanoparticles, which enhance osmotic adjustment and antioxidant capacity. Additionally, these nanomaterials can serve as carriers for the transient delivery of nucleic acids, proteins, or elicitors to modulate stress-responsive gene expression without stable genetic modification (Yang et al. 2025). Recent reviews have highlighted increased nutrient-use efficiency and improved physiological resilience under abiotic stress in climate-resilient wheat production through nano-enabled approaches (Demeke et al. 2025). Nano-fertilizers, nano-encapsulated plant growth regulators, and nano-pesticides are engineered to release their cargo in a controlled manner, enhance foliar or root uptake, and reduce off-target losses. Advances in polymeric carriers, mesoporous silica, lipid nanocarriers, and stimulus-responsive matrices enable tailored release profiles, such as pH- or moisture-triggered delivery, thereby improving efficacy during stress and potentially reducing input quantities and greenhouse gas emissions from fertilizer loss. Although the initial formulation costs of these nano-formulations may be higher, they can be offset by enhanced nutrient uptake and reduced need for repeat applications at the farm scale. However, outcomes may vary depending on region, scale, and crop type (Mim et al. 2025).

A significant translational challenge in nanotechnology arises from the potential for nanomaterials to exhibit dose-dependent phytotoxicity, induce oxidative stress, genotoxicity, or cause membrane damage in plants and nontarget organisms (Rajpal et al. 2025). Additionally, nanomaterials may adversely impact soil microbial diversity, biomass, and enzyme activities, thereby impairing soil health and long-term productivity (Chen et al. 2023b). Nanoparticles have also demonstrated toxicity in mammalian systems, raising concerns regarding human exposure through food chains and occupational pathways (Kumah et al. 2023). Consequently, it is essential to establish material-specific toxicity thresholds and realistic exposure scenarios, including soil, foliar, and seed applications, rather than relying solely on laboratory-scale or short-term greenhouse trial data. Comprehensive multi-season, multi-site field trials and long-term fate studies are necessary prior to widescale deployment.

Nanoparticle persistence, transformation, and trophic transfer present significant concerns for ecosystems. Metal and metal-oxide nanoparticles may remain as reactive particulate phases or dissolve into bioavailable ions, both of which increase the risk of accumulation in soils and subsequent entry into the food chain. Key issues, including unequal access to nanotechnologies, disposal of nano-waste, and the need for consent and labelling of nano-treated seeds or food, require policy decisions informed by government action, public engagement, and transparent risk communication (Singh et al. 2025). A comprehensive approach that balances the benefits of food security with potential long-term ecosystem impacts is necessary.

The European food safety authority (EFSA) has issued technical guidance for assessing small particles and nanomaterials in food and feed applications, while the OECD and national agencies are developing standardized testing and reporting protocols. However, global regulatory heterogeneity persists due to variations in definitions of nanomaterials and differing requirements for characterization and safety testing, such as toxicity, residue, and environmental fate. This regulatory inconsistency complicates product development and international trade. Researchers recommend prerequisites for robust risk assessment and responsible commercialization of nanotechnology, including standardized characterization methods, comprehensive reporting of particle number, surface area, and mass, implementation of real-world ecotoxicology protocols, and harmonized data-sharing (EFSA Scientific Committee 2021; Schoonjans et al. 2023). There is also a recognized need for nanosensors and molecular biomarkers to detect early ecosystem perturbations, such as microbiome shifts and bioaccumulation, and to monitor the efficacy of nanomaterials *in situ*. These data will support the establishment of evidence-based thresholds for safe use (Bouhadi et al. 2025; Demeke et al. 2025).

To achieve meaningful impact, cost-effective, scalable, and locally adaptable nano-agricultural solutions are necessary across diverse wheat-growing regions. Green synthesis methods for nanoparticles can reduce production costs and enhance biocompatibility, while formulation strategies that minimize active ingredient load, such as seed nanopriming, can decrease per hectare input and environmental burden. Successful implementation of nanotechnology will require business models that mitigate risks for smallholder farmers, robust training, and extension infrastructure to ensure proper handling and application, and comprehensive lifecycle cost–benefit analyses, including environmental externalities, to inform policy incentives and adoption. Regional validation of all nanotechnologies is essential (Mim et al. 2025; Demeke et al. 2025).

The implementation of responsible nanotechnology for climate-resilient wheat requires standardized testing and reporting, interdisciplinary research involving scientists from materials science, plant physiology, soil ecology, and toxicology, as well as economists and social scientists. Open data platforms that include both negative and positive results, along with phased field demonstrations supported by independent environmental monitoring, are essential. International regulatory guidelines are necessary to safeguard human and ecological health. Furthermore, all nanomaterial design strategies should emphasize safe-by-design materials, stimulus-responsive formulations that reduce the quantity of active ingredients, and seed-based deployment methods such as nanopriming, which provide low environmental loading and maximize benefit per unit material (EFSA Scientific Committee 2021; Rajpal et al. 2025; Singh et al. 2025).

## 15.8 Conclusion

Nanotechnology presents a transformative opportunity to enhance wheat resilience to both abiotic and biotic stresses (Fig. 15.6). Nanomaterials have been shown to modulate physiological, biochemical, and molecular pathways, thereby improving plant growth and productivity under adverse conditions. Laboratory and greenhouse studies provide evidence of increased stress tolerance following nano-enabled interventions; however, large-scale field trials and comprehensive biosafety assessments remain essential before these technologies can be adopted in mainstream wheat cultivation. Realizing the full potential of nanotechnology for sustainable wheat improvement will require a multidisciplinary approach that integrates materials science, plant biology, environmental safety, policy, and farmer engagement. This

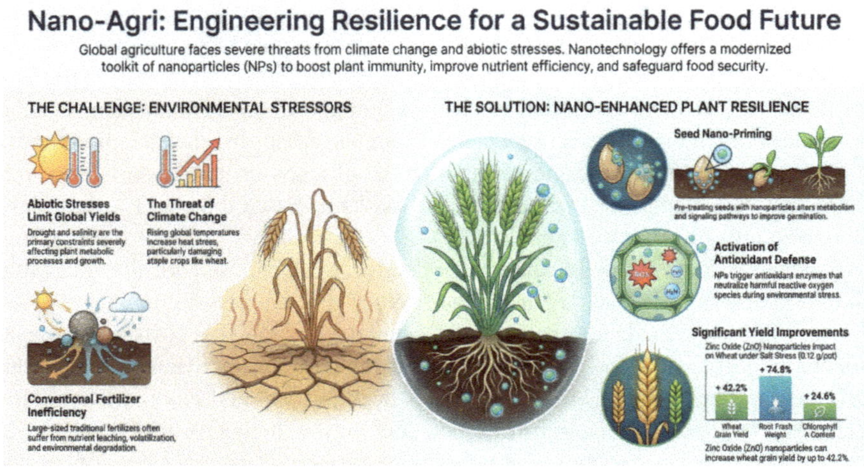


**Fig. 15.6** The transformative potential of agri-nanotechnology

convergence will facilitate the responsible deployment of nanotechnologies and ensure that their benefits are delivered to diverse wheat-growing regions in a safe, scalable, and equitable manner.

## References


Adil M, Bashir S, Bashir S et al (2022) Zinc oxide nanoparticles improved chlorophyll contents, physical parameters, and wheat yield under salt stress. Front Plant Sci 13:932861. https://doi.org/10.3389/fpls.2022.932861

Afzal S, Singh NK, Lal AF et al (2024) Nanostructure and plant uptake: assessing the ecological footprint and root-to-leaf dynamics. Plant Nano Biol 10:100122. https://doi.org/10.1016/j.plana.2024.100122

Akhter Bhat T, Rather RA, Bashir S et al (2025) Nanoparticles in plant systems: omics-based perspectives on stress adaptation and toxicological implications. Plant Nano Biol 13:100181. https://doi.org/10.1016/j.plana.2025.100181

Alam MN, Islam MZ, Farukh MA et al (2025) Detrimental effects of abiotic stress on wheat and its management techniques. Cereal Res Commun 53:873–886. https://doi.org/10.1007/s42976-024-00588-2

Altammar KA (2023) A review on nanoparticles: characteristics, synthesis, applications, and challenges. Front Microbiol 14:1155622. https://doi.org/10.3389/fmicb.2023.1155622

Aryal S, Shiksha KC, Dahal M et al (2021) Introduction to abiotic stress and its effect in wheat: a review. Plant Physiol Soil Chem 1:04–07. https://doi.org/10.26480/ppsc.01.2021.04.07

Asseng S, Ewert F, Martre P et al (2015) Rising temperatures reduce global wheat production. Nat Clim Chang 5:143–147. https://doi.org/10.1038/nclimate2470

Ateş Sönmezoğlu Ö, Çevik E, Terzi A et al (2022) Assessment of some bread wheat (*Triticum aestivum* L.) genotypes for drought tolerance using SSR and ISSR markers. Biotechnol Stud 31:45–52. https://doi.org/10.38042/biotechstudies.1159128

Ayman M, Fahmy MA, Elnahal ASM et al (2024) Enhancing wheat tolerance to salinity using nanomaterials, proline, and biochar-inoculated with *Bacillus subtilis*. Heliyon 10(17):e37160. https://doi.org/10.1016/j.heliyon.2024.e37160

Balali-Mood M, Naseri K, Tahergorabi Z et al (2021) Toxic mechanisms of five heavy metals: mercury, lead, chromium, cadmium, and arsenic. Front Pharmacol 12:643972. https://doi.org/10.3389/fphar.2021.643972

Bayda S, Adeel M, Tuccinardi T et al (2019) The history of nanoscience and nanotechnology: from chemical-physical applications to nanomedicine. Molecules 25(1):112. https://doi.org/10.3390/molecules25010112

Bouhadi M, Javed Q, Jakubus M et al (2025) Nanoparticles for sustainable agriculture: assessment of benefits and risks. Agronomy 15(5):1131. https://doi.org/10.3390/agronomy15051131

Budak H, Hussain B, Khan Z et al (2015) From genetics to functional genomics: improvement in drought signaling and tolerance in wheat. Front Plant Sci 6:1012. https://doi.org/10.3389/fpls.2015.01012

Cano PB, Carcedo AJP, Hernandez CM et al (2025) Trends in agricultural technology: a review of US patents. Precis Agric 26:59. https://doi.org/10.1007/s11119-025-10257-x

Chen S, Liu H, Yangzong Z et al (2023a) Seed priming with reactive oxygen species-generating nanoparticles enhanced maize tolerance to multiple abiotic stresses. Environ Sci Technol 57(48):19932–19941. https://doi.org/10.1021/acs.est.3c07339

Chen S, Teng Y, Luo Y et al (2023b) Threats to the soil microbiome from nanomaterials: a global meta and machine-learning analysis. Soil Biol Biochem 188:109248. https://doi.org/10.1016/j.soilbio.2023.109248

Das D, Ingti B, Paul P et al (2025) Soil microbial dynamics in response to the impact of nanoparticles on agricultural implications. Discov Soil 2:58. https://doi.org/10.1007/s44378-025-00087-8

Demeke ED, Benti NE, Terefe MG et al (2025) A comprehensive review on nano-fertilizers: preparation, development, utilization, and prospects for sustainable agriculture in Ethiopia. Nanoscale Adv 7(8):2131–2144. https://doi.org/10.1039/d4na01068j

Dhakal A, Adhikari C, Manandhar D et al (2021) Effect of abiotic stress in wheat: a review. Rev Food Agric 2(2):69–72. https://doi.org/10.26480/rfna.02.2021.69.72

Ding Y, Nadeem M, Zhang T et al (2025) Alleviation role of $MnFe_2O_4$ nanomaterials for soybean growth under drought stress. J Plant Growth Regul 44:6098–6111. https://doi.org/10.1007/s00344-025-11815-3

Djanaguiraman M, Anbazhagan V, Dhankher OP et al (2024) Uptake, translocation, toxicity, and impact of nanoparticles on plant physiological processes. Plants (Basel) 13(22):3137. https://doi.org/10.3390/plants13223137

Dong B, Zheng X, Liu H et al (2017) Effects of drought stress on pollen sterility, grain yield, abscisic acid and protective enzymes in two winter wheat cultivars. Front Plant Sci 8:1008. https://doi.org/10.3389/fpls.2017.01008

Donia DT, Carbone M (2023) Seed priming with zinc oxide nanoparticles to enhance crop tolerance to environmental stresses. Int J Mol Sci 24(24):17612. https://doi.org/10.3390/ijms242417612

EFSA Scientific Committee et al (2021) Guidance on technical requirements for regulated food and feed product applications to establish the presence of small particles including nanoparticles. EFSA J 19(8):e06769. https://doi.org/10.2903/j.efsa.2021.6769

El Sabagh A, Islam MS, Skalicky M et al (2021) Salinity stress in wheat (*Triticum aestivum* L.) in the changing climate: adaptation and management strategies. Front Agron 3:661932. https://doi.org/10.3389/fagro.2021.661932

El-Saadony MT, Saad AM, Soliman SM et al (2022) Role of nanoparticles in enhancing crop tolerance to abiotic stress: a comprehensive review. Front Plant Sci 13:946717. https://doi.org/10.3389/fpls.2022.946717

Faizan M, Sharma P, Sultan H et al (2024) Nano-priming: improving plant nutrition to support the establishment of sustainable agriculture under heavy metal stress. Plant Nano Biol 10:100096. https://doi.org/10.1016/j.plana.2024.100096

Fan N, Zhao C, Chang Z et al (2023) Silicon quantum dots promote radish resistance to root herbivores without impairing rhizosphere microenvironment health. Environ Sci Nano 10(9):2232–2244. https://doi.org/10.1039/D3EN00368J

García-Sánchez S, Gala M, Žoldák G (2021) Nanoimpact in plants: lessons from the transcriptome. Plants (Basel) 10(4):751. https://doi.org/10.3390/plants10040751

Ghosh D, Das T, Paul P et al (2024) Zinc-loaded mesoporous silica nanoparticles mitigate salinity stress in wheat seedlings through silica–zinc uptake, osmotic balance, and ROS detoxification. Plant Physiol Biochem 211:108693. https://doi.org/10.1016/j.plaphy.2024.108693

Gomes D, Pieretti J, Lange C et al (2024) Seed nanopriming with green-synthesized Cu-based nanoparticles for maize plants with improved tolerance to hydric stress. ACS Appl Nano Mater 7. https://doi.org/10.1021/acsanm.4c02607

Goyal A, Chavan SS, Mohite RA et al (2025) Emerging trends and perspectives on nano-fertilizers for sustainable agriculture. Discov Nano 20(1):97. https://doi.org/10.1186/s11671-025-04286-8

Hosseinpour A, Ilhan E, Özkan G et al (2022) Plant growth-promoting bacteria (PGPBs) and copper (II) oxide (CuO) nanoparticle ameliorate DNA damage and DNA methylation in wheat (*Triticum aestivum* L.) exposed to NaCl stress. J Plant Biochem Biotechnol 31:751–764. https://doi.org/10.1007/s13562-021-00713-w

Huang F, Chen L, Zeng Y et al (2024) Unveiling influences of metal-based nanomaterials on wheat growth and physiology: from benefits to detriments. Chemosphere 364:143212. https://doi.org/10.1016/j.chemosphere.2024.143212

Hussain B, Akpınar BA, Alaux M et al (2021) Wheat genomics and breeding: bridging the gap. AgriRxiv. https://doi.org/10.31220/agriRxiv.2021.00039

Ijaz I, Gilani E, Nazir A et al (2020) Detail review on chemical, physical and green synthesis, classification, characterizations and applications of nanoparticles. Green Chem Lett Rev 13(3):223–245. https://doi.org/10.1080/17518253.2020.1802517

Ilya A, Ahmad I, Romanovna KM et al (2025) Assessing the exposure of lead, cadmium, and arsenic on growth parameters and antioxidant defense system in wheat. Biodegradation 36:35. https://doi.org/10.1007/s10532-025-10138-0

Kajla M, Yadav V, Khokhar J et al (2015) Increase in wheat production through management of abiotic stresses: a review. J Appl Nat Sci 7:1070–1080. https://doi.org/10.31018/jans.v7i2.733

Kannaujia R, Singh P, Prasad V et al (2021) Evaluating impacts of biogenic silver nanoparticles and ethylenediurea on wheat (*Triticum aestivum* L.) against ozone-induced damages. Environ Res 203:111857. https://doi.org/10.1016/j.envres.2021.111857

Kaur T, Tiwari A, Mathur M et al (2025) Nanomaterials for wheat stress tolerance: trade-off on underlying mechanisms and latest breakthroughs under changing climate. J Agron Crop Sci 211(4):e70096. https://doi.org/10.1111/jac.70096

Kaymak HÇ, Sevim M, Metin Ö (2022) Graphene oxide: a promising material for the germination of melon seeds under salinity stress. Turk J Agric For 46:863–874. https://doi.org/10.55730/1300-011X.3048

Khan MN, Fu C, Li J et al (2023) Seed nanopriming: how do nanomaterials improve seed tolerance to salinity and drought? Chemosphere 310:136911. https://doi.org/10.1016/j.chemosphere.2022.136911

Kulkarni M, Soolanayakanahally R, Ogawa S et al (2017) Drought response in wheat: key genes and regulatory mechanisms controlling root system architecture and transpiration efficiency. Front Chem 5:106. https://doi.org/10.3389/fchem.2017.00106

Kumah EA, Fopa RD, Harati S et al (2023) Human and environmental impacts of nanoparticles: a scoping review of the current literature. BMC Public Health 23:1059. https://doi.org/10.1186/s12889-023-15958-4

Kumari A, Gupta AK, Sharma S et al (2024) Nanoparticles as a tool for alleviating plant stress: mechanisms, implications, and challenges. Plants (Basel) 13(11):1528. https://doi.org/10.3390/plants13111528

Li L, Mao X, Wang J et al (2019) Genetic dissection of drought and heat-responsive agronomic traits in wheat. Plant Cell Environ 42(9):2540–2553. https://doi.org/10.1111/pce.13577

Liu Z, Faizan M, Zheng L et al (2023) Nanoparticles enhance plant resistance to abiotic stresses: a bibliometric statistic. Agronomy 13(3):729. https://doi.org/10.3390/agronomy13030729

Luo X, Li J, Guo S et al (2025) Analysis of research status and trends on nano-agricultural application: a bibliometric study. Front Plant Sci 16:1530629. https://doi.org/10.3389/fpls.2025.1530629

Maaz TM, Dobermann A, Lyons SE et al (2025) Review of research and innovation on novel fertilizers for crop nutrition. npj Sustain Agric 3:25. https://doi.org/10.1038/s44264-025-00066-0

Malejko J, Godlewska-Żyłkiewicz B, Vanek T et al (2021) Uptake, translocation, weathering and speciation of gold nanoparticles in potato, radish, carrot and lettuce crops. J Hazard Mater 418:126219. https://doi.org/10.1016/j.jhazmat.2021.126219

Malipatil SS, Biradar SS, Desai SA et al (2023) Studies on genetic diversity and stress indices for salinity tolerance in *Triticum dicoccum*. Int J Environ Clim Change 13(11):2609–2621. https://doi.org/10.9734/ijecc/2023/v13i113429

Mao H, Jiang C, Tang C et al (2023) Wheat adaptation to environmental stresses under climate change: molecular basis and genetic improvement. Mol Plant 16(10):1564–1589. https://doi.org/10.1016/j.molp.2023.09.001

Mao T, Bao L, Zhang H et al (2024) $Mn_3O_4$ nanoenzyme seed soaking enhanced salt tolerance in soybean through modulating homeostasis of reactive oxygen species and ATPase activities. Plants (Basel) 13(21):3011. https://doi.org/10.3390/plants13213011

Maulana F, Ayalew H, Anderson JD et al (2018) Genome-wide association mapping of seedling heat tolerance in winter wheat. Front Plant Sci 9:1272. https://doi.org/10.3389/fpls.2018.01272

Mim JJ, Rahman SM, Khan F et al (2025) Towards smart agriculture through nano-fertilizer: a review. Mater Today Sustain 30:101100. https://doi.org/10.1016/j.mtsust.2025.101100

Mondal S, Rutkoski JE, Velu G et al (2016) Harnessing diversity in wheat to enhance grain yield, climate resilience, disease and insect pest resistance and nutrition through conventional and modern breeding approaches. Front Plant Sci 7:991. https://doi.org/10.3389/fpls.2016.00991

Morales-Espinoza MC, Cadenas-Pliego G, Pérez-Alvarez M et al (2019) Se nanoparticles induce changes in the growth, antioxidant responses, and fruit quality of tomato developed under NaCl stress. Molecules 24(17):3030. https://doi.org/10.3390/molecules24173030

Muhammad F, Raza MAS, Iqbal R et al (2022) Ameliorating drought effects in wheat using an exclusive or co-applied rhizobacteria and ZnO nanoparticles. Biology (Basel) 11(11):1564. https://doi.org/10.3390/biology11111564

Pais IP, Moreira R, Semedo JN et al (2023) Wheat crop under waterlogging: potential soil and plant effects. Plants (Basel) 12(1):149. https://doi.org/10.3390/plants12010149

Pandey K, Anas M, Hicks VK et al (2019) Improvement of commercially valuable traits of industrial crops by application of carbon-based nanomaterials. Sci Rep 9:19358. https://doi.org/10.1038/s41598-019-55903-3

Porras R, Miguel-Rojas C, Lorite IJ et al (2023) Characterization of durum wheat resistance against leaf rust under climate change conditions of increasing temperature and $CO_2$. Sci Rep 13:22001. https://doi.org/10.1038/s41598-023-49118-w

Pramanick B, Dubey R, Kesarwani A et al (2023) Advancement in mitigating the effects of waterlogging stress in wheat. Advances in Wheat Research. https://doi.org/10.1016/B978-0-323-95368-9.00013-8

Rajpal VR, Nongthongbam B, Bhatia M et al (2025) The nano-paradox: addressing nanotoxicity for sustainable agriculture, circular economy and SDGs. J Nanobiotechnol 23:314. https://doi.org/10.1186/s12951-025-03371-5

Rashid A, Schutte BJ, Ulery A et al (2023) Heavy metal contamination in agricultural soil: environmental pollutants affecting crop health. Agronomy 13(6):1521. https://doi.org/10.3390/agronomy13061521

Roy C, He X, Gahtyari NC et al (2023) Managing spot blotch disease in wheat: conventional to molecular aspects. Front Plant Sci 14:1098648. https://doi.org/10.3389/fpls.2023.1098648

Savary S, Willocquet L, Pethybridge SJ et al (2019) The global burden of pathogens and pests on major food crops. Nat Ecol Evol 3:430–439. https://doi.org/10.1038/s41559-018-0793-y

Schoonjans R, Castenmiller J, Chaudhry Q et al (2023) Regulatory safety assessment of nanoparticles for the food chain in Europe. Trends Food Sci Technol 134:98–111. https://doi.org/10.1016/j.tifs.2023.01.017

Sharma RC, Duveiller E (2004) Effect of *Helminthosporium* leaf blight on performance of timely and late-seeded wheat under optimal and stressed levels of soil fertility and moisture. Field Crop Res 89:205–218. https://doi.org/10.1016/j.fcr.2004.02.002

Shelar A, Singh AV, Chaure N et al (2024) Nanoprimers in sustainable seed treatment: molecular insights into abiotic–biotic stress tolerance mechanisms for enhancing germination and improved crop productivity. Sci Total Environ 951:175118. https://doi.org/10.1016/j.scitotenv.2024.175118

Shewry PR (2025) Wheat in 2025: challenges for sustainable production and consumption. J Exp Bot eraf514. https://doi.org/10.1093/jxb/eraf514

Shewry PR, Hey SJ (2015) The contribution of wheat to human diet and health. Food Energy Secur 4:178–202. https://doi.org/10.1002/fes3.64

Sidhu AK, Verma N, Kaushal P et al (2025) Green synthesis of polymeric nanoparticles: agricultural applications and toxicological implications. Discov Appl Sci 7:505. https://doi.org/10.1007/s42452-025-06877-7

Silva S, Dias MC, Silva AMS (2022) Titanium and zinc-based nanomaterials in agriculture: a promising approach to deal with (a)biotic stresses? Toxics 10:172. https://doi.org/10.3390/toxics10040172

Singh K, Malla MA, Kumar A et al (2025) Toxicological concerns of nanomaterials in agroecosystem: risk, fate, and their analytical and regulatory assessment. Environ Qual Manag 35:e70136. https://doi.org/10.1002/tqem.70136

Stefanello R, da Silva Garcia WJ, Rossato Viana A et al (2024) Graphene oxide decreases the effects of salt stress on Persian clover seed germination. J Toxicol Environ Health A 87:47–56. https://doi.org/10.1080/15287394.2023.2274338

Tamindžić G, Azizbekian S, Miljaković D et al (2024) Assessment of various nanoprimings for boosting pea germination and early growth in both optimal and drought-stressed environments. Plants 13:1547. https://doi.org/10.3390/plants13111547

Tang Y, Zhao W, Zhu G et al (2023) Nano-pesticides and fertilizers: solutions for global food security. Nanomaterials 14:90. https://doi.org/10.3390/nano14010090

Türkoğlu A, Haliloğlu K, Ekinci M et al (2024) Zinc oxide nanoparticles: an influential element in alleviating salt stress in quinoa (*Chenopodium quinoa* L. cv. Atlas). Agronomy 14:1462. https://doi.org/10.3390/agronomy14071462

Ullah I, Toor MD, Basit A et al (2023) Nanotechnology: an integrated approach towards agriculture production and environmental stress tolerance in plants. Water Air Soil Pollut 234:666. https://doi.org/10.1007/s11270-023-06675-0

Venzhik Y, Deryabin A, Popov V et al (2022) Gold nanoparticles as adaptogens increasing the freezing tolerance of wheat seedlings. Environ Sci Pollut Res 29:55235–55249. https://doi.org/10.1007/s11356-022-19759-x

Wang B, Meng T, Xiao B et al (2023a) Fighting wheat powdery mildew: from genes to fields. Theor Appl Genet 136:196. https://doi.org/10.1007/s00122-023-04445-4

Wang X, Xie H, Wang P et al (2023b) Nanoparticles in plants: uptake, transport and physiological activity in leaf and root. Materials 16:3097. https://doi.org/10.3390/ma16083097

Wang Z, Wang S, Ma T et al (2023c) Synthesis of zinc oxide nanoparticles and their applications in enhancing plant stress resistance: a review. Agronomy 13:3060. https://doi.org/10.3390/agronomy13123060

Wei L, Liu J, Jiang G (2024) Nanoparticle-specific transformations dictate nanoparticle effects associated with plants and implications for nanotechnology use in agriculture. Nat Commun 15:7389. https://doi.org/10.1038/s41467-024-51741-8

Yang Y, Ye C, Zhao M et al (2025) Nanoparticles in sustainable agriculture: enhancing nutrient use efficiency and abiotic stress resilience under climate change. Plant Stress 17:100982. https://doi.org/10.1016/j.stress.2025.100982

Yashavanthakumar KJ, Pawar P, Basavaraj PS et al (2026) Climate-smart strategies for mitigating abiotic stresses in cereal crops: a review. J Cereal Res 17:221–263. https://doi.org/10.25174/2582-2675/2025/166751

Yu M, Luobu Z, Zhuoga D et al (2025) Advances in plant response to low-temperature stress. Plant Growth Regul 105:167–185. https://doi.org/10.1007/s10725-024-01253-8

Zhang T, Wang Q, Yukui R (2025) The impact of nanomaterials on plant health: a review of exposure, toxicity, and control. Environ Sci Nano 12:2965–2982. https://doi.org/10.1039/D5EN00037H

# Exploring the Potential of Artificial Intelligence in Managing Biotic and Abiotic Stresses in Wheat

# 16

Tilak Chandra, M. A. Iquebal, and Sarika Jaiswal

**Abstract**

Wheat plays a critical role in global food security. However, its productivity and quality are increasingly threatened by biotic and abiotic stresses driven by climate variability, pathogen evolution, and soil degradation. Conventional breeding and management approaches are constrained by slow data acquisition, subjective field evaluations, and delayed responses. Recent advances in artificial intelligence (AI), encompassing machine learning, deep learning, computer vision, and advanced sensor networks, offer transformative opportunities for stress detection, prediction, and mitigation in wheat systems. AI-driven platforms facilitate high-throughput phenotyping, real-time monitoring of disease and pests, early prediction of climate-induced stress, and accelerated genomic analysis, collectively advancing wheat improvement and precise postharvest management. Progress in modelling major biotic threats, including rust, powdery mildew, viral diseases, and insect pests, as well as abiotic challenges such as drought, heat, salinity, and nutrient deficiencies, highlights AI's capacity to integrate diverse datasets into robust, scalable decision-support tools. Despite ongoing challenges related to data inconsistency, model interpretability, infrastructure limitations, and restricted field deployment, the outlook for AI-enabled wheat improvement in managing multiple stresses remains optimistic. Emerging innovations, such as large-scale foundation models, digital twins, integration of Internet of Things (IoT) with AI, and AI-assisted gene editing, are poised to enhance wheat resilience, promote long-term sustainability, and ensure food security.

**Keywords**

Wheat · Artificial intelligence · Machine learning · Wheat productivity · Biotic and abiotic stress

T. Chandra · M. A. Iquebal · S. Jaiswal (✉)
ICAR-Indian Agricultural Statistics Research Institute, New Delhi, India



R. Tiwari et al. (eds.), *Wheat Stress Management*,
https://doi.org/10.1007/978-981-92-0902-6_16

## 16.1 Introduction

Wheat (*Triticum aestivum* L.) is among the world's most important staple food crops, supplying a substantial share of global dietary calories and proteins (Chandra et al. 2024). Stable wheat production is essential to satisfy the demands of a growing population. However, wheat productivity is constrained by a diverse array of biotic and abiotic stressors, such as fungal diseases, insect pests, viral pathogens, drought, terminal heat, salinity, nutrient imbalances, and flooding (Mushtaq et al. 2024; Lin et al. 2025). The impacts of climate change further intensify these stressors, resulting in yield instability, production losses, quality deterioration, and economic setbacks on a global scale (Chandra et al. 2024). While traditional breeding, agronomic practices, and integrated pest management have mitigated some of these challenges, they often do not fully address the complexity and rapid evolution of environmental threats. Recent advancements in artificial intelligence (AI), particularly in machine learning (ML) and deep learning (DL), are transforming crop science by enabling more precise, data-driven decision-making throughout the wheat growth and management cycles (Mushtaq et al. 2024). The application of AI in wheat research has expanded rapidly, supported by high-throughput phenotyping platforms, drone and satellite-based remote sensing, hyperspectral imaging, and the analysis of large-scale multi-omics datasets. These complex data streams frequently exceed the capabilities of traditional analytical methods. AI models, including convolutional neural networks (CNNs), transformers, decision trees, and ensemble approaches, can effectively identify patterns, predict outcomes, and classify stress signatures with high accuracy (Patrício and Rieder 2018). In the context of biotic stress management, predictive AI frameworks that integrate weather, crop growth, and pathogen surveillance data can forecast disease and pest outbreaks, thereby enhancing early warning systems (Ashfaq et al. 2025). For abiotic stresses, AI enables precise mapping and prediction of drought severity, heatwave impacts, soil salinity, and nutrient deficiencies using remote sensing and environmental data. AI-driven genomic selection and phenome–genome integration are accelerating the identification and breeding of wheat varieties resilient to drought, heat, salinity, and combined stresses (Mushtaq et al. 2024; Barbedo 2025; Lin et al. 2025). Despite these advancements, several challenges impede the widespread adoption of AI in wheat improvement, including limited availability of labelled datasets, inconsistent environmental conditions, issues with model transferability, and the need for significant computational resources and user-friendly tools (Mushtaq et al. 2024; Lin et al. 2025). Overcoming these challenges is critical for translating AI innovations into practical field solutions. This chapter synthesizes recent progress in AI-driven strategies for managing biotic and abiotic stresses in wheat, emphasizing their applications, benefits, limitations, and prospects. By integrating advances from breeding, phenotyping, remote sensing, and field management, these technologies offer significant potential to advance climate-resilient wheat production.

## 16.2 Abiotic and Abiotic Threats and Their AI-Mediated Management: An Overview

A variety of biotic stressors, including fungal, bacterial, and viral pathogens, insect pests, and invasive weeds, constrain wheat production. Diseases such as rust, powdery mildew (PM), Fusarium head blight (FHB), and septoria cause substantial yield losses and compromise grain quality through toxin contamination (Mushtaq et al. 2024). Viral infections, including wheat streak mosaic virus and barley yellow dwarf virus, which are frequently transmitted by aphids, further diminish plant vigour and productivity. Insect pests such as aphids, hessian fly, armyworms, and termites inflict direct damage through feeding and facilitate the spread of pathogens. Weeds like *Phalaris minor* and *Avena fatua* compete for essential nutrients and impair canopy health (Mushtaq et al. 2024; Barbedo 2025). Artificial intelligence (AI) has emerged as a transformative tool for addressing these challenges, delivering accurate, rapid solutions for disease and pest management (Mushtaq et al. 2024). AI-powered computer vision models, particularly convolutional neural networks (CNNs), enable early detection of leaf diseases and pest damage using images collected from smartphones, drones, and field sensors. Machine learning algorithms, including random forest (RF), support vector machine (SVM), and deep learning architectures, integrate real-time weather, soil, and crop data to forecast outbreaks of rusts, aphids, and other pests, thereby facilitating timely interventions (Barbedo 2025). AI also provides robust strategies for managing abiotic stresses in wheat by supporting more effective monitoring and climate-adaptive decision-making. By using sensor data, satellite imagery, and predictive models, AI can anticipate heat, drought, salinity, and nutrient stress before significant yield reductions occur. These algorithms inform timely irrigation, optimize nutrient application, and support the development of climate-resilient varieties (Lin et al. 2025). In precision agriculture, AI advances the deployment of automated sprayers and robotic platforms that identify and treat infected areas or weed patches, thereby reducing pesticide usage and operational costs. In breeding programs, AI accelerates the development of disease-resistant wheat by analyzing genomic data, identifying resistance genes, and enhancing the accuracy of genomic selection (Mushtaq et al. 2024; Barbedo 2025; Lin et al. 2025). Collectively, AI-based technologies provide an efficient, proactive, and sustainable approach to safeguarding wheat from biotic threats and promoting stable global food production.

## 16.3 Application of AI Tools and Technologies Relevant to Biotic Stress Management

Fungal diseases such as leaf rust, stem rust, and stripe rust significantly reduce wheat yield quality and quantity. Recent advances in deep learning have improved disease detection accuracy across diverse agroecological settings. For instance, InceptionV3, ResNet50, and VGG16/19 architectures have been evaluated for rust diagnosis, with VGG19 achieving the highest classification accuracy of 99.38%, indicating superior feature extraction capabilities for wheat pathology (Aboneh et al. 2021). Automated yellow rust detection using high-resolution hyperspectral UAV imagery has demonstrated the benefits of integrating spectral and spatial data. The resulting model achieved an overall accuracy of 0.85, surpassing traditional random forest classifiers and maintaining robust detection throughout the crop growth cycle, particularly during late-stage disease development (Zhang et al. 2019a). Additionally, hyperspectral imaging combined with deep neural networks (DNN) has proven effective for Fusarium head blight (FHB) assessment, with UAV-based spectral and texture analyses further supporting precise detection under field conditions (Xiao et al. 2021). Deep learning approaches have also integrated feature extraction and metric learning to reliably identify wheat leaf diseases. Xu et al. (2023) developed a dual-parallel CNN with feedback blocks, which facilitates optimized feature learning and improves classification performance. Furthermore, a neural network-based system for measuring FHB severity in field environments achieved high accuracy for single-ear segmentation and near-perfect accuracy for disease-spot segmentation, with processing times under 1 s, highlighting its potential for real-time disease monitoring (Zhang et al. 2019b).

Advances in deep learning and sensor-based analytics have enhanced the detection and monitoring of Fusarium head blight (FHB) and rust diseases in wheat. Surface-enhanced Raman spectroscopy, when combined with deep learning, enables rapid and sensitive detection of deoxynivalenol (DON) in FHB-infected kernels. In this approach, Cit-AuNRs serve as effective Raman substrates, and liquid–liquid interface self-extraction enables rapid DON separation. In the context of smart farming, UAV-based imaging has emerged as a valuable tool for disease surveillance. Aerial visual perception systems that capture both spectral and spatial information outperform traditional spectral-only classifiers, such as random forests, particularly for yellow rust monitoring (Su et al. 2020). Further progress includes a dual Mask R-CNN framework for rapid identification of wheat spikes and infected regions, achieving detection accuracies of 98.81% for diseased areas, 77.76% for spike recognition, and 77.19% for FHB severity assessment (Su et al. 2020). Deep residual networks have also shown promise for early detection of stripe rust. Schirrmann et al. (2021) reported a classifier with 77% overall accuracy, detecting 57% of cases at very early infection levels (0.5% incidence) and 76% during the initial spread (2%–4% incidence), highlighting the potential of deep learning for early warning systems. New methods for canopy-level FHB detection, such as pixel-based prediction in field canopy data, show potential for large-scale operational monitoring. Similarly, colour-image-based DNNs effectively identify wheat spikes and symptomatic regions, increasing the efficiency of field-level FHB assessment (Qiu et al. 2019). Weakly supervised learning further addresses the challenge of analysing large-scale UAV imagery.

Innovations in deep learning have markedly improved the accuracy and practicality of wheat disease detection, particularly in field applications. The integration of attention mechanisms into mobile-captured imagery represents a significant advancement. For example, Mi et al. (2020) demonstrated that C-DenseNet achieved a test accuracy of 97.99% for stripe rust grading, surpassing classical DenseNet

(92.53%) and ResNet (73.43%). GradCAM++ visualizations confirmed that C-DenseNet localizes disease-critical regions more effectively, thereby enhancing model interpretability. To address the challenge of low-resolution imagery from edge devices, super-resolution generative adversarial networks (SRGANs) have improved classification performance. Maqsood et al. (2021) reported that up-sampling increased accuracy to 83%, compared to 75% with native low-resolution data, indicating SRGANs' utility as a preprocessing tool in field scenarios. Mobile-enabled diagnostic tools, such as GranoScan, continue to advance; this AI-powered application achieved 77% accuracy for leaf diseases, 95% for spike, stem, and root diseases, and 94% for pest recognition. Ensemble learning approaches also demonstrate potential. Beyond rust diseases, spectral-based methods have been applied to detect wheat powdery mildew. Liu et al. (2024) found that a three-band texture index enabled early detection, and that combining spectral, textural, and canopy-temperature features improved the estimation of powdery mildew severity. Advanced machine learning systems have further enhanced diagnostic performance. Khan et al. (2022) introduced an automated wheat disease classification system that achieved 99.8% accuracy, outperforming traditional machine learning techniques across accuracy, precision, recall, and F1-score metrics. Classical machine learning methods remain relevant; for example, support vector machines (SVMs) using morphological traits have been applied to identify diseased plants, demonstrating the continued value of traditional approaches in breeding and pathology workflows (Azimi et al. 2020).

An infield automatic wheat disease diagnosis system (DMIL-WDDS) was first proposed to enable simultaneous identification and localization of wheat diseases, demonstrating superior recognition accuracy compared to conventional CNN architectures, supported by the newly collected WDD2017 field dataset; the system has also been deployed as a real-time mobile application (Lu et al. 2017). A lightweight CNN incorporating an attention mechanism was developed for wheat ear disease detection under complex field conditions, achieving approximately 94% accuracy. Comparative analyses further showed that a hybrid neural network integrating a 2D-CNN and a BidGRU achieved the best generalization performance and minimized overfitting when classifying hyperspectral pixels of healthy and Fusarium-infected wheat heads under field conditions (Jin et al. 2018). Additionally, a unified matrix-based convolutional neural network (M-bCNN) designed for fine-grained leaf disease classification achieved substantial improvements over AlexNet and VGG16, with validation and test accuracies of 96.5% and 90.1%, respectively (Lin et al. 2019). CNN-based comparisons of septoria- and stripe rust-infected leaves against healthy counterparts using MobileNet and EfficientNet-B3 further demonstrated progressive improvements in wheat leaf disease detection accuracy (Akbar et al. 2022). The Otsu segmentation algorithm was refined by optimizing combinations of colour and texture features, yielding an E-MMC metric-learning-based approach that improved the identification and severity assessment of wheat leaf diseases. A novel multi-temporal and multimodal UAV dataset tailored for wheat stripe rust enabled the development of an AI-driven multimodal regression network for quantitative wheat stripe rust inversion, demonstrating significantly enhanced performance relative to unimodal approaches (Deng et al. 2024). A lightweight multiscale CNN combining residual and inception modules, enriched with CBAM and ECA attention mechanisms, was further shown to improve disease sensitivity while reducing background interference in complex field environments (Fang et al. 2023). Moreover, a tandem dual-BlendMask deep learning framework enabled simultaneous segmentation of wheat spikes and diseased regions, supporting rapid and accurate assessment of disease severity under field conditions (Gao et al. 2022).

A multispecies dataset containing annotated soybean, wheat, and coffee leaves exhibiting disease symptoms was used to train six CNN architectures for lesion segmentation, with three architectures delivering highly accurate predictions and strong agreement between estimated and reference lesion areas (Goncalves et al. 2021). An improved deep convolutional architecture for wheat leaf and spike disease detection showed significant performance gains, increasing accuracy by 7.01% and 15.92%, respectively. Ninety-two percentage over VGG 16 and ResNet 50, respectively (Goyal et al. 2021). A generic identification and classification approach combined decision trees with deep learning models, with expert verification enhancing decision-tree rules and integrating them into a knowledge-based system for wheat disease diagnosis (Haider et al. 2021). CNNs were also employed to estimate yellow rust severity, and the newly developed Yellow-Rust-Xception network, trained on a dedicated yellow-rust dataset, achieved a test accuracy of 91%, comparable to or better than human assessment in task-specific evaluation (Hayit et al. 2021). A multitask learning framework was further proposed to simultaneously recognize rice and wheat leaf diseases, in which VGG-16 was adapted to enable multitask training for dual-crop diagnosis (Jiang et al. 2021). For FHB detection, optimal spectral and imaging features were selected via a decision-tree feature-selection process, and their fusion enabled deep learning models to achieve superior performance. Digital imaging was also used to evaluate the efficacy of fungicides for controlling FHB in wheat, demonstrating the utility of imaging-based quantification (Zhang et al. 2020a). In UAV-assisted monitoring, the Ir-UNet architecture demonstrated that using only three selected multispectral bands (Red, NIR, and Red-Edge) achieved high performance with an overall accuracy of 96.83%, while significantly reducing computational load (Zhang et al. 2021). Therefore, AI is redefining biotic stress management in wheat by enabling early detection, accurate prediction, and precise intervention. With innovative imaging, predictive models, and AI-driven precision tools, farmers can respond to diseases, pests, and weeds before they cause significant damage. AI not only boosts wheat productivity but also supports sustainable, low-input crop protection. It is fast becoming an essential ally for resilient and future-ready wheat production. Table 16.1 delineates recent applications of AI tools and technologies relevant to biotic stress management to accelerate productivity amidst variable pathogenic events.

**Table 16.1** Tabulation of recent applications of AI tools and technologies relevant to biotic stress management to accelerate productivity amidst variable pathogenic events

| AI/ML approaches | Model/algorithm | Results accuracy (%) | Inferences from the studies | References |
|---|---|---|---|---|
| CNN | Inceptionv3, Resnet50, and VGG16/19 | 99.38 | Leaf stem and stripe rust detection | Aboneh et al. (2021) |
| CNN | DenseNet, ResNet, and GradCAM++ | 97 | Wheat stripe rust grading | Mi et al. (2020) |
| CNN | DT, KNN | 97 | Diseases classification | Khan et al. (2022) |
| CNN | Mobilenet and efficient net-b3 | 97 | Septoria and stripe rust detection | Akbar et al. (2022) |
| CNN | ZFNet and GoogLeNet | 99 | Leaf disease identification | Xu et al. (2023) |
| SVM | DT, kNN, and NB | 98–100 | Fusarium head blight detection | Azimi et al. (2020) |
| CNN | SRGAN | 75–83 | Stripe rust classification | Maqsood et al. (2021) |
| – | RustQNet | 80 | Stripe rust identification | Deng et al. (2024) |
| FL and TL | ConvNeXtBase, ConvNeXtLarge, etc. | 99 | Leaf disease classification | Fahim-Ul-Islam et al. (2024) |
| CNN | MobileNetV3 and EfficientNetb0 | 98 | Disease detection | Fang et al. (2023) |
| DL | ResNet-50 and BlendMask | 92 | Severity assessment of fusarium head blight | Gao et al. (2022) |
| SVM, CNN | 2D-CNN-BidGRU | 75 | Healthy heads and fusarium head blight classification | Jin et al. (2018) |
| CNN | VGG-FCN-VD16 and VGG-FCN-S | 98 | Disease diagnosis system | Lu et al. (2017) |
| CNN | FPN, Unet, and DeepLabv3 + (Xception) | 97 | Severity of foliar symptoms | Goncalves et al. (2021) |
| CNN | AlexNet and VGG-16 | 90 | Leaf disease classification | Lin et al. (2019) |
| CNN | VGG16 and RESNET50 | 98 | Leaf and spike disease detection | Goyal et al. (2021) |
| Regression | GLCM | 90 | Head blight detection | Xiao et al. (2021) |
| RF | U-Net | 90 | Yellow rust monitoring | Su et al. (2020) |
| CNN | Decision trees (DT) | 97 | Disease classification and verification | Haider et al. (2021) |
| CNN | ConvNet | 91 | Severity level of yellow rust disease | Hayit et al. (2021) |

(continued)

**Table 16.1** (continued)

| AI/ML approaches | Model/algorithm | Results accuracy (%) | Inferences from the studies | References |
|---|---|---|---|---|
| MLR | RMSE | 90 | Powdery mildew comparison | Liu et al. (2024) |
| CNN | resnet50 and densenet121 | 98 | Detection of leaf diseases | Jiang et al. (2021 |
| CNN | ResNet | 90 | Early detection of stripe rust | Schirrmann et al. (2021) |
| CNN | Inception-Resnet | 85 | Yellow rust detection | Zhang et al. (2019a) |
| RF | – | 86 | Efficacy of fungicides for control of fusarium head blight | Zhang et al. (2020b) |
| R-CNN | ResNet 50 and ResNet 101 | 80 | Detection of fusarium head blight | Qiu et al. (2019) |
| R-CNN | ResNet-101 | 77 | Resistance to fusarium head blight | Su et al. (2020) |
| CNN | – | 97 | Detect fusarium head blight | Zhang et al. (2020a) |
| FCN, PCNN | U-Net | 98 | Severity of the fusarium head blight in the field | Zhang et al. (2019b) |
| Ir-UNet | – | 97 | Yellow rust detection | Zhang et al (2021) |

## 16.4 Application of AI Tools and Technologies Relevant to Abiotic Stress Management

Wheat is a staple food crop worldwide, and research into its drought and salinity tolerance is essential for global food security. The advancement of artificial intelligence (AI) in agriculture has introduced innovative methods to enhance wheat stress resistance (Lin et al. 2025). Machine learning (ML) has shown considerable promise in improving drought-stress detection and informing water-management decisions in wheat cultivation. For example, Gupta et al. (2023) performed a comprehensive grid-search hyperparameter optimization with tenfold cross-validation across nine ML algorithms, identifying the random forest model as the most robust, achieving a global diagnostic accuracy of 91.16% for drought stress detection. Genotype screening for drought resilience has also advanced through the integration of ML. Benlioğlu et al. (2024) utilized ML algorithms to assess drought tolerance in tetraploid wheat during germination, demonstrating the value of data-driven approaches for early-stage selection. Deep learning techniques have further improved the modelling of long-term evapotranspiration. Elbeltagi et al. (2020)

found that a deep neural network (DNN) achieved strong correlations between observed and predicted ETc values using limited meteorological data in semi-arid environments. Advances in computer vision have also enabled precise phenotyping of germination traits. Among the VGG16+ Faster R-CNN, ResNet50 + Faster R-CNN, and YOLOv4 models evaluated for wheat seed germination detection, YOLOv4 achieved the highest performance, with a mean average precision (mAP) of 97.59%, a recall of 97.35%, and a detection speed of 6.82 FPS, confirming its suitability for automated germination counting (Fu et al. 2022).

Deep learning methods have substantially improved drought and stress monitoring in wheat by integrating various remote sensing data sources. For instance, a deep learning-based drought monitoring model using multisource remote sensing data demonstrated strong applicability for both meteorological and agricultural drought assessments, effectively capturing actual drought events (Shen et al. 2019). Cloud-based AI, combined with small unmanned aerial systems (sUAS) and spectral imaging, has shown promise in detecting early signs of water stress and supporting timely irrigation decisions (Freeman et al. 2019). The integration of AI with geographic information systems (GIS) has enabled precise land suitability assessments for wheat cultivation in arid regions, thereby enhancing food security planning (El Behairy et al. 2023). High-resolution thermal and red, green, and blue (RGB) imagery, when paired with advanced AI techniques, has improved water-stress identification in winter wheat. Thermal imagery outperformed RGB data, and a deep learning long short-term memory (DL-LSTM) model achieved the highest stress-classification accuracy of 96.7% (Chandel et al. 2022). In addition to drought assessment, AI has advanced the detection of lodging. A lightweight Mobile U-Net model that integrates RGB and digital surface model (DSM) data achieved high accuracy and robustness in extracting wheat lodging regions (Yang et al. 2021). Furthermore, combining transfer learning with the DeepLabv3+ network provided an effective approach for lodging area extraction across multiple wheat growth stages, surpassing the performance of the traditional U-Net model (Zhang et al. 2020c).

Machine learning and advanced sensing technologies have significantly improved the quantification and monitoring of salinity stress in wheat. The combination of Raman spectroscopy and machine learning enables precise measurement of salt stress in wheat leaves, supporting informed decision-making to enhance efficiency and reduce crop losses (Kecoglu et al. 2022). Artificial neural network (ANN) models have been effectively used to assess salinity indices, with the yield stability index (YSI), mean productivity (MP), geometric mean productivity (GMP), and stress tolerance index (STI) identified as the most reliable predictors of salinity-tolerant wheat varieties (Ravari et al. 2016). High-throughput phenotyping platforms such as PhenoField® provide detailed characterization of genotype performance throughout the growth cycle and facilitate the refinement of crop models and sensor evaluation (Beauchêne et al. 2019). AI applications have also advanced the study of drought effects on root traits in both modern and ancient wheat varieties, offering deeper insights into stress-responsive root architectures (Licaj et al. 2023). The integration of thermal and RGB imaging with ANN models has proven highly effective for noninvasive assessment of salt tolerance in wheat genotypes under field conditions, enabling rapid screening of large germplasm collections (Author/year needed). Additionally, UAV LiDAR metrics have demonstrated strong potential for monitoring crop height, biomass, and nitrogen uptake in winter wheat, with the 60th and 90th percentiles of reflectance serving as reliable proxies for plant structural and nutritional status (Hütt et al. 2023).

Recent advances demonstrate the increasing effectiveness of integrating remote sensing technologies with AI for monitoring water stress in wheat. For instance, the combination of UAV-based multispectral and thermal infrared imagery with machine learning models has enabled accurate prediction of water stress in winter wheat, efficiently capturing spatial heterogeneity and supporting precision irrigation practices (Mali et al. 2025). Building on these innovations, state-of-the-art AI-enabled mobile devices have been developed to facilitate real-time, on-field detection of water stress in wheat and other field crops, providing a highly accessible and rapid decision-support tool for farmers (Chandel et al. 2024). Additionally, hyperspectral imaging integrated with multivariate modelling and machine learning techniques has been successfully employed to phenotype drought stress and to understand the dynamics of the drought–nitrogen interaction in wheat, providing a robust framework for high-throughput physiological assessment (Okyere et al. 2024). Future advancements will focus on integrating multimodal sensing platforms with advanced AI models to deliver more precise, real-time assessments of water stress in wheat. The adoption of explainable AI will enhance trust and interpretability in stress-prediction outputs, while low-cost, scalable AI-sensor systems will expand accessibility for smallholder farmers. Moreover, coupling AI-driven diagnostics with automated irrigation technologies and genotype–environment modelling will accelerate the development of drought-resilient wheat varieties and promote more innovative, resource-efficient crop management. Table 16.2 presents recent applications of AI tools and technologies relevant to abiotic stress management to accelerate productivity amid variable environments, including extreme heat, drought, and salt stress, as well as other environmental factors that negatively affect wheat production and quality.

## 16.5 Conclusion

Artificial intelligence (AI) has significant potential to transform wheat production by enabling precise, rapid, and scalable management of both biotic and abiotic stresses. Applications include disease diagnosis, stress prediction, genomic selection, phenotyping, and climate modelling. Although challenges such as limited data availability, interpretability, and adoption barriers persist, integrating AI with genomics, remote sensing, robotics, and gene editing is expected to accelerate progress toward resilient, sustainable wheat production. The advancement of wheat improvement will increasingly depend on interdisciplinary, AI-driven innovations. Foundation models and large-scale agricultural AI systems can be trained on satellite imagery, sequence data, and climate data to facilitate generalized stress

**Table 16.2** The tabulation of recent applications of AI tools and technologies relevant to abiotic stress management to accelerate productivity amid variable environments, such as extreme heat, drought, and salt stress, as well as other environmental factors that negatively influence wheat production and quality

| AI/ML approaches | Model/ algorithm | Results accuracy | Inferences from the studies | References |
|---|---|---|---|---|
| DNN | NetCDF | 99% | Crop evapotranspiration in a semi-arid environment | Aboneh et al. (2021) |
| DFNN | – | 90% | Drought monitoring on multisource remote sensing data | Shen et al. (2019) |
| FCN | U-net | 89% | Accurate lodging extraction | Yang et al. (2021) |
| U-net | deeplabv3+ | 92% | Automatic extraction of the lodging area | Zhang et al. (2020c) |
| CNN | GoogLeNet | 93% | Lodging detection from UAS images | Zhang et al. (2020d) |
| – | PhénoField® | 99% | Characterization of abiotic stress | Beauchêne et al. (2019) |
| – | UAV LiDAR | 95% | Monitoring height and biomass | Hütt et al. (2023) |
| ANN | BPNN | – | Assessing salt tolerance | El-Hendawy et al. (2024) |
| AI | – | – | Impact of drought stress on root traits | Licaj et al. (2023) |
| ANN | – | – | Assessment of salinity indices | Ravari et al. (2016) |
| ML | GPR, NN | 99% | Quantification of salt stress | Kecoglu et al. (2022) |
| ANN, KNN | AlexNet, GoogLeNet | 97% | Water stress identification | Chandel et al. (2022) |
| GIS, FIS | – | – | Land suitability assessment | El Behairy et al. (2023) |
| YOLOv4 | ResNet50 | 97% | Detecting salt tolerance | Fu et al. (2022) |
| TV-L1 | Cfit-kmeans | 91% | Multiple stress analysis | Gupta et al. (2023) |
| CNN | – | 98% | Identification of the early indicators of water stress | Freeman et al. (2019) |
| SVM | ELANET | 91% | Insights into drought tolerance | Benlioğlu et al. (2024) |
| ANN | Nnet | 90% | Predicting water stress | Mali et al. (2025) |
| DCNN | GoogLeNet | 97% | Real-time water stress detection | Chandel et al. (2024) |
| DNN | – | 87% | Phenotyping drought stress and nitrogen interactions | Okyere et al. (2024) |

detection. Digital twins of wheat simulate field conditions in virtual environments to model stress responses and optimize management strategies. AI and robotics also enable automation in precision spraying, disease scouting, and targeted irrigation. Integration with advanced gene-editing platforms enables AI-guided CRISPR design to introduce stress-resilience traits and rapidly develop climate-smart wheat ideotypes. Farmer-focused decision tools and AI-powered mobile applications for real-time disease diagnosis, irrigation scheduling, and fertilizer optimization deliver substantial stakeholder benefits and support market forecasting, thereby enhancing financial returns.

## References


Aboneh T, Rorissa A, Srinivasagan R et al (2021) Computer vision framework for wheat disease identification and classification using Jetson GPU infrastructure. Technologies 9:47. https://doi.org/10.3390/technologies9030047

Akbar S, Ahmad KT, Abid MK et al (2022) Wheat disease detection for yield management using IoT and deep learning techniques. VFAST Trans Softw Eng 10:80–89. https://doi.org/10.21015/vtse.v10i3.1108

Ashfaq M, Khan I, Afzal RF et al (2025) Enhanced wheat yield prediction through integrated climate and satellite data using advanced AI techniques. Sci Rep 15:18093. https://doi.org/10.1038/s41598-025-02700-w

Azimi N, Sofalian O, Davari M et al (2020) Statistical and machine learning-based FHB detection in durum wheat. Plant Breed Biotechnol 8:265–280. https://doi.org/10.9787/PBB.2020.8.3.265

Barbedo JGA (2025) A review of artificial intelligence techniques for wheat crop monitoring and management. Agronomy 15:1157. https://doi.org/10.3390/agronomy15051157

Beauchêne K, Leroy F, Fournier A et al (2019) Management and characterization of abiotic stress via PhénoField®, a high-throughput field phenotyping platform. Front Plant Sci 10:904. https://doi.org/10.3389/fpls.2019.00904

Benlioğlu B, Demirel F, Türkoğlu A et al (2024) Insights into drought tolerance of tetraploid wheat genotypes in the germination stage using machine learning algorithms. Agriculture 14:206. https://doi.org/10.3390/agriculture14020206

Chandel NS, Rajwade YA, Dubey K et al (2022) Water stress identification of winter wheat crop with state-of-the-art AI techniques and high-resolution thermal-RGB imagery. Plants 11:3344. https://doi.org/10.3390/plants11233344

Chandel NS, Chakraborty SK, Chandel AK et al (2024) State-of-the-art AI-enabled mobile device for real-time water stress detection of field crops. Eng Appl Artif Intell 131:107863. https://doi.org/10.1016/j.engappai.2024.107863

Chandra T, Sahu J, Jaiswal S et al (2024) Current research status and emerging trends in wheat: an integrated scientometric analysis based on ploidy uncovers hidden footprints in the scientific landscape. Heliyon 10:e36375. https://doi.org/10.1016/j.heliyon.2024.e36375

Deng J, Hong D, Li C et al (2024) RustQNet: multimodal deep learning for quantitative inversion of wheat stripe rust disease index. Comput Electron Agric 225:109245. https://doi.org/10.1016/j.compag.2024.109245

El Behairy RA, Arwash HME, El Baroudy AA et al (2023) Artificial intelligence integrated GIS for land suitability assessment of wheat crop growth in arid zones to sustain food security. Agronomy 13:1281. https://doi.org/10.3390/agronomy13051281

Elbeltagi A, Deng J, Wang K et al (2020) Modeling long-term dynamics of crop evapotranspiration using deep learning in a semi-arid environment. Agric Water Manag 241:106334. https://doi.org/10.1016/j.agwat.2020.106334

El-Hendawy S, Tahir MU, Al-Suhaibani N et al (2024) Potential of thermal and RGB imaging combined with artificial neural networks for assessing salt tolerance of wheat genotypes grown in real-field conditions. Agronomy 14:1390. https://doi.org/10.3390/agronomy14071390

Fahim-Ul-Islam M, Chakrabarty A, Ahmed ST et al (2024) A comprehensive approach toward wheat leaf disease identification leveraging transformer models and federated learning. IEEE Access 12:109128–109156. https://doi.org/10.1109/ACCESS.2024.3438544

Fang X, Zhen T, Li Z (2023) Lightweight multiscale CNN model for wheat disease detection. Appl Sci 13:5801. https://doi.org/10.3390/app13095801

Freeman D, Gupta S, Smith DH et al (2019) Watson on the farm: using cloud-based artificial intelligence to identify early indicators of water stress. Remote Sens 11:2645. https://doi.org/10.3390/rs11222645

Fu X, Han B, Liu S et al (2022) WSVAS: a YOLOv4-based phenotyping platform for automatically detecting the salt tolerance of wheat based on seed germination vigour. Front Plant Sci 13:1074360. https://doi.org/10.3389/fpls.2022.1074360

Gao Y, Wang H, Li M et al (2022) Automatic tandem dual blend mask networks for severity assessment of wheat fusarium head blight. Agriculture 12:1493. https://doi.org/10.3390/agriculture12091493

Goncalves JP, Pinto FA, Queiroz DM et al (2021) Deep learning architectures for semantic segmentation and automatic estimation of severity of foliar symptoms caused by diseases or pests. Biosyst Eng 210:129–142. https://doi.org/10.1016/j.biosystemseng.2021.08.011

Goyal L, Sharma CM, Singh A et al (2021) Leaf and spike wheat disease detection and classification using an improved deep convolutional architecture. Inf Med Unlocked 25:100642. https://doi.org/10.1016/j.imu.2021.100642

Gupta A, Kaur L, Kaur G (2023) Drought stress detection technique for wheat crop using machine learning. PeerJ Comput Sci 9:e1268. https://doi.org/10.7717/peerj-cs.1268

Haider W, Rehman AU, Durrani NM et al (2021) A generic approach for wheat disease classification and verification using expert opinion for knowledge-based decisions. IEEE Access 9:31104–31129. https://doi.org/10.1109/ACCESS.2021.3058582

Hayit T, Erbay H, Varçın F et al (2021) Determination of the severity level of yellow rust disease in wheat by using convolutional neural networks. J Plant Pathol 103:923–934. https://doi.org/10.1007/s42161-021-00886-2

Hütt C, Bolten A, Hüging H et al (2023) UAV LiDAR metrics for monitoring crop height, biomass and nitrogen uptake: a case study on a winter wheat field trial. PFG 91:65–76. https://doi.org/10.1007/s41064-022-00228-6

Jiang Z, Dong Z, Jiang W, Yang Y (2021) Recognition of rice leaf diseases and wheat leaf diseases based on multi-task deep transfer learning. Comput Electron Agric 186:106184. https://doi.org/10.1016/j.compag.2021.106184

Jin X, Jie L, Wang S et al (2018) Classifying wheat hyperspectral pixels of healthy heads and fusarium head blight disease using a deep neural network in the wild field. Remote Sens 10:395. https://doi.org/10.3390/rs10030395

Kecoglu I, Sirkeci M, Unlu MB et al (2022) Quantification of salt stress in wheat leaves by Raman spectroscopy and machine learning. Sci Rep 12:7197. https://doi.org/10.1038/s41598-022-10767-y

Khan H, Haq IU, Munsif M et al (2022) Automated wheat diseases classification framework using advanced machine learning technique. Agriculture 12:1226. https://doi.org/10.3390/agriculture12081226

Licaj I, Felice D, Germinario C et al (2023) An artificial intelligence-integrated analysis of the effect of drought stress on root traits of "modern" and "ancient" wheat varieties. Front Plant Sci 14:1241281. https://doi.org/10.3389/fpls.2023.1241281

Lin Z, Mu S, Huang F et al (2019) A unified matrix-based convolutional neural network for fine-grained image classification of wheat leaf diseases. IEEE Access 7:11570–11590. https://doi.org/10.1109/ACCESS.2019.2891739

Lin J, Zhao Y, Tang J et al (2025) The application and progress of artificial intelligence in improving drought and salinity tolerance in wheat. Adv Resour Res 5:736–750. https://doi.org/10.50908/arr.5.2_736

Liu Y, Liu G, Sun H et al (2024) Exploring multi-features in UAV-based optical and thermal infrared images to estimate disease severity of wheat powdery mildew. Comput Electron Agric 225:109285. https://doi.org/10.1016/j.compag.2024.109285

Lu J, Hu J, Zhao G et al (2017) An in-field automatic wheat disease diagnosis system. Comput Electron Agric 142:369–379. https://doi.org/10.1016/j.compag.2017.09.012

Mali SS, Scobie M, Baillie J et al (2025) Integrating UAV-based multispectral and thermal infrared imageries with machine learning for predicting water stress in winter wheat. Precis Agric 26:44. https://doi.org/10.1007/s11119-025-10239-z

Maqsood MH, Mumtaz R, Haq IU et al (2021) Super resolution generative adversarial network (SRGANs) for wheat stripe rust classification. Sensors 21:7903. https://doi.org/10.3390/s21237903

Mi Z, Zhang X, Su J et al (2020) Wheat stripe rust grading by deep learning with attention mechanism and images from mobile devices. Front Plant Sci 11:558126. https://doi.org/10.3389/fpls.2020.558126

Mushtaq MA, Ahmed HGM-D, Zeng Y (2024) Applications of artificial intelligence in wheat breeding for sustainable food security. Sustainability 16:5688. https://doi.org/10.3390/su16135688

Okyere FG, Cudjoe DK, Virlet N et al (2024) Hyperspectral imaging for phenotyping plant drought stress and nitrogen interactions using multivariate modeling and machine learning techniques in wheat. Remote Sens 16:3446. https://doi.org/10.3390/rs16183446

Patrício DI, Rieder R (2018) Computer vision and artificial intelligence in precision agriculture for grain crops: a systematic review. Comput Electron Agric 153:69–81. https://doi.org/10.1016/j.compag.2018.08.001

Qiu R, Yang C, Moghimi A et al (2019) Detection of fusarium head blight in wheat using a deep neural network and color imaging. Remote Sens 11:2658. https://doi.org/10.3390/rs11222658

Ravari SZ, Dehghani H, Naghavi H (2016) Assessment of salinity indices to identify Iranian wheat varieties using an artificial neural network. Ann Appl Biol 168:185–194. https://doi.org/10.1111/aab.12254

Schirrmann M, Landwehr N, Giebel A et al (2021) Early detection of stripe rust in winter wheat using deep residual neural networks. Front Plant Sci 12:469689. https://doi.org/10.3389/fpls.2021.469689

Shen R, Huang A, Li B et al (2019) Construction of a drought monitoring model using deep learning based on multi-source remote sensing data. Int J Appl Earth Obs Geoinf 79:48–57. https://doi.org/10.1016/j.jag.2019.03.006

Su J, Yi D, Su B et al (2020) Aerial visual perception in smart farming: field study of wheat yellow rust monitoring. IEEE Trans Ind Inform 17:2242–2249. https://doi.org/10.1109/TII.2020.2979237

Xiao Y, Dong Y, Huang W et al (2021) Wheat fusarium head blight detection using UAV-based spectral and texture features in optimal window size. Remote Sens 13:2437. https://doi.org/10.3390/rs13132437

Xu L, Cao B, Zhao F et al (2023) Wheat leaf disease identification based on deep learning algorithms. Physiol Mol Plant Pathol 123:101940. https://doi.org/10.1016/j.pmpp.2022.101940

Yang B, Zhu Y, Zhou S (2021) Accurate wheat lodging extraction from multi-channel UAV images using a lightweight network model. Sensors 21:6826. https://doi.org/10.3390/s21206826

Zhang X, Han L, Dong Y et al (2019a) A deep learning-based approach for automated yellow rust disease detection from high-resolution hyperspectral UAV images. Remote Sens 11:1554. https://doi.org/10.3390/rs11131554

Zhang D, Wang D, Gu C et al (2019b) Using neural network to identify the severity of wheat fusarium head blight in the field environment. Remote Sens 11:2375. https://doi.org/10.3390/rs11202375

Zhang DY, Chen G, Yin X et al (2020a) Integrating spectral and image data to detect fusarium head blight of wheat. Comput Electron Agric 175:105588. https://doi.org/10.1016/j.compag.2020.105588

Zhang D, Wang Z, Jin N et al (2020b) Evaluation of efficacy of fungicides for control of wheat fusarium head blight based on digital imaging. IEEE Access 8:109876–109890. https://doi.org/10.1109/ACCESS.2020.3001652

Zhang D, Ding Y, Chen P et al (2020c) Automatic extraction of wheat lodging area based on transfer learning method and Deeplabv3+ network. Comput Electron Agric 179:105845. https://doi.org/10.1016/j.compag.2020.105845

Zhang Z, Flores P, Igathinathane C et al (2020d) Wheat lodging detection from UAS imagery using machine learning algorithms. Remote Sens 12:1838. https://doi.org/10.3390/rs12111838

Zhang T, Xu Z, Su J et al (2021) Ir-UNet: irregular segmentation U-shape network for wheat yellow rust detection by UAV multispectral imagery. Remote Sens 13:3892. https://doi.org/10.3390/rs13193892

GPSR Compliance
The European Union's (EU) General Product Safety Regulation (GPSR) is a set of rules that requires consumer products to be safe and our obligations to ensure this.

If you have any concerns about our products, you can contact us on

ProductSafety@springernature.com

In case Publisher is established outside the EU, the EU authorized representative is:

Springer Nature Customer Service Center GmbH
Europaplatz 3
69115 Heidelberg, Germany

www.ingramcontent.com/pod-product-compliance
Ingram Content Group UK Ltd.
Pitfield, Milton Keynes, MK11 3LW, UK
UKHW020631260926
475829UK00003BA/18
* 9 7 8 9 8 1 9 2 0 9 0 1 9 *

For Product Safety Concerns and Information please contact our EU
representative GPSR@taylorandfrancis.com
Taylor & Francis Verlag GmbH, Kaufingerstraße 24, 80331 München, Germany

www.ingramcontent.com/pod-product-compliance
Ingram Content Group UK Ltd.
Pitfield, Milton Keynes, MK11 3LW, UK
UKHW020316270926
475702UK00050B/908

* 9 7 8 9 0 4 8 5 7 7 2 7 9 *